Turbulent Combustion Modeling

FLUID MECHANICS AND ITS APPLICATIONS
Volume 95

Series Editor: R. MOREAU
MADYLAM
Ecole Nationale Supérieure d'Hydraulique de Grenoble
Boîte Postale 95
38402 Saint Martin d'Hères Cedex, France

Aims and Scope of the Series
The purpose of this series is to focus on subjects in which fluid mechanics plays a fundamental role.
As well as the more traditional applications of aeronautics, hydraulics, heat and mass transfer etc., books will be published dealing with topics which are currently in a state of rapid development, such as turbulence, suspensions and multiphase fluids, super and hypersonic flows and numerical modeling techniques.
It is a widely held view that it is the interdisciplinary subjects that will receive intense scientific attention, bringing them to the forefront of technological advancement. Fluids have the ability to transport matter and its properties as well as to transmit force, therefore fluid mechanics is a subject that is particularly open to cross fertilization with other sciences and disciplines of engineering. The subject of fluid mechanics will be highly relevant in domains such as chemical, metallurgical, biological and ecological engineering. This series is particularly open to such new multidisciplinary domains.
The median level of presentation is the first year graduate student. Some texts are monographs defining the current state of a field; others are accessible to final year undergraduates; but essentially the emphasis is on readability and clarity.

For further volumes:
www.springer.com/series/5980

Tarek Echekki • Epaminondas Mastorakos
Editors

Turbulent Combustion Modeling

Advances, New Trends and Perspectives

Springer

Editors
Tarek Echekki
Dept. Mechanical & Aerospace Engineering
North Carolina State University
911 Oval Drive
27695 Raleigh
North Carolina
USA
techekk@ncsu.edu

Epaminondas Mastorakos
Department of Engineering
University of Cambridge
CB2 1PZ Cambridge
UK
em257@eng.cam.ac.uk

ISSN 0926-5112
ISBN 978-94-007-0411-4 e-ISBN 978-94-007-0412-1
DOI 10.1007/978-94-007-0412-1
Springer Dordrecht Heidelberg London New York

Cover design: VTEX, Vilnius

Printed on acid-free paper

Springer is part of Springer Science+Business Media (www.springer.com)

To our families and mentors

Preface

The field of turbulent combustion has undergone significant progress since the first original paradigms for modeling turbulent combustion flows emerged more than 60 years ago. In the seventies, the emergence of computational fluid dynamics (CFD) and access to more advanced non-intrusive laser-based techniques for combustion measurements have enabled further development in the field. More recently, rapid progress in the modeling and simulation of turbulent flows has occurred. This progress may be attributed to different factors. First, we have access to more advanced computational and experimental resources. Advanced computational resources enable the computations of more realistic combustion flows with better description of the flow and chemical reactions. The higher access to computational resources also enabled the emergence of new paradigms in turbulent combustion simulation that address direct computations of unresolved physics. Turbulent combustion has long been considered a paradigm for multiscale problems and has long been identified as one of the important problems to solve, hence the increasing interest from the computational and applied mathematics communities.

Advanced experimental resources enabled measurements that serve as standards for validation. We cite here the great synergy created by the Turbulent Nonpremixed Flames (TNF) workshop and its counterpart, Turbulent Premixed Flames workshop. More experimental data is being made available online for validation, where in the past they can be gleamed only through plots from archival journals and reports.

A second impetus is the increasing requirements to design both efficient and clean combustion technologies. These technologies no longer represent operation at prescribed modes (e.g. premixed, non-premixed) or regimes (e.g. the flamelet regime). Computer simulation of turbulent combustion flows can potentially reduce the turn-around time for the expensive design cycle.

To cover all the progress in the field of turbulent combustion may be beyond the scope of a single volume; and chances are, revisions on such a volume may have to be started once it is published. However, some of the basic foundations that every researcher in the field relies on remain timeless and are covered by some recent textbooks and monographs. We cite here, for example, *Turbulent Combustion* by Peters [1]; *Theoretical and Numerical Combustion* by Poinsot and Veynante [2]; *Computational Models for Turbulent Reacting* Flows by Fox [3]; and *An Introduction to Turbulent Reacting Flows* by Cant and Mastorakos [4]. There are also books that have reviewed on a regular basis progress in the field. They include in particular the series *Turbulent Reacting Flows* edited by Libby and Williams. [5] [6] Extensive reviews in aspects of turbulent combustion models are also available in a number of journals and proceedings, including *Progress in Energy and Combustion Science*, *Proceedings of the Combustion Institute*, and *Annual Reviews of Fluid Mechanics*.

Having recognized the progress that has been achieved recently in the turbulent combustion field, we have adopted a two-pronged approach in this book. First, we have attempted to present to the reader the current state-of-the-art in advanced models in turbulent combustion. Here, we have attempted to avoid duplication of topics that are already covered in the aforementioned textbooks. Instead, we have emphasized more recent progress and have identified the current needs and trends associated with the state-of-the-art models in turbulent combustion.

Perhaps an important distinction between the present volume and the now classic Libby and Williams series is a greater emphasis on the topic of computation of turbulent combustion flows. This emphasis represents the second scope of the present volume.

The primary audience for this book is graduate students in engineering, applied and computational mathematics, and researchers in both academia and industry. It is assumed that readers have a good knowledge of the state-of-the-art models in turbulent reacting flows. Therefore, the book can serve as a graduate text or desk reference.

The book is divided into four major parts and includes 19 chapters. The first introductory part includes two chapters. The first chapter attempts to reassert the role of combustion science in the current energy debate and identifies the current challenges and requirements to advance the field forward. The second chapter summarizes the governing equations for turbulent reacting flows and motivates the various approaches needed to address the closure problem.

In Part II, the state-of-the-art and current trends of advanced turbulent combustion models are presented. The discussion addresses the flamelet approach (Chapter 3), models for premixed combustion (Chapter 4), the conditional moment closure model (Chapter 5), transported probability density function function meth-

---

[1] Peters, N.: Turbulent Combustion, Cambridge University Press, Cambridge, UK (2000)

[2] Poinsot, T., Veynante, D.: Theoretical and Numerical Combustion, 2nd Ed., R.T. Edwards, Philadelphia, USA (2005)

[3] Fox, R.O.: Computational Models for Turbulent Reacting Flows, Cambridge University Press, Cambridge, UK (2003)

[4] Cant, R.S., Mastorakos, E.: An Introduction to Turbulent Reacting Flows, Imperial College Press, London, UK (2008)

[5] Libby, P.A., Williams, F.A. (Eds): Turbulent Reacting Flows, Springer-Verlag, Berlin Heidelberg, Germany (1980)

[6] Libby, P.A., Williams, F.A. (Eds): Turbulent Reacting Flows, Academic Press, London, UK (1994)

ods (Chapter 6), and the multiple mapping conditioning approach (Chapter 7). These chapters introduce aspects of the model formulations, illustrate the models through examples and identify challenges and trends in advancing these modeling approaches.

Part III addresses multiscale approaches in turbulent combustion. Chapter 8 motivates requirements for multiscale scale models in combustion and summarizes current approaches. Chapter 9 addresses methods and strategies for the integration and acceleration of chemistry in reacting flow computations. Chapters 10, 11 and 12 present the linear-eddy, the one-dimensional turbulence and the unsteady flame-embedding approaches as multiscale strategies based on hybrid solutions combining coarse-grained and fine-grained approaches. The following two chapters illustrate multiscale strategies through mesh-adaptivity based on the adaptive mesh refinement approach (Chapter 13) and the wavelet approach (Chapter 14).

The final part of the book, Part IV, presents what is termed 'cross-cutting science'. It samples important disciplines that are relevant to advancing the field of turbulent combustion. The first discipline is associated with the topic of validation and verification. Chapter 15 reasserts the role of experiment in advancing turbulent combustion models. The second discipline is associated with requirements to manage large-scale computations associated with turbulent combustion. This discipline is represented by two chapters. Chapter 16 reviews recent progress and trends in uncertainty quantification. Chapter 17 discusses a computational framework, based on the common component architecture, as a strategy to efficiently develop computational tools for advancing science. Examples are presented in this chapter for combustion flows. The third discipline is multiscale science. An important role of multiscale mathematics is the construction of frameworks to couple models designed for disparate scales. Chapter 18 illustrates the development and implementation of such frameworks for combustion using the heterogeneous multiscale method. The fourth discipline is associated with the choices of the governing equations for turbulent reacting flows. The Navier-Stokes equations in their instantaneous or filtered forms are the most popular forms of representing turbulent combustion flows. Chapter 19 discusses progress on an alternative representation of the governing equations based on the lattice-Boltzmann method. The method may serve as an alternative to the continuum-based Navier-Stokes equations where potentially closer coupling with atomistic scales is needed.

This book is a collaborative effort that has involved researchers/experts from different disciplines. This is a reflection of how complex the theme of turbulent combustion has become. It could not have been completed without the expert opinion, the patience and the diligence of the 31 contributors.

Raleigh, NC, USA and Cambridge, UK, *Tarek Echekki*
July 2010 *Epaminondas Mastorakos*

Contents

Part II Recent Advances and Trends in Turbulent Combustion Models

List of Contributors

Benjamin A. Allan
Sandia National Laboratories, Livermore, CA, USA, e-mail: baallan@sandia.gov

Rob Armstrong
Sandia National Laboratories, Livermore, CA, USA, e-mail: rob@sandia.gov

John B. Bell
Center for Computational Sciences and Engineering, Lawrence Berkeley National Laboratory, Berkeley, CA, USA, e-mail: jbbell@lbl.gov

Robert W. Bilger
School of Aerospace, Mechanical and Mechatronic Engineering, The University of Sydney, NSW 2006, Australia, e-mail: robert.bilger@sydney.edu.au

R. Stewart Cant
Department of Engineering, University of Cambridge, Cambridge, UK, e-mail: rsc10@eng.cam.ac.uk

Matthew J. Cleary
The University of Queensland, School of Mechanical and Mining Engineering, St Lucia, Queensland, 4072, Australia, e-mail: m.cleary@uq.edu.au

Bénédicte Cuenot
Centre Européen de Recherche et Formation Avancées en Calcul Scientifique (CERFACS), Toulouse, France, e-mail: Benedicte.Cuenot@cerfacs.fr

Marcus S. Day
Center for Computational Sciences and Engineering, Lawrence Berkeley National Laboratory, Berkeley, CA, USA, e-mail: msday@lbl.gov

Bert Debusschere
Sandia National Laboratories, Livermore, CA, USA, e-mail: bjdebus@sandia.gov

Weinan E
Department of Mathematics and Program in Applied and Computational Mathematics, Princeton University, Princeton, NJ, USA, e-mail: weinan@math.princeton.edu

Tarek Echekki
Department of Mechanical and Aerospace Engineering, North Carolina State University, NC, USA, e-mail: techekk@ncsu.edu

Björn Engquist
Department of Mathematics, University of Texas at Austin, Autsin, TX, USA, e-mail: engquist@math.utexas.edu

Christos E. Frouzakis
Aerothermochemistry and Combustion Systems Laboratory, Swiss Federal Institute of Technology, Zurich, Switzerland, e-mail: frouzakis@lav.mavt.ethz.ch

Ahmed F. Ghoniem
Department of Mechanical Engineering, Massachusetts Institute of Technology, Cambridge MA, USA, e-mail: ghoniem@mit.edu

Dimitris A. Goussis
School of Applied Mathematics & Physical Sciences, National Technical University of Athens, Greece, e-mail: dagoussi@mail.ntua.gr

Daniel C. Haworth
Department of Mechanical and Nuclear Engineering, The Pennsylvania State University, University Park, PA, USA, e-mail: dch12@psu.edu

Hossam A. El-Asrag
Department of Mechanical Engineering, Massachusetts Institute of Technology, Cambridge MA 02139, e-mail: helasrag@mit.edu

Alan R. Kerstein
Sandia National Laboratories, Livermore, CA, USA, e-mail: arkerst@sandia.gov

Alexander Klimenko
The University of Queensland, School of Mechanical and Mining Engineering, St Lucia, Queensland, 4072, Australia, e-mail: a.klimenko@uq.edu.au

Andreas Kronenburg
Institut für Technische Verbrennung, University of Stuttgart, 70569 Stuttgart, Germany, e-mail: kronenburg@itv.uni-stuttgart.de

Ulrich Maas
Institut für Technische Thermodynamik, Karlsruhe Institute of Technology, Karlsruhe, Germany, e-mail: ulrich.maas@kit.edu

Assaad R. Masri
School of Aerospace, Mechanical and Mechatronic Engineering, The University of Sydney, NSW 2006, Australia, e-mail: masri@aeromech.usyd.edu.au

Epaminondas Mastorakos
Department of Engineering, University of Cambridge, Cambridge, UK, e-mail: em257@eng.cam.ac.uk

Suresh Menon
School of Aerospace Engineering, Georgia Institute of Technology, e-mail: suresh.menon@aerospace.gatech.edu

Habib N. Najm
Sandia National Laboratories, Livermore, CA, USA, e-mail: hnnajm@sandia.gov

Stephen B. Pope
Sibley School of Mechanical and Aerospace Engineering, Cornell University, Ithaca, N.Y., USA, e-mail: s.b.pope@cornell.edu

Robert Prosser
School of Mechanical, Aerospace and Civil Engineering, University of Manchester, Manchester, UK, e-mail: robert.prosser@manchester.ac.uk

Jaideep A. Ray
Sandia National Laboratories, Livermore, CA, USA, e-mail: jairay@sandia.gov

Cosmin Safta
Sandia National Laboratories, Livermore, CA, USA, e-mail: csafta@sandia.gov

Yi Sun
Statistical and Applied Mathematical Sciences Institute (SAMSI), Research Triangle Park, NC, USA, e-mail: yisun@samsi.info

James C. Sutherland
Department of Chemical Engineering, The University of Utah, Salt Lake City, UT, USA, e-mail: James.Sutherland@utah.edu

Part I
Introductory Concepts

Chapter 1
The Role of Combustion Technology in the 21st Century

R.W. Bilger

Abstract Continuing support for combustion research is threatened by those who argue that continued use of fossil fuels is not sustainable. Reserves of fossil fuels are in fact sufficient for several generations and their use can be continued if their greenhouse gas emissions can be captured and sequestered or otherwise offset. Forecasts for future energy technologies foresee considerable reductions in the role for combustion technologies. It is important that combustion researchers become involved in such forecasting so that viable combustion technologies such as carbon capture and sequestration do not become sidelined by over-optimistic projections for photovoltaics and the hydrogen economy. It is evident that many energy technologies will undergo at least two stages in transition to achieving the goals that are needed by 2050. Furthermore, there will be many factors such as geography and economic and policy changes that will have drastic effects on the marketability of energy systems. Combustion systems will need to be able to respond quickly to these rapidly changing markets. There is a vitally important role for advanced computer modelling in meeting this challenge. Current combustor modelling capability is of only peripheral use in the development of new combustion systems. The main problem is in the modelling of turbulence chemistry interactions. Considerable investment is needed in the development of advanced modelling approaches and in high quality measurements of an hierarchy of experimental data bases that will provide physical insights and become a basis for model validation.

1.1 Introduction

For tens of thousands of years combustion has been the main means of providing heating for human comfort and for the cooking of food. For about five thousand years combustion has been the means for materials processing - producing pottery,

---

Department of Mechanical and Mechatronic Engineering, University of Sydney, NSW 2006, Australia, e-mail: robert.bilger@sydney.edu.au

T. Echekki, E. Mastorakos (eds.), *Turbulent Combustion Modeling*,
Fluid Mechanics and Its Applications 95, DOI 10.1007/978-94-007-0412-1_1,

bronze, cement, iron, steel, and many other materials associated with the rise of human civilization. It is only a few hundred years since mechanical power from humans, animals, wind and water has been surpassed by the mechanical power from combustion engines - steam, spark-ignition, diesel and gas turbine. Much of this mechanical power is now used for generation of electricity, transportation of materials and for personal mobility.

Combustion is a means of transforming primary energy into useful secondary energy such as heating, processed materials and mechanical and electrical power. For most of history the primary energy consumed in combustion has been wood, charcoal and other so-called bio-fuels. Only after the steam engine was stripping bare the forests of Europe did coal become the main source of primary energy in the industrialized world, later to be augmented by oil and natural gas. For some two centuries now the main sources of primary energy have been these fossil fuels.

Fossil fuels are so named as the energy they embody is energy from the sun - solar energy - that was originally captured in plants of various kinds and has been stored for millions of years in fossil form. This source of primary energy has proved to be a great boon to mankind. It soon became evident that fossil energy was far less costly to obtain than wood, charcoal and other bio-fuels. The technology needed to combust fossil fuels was quickly found - no doubt because these fuels are far more energy intensive than wood, charcoal and other bio-fuels.

Coal and other fossil fuels contain sulphur and the fumes from their combustion contain carcinogens and toxins not found in wood smoke. Wood smoke contains particulates and tars that seemingly should be very harmful - but perhaps over 100,000 years the genes of humans have been selected to provide most people with some tolerance to this cocktail. Concerns about air pollution in the troposphere affecting human health and amenity have largely been associated with fossil fuels. In the 20th century laws based on torts between individuals became surpassed by air pollution laws aimed at preserving human health and amenity in urban air basins.

From their very nature, it is certain that only finite amounts of fossil fuels have been stored in the earth's crust. Our consumption of fossil fuels has been rising exponentially. An important question arises as to whether our consumption will soon deplete these storages by significant amounts. This question is compounded by the fact that the geographical locations of the storages as currently known are some distance from the locations where their consumption is highest. This fact immediately transforms questions of global sufficiency into ones of national energy security.

The dominant constituent of fossil fuels is the element carbon. Combustion of fossil fuels turns essentially all of this carbon into carbon dioxide. Carbon dioxide mixed into the local atmosphere has essentially no direct effect on human health or on the local amenity of humans or of their immediate environment. Carbon dioxide levels in the global atmosphere have been rising. There is now widely-accepted evidence that human caused emissions of carbon dioxide are directly linked to increases in global average temperature levels, depletion of arctic ice floes and many other climate and environmental changes of serious concern. This conclusion is strongly disputed by a small group of climate change skeptics or deniers.

It is seen that the use of fossil fuels is directly and intimately connected with tropospheric air pollution, energy security and climate change. These are three of the most important challenges for the world in the 21st century.

A common facile response to these challenges is for proposals to ban the use of fossil fuels. The world has agreed that we must only have development that is sustainable [24]. In this facile response it is argued that 'since fossil fuels are being depleted they cannot be sustainable - only renewable sources of energy can be sustainable'. In reality, many renewable sources of energy are not sustainable solutions to the world's energy needs and fossil fuels can be the preferred sustainable option for the way ahead. This is examined in more detail in the next section.

Our main concern here is with combustion: its technology and its underlying science. With some 100,000 years of development, it might be expected to be a mature technology. In fact, it is the least developed technology of modern engineering systems. Engineering systems with mature technology have computer models that are routinely used in analysis and can be used in the optimization of the geometry and operating conditions of new designs. Thus the wings of new models of jet aircraft have subtle geometry contours that greatly increase their efficiency. The same can be said for the new generation of fans in turbo-fan engines. Computer modeling of combustors is still at a fledgling stage and is only used peripherally in the development of new combustors. Combustor development is still largely by cut-and-try testing in experimental rigs and in prototype and in-service engines and power plants. Mongia [21] reviews important concepts and methodologies for usefully integrating CFD into the design process for gas-turbine combustors in a commercial engineering environment.

The fledgling status of computer modelling of combustors comes not from the want of trying. More than forty years of effort by some of the leading scientists in the world has still not got us within close sight of achieving the goal of having soundly based predictive models for combustors that can be used in industry with confidence on a routine basis. This is because it is a multi-scale problem requiring advanced models for highly turbulent recirculating and swirling fluid flows, for gaseous and particulate fuel dispersion and mixing and for the extremely complex chemical kinetics involved in the oxidation of hydrocarbon fuels and the formation and burnout of pollutants such as soot, polycyclic hydrocarbons, carbon monoxide and oxides of nitrogen. Furthermore, these highly non-linear processes are inherently and strongly coupled together. A particularly difficult coupling is that between turbulence and chemical reaction [7]. The main chapters in this book summarize the variety of approaches that are being developed to overcome this particular difficulty in the modeling of practical combustion systems.

The question arises as to whether efforts to improve the capabilities for computer modeling of combustors are a worthwhile use of scientific resources. Some authorities argue that combustion will quickly lose the dominant role that it has historically had due to the widely accepted need for decarbonization of energy processes to mitigate the causes of climate change - investment in improving such tools for combustion systems is wasting money on a technology that has already passed its potential usefulness. A main purpose of this introductory chapter is to explore the nature of such technology forecasting and whether it is prudent to make such summary dismissals of combustion technology.

Another main purpose is to outline the prospects for advanced computational modeling of combustors in practical systems and to identify the major thrusts needed in research.

1.2 Sustainable Energy

All of the significant nations in the world have ratified the United Nations Framework Convention on Climate Change [24] the so-called Rio Convention. They are committed to 'stabilization of greenhouse gas concentrations in the atmosphere at a level that would prevent dangerous interference with the climate system'. The Rio convention also defines 'sustainable development' as 'development that meets the needs of the present without compromising the ability of future generations to meet their own needs'.

In the 1970s we had two energy crises associated with wars in the Middle East. Fear was engendered that we were running out of oil as proven reserves were only 30 times the then current rates of consumption. The movement for renewable sources of energy was born, including solar thermal, solar photovoltaic, hydroelectricity, wind turbines, and combustion of biomass. In the 1990s renewable energy was widely touted as being the only truly sustainable source of energy and should play the major role in the mitigation of climate change.

Today proven reserves of oil have not been exhausted as feared in the 1970s but are now 40 times our much increased current rates of consumption [26]. Proven reserves for oil, natural gas, coal, oil shale, tar sands and other fossil fuels are such that they could provide the world's energy needs for more than a century at current rates of consumption [26]. The cost of exploration for oil is high and it is not profitable to spend money today to prove reserves that will not be producing income for more than a few years ahead. Storages of fossil fuels as yet unproven are likely to be many times those already proven. Even with the industrialization of China, India, Brazil and other emerging economies it is certain that there is enough energy from fossil fuels to meet world needs for several generations to come. It may be that oil will be more short-lived, but liquid hydrocarbon fuels can readily be obtained from coal, oil shale and tar sands at modest cost. It can be firmly concluded that fossil fuels are not to be disqualified as being sustainable on the grounds that they will be exhausted within a few generations.

It is clear, however, that continued use of fossil fuels is un-sustainable if the carbon dioxide produced upon combustion needs to be contained or offset. Without such containment or offsetting, it is widely accepted that significant interference with the climate system will occur and the ability of future generations to meet their own needs will be greatly compromised. It is also clear that use of fossil fuels can qualify as being sustainable where carbon capture and storage technologies [5, 14] are used or where their carbon dioxide emissions are offset by black carbon sequestration [11] or other such means of extracting and storing carbon from the atmosphere. On the other hand, if the climate change skeptics are correct, the continued use of fossil fuels in this century will be sustainable but policy will be largely focused on energy security issues.

Renewable energy sources appear to easily qualify as being sustainable on the basis of their carbon dioxide emission contributions to climate change. Several of them may not be sustainable on the basis of other aspects of sustainability associated with 'meeting the needs of the present' and 'compromising the ability of future generations to meet their own needs'. Significant early investment in photovoltaic generation of electricity will require huge amounts of extra capital that may be more sorely needed to fight disease and starvation in under-developed parts of the world. Bio-fuel production requires land and water resources that may be better used for food production or for conservation of ecosystems that sustain endangered plant and animal species that would be lost to future generations. It has long been recognized that hydroelectric dams do great damage to local ecosystems and even to estuarine and coastal systems that may be many hundreds of kilometers downstream. It is clear that many renewable sources of energy may be less sustainable than using fossil fuels with carbon capture and storage or offset technologies.

1.3 Technology Forecasts

In 2006 the US Department of Energy held a workshop on the 'Basic Research Needs for Clean and Efficient Combustion of 21st Century Transportation Fuels' [19]. The workshop focused on new fuels and engines needed to meet the challenge of energy security. The new fuels considered included oil from coal, shale and tar sands. New engines considered included the homogeneous charge compression ignition (HCCI) engine. There was no direct consideration given to needs related to severe mitigation of greenhouse gases (GHGs). With a change of administration in the USA much more emphasis is now placed on mitigation of GHGs. A recent DOE workshop [12] recognizes this need for GHG mitigation and places considerable emphasis on the prospects for advanced bio-fuels.

Technology forecasts are very dependent on whether the already expressed government policies of containing carbon emissions will be followed through. The UNCCC meeting in Copenhagen in December, 2009 failed to come up with a binding resolution extending the Kyoto protocol. Much of this failure of outcome can perhaps be attributed to the campaign mounted by the climate change skeptics. It seems that this campaign may have greatly defused international attempts to put world-wide regulation of carbon emissions in place.

There is small list of scientists who dispute the conclusions of the IPCC reports on climate change[1]. Essentially all of these doubts have been put forward in news-

[1] Wikipedia lists some 40 or more of them (`http://en.wikipedia/wiki/:26/03/2010`). Their doubts are grouped there under the following headings: 1) Global warming is not occurring (3 scientists), 2) Accuracy of IPCC climate projections is questionable (4 scientists), 3) Global

papers, books and the general electronic media rather than in anonymously peer-reviewed scientific journals.[2]

It is evident that much confusion about the validity of projections of climate change science has been engendered in the USA, Australia and many other countries. It is far from certain that international agreement will be reached on carbon emission targets and the international governance needed to implement them. Of perhaps more importance is that emission trading schemes (ETS) or cap-and-trade (CaT) legislation and other such measures intended to provide a level playing field for the the development of GHG control technologies may not succeed. It seems that, although they are the most sensible economic measures, they are too easily branded as 'high tax' in Western democracies. The failure to implement ETS or CaT is likely to severly affect the prospects for carbon capture and storage (CCS) technologies that need such legislation to underwrite the large capital investment involved.

The International Energy Agency (IEA) of the OECD has made a comprehensive study of the energy technologies that are likely to succeed in the transition to the low carbon emissions regime that has been projected to be needed by the year 2050 [13]. They consider two scenarios: the ACT scenario uses currently available technologies and brings world greenhouse gas emissions down to the 2005 level of 27 GtCO2e/yr; and the BLUE scenario that brings GHG emissions down to half this level or 14 GtCO2e/yr but requires the development of new technologies that are little beyond proof of feasibility at this stage. World growth in energy demand by 2050 would result in baseline emissions of 62 GtCO2e/yr if the technologies of 2005 were kept in place. The study was requested by the G8 members of OECD and has taken into account the capital investment needed, the availability of the resources needed and feasible rates of commercial deployment. Figure 1.1 shows the marginal emission reduction cost in USD/tCO2 as a function of the emissions reduction relative to baseline. It is seen that improvements in end use efficiency can, in fact, be profitable, and that reducing emissions from the electrical power generation sector will cost up to 30 USD/tCO2 which is about the current European carbon trading price.

Controlling GHG emissions from industry and the transport sector are seen to be very expensive at 50–500 USD/tCO2. It should be noted that GHG emission reductions from the transport sector by downsizing cars and their engines are seen as belonging to the 'increasing end use efficiency' sector. The predictions for the power generation sector see little contribution from carbon capture and storage and from combustion of biomass as the sector becomes dominated by wind, photo-voltaic

warming is primarily caused by natural processes (24 scientists), 4) Cause of global warming is unknown (7 scientists), 5) Effects of global warming will not be significantly negative (3 scientists)

[2] There are many web sites that refute these claims - see for example: 1) http://www.grist.org/article/series/skeptics, http://www.astronomynotes.com/solarsys/s11b.htm, http://scientificamerican.com/article.cfm?id=seven-answers-to-climate-contrarian-nonsense. It seems that such refutation is also confined to the public media rather than to the anomynously peer-reviewed scientific literature. Meehl et al. [20] is, however, quoted in the Wikipedia summary as the widely accepted response on the effects of 'natural causes'.

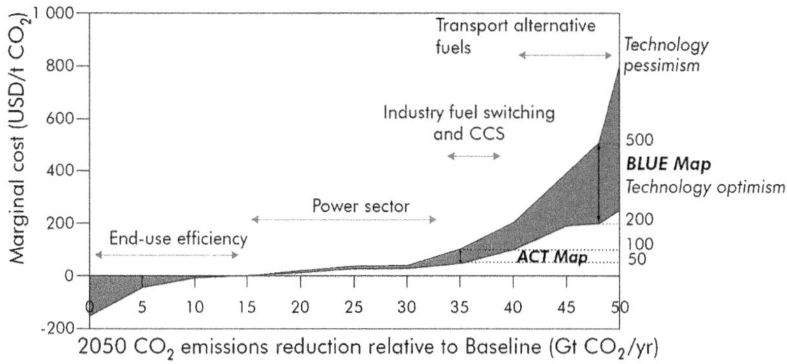

Fig. 1.1: Marginal emission reduction costs for the global energy system (2050). Reprinted from [13] with permission. Copyright © 2010, OECD/IEA.

electricity and nuclear power. In the transport sector, it is noteworthy that bio-fuels are seen as being reserved for large trucks, ships and aircraft. Fuel cells and battery electric vehicles are seen as being the power systems for cars. It is evident that this reflects the limitations on land, water and other resources in competition with needs for food supply. Overall it is predicted that use of combustion technologies in energy transformation will decline from present levels of about 80% on a world wide basis to less than 30%. The predictions are thus supportive of those that question further investment in improving computational aids for combustor design.

Such studies of the potential of various energy technologies are of great interest and importance for government policy makers. The UK Energy Research Centre's 'mission is to be the UK's pre-eminent centre of research, and source of information and leadership on sustainable energy systems. The Centre takes a whole systems approach to energy research, incorporating economics, engineering and the physical, environmental and social sciences while developing and maintaining the means to enable cohesive research in energy' [25]. The Centre is based at the University of Edinburgh with satellite contributors, mainly from other universities. Their major project is the UKERC Energy 2050 project which 'is exploring how the UK can move towards a resilient low-carbon energy system over the next forty years.' Their second report on this project [25] forecasts energy technologies for 2050 that are similar in nature to those of the IEA predictions. They also predict a low contribution from combustion technologies.

It is apparent that there are mounting predictions against combustion having a major role in energy systems in the middle and latter half of this century. It is evident that the combustion research community will, in the not too distant future, have a challenging job to keep university and government administrators convinced that this difficult area of engineering science merits further investment in research. It is important that the combustion research community engage itself in this emerging research area of energy technology prediction so that we do not get sidelined by enthusiastic protagonists for energy systems that are inherently more costly to develop to commercial viability and require distorted subsidies or market guarantees to help them to get into the marketplace.

On a more detailed reading it seems that the IEA and UKERC studies may have fallen into this trap. Figure 1.2 shows the projections assumed for the capital cost of photo-voltaic systems in the UKERC study [25]. It is seen that an order of magnitude decrease in capital costs is assumed, similar in magnitude for the semi-conductor industry. It seems to the author that the photo-voltaics community have been making such projections for several decades now and have not been able to show such 'learning curves' despite large government investment. Kazmerski [15] argues that such large advances in the capital cost of solar photovoltaics will require a break-through or 'tipping point' in the fundamental science of photovoltaics.

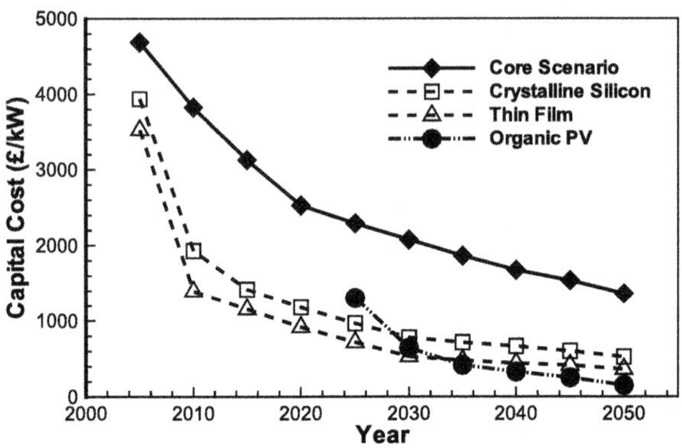

Fig. 1.2: Capital costs of photo-voltaic systems. Adapted from [25].

In the IEA and UKERC studies, photo-voltaic generation of electricity is projected to replace carbon capture and storage (CCS) for power generation. Assumptions for CCS on the other hand see little reduction in costs over this period and seem to be lumbered with a further assumption that they will be limited to only 90% capture which is a severe penalty in reaching the IEA BLUE scenario and the UKERC 80% reduction scenario. They foresee no dramatic advances in CCS technology over the next 40 years.

Projections of this sort are fraught with much uncertainty beyond a decade or two. Optimistic projections for photo-voltaic systems, batteries and fuel cells may not be realized. There is always the prospect that technologies yet to be discovered or now in early stages of development will emerge as winners and play a very important role in two or three decades from now. Novel technologies for carbon capture in the CCS approach have, in my view, a high potential for such success. These technologies are likely to involve the development of entirely new combustion systems and thus provide a challenge to combustion researchers.

Oxy-fuel firing is a combustion technology that has promise [14] of providing carbon dioxide capture at costs much lower than those for systems involving scrubbing of the power plant flue gases. Most proposals for oxy-fuel firing involve recycle of CO_2 to control combustion temperatures. In the Clean Energy Systems realization of this technology [2, 14] it is H_2O that is recycled. Thermodynamically this system has prospects for much higher cycle efficiencies as heat addition temperatures are not limited by the need to transfer heat through a boiler tube wall and heat rejection temperatures can be low due to condensation of the steam at sub-atmospheric pressure. As in the Rankine cycle, the costs of compression of the H_2O as water are much reduced as compared with the costs of compression of N_2 or CO_2 as the diluent in the Brayton cycle part of a Gas Turbine Combined Cycle (CTCC) system. The author has given the name of the internal combustion Rankine cycle (ICRC) to this cycle [6]. Even accounting for the energy needed for air separation, cycle efficiencies can be expected to be not much less than those of current day GTCCs. Furthermore, capital costs will be significantly lowered due to the use of a pump instead of a N_2 compressor and the elimination of the waste heat boiler. It seems clear that this system should be capable of providing CCS at much less than \$30/tCO2 when fully developed. Novel combustion systems are used for the primary 'steam generator' combustor and for the reheat combustor. Current designs of these combustors appear to be quite rudimentary and can be expected to evolve considerably with the use of advanced combustion modeling tools.

Reciprocating engine versions of the ICRC cycle have been studied [6, 8]. Such engines have application at lower power ratings than for the turbine system and could be suitable for stationary power plants up to 10MW, for combined heat and power (CHP) applications in buildings and industry, for ships, railway locomotives, trucks, buses, and even cars. Bilger & Wu [8] consider the application to a small car with a 50 kW power plant. Oxygen is carried in a high pressure storage tank which is also used to collect the CO_2 that is captured. Conventional gasoline or diesel is used, or liquids from coal, shale or tar sands or bio-fuels. Increases in power plant volume weight and cost are predicted to be only moderate compared with those for current uncontrolled cars and range and acceleration performance are little affected. They will far out-perform fuel-cell and electric vehicles. Control of GHG emissions will be close to 100% even using fossil fuels. Control costs are estimated to be about \$7/tCO2 which is almost two orders of magnitude less than is forecast for fuel cells and electrics (see Fig. 1, above). So far these are only paper studies. A demonstration of the engine is planned at Tongji University using a modified motor cycle engine. The design of the combustion system involving ignition of a fuel oxygen mixture and injection of a water spray is very challenging. Development of the engine will be greatly aided by advances in combustion modeling tools.

It is apparent that there will be major changes in the energy technologies within the next 40 years driven by the need to drastically reduce GHG emissions but also for improving energy security and decreasing tropospheric air pollution. It is also apparent that this will not be a single smooth transition from current technologies to the technologies needed by 2050 to provide a long-term sustainable energy system mix. This is because near-term reductions in emissions are needed as well as the achievement of a low long-term goal [22]. The near-term reductions can be achieved with existing technologies and technologies that are well-advanced in their development. Many of these technologies will not survive in the long term as they cannot provide sufficient control, will not be cost effective compared with newly developed technologies or there will not be sufficient availability of input resources.

An illustration of this can be seen for the transportation sector. This sector amounts to some 40% of current total energy consumption and early curtailment of GHG emissions from this sector is important. This early containment can most readily be obtained by regulating the fleet average emissions from new cars. The EU has proposed a limit of 130 gCO2/km for new vehicles sold in 2012 and 95 gCO2/km for 2020. These limits can be met by down-sizing of vehicle mass and engine power, using advanced engines such as the HCCI engine [10], using hybrids, and with the use of bio-fuel blends. The 2020 goal is only about a 35% reduction below current total fleet, old and new, of European emissions. The 2050 goal is for an 80% decrease overall on a world wide fleet double that of the current size, implying a per vehicle reduction of 90% averaged over the then fleet, new and old. It is apparent that by 2050 new cars sold must be essentially zero emission vehicles. By 2050 it is also apparent that bio-fuels will not be able to provide this level of control for the world-wide transportation fleet. The IEA study [13] projects that bio-fuels will only be viable for aeroplanes, ships and large trucks. There will be insufficient availability of land and water resources to produce more. So, bio-fuels and new conventional engine technologies will not have a major role in the 2050 transportation energy mix. Hydrogen fuel cells and electric vehicles, both deriving their energy from nuclear, wind and photo-voltaic systems are thus seen as the end-of-transition mix. It is clear that the transition to the long-term energy mix for transport will be at least a two stage process.

Multi-stage transition of energy technologies in other sectors such as power generation can also be expected. In the near term co-firing of biomass will increase together with increasing contributions from wind power. In the mid term nuclear power will play an increasing role together with ‘conventional’ capture for CCS. In the long term photo-voltaics may supplant advanced CCS systems, but this is far from certain.

1.4 Implications for Combustion Technology

It is seen that a variety of market forces and government regulations will strongly influence combustion system design in the near term and in the longer term. This environment can be expected to be highly volatile due to strong fluctuations in prices of primary energy, economic and political cycles and a host of other factors. Furthermore, geographical factors are likely to be important with technologies suitable for Europe or the USA possibly not being appropriate for China or Japan. It is apparent that strategies for survival of commercial enterprises in the energy technology sector must include a broad scope of inherently different products that are each being

Table 1.1: Status of current use of computational modelling in thermal systems in commercial engineering enterprises. The number of ($\star$) indicates the confidence in computing certain physics. For example, 5 ($\star$)'s indicate everyday use with very high confidence on quantitative prediction; 3 ($\star$)'s indicates frequent use with engineering value for qualitative information on trends; and 1 ($\star$) indicates occasional use with only marginal input into the design and development process.

Status	**Physics**
$\star\star\star\star\star$	Stress and deformation prediction.
$\star\star\star\star$	Heat transfer and solid temperature prediction.
$\star\star\star$	Fluid flow and convective heat transfer prediction in the absence of combustion.
$\star\star$	General flow and mixing prediction of combustion systems.
$\star$	Prediction of finite-rate chemistry effects, such as ignition, extinction, and pollutant formation of CO, NOx, soot, etc.

rapidly developed to keep pace with the competition. It will not be prudent to 'bet the firm' on a single solution, either in the near term or in the longer term.

For energy technologies involving combustion this will focus attention on the ability to rapidly develop combustors. This will include evolutions of existing designs to meet increasingly stringent regulations and market changes and for the development of entirely new concepts such as for CCS using oxy-fuel firing in power generation [2] and transportation [8].

The focus on combustor development will be intensified by the fact that most other elements of thermal systems are much further advanced in being able to be successfully modeled computationally. This can be seen from the assessment given in Table 1.1.

This assessment is the author's based on numerous contacts with colleagues in industry, academe and government research laboratories. It is seen that computational modeling of combustors is in only a fledgling state and its role is a secondary one in combustor development in commercial enterprises. Combustor development is still largely by cut-and-try testing in experimental rigs and in prototype and in-service engines and power plants.

It is evident that the fledgling status of computational modeling of combustors will be a severe handicap for the development of energy technologies involving combustion in the 21st century. The 2006 DOE workshop [19] "identified a single, overarching grand challenge: the development of a validated, predictive, multi-scale, combustion modeling capability to optimize the design and operation of evolving fuels in advanced engines for transportation applications." Such modeling capability is also sorely needed in the power generation and materials processing sectors of energy technologies.

1.5 Prospects for Advanced Computer Modeling of Combustors

As has already been stated in the Introduction, the fledgling status of computer modeling of combustors comes despite much effort over many years. It is an extremely complex multi-scale problem.

In terms of prescribed input length scales: overall dimensions of combustors are of order of 0.1 to 10 meters; details of their geometry affecting the flow are of order of 10^{-2} of this; and the dimensions of liquid or solid fuel particles are of order a further factor of 10^{-2} lower, making them 10^{-4} of the overall dimension. Fluid dynamic length scales, at the Reynolds numbers of practical systems, range from these dimensions down to a further factor of 10^{-2} for each of these prescribed input length scales. For example the boundary layers on injection orifices will be of order 10^{-2} of the orifice size, and the boundary layers on fuel particles will be of 10^{-2} of the particle size. Reaction zone length scales at the high Damkohler numbers of practical combustion systems can be even thinner than these fluid dynamic length scales, say 10^{-1}. This amounts to 10^{-7} overall, so far. Length scales associated with formation of particulates such as soot are of this order or even smaller.

In terms of prescribed input time scales: the residence time in combustors is of order 10 ms to 1 s; acoustic time scales can be of order 10^{-2} to 10 times this; and particulate fuel combustion time scales are of order 10^{-2} of this. At the Reynolds numbers of practical systems, fluid dynamic time scales range from these dimensions down to a further factor of 10^{-2} for each of these prescribed input time scales. Chemical time scales associated with ignition and extinction phenomena are of order 10^{-2} of the residence time for the combustor as a whole and also 10^{2} of particulate fuel combustion time scales for phenomena associated with ignition/extinction phenomena associated with individual particles. Chemical time scales associated with pollutant formation and burnout can be of order 10^{2} of the combustor residence time.

Important physics, including chemical physics, can be associated with any combination of these length and time scales.

Advances in computational power can be expected to continue. Even if the advances continue at present exponential rates, it is evident that full direct numerical simulation (DNS) is not possible for practical combustion systems in the foreseeable future; certainly not for the next two decades of the developments needed for combustion technology responses to GHG reduction for climate change and energy security in the near-term. It should be remembered that, to be an effective design tool in industry, a 20-hr turnaround time, or better, is needed. Perhaps new developments in computers, such as quantum computing [1], will become available in the longer term and be available to provide the computational tools in DNS that are needed before 2050. Such developments cannot be counted on, however.

Advanced computer methods for combustors will thus, inevitably, involve modeling of many of the important physical aspects of the flow that we cannot afford to resolve in our computations. The relative importance of these physical aspects to be modeled will depend on the application and on the information being sought. Thus predicting the altitude ignition behavior of a helicopter gas turbine engine will require a different modeling focus from that needed to predict re-heater tube fouling in a boiler co-firing coal and biomass. The prediction of NO_x emissions from the same helicopter engine will require a different modeling focus from that needed for ignition prediction: as will the prediction of flame stability in the co-fired boiler from that needed for tube fouling.

This diversity in application and purpose of combustion modeling of practical systems, and the perceived economic importance of this research area, is reflected in the very diverse range of research efforts directed at advancing combustion modeling. Table 1.2 lists a categorization of those involved.

Table 1.2: Current provenenance of turbulent combustion codes.

Code categories	**Examples**
Public domain codes	KIVA.
Government lab codes	NASA, CERFACS, IFP, CSIRO, DSTO.
Commercial codes	Fluent/CFX, Adapco/STAR-CD.
Corporate codes	Ferrari, GE, Rolls-Royce, AVL, Ricardo.
Academic codes	Furnace, BOFFIN.

It is apparent that there are several hundred code developers around the world involved in advancing computer predictions for combustion flows. It is also apparent that most of these efforts are of a 'man and boy' nature and there is no overall co-ordination.

From the wider perspective of the challenges facing the world in combustion technology in the 21st century, it would seem that a more concerted international effort is needed. Computational modeling is a 'tool' and so in the hierarchy of research and development would seem to fall in the 'pre-competitive' range of this path to innovation in energy systems. If this is so, then it seems that the world would be better served if there was open and co-operative interchange - world wide - on developments in codes. And yet this is far from the case: computer codes are viewed as intellectual property, are subject to copyright law and in most cases are jealously protected and only communicated under stringent contractual arrangements at quite high prices.

A closer look at the situation, however, suggests that efficient coding and fast numerical algorithms are not the main limitations to advancement of combustion codes. These aspects can be quite significant. Of more importance, however, is the validity of the modeling approach and the sub-models that are used for the essential physics (including chemical physics) involved in the particular application. The fledgling status of the use of combustion prediction in industry depicted in Table 1.1 above is widely attributed to inadequacies in capturing the right physics in the modeling. This is often attributed to computational limitations that prohibit the inclusion of more detailed modeling of important physical processes. More often, it is acknowledged that the modeling approach is lacking in validity.

An important problem area in such modeling is that associated with turbulence-chemistry interactions. Bilger et al. [7] survey the modeling approaches and combustion sub-models that were of major interest at that time. The emphasis was on gaseous combustion - non-premixed, premixed and partially premixed. Here, the main chapters in this book outline and update these approaches and others that may have potential usefulness. It is noted that it is unlikely that a single modeling approach will be able to be adequate for all applications: it is more likely that different approaches will be needed for different applications.

The most important area is undoubtedly that of validation. All too often computer model predictions are 'validated' against sparse experimental data for full-scale combustors or for laboratory combustors that have some of the flow and other complexities of practical systems. Such data can often be adequately matched by choices made for boundary conditions that are unfettered by experimental measurement. Little information is gained about the adequacy of the modeling approach or of the combustion models used. A more scientific approach is to devise experiments in which the flow is simple enough for the boundary conditions to be adequately measured and for which the crucial physics that need modeling are dominant.

The Sandia flame series of experiments [4] using the Sydney [16, 23] burner for piloted jet diffusion flames are an outstanding example of the sort of validation experiments that are needed. The burner was designed to provide clear optical access to the regions of interest in the flame. The pilot was designed to be large enough so that the region of strong coupling between mixing and chemical reaction would be delayed beyond the near field where transitional turbulent flow associated with 'coherent structures' would be located. The upstream inflow boundary conditions could be adequately measured. The data from these experiments has proved to be a widely-recognized stringent test of the adequacy of the modeling of the coupling between turbulence and chemistry in such flows. It has led to a series of international workshops on turbulent non-premixed flames [3] in which there has been a robust interaction between modelers and experimentalists that has led to significant advances in both experimentation and modeling. Lifted flame stabilization [9], auto-ignition and forced ignition [18] in such flows are further steps in this hierarchy. These more complex problems still have relatively simple geometry and flow structure. Their study builds on the gains in experimental accuracy and improvements in modeling arising from the jet-flow experiments and provides an important stepping stone to the modeling of practical combustors. Bluff-body and swirling flames have been recognized as next steps in an hierarchy of such test problems and measurements have been made and some modeling predictions are becoming available [3]. Such flows where there is not a dominant advective time scale giving a direct analogue for the chemical time scales provide an important challenge to unsteady flamelet models and several of the new multi-scale modeling approaches.

It is only with such an hierarchical approach to validation that sure progress on advanced computational modeling for practical combustors can be made. Such an approach for systems with premixed combustion is yet to be satisfactorily defined [7]. The experiments at Sydney University [17] on spray jet mixing and combustion appear to be a firm basis for such an hierarchy for the combustion of particulate fuels.

1.6 Concluding Remarks

It is apparent that the dominant role of combustion technology in conversion of primary energy into secondary energy may be greatly reduced by the middle of the 21st century. It is important that the combustion community become actively engaged in the new research area of energy technology forecasting so that viable combustion technologies such as CCS do not become sidelined by over-optimistic projections for photo-voltaics and the hydrogen economy.

The fledgling status of computational modeling of practical combustion systems will be a seriously limiting factor in the ability of combustion technology to meet the 21st century challenges of climate change, energy security and tropospheric pollution. The advances needed in such computational modeling for practical combustion systems will require considerable investment in modeling research in this area, including the new paradigms of multi-scale modeling. It will also need an increased emphasis on research directed at improving experimental data bases for model validation that are of an hierarchical nature and start with relatively simple problems that encapsulate the most important physics.

Acknowledgements

This research is supported by the Australian Research Council.

References

1. Aharonov, D.: Quantum computing. In Annual Review of Computational Physics, VI, Singapore, World Scientific (1998)
2. Anderson, R.E., MacAdam, S., Viteri, F., Davies, D.O., Downs, J.P., Paliszewski, A.: Adapting gas turbines to zero emission oxy-fuel power plants. Proc. of ASME Turbo Expo 2008: Power for Land, Sea and Air, GT 2008, June 9–13, Berlin, Germany, GT2008-51377 (2008)
3. Barlow, R.S.: Proceedings of the TNF9 Workshop: http://public.ca.sandia.gov/TNF/9thWorkshop/TNF9.html (2009)
4. Barlow, R.S., Frank, J.H.: Effects of turbulence on species mass fractions in methane-air jet flames. Proc. Combust. Inst. **27**, 1087–1095 (1998)
5. Bilger, R.W.: The future for energy from the combustion of fossil fuels. 5th International Conference on Technologies and Combustion for a Clean Environment, Lisbon, Portugal, pp. 617–623, July (1999)
6. Bilger, R.W.: Zero release combustion technologies and the oxygen economy, 5th International Conference on Technologies and Combustion for a Clean Environment, Lisbon, Portugal, pp. 1039–1046, July (1999)

7. Bilger, R.W., Pope, S.B., Bray, K.N.C., and Driscoll, J.M.: Paradigms in turbulent combustion. Proc. Combust. Inst. **30**, 21–42 (2005)
8. Bilger, R.W., Wu, Z-J.: Carbon capture for automobiles using internal combustion rankine cycle engines. ASME J. Eng. Gas Turb. Power **131** 034502-1-4 (2009)
9. Cabra, R., Chen, J.-Y., Dibble, R.W., Karpetis, A.N., Barlow, R.S.: Lifted methane-air jet flame in vitiated coflow. Combust. Flame **143**, 491–506 (2005)
10. Chen, K., Karim, G.A., Watson, H.C.: Experimental and analytical examination of the development of inhomogeneities and autoignition during rapid compression of hydrogen-oxygen-argon mixtures. ASME J. Eng. Gas Turb. Power **125**, 458–465 (2003)
11. Fowles, M.: Black carbon sequestration as an alternative to bioenergy. Biomass Bioenergy **31**, 426–432 (2007)
12. HITEC Workshop: Next Generation Biofuels and Advanced Engines for Tomorrow's Transportation Needs, Workshop, November 17–18 (2009) Presentations available from (accessed on May 2010): `http://hitectransportation.org/events/`
13. IEA: Energy Technology Perspectives: Scenarios and Strategies to 2050. OECD/IEA, 75739 Paris, Cedex 15, France (2008)
14. IPCC Carbon Dioxide Capture and Storage, Intergovernmental on Climate Change, Cambridge University Press (2005)
15. Kazmerski, L.L.: Solar photovoltaics R& D at the tipping point: A 2005 technology review. J. Electron Spect. Related Phen. **150**, 105–135 (2006)
16. Masri, A.R., Dibble, R.W., Bilger, R.W.: Turbulent nonpremixed flames of methane near extinction: Mean structure from Raman measurements. Combust. Flame **71**, 245–266 (1988)
17. Masri, A.R.: International Workshop on Turbulent Combustion in Sprays, `http://www.FloHeaCom.UGent.be/` (2010)
18. Mastorakos, E.: Ignition of turbulent non-premixed flames. Progr. Energy Combust. Sci. **35**, 57–97 (2009)
19. McIlroy, A., McRae, G., Sick, V., Siebers, D.L., Westbrook, C.K., Smith, P.J., Taatjes, C., Trouve, A., Wagner, A.E., Rohlfing, E., Manley, D., Tully, F., Hilderbrandt, R., Green, W., Marceau, D., O'Neal, J., Lyday, M., Cebulski, F., Garcia, T.R., Strong, D.: Basic research needs for clean and efficient combustion of 21st century transportation fuels. Department of Energy Office of Science Report (2006)
20. Meehl, G.A., Washington, W.M., Ammann, C.M., Arbalaster, J.M., Wigley, T.M.L., Tebaldi, C.: Combinations of natural and anthropogenic forcings in the twentieth-century climate. J. Climate **17**, 3723–3277 (2004)
21. Mongia, H.C.: Recent progress in comprehensive modeling of gas turbine combustion. AIAA Paper 2008-1445 (2008)
22. Socolow, R.H., Lam, S.H.: Good enough tools for global warming policy making. Phil. Trans. R. Soc. A **365**, 897–934 (2006)
23. Starner, S.H., Bilger, R.W.: Characteristics of a piloted diffusion flame designed for study of combustion turbulence interactions. Combust. Flame **61**, 29–38 (1985)
24. UNFCC: United Nations Framework Convention on Climate Change. `http://www.unfcc.de` (1992)
25. Winskel, M., Markusson, N., Moran, B., Jeffrey, H., Anandarajah, G., Hughes, N., Candelise, C., Clarke, D., Taylor, G.,Chalmers, H., Dutton, G., Howarth, P., Jablonski, S., Kalyvas, C., Ward, D.: Decarbonising the UK energy system: Accelerated development of low carbon energy supply technologies, UK Energy Research Centre, Edinburgh University (2009)
26. World Energy Outlook 2008, International Energy Agency, Paris, France (2008)

Chapter 2
Turbulent Combustion: Concepts, Governing Equations and Modeling Strategies

Tarek Echekki and Epaminondas Mastorakos

Abstract The numerical modeling of turbulent combustion problems is based on the solution of a set of conservation equations for momentum and scalars, plus additional auxiliary equations. These equations have very well-defined foundations in their instantaneous and spatially-resolved forms and they represent a myriad of problems that are encountered in a very broad range of applications. However, their practical solution poses important problems. First, models of turbulent combustion problems form an important subset of models for turbulent flows. Second, the reacting nature of turbulent combustion flows imposes additional challenges of resolution of all relevant scales that govern turbulent combustion and closure for scalars. This chapter attempts to review the governing equations from the perspective of modern solution techniques, which take root in some of the classical strategies adopted to address turbulent combustion modeling. We also attempt to outline common themes and to provide an outlook where present efforts are heading.

2.1 Introduction

The subject of turbulent combustion spans a broad range of disciplines. The combination of the subject of turbulence on one hand and that of combustion already reveals the daunting task of predicting turbulent combustion flows. At the heart of the challenge is the presence of a broad range of length and time scales spanned by the various processes governing combustion and the degree of coupling between these processes across all scales.

Bilger et al. [4] have discussed the various paradigms that have evolved over the years to address the turbulent combustion problem. A running theme among these

Tarek Echekki
North Carolina State University, Raleigh NC 27695-7910, USA, e-mail: techekk@ncsu.edu

Epaminondas Mastorakos
Cambridge University, Cambridge, CB2 1PZ, UK, e-mail: em257@eng.cam.ac.uk

T. Echekki, E. Mastorakos (eds.), *Turbulent Combustion Modeling*,
Fluid Mechanics and Its Applications 95, DOI 10.1007/978-94-007-0412-1_2,

paradigms is the separation of scales to overcome the coupled multiscale complexity of turbulent combustion flows. In many respects, these strategies have been successful for a large class of problems and enabled the use of computational fluid dynamics (CFD) for the prediction and design of combustion in practical devices. The review by Bilger et al. [4] also identified recent trends in turbulent combustion modeling. These trends are motivated and enabled by the need to represent important finite-rate chemistry effects and non-equilibrium chemistry effects in combustion. Requirements for combustion technologies only 20 years ago are not the same as the requirements we dictate now. A variety of alternative fuels are explored in addition to high grade fossil fuels. Pollution mitigation also enforces additional requirements on the choice of the fuel, its equivalence ratio and mixture control (e.g. homogeneity of the charge).

Additional qualitative changes in the scope of turbulent combustion models can be gaged from two seminal contributions in the field of turbulent combustion. They correspond to two contributed volumes entitled 'Turbulent Reacting Flows', which were edited by Libby and Williams in 1980 and 1994 [27, 28]. A comparison of the topics covered in the two books and the present volume illustrates important expansions in the scope of the field of turbulent combustion. The key areas of expansion are outlined here:

- The role of chemistry in turbulent combustion simulations has seen a tremendous growth since the Libby and Williams [27, 28] volumes. Already in the 1980's software packages, such as Sandia's Chemkin [20] chemistry and transport libraries and associated zero-dimensional and one-dimensional applications, have enabled important advances in the prediction of the role of finite-rate chemistry effects in combustion [9]. Because of the disparity of chemical scales, stiff-integration software were becoming available for the integration of chemistry, such as the DASSL [41] and VODE [5] software packages. These packages played an essential role in the implementation of chemistry in combustion problems. Beyond the traditional strategies of quasi-steady state assumptions (QSSA) for species and partial equilibrim (PE) for reactions and sensitivity analysis, novel numerical tools have contributed to efficient strategies for the acceleration of chemistry in numerical codes. Examples of such strategies include mechanism automation strategies based on QSSA and PE [6], systematic eigenvalue based approaches, including the computational singular perturbation (CSP) [23] approach and the intrinsic low-dimensional manifold (ILDM) [47] approach, and direct relation graph [24]. Chapter 9 provides ample discussion on chemistry reduction and integration.
- Large-eddy simulation (LES) has emerged as as an alternative mathematical framework for the solution of transport equations for momentum and scalars. The traditional strategy, which is more common, is based on Reynolds-averaged Navier-Stokes (RANS) and associated equations for scalar transport, is not always sufficient for complex flows. LES has seen tremendous growth in the 1980's for turbulent non-reacting canonical flows; but, it is increasingly becoming a viable modeling framework for practical combustion flows. LES potentially enables accurate solutions of combustion flows incorporating unsteady flow effects.

Successful LES simulations with advanced combustion models are increasingly being used to model practical combustion devices [49].

- A broader range of combustion modes (e.g. premixed, non-premixed, stratified) and combustion regimes (e.g. thin or relatively thick reaction zones) are being explored in current and novel combustion technologies. The strict classification of combustion modes as either premixed or non-premixed, while powerful for the development of physical models of turbulent combustion, may not be separately adequate to represent partially-premixed combustion modes. Combustion in stratified mixture plays key role in a number of practical combustion applications, including diesel, gas turbine and homogeneous charge compression ignition (HCCI) combustion. In his textbook, Peters [40] dedicates an entire chapter to partially-premixed combustion with the recognition of the role of this combustion mode in a broad range of combustion problems, and novel modeling strategies have been developed for this combustion regime.
- Moreover, earlier important advances in turbulent combustion concerned primarily phenomena in which the separation of scales can be justified, such as in the cases of fast chemistry and in the flamelet regime. For example, both the eddy-dissipation model (EDM) [31] and the flamelet model [39] demonstrated a broad range of applicability in predicting combustion in practical combustion devices. However, combustion in other regimes where both chemistry and mixing are competitive during ignition (e.g. HCCI combustion) and flame-based combustion (e.g. distributed reaction, corrugated flames), are more challenging.
- Both books by Libby and Williams [27, 28] adopt the traditional view that turbulent combustion modeling primarily is a physical modeling challenge. However, the increasing availability of computational resources has enabled further and accelerated development of direct numerical simulation (DNS) techniques for combustion. In a recent paper, Valorani and Paolucci [53] make the observation 'No longer than 10 years ago, a direct numerical simulation (DNS) [11] of a turbulent flame with a four-step kinetics mechanism on a 10 mm box constituted the state-of-the-art in combustion simulation. Nowadays, the targets are DNSs of turbulent combustion of surrogate fuels, in half-a-meter domains.' As stated in Chapter 1 and elsewhere in this book, DNS may not be applicable to practical combustion devices for some time to come. However, other DNS-like techniques have been used to model laboratory-scale burners, such as recent simulations based on adaptive mesh refinement (AMR) [1, 2].

Our current understanding of the fundamental laws governing reacting flows enables us to formulate detailed physical models, with minimum empiricism, for a large number of the processes underlying turbulent combustion. For example, atomistic simulations may be used to construct databases for rate constants and thermochemical and molecular transport properties of reacting species. But, atomistic approaches alone may not extend to the scales relevant to practical combustion problems; yet, with the help of constitutive relations derived for molecular processes, continuum-based formulations for reacting flows are a good starting point.

Even within the continuum limit, various strategies may be adopted. These strategies may reflect the formulation of the mathematical models for the governing equations as well as their numerical solution in addition to inherent simplification of these equations due to the flow regime (e.g. low Mach number formulations). They also reflect the scope of the modeler whether she/he is interested in statistical results or fully-resolved (spatially and temporally) results. The latter scope belongs to the realm of direct numerical simulations (DNS) where the governing equations are solved without filtering or averaging of the solution vector and with a full account of the required spatial and temporal resolution within the continuum limit. However, recourse to unsteady information is progressively seen also as one of the reasons for moving towards LES as in, for example, the effort to capture ignition or extinction phenomena [52]. The governing equations for DNS will be the starting point for discussing the different strategies adopted to address the mathematical models in turbulent combustion and their numerical solutions. Our emphasis is on two mathematical frameworks for representing the solution vector based on RANS and LES. Following effort in the turbulence community, other mathematical frameworks may be feasible as well, but RANS and LES are the most common approaches in modern turbulent combustion modeling and will form the focus of this book.

2.2 Governing Equations

2.2.1 Conservation Equations

The governing equations for turbulent combustion flows may be expressed in different forms; however, they normally are represented as transport equations for overall continuity, momentum and additional scalars that can be used to spatially- and temporally- resolve the thermodynamic state of the mixture. These equations are augmented by initial and boundary conditions, as well as constitutive relations for atomistic processes (e.g. reaction, molecular diffusion, equations of state). Therefore, in addition to density, transport equations for the evolving momentum and composition (e.g. mass or mole fractions, species densities or concentrations) and a scalar measure of energy (e.g. internal energy, temperature, or enthalpy). For illustration purposes, we present the compressible form of the instantaneous governing equations in non-conservative form for the mass density, momentum, species mass fractions and internal energy. A more detailed discussion on the various forms and their equivalence, especially for the energy equation can be found in the textbooks by Williams [57] or Poinsot and Veynante [44].

- Continuity

$$\frac{\partial \rho}{\partial t} + \nabla \cdot \rho \mathbf{u} = 0, \tag{2.1}$$

- Momentum

$$\rho \frac{D\mathbf{u}}{Dt} = \rho \frac{\partial \mathbf{u}}{\partial t} + \rho \mathbf{u} \cdot \nabla \mathbf{u} = -\nabla p + \nabla \cdot \tau + \rho \sum_{k=1}^{N} Y_k \mathbf{f}_k, \tag{2.2}$$

- Species continuity ($k = 1, \cdots, N$)

$$\rho \frac{DY_k}{Dt} = \rho \frac{\partial Y_k}{\partial t} + \rho \mathbf{u} \cdot \nabla Y_k = \nabla \cdot (-\rho \mathbf{V}_k Y_k) + \omega_k, \tag{2.3}$$

- Energy

$$\rho \frac{De}{Dt} = \rho \frac{\partial e}{\partial t} + \rho \mathbf{u} \cdot \nabla e = -\nabla \cdot \mathbf{q} - p \nabla \cdot \mathbf{u} + \tau : \nabla \mathbf{u} + \rho \sum_{k=1}^{N} Y_k \mathbf{f}_k \cdot \mathbf{V}_k. \tag{2.4}$$

In the above equations, ρ is the mass density; $\mathbf{u}$ is the velocity vector; p is the pressure; $\mathbf{f}_k$ is the body force associated with the kth species per unit mass; τ is the viscous stress tensor; $\mathbf{V}_k$ is the diffusive velocity of the kth species, where the velocity of the kth species may be expressed as the sum of the mass-weighted velocity and the diffusive velocity, $\mathbf{u} + \mathbf{V}_k$; ω_k is the kth species production rate; e is the mixture internal energy, which may be expressed as $e = \sum_{k=1}^{N} h_k Y_k - p/\rho$; $\mathbf{q}$ is the heat flux, which represents heat conduction, radiation, and transport through species gradients and the Soret effect. The solution vector, Ξ represented by the above governing equations (2.1)–(2.4) is $\Xi = (\rho, \rho\mathbf{u}, \rho\mathbf{Y}, \rho e)$ in its conservative form or $\Xi = (\rho, \mathbf{u}, \mathbf{Y}, e)$ in its non-conservative form. The governing equations may be expressed in a more compact form as follows:

$$\frac{D\Xi}{Dt} = \mathbf{F}(\Xi) \tag{2.5}$$

where $\mathbf{F}(\Xi)$ represents the right-hand side of the governing equations and features terms with spatial derivatives (e.g. diffusive fluxes for mass and heat) and source terms (e.g. reaction source terms). The material derivative $D\Xi/Dt$ includes both the unsteady term and the advective term in the Eulerian representation such that: $D\Xi/Dt \equiv \partial\Xi/\partial t + \mathbf{u} \cdot \nabla\Xi$. As can be seen, a number of terms in the governing equations are not explicitly expressed in terms of the solution vector and must rely on constitutive relations, equations of state or any additional auxiliary relations. These terms include expressions for the viscous stress, the species diffusive velocities, the body forces, the species reaction rate and the heat flux. The bulk of these terms have their origin in the molecular scales, and therefore, the role of constitutive relations is to represent them in continuum models. In fact, the use of constitutive equations is the first level of multiscale treatment for the modeling of turbulent combustion flows. Alternative, but significantly more costly approaches, involve their determination using atomistic models coupled 'on the fly' with continuum models. However, cases where such approaches are needed are very limited.

2.2.2 *Constitutive Relations, State Equations and Auxiliary Relations*

2.2.2.1 Constitutive Relations, Transport Properties and State Equations

The constitutive relations for the conversation equations outlined above represent primarily relations between transport terms for momentum, energy and species and the solution vector as well as relations that describe the rate of chemistry source terms in the species equations. They are designed to represent atomistic scale effects of transport and reaction. Below we outline the principal terms that are represented by constitutive relations.

- The pressure and viscous stress tensor: In gas-phase flows applicable to combustion problems, the Newtonian fluid assumption is reasonably valid, and the viscous stress tensor may be represented through the following relation:

$$\tau = \mu\left[(\nabla \mathbf{u}) + (\nabla \mathbf{u})^T\right] + \left(\frac{2}{3}\mu - \kappa\right)(\nabla \cdot \mathbf{u})\,\mathbf{I} \tag{2.6}$$

  In this expression, μ is the dynamic viscosity; κ is the bulk viscosity; and $\mathbf{I}$ is the identity matrix. The principle of corresponding states provides generalized curves for the viscosity of gases, liquids and supercritical fluids for a broad range of temperature and pressure conditions. The principle states that a reduced viscosity, based on the ratio of the dynamic viscosity to that at critical conditions, may be uniquely defined in terms of a reduced temperature and pressure, both reduced values result from the normalization of temperature and pressure with their corresponding critical values.
- The diffusive mass flux, $\rho Y_k \mathbf{V}_k$: The diffusive mass flux represents the transport of species in addition to their transport with the bulk flow, $\mathbf{u}$. Diffusive mass transport may be associated with gradients in mass or species concentration, the so-called Fickian diffusion, temperature gradients, or the so-called Dufour effect, and pressure gradient. A hierarchy of models for the diffusive mass flux may be adopted. The first is based on adopting a Fick's law model using mixture-averaged transport coefficients:

$$X_k \mathbf{V}_k = -D_k^m \nabla X_k \tag{2.7}$$

  where D_k^m is the mixture-averaged mass diffusion coefficient for species k. The mixture-averaged mass diffusion coefficient is derived, in general, using mixture weighting rules and multi-component diffusion coefficients. A simple form of the mixture-averaged diffusivity is based on the assumption of constant diffusion coefficients' ratios (e.g. fixed Lewis numbers or Schmidt numbers), such that the mixture-averaged mass diffusion coefficient is expressed as follows [50]:

$$D_k^m = \frac{\lambda}{\rho c_p \mathrm{Le}_k} \tag{2.8}$$

where c_p is the mixture specific heat; λ is the mixture thermal conductivity; and Le_k is the kth species Lewis number.
The second approach is based on a multi-component formulation [8]:

$$\mathbf{V}_k = \frac{1}{X_k W} \sum_{j=1, j\neq k}^{N} W_j D_{kj} \mathbf{d}_j - \frac{D_{T,k}}{\rho Y_k} \frac{1}{T} \nabla T \tag{2.9}$$

In this expression, D_{kj} and $D_{T,k}$ are the binary mass diffusivity between species k and j and the thermal diffusion coefficient, respectively; $\mathbf{d}_j$ is the concentration and pressure gradients for species j:

$$\mathbf{d}_j = \nabla X_j + (X_j - Y_j) \frac{\nabla p}{p} \tag{2.10}$$

Detailed formulations for D_{kj} and $D_{T,k}$ may be found in various textbooks (see for example, Kee et al. [21]).

- The heat flux vector, $\mathbf{q}$: The heat flux vector $\mathbf{q}$ features contributions from different modes of heat transfer, including heat conduction, heat diffusion by mass diffusion of the various species, thermal diffusion (Dufour effect), and radiative heat transfer. A general form of the heat flux featuring the contribution of these different heat transfer modes may be written as follows:

$$\mathbf{q} = -\lambda \nabla T + \rho \sum_{i=1}^{N} h_i Y_i \mathbf{V}_i + R_u T \sum_{i=1}^{N} \sum_{j=1}^{N} \left(\frac{X_j D_{T,i}}{W_i D_{ij}} \right) (\mathbf{V}_i - \mathbf{V}_j) + \mathbf{q}_{rad} \tag{2.11}$$

In this equation, λ is the mixture thermal conductivity and $\mathbf{q}_{rad}$ is the radiative heat flux.

- The chemical reaction term, ω_k: This term is derived from the law of mass action, which dictates that the rate of chemical reactions is proportional to the concentrations of the contributing species. The proportionality factor is primarily a function of temperature and is denoted as the reaction rate constant. Contributions to this term include statistical information about the rates of collisions, and the fraction of collisions resulting in reactions as well as steric factor, which take into consideration the shapes of molecules during collisions. The following equation represents the rate of production of species k due to its involvement in R reversible reactions:

$$\omega_k = W_k \sum_{r=1}^{R} \left\{ \left(\nu''_{k,r} - \nu'_{k,r} \right) \left[k_{f,r} \prod_{j=1}^{N} \left(\frac{X_j p}{R_u T} \right)^{\nu'_{k,r}} - k_{b,r} \prod_{j=1}^{N} \left(\frac{X_j p}{R_u T} \right)^{\nu''_{k,r}} \right] \right\}, \tag{2.12}$$

where

$$k_{f,r}(T) = A_r T^{\alpha_r} \exp\left(\frac{-E_{a,r}}{R_u T} \right), k_{b,r} = \frac{k_{f,r}}{K_{C,r}} \tag{2.13}$$

In these expressions, W_k is the molecular weight for species k; $\nu'_{j,r}$ and $\nu''_{j,r}$ are the rth reaction stoichiometric coefficients on the reactants and the products sides, respectively; $k_{f,r}$ and $k_{b,r}$ are the forward and backward rate constants for the reversible reaction, r. The backward reaction rate constant is related to the forward rate constant through the concentration-based equilibrium constant, $K_{C,r}$ for reaction r. In the Arrhenius form for the forward rate constant expression, A_r and α_r are the pre-exponential coefficients, and $E_{a,r}$ is the activation energy for the forward reaction, r. An elementary reaction, r, is prescribed as follows:

$$\sum_{j=1}^{N} \nu'_{j,r} \mathrm{A}_j \rightarrow \sum_{j=1}^{N} \nu''_{j,r} \mathrm{A}_j \tag{2.14}$$

where A_j is the jth species chemical symbol.
The integration of the chemical source term in the species equation (as well as in the temperature or sensible enthalpy forms of the energy equation) poses important and limiting challenges in computational combustion, as discussed below.

The determination of transport properties for momentum, mass and energy remains an understated challenge. Various software packages for the evaluation of transport properties are available, including MIXRUN [56], TRANLIB [19], EGLIB [13] and DRFM [38]. A first challenge is to assemble reliable data for potential parameters that contribute to the evaluation of the collision integrals. Paul [38] find that special attention needs to be made in determining the transport properties for molecules with dipole moments (e.g. H atom, H_2 molecule) and indeed numerical simulations with different levels of modeling transport can lead to different results.

2.2.2.2 Mixture Properties and State Equations

State equations enable to evaluate thermodynamic properties from known properties. A common relation involves the ideal gas law:

$$p = \rho R_u T \sum_{j=1}^{N} \left(\frac{Y_j}{W_j} \right) \tag{2.15}$$

The caloric equation of state may be used to relate a species enthalpy or internal energy to temperature as follows:

$$h_k(T) = h_{k,chem} + \int_{T^\circ}^{T} c_{p,k} dT \tag{2.16}$$

and

$$e_k(T) = h_{k,chem} + \int_{T^\circ}^{T} c_{v,k} dT \tag{2.17}$$

where h_k and e_k are the kth species total enthalpies and internal energies; T° is a reference temperature for the sensible enthalpy. Here, $h_{k,chem}$ corresponds to the chemical enthalpy of the kth species, and the second terms on the right hand-sides of the two above equations corresponds to the sensible contributions; $c_{v,k}$ and $c_{p,k}$ are the specific heats for species k at constant volume and pressure, respectively.

2.2.2.3 Other Transport Equations

Along with, or instead of, the scalar transport equations, transport equations for additional scalars may be used. These include conserved scalars (e.g. mixture fraction, total enthalpy), normalized reaction progress variables and flame surface variables (e.g. flame surface density).

Conserved scalars may be found in different aspects of combustion analysis from theory to experiment. They offer the convenience that their transport equations are devoid of source terms. Therefore, their integration is not subject to the steep time constraints of integrating chemistry. The Shvab-Zeldovich [57] formulation offers an early example of the use of conserved scalars in the limit of fast chemistry in terms of the so-called 'coupling functions'. The same concept based on this formulation resulted in one of the classic analytical solutions in combustion based on the Burke-Schumann jet flame model [57]. However, the concepts of elemental mass fractions and mixture fractions have offered significantly more insight into processes in turbulent combustion, especially in non-premixed combustion. From a mixture composition, it is possible to construct an elemental mass fraction, Z_l, for element l, which may be prescribed as:

$$Z_l = \sum_{j=1}^{N} \mu_{j,l} Y_j \tag{2.18}$$

where $\mu_{j,l}$ is the mass fraction of element l in species j. The elemental mass fraction is unaltered by reaction; and therefore, there is no source term associated with its transport equation:

$$\rho \frac{DZ_l}{Dt} = \rho \frac{\partial Z_l}{\partial t} + \rho \mathbf{u} \cdot \nabla Z_l = \nabla \cdot (-\rho \mathbf{V}_l Z_l). \tag{2.19}$$

Here, the diffusive velocity associated with the elemental mass fraction is expressed as follows:

$$\sum_{j=1}^{N} \mathbf{V}_j \mu_{j,l} Y_j = \mathbf{V}_l Z_l \tag{2.20}$$

The mixture fraction represents a normalized form of the elemental mass fraction, and it is a parameter of great value for non-premixed chemical systems. It measures the fraction by mass in the mixture of the elements, which originates in the fuel. When derived from elemental mass fractions, it may be expressed in normalized form as:

$$F_l = \frac{Z_l - Z_{l,o}}{Z_{l,f} - Z_{l,o}} \tag{2.21}$$

where the subscripts o and f refer to the oxidizer and the fuel mixture conditions, respectively. In a mixing system of fuel and oxidizer streams, values of the mixture fractions based on different elements may be different because of differential diffusion effects. Element-averaged mixture fractions, such as the Bilger mixture fraction [3], may be adopted:

$$F_{\text{Bilger}} = \frac{2\left(Z_C - Z_{C,o}\right)/W_C + \left(Z_H - Z_{H,o}\right)/(2W_H) - \left(Z_O - Z_{O,o}\right)/W_O}{2\left(Z_{C,f} - Z_{C,o}\right)/W_C + \left(Z_{H,f} - Z_{H,o}\right)/(2W_H) - \left(Z_{O,f} - Z_{O,o}\right)/W_O} \tag{2.22}$$

where the subscripts C, H and O correspond to the elements C, H and O, respectively, and the symbol W refers to their corresponding molar masses. The coefficients in front of the different elemental contributions serve the important role of maintaining the stoichiometric Bilger mixture fraction value identical to the elemental mixture fractions. As stated earlier, the mixture fraction is an important parameter for the modeling of non-premixed systems [3, 40, 57].

2.3 Conventional Mathematical and Computational Frameworks for Simulating Turbulent Combustion Flows

Within the continuum limit, there are different mathematical and computational frameworks to model turbulent combustion flows. These frameworks address the way the governing equations are modified and the way the solutions are implemented computationally (e.g. discretization). Strategies to overcome the limitations of resolving all the time and length scales even within the continuum limit motivates two principal classes of modeling frameworks associated with model-adaptivity or mesh-adaptivity or both. Model adaptivity refers to a class of models in which the governing equations, and accordingly the solution vector Ξ, are modified to a reduced order, a reduced dimension, or a statistical form, which effectively decouples or eliminates ranges of scales from the solution vector. Mesh adaptivity refers to a class of models in which the solution vector, Ξ, is resolved by adapting the grid hierarchy or the resolution hierarchy where it is needed to meet prescribed error criteria.

As indicated above, model adaptivity is concerned with modifying the governing equation and the solution vector. For combustion flows, three principal strategies have been implemented for model adaptivity; while, potentially other approaches may be considered. They correspond to DNS, RANS and LES.

2.3.1 Direct Numerical Simulation (DNS)

DNS corresponds to the solution of the 3D unsteady governing equations (Eqs. 2.1–2.4) with the necessary resolution required to accurately integrate the solution in time and predict the details of velocity and scalar fields. Therefore, DNS offers the best resolved framework for the study of turbulent combustion flows. A typical 3D unsteady DNS in combustion must span the ranges of time and length scales discussed above (approximately 4-5 decades in length scales within the continuum regime in a given direction), which entail resolution requirements of the order of trillions of grid points or higher and tens of millions of time steps. Yet, the state-of-the-art DNS of combustion have been limited to computational domains that are approximately one order of magnitude smaller in linear scale (or three orders of magnitude in volume) than laboratory flames or practical combustion devices. However, achieving these length scales is fast approaching with petascale capabilities and beyond. Nonetheless, DNS remains a powerful tool to understand important turbulence-chemistry interaction processes and formulate closure models in turbulent combustion [43, 54, 55]. Computational requirements for DNS may vary depending on the level of description of the chemistry, molecular transport and radiation as well as the representation of the governing equations (e.g. low-Mach number formulation vs. compressible formulations). Examples of computational requirements may be found in a recent paper by Chen et al. [7].

A principal challenge for DNS remains the temporal integration of the conservation equations, especially those pertaining to the integration of the reactive scalars. A temporal resolution from the fastest reactions (of the order of 10^{-9} s for hydrocarbon chemistry) to integral scales of the flow results in hundreds of thousands to million time steps with explicit integration schemes; accordingly, DNS simulations remain largely CPU-limited. Lu and Law [29] present an analysis of the cost of integrating chemistry within DNS. Their analysis shows that:

- The size of a chemical mechanism (i.e. the number of reactions) increases approximately linearly with the number of chemical species considered; the scaling factor is approximated as 5 between the chemical mechanism size and the number of species involved. This scaling is presented for hydrocarbon fuels. However, it is clear that as DNS applications are extended from hydrogen and simple hydrocarbon fuels to more common fuels (e.g. gasoline, diesel, kerosene), additional cost is associated with both the transport of more scalar equations as well as in the evaluation and the integration of non-linear reaction rate terms. The task is daunting given that more than one order of magnitude separates the size of simple and more complex fuel chemistries.
- The computational cost of DNS at each grid, C_{DNS}, also scales approximately linearly with the number of species, N, involved: $C_{DNS} \propto N$. The proportionality factor subsumes contributions associated with the spatial resolution and the cost of advancing the scalar transport equations, including the evaluation of transport properties and chemical reaction rates.

Therefore, aggressive strategies for chemistry reduction are warranted and may need to go beyond the development of skeletal mechanisms. Additional strategies for chemistry calculation acceleration are needed equally to overcome the stiff chemistry. These strategies have been pursued and significant progress has been achieved in recent years. An additional challenge is to account for the transport of tens to hundreds of species in mechanisms that range in size from tens to thousands of reactions. The subject of chemistry reduction and acceleration has received increasing interest in recent years (see for example the recent reviews by Lu and Law [29] and Pope and Ren [48]). Chapter 9 details further strategies for chemistry reduction and acceleration.

Another equally important effort is to address spatial resolution requirements. Spatial resolution must resolve the thinnest layers of reaction zone structures; these layers represent the balance of reaction and diffusion and at their thinnest may be of the order of 10 μm or smaller. It is difficult to justify not to resolve these layers if they serve a role in the combustion process; and often strategies to address the resolution of these layers may be implemented through adapting the spatial resolutions where these layers are present and coarsening the resolution when such fine resolutions are not needed. Adaptive resolution strategies offer the most promising strategies for addressing spatial resolution and often result in almost an order of magnitude gain in the size of problems to be solved in comparison with DNS. Typical examples of mesh adaptivity include adaptive mesh refinement (AMR), which is discussed in Chapter 13 and wavelet-based adaptive multiresolution strategies, which are discussed in Chapter 14.

2.3.2 Reynolds-Averaged Navier-Stokes (RANS)

The Reynolds-averaged Navier-Stokes (RANS) formulation is based on time or ensemble averaging of the instantaneous transport equations for mass, momentum and reactive scalars. Within the context of scale separation, the RANS approach indiscriminately impacts all scales. Consequently, all unsteady turbulent motion and its coupling with combustion processes are unresolved over the entire range of length and time scales of the problem, and closure models are needed to represent the unresolved physics. An additional complexity introduced by averaging is that non-linear terms in the governing equations result in unclosed terms. The closure problem is particularly critical for the reaction source terms in the species and some forms of the energy equations. The treatment of these terms has been the scope of the moment-based methods. We illustrate the closure problem using the transport equation for the conservation equations above (Eqs. 2.1–2.4). Before listing the conservation equations, we briefly address the advantages of density-weighted averaging or the so-called Favre-averaging [14, 18]. With Favre-averaging, all momentum and scalars, at the exception of the density, the pressure and diffusive fluxes, are decomposed using a Favre-averaged means and fluctuations:

$$\Xi = \widetilde{\Xi} + \Xi' \tag{2.23}$$

The Favre average, $\widetilde{\Xi}$ may be expressed in terms of the non-weighted average as:

$$\widetilde{\Xi} \equiv \frac{\overline{\rho \Xi}}{\overline{\rho}} \tag{2.24}$$

The contribution Ξ' corresponds to the fluctuating components of the solution vector Ξ relative the Favre mean. The overbar denotes an unweighted ensemble average over a statistically-meaningful set of realizations; the symbol '$\sim$' denotes a density-weighted ensemble average. Density-weighted averaging eliminates the need to explicitly represent the density-momentum and density-scalar correlations, which when kept in the governing equations generate additional closure terms. Based on the above conservation equations, the Favre-averaged continuity, momentum and scalar equations are expressed as follows:

- Continuity

$$\frac{\partial \overline{\rho}}{\partial t} + \nabla \cdot \overline{\rho} \widetilde{\mathbf{u}} = 0, \tag{2.25}$$

- Momentum

$$\overline{\rho} \frac{\partial \widetilde{\mathbf{u}}}{\partial t} + \overline{\rho} \widetilde{\mathbf{u}} \cdot \nabla \widetilde{\mathbf{u}} = -\nabla \overline{p} + \nabla \cdot \overline{\tau} + \overline{\rho} \sum_{k=1}^{N} \widetilde{Y_k \mathbf{f}_k} - \nabla \cdot \left(\overline{\rho} \widetilde{\mathbf{u}'\mathbf{u}'} \right), \tag{2.26}$$

- Species continuity ($k = 1, \cdots, N$)

$$\overline{\rho} \frac{\partial \widetilde{Y}_k}{\partial t} + \overline{\rho} \widetilde{\mathbf{u}} \cdot \nabla \widetilde{Y}_k = \nabla \cdot (-\overline{\rho} \widetilde{\mathbf{V}_k Y_k}) + \overline{\omega}_k - \nabla \cdot (\overline{\rho} \widetilde{\mathbf{u}' Y_k'}), \tag{2.27}$$

- Energy

$$\overline{\rho} \frac{\partial \widetilde{e}}{\partial t} + \overline{\rho} \widetilde{\mathbf{u}} \cdot \nabla \widetilde{e} = -\nabla \cdot \widetilde{\mathbf{q}} - \overline{p \nabla \cdot \mathbf{u}} + \overline{\tau : \nabla \mathbf{u}} + \overline{\rho} \sum_{k=1}^{N} \widetilde{Y_k \mathbf{f}_k \cdot \mathbf{V}_k} - \nabla \cdot \left(\overline{\rho} \widetilde{\mathbf{u}' e'} \right). \tag{2.28}$$

In the RANS formulation, the solution vector is expressed as $\overline{\Xi} = (\overline{\rho}, \widetilde{\mathbf{u}}, \widetilde{\mathbf{Y}}, \widetilde{e})$. Both the source term and the advective term are non-linear contributions to the governing equations for the species, and invariably these terms will be unclosed since there is no explicit transport equation used to solve them. Additional new terms in the governing equation correspond to the Reynolds stresses and fluxes: $\overline{\rho} \widetilde{\mathbf{u}'\mathbf{u}'}$, $\overline{\rho} \widetilde{\mathbf{u}' Y_k'}$ and $\overline{\rho} \widetilde{\mathbf{u}' e'}$ which are also unclosed and must be modeled. It is quite common to treat this term as a turbulent diffusion term with a gradient model. The molecular diffusion term is also unclosed; but, its description may depend largely on how it is modeled and how its effects are prescribed with the reaction source term. It is also common to assume that the turbulent diffusion term is much larger than the molecular diffusion term in the governing equations and that the molecular diffusion term is often dropped from the above governing equations. This is not strictly true because these two transport terms may act on different scales. Therefore, the effects of molecular diffusion may still have to be represented (typically through the chemistry closure). Nonetheless, the most critical closure arises from modeling the reaction source term, $\overline{\omega}_k$.

To motivate the strategies adopted for the closure for $\overline{\omega}_k$, the statistical representation of this term is expressed as follows:

$$\overline{\omega}_k = \int_{\Xi} \omega_k(\psi) f(\psi) d\psi. \tag{2.29}$$

In this expression, $f(\psi)$ is the joint scalar probability density function (PDF). The vector ψ represents components of the thermodynamic state vector, which may include for example, pressure, temperature and composition. Therefore, the vector, ψ, may be a subset of the solution vector, Ξ, which also includes the momentum components. The joint PDF contains the complete statistical information about all scalars. Therefore, the averaged scalar field, its moments and any related functions of the field may be constructed using this joint scalar PDF:

$$\overline{\Xi} = \int_{\psi} \Xi(\psi) f(\psi) d\psi, \tag{2.30}$$

and

$$\overline{\Omega(\Xi)} = \int_{\psi} \Omega(\psi) f(\psi) d\psi. \tag{2.31}$$

A density-weighted PDF may be defined as well, which may be written as follows:

$$\widetilde{f}(\psi) = \frac{\rho(\psi)}{\overline{\rho}} f(\psi) \tag{2.32}$$

These expressions offer an important window into closure strategies in turbulent combustion within the context of RANS. An accurate description of averaged scalars, their moments and functions of these moments must involve an accurate account of the state-vector solution (i.e. the instantaneous correlation, $\Xi(\psi)$ inside the integral) as well as an accurate account of the statistical distribution, $f(\psi)$. Often, the modeling of the two contributions is coupled, and the choice of the combustion mode or regime may enable strategies for the modeling of the state-vector solutions as well as the joint scalar PDF.

2.3.3 Large-Eddy Simulation (LES)

The third approach is based on spatially filtering the instantaneous equations to capture the contribution of large scales, resulting in transport equations for spatially filtered mass, momentum and scalars, while the effects of smaller scales are modeled. This approach is known as large-eddy simulation [42]. LES relies on scale separation between (kinetic) energy containing eddies and small scales responsible for its dissipation (the so-called subgrid scale, or SGS). The approach is rooted in the traditional view of turbulent flows where the bulk of turbulent kinetic energy originates at the large scales; however, this choice has limited justification in combustion flows: important physics may reside and originate at small scales.

From a modeling standpoint, LES provides a number of important advantages towards the prediction of turbulent combustion flows over RANS. First, LES captures large scale information in both the momentum and scalar fields. Therefore, it is able to capture the role of large flow structures on mixing and, therefore, on combustion. These flow structures are inherently unsteady, and capturing their interactions with combustion chemistry is very crucial in a broad range of practical applications. For example, tumble and swirl in internal combustion engines serve to enhance the volumetric rate of heat release and contribute to cycle-to-cycle variations. Moreover, the lift-off and blow-out of lifted flames in practical burners is dependent on the unsteady flow dynamics around the flame leading edge and the inherent instabilities in the presence of shear. Another example is associated with the prediction of thermo-acoustic and other flow-associated instabilities in gas turbine combustors.

Second, because LES is designed to capture the role of a band of scales, it can naturally be implemented within the context of multiscale frameworks. In these frameworks, simulations of the subgrid scale physics is implemented along with LES to capture the contributions from all relevant scales. Hybrid approaches of LES with low-dimensional stochastic models, such as the LEMLES and the ODTLES frameworks discussed in Chapters 10 and 11 illustrate the implementation of LES for combustion within the context of multiscale approaches. However, as outlined in this book and briefly discussed above, LES is a promising strategy within the context of moment-based approaches, such as non-premixed and premixed flames (Chapters 3 and 4), the conditional moment closure (CMC) model (Chapter 5), the transported PDF (Chapter 6), and the multiple mapping conditioning (MMC) approach (Chapter 7).

Third, since closure in LES targets primarily subgrid scale physics, a higher degree of universality in statistics may be achieved when the contribution from geometry-dependent large scales are eliminated from consideration.

We consider a filtering operation applied to the conservation equations. The filtering operation corresponds to the implementation of a low-pass filter, which is expressed as follows for a solution vector component, Ξ:

$$\overline{\Xi}(\mathbf{x},t) = \int_{\Delta} G\left(\mathbf{x},\mathbf{x}'\right) \Xi\left(\psi;\mathbf{x}',t\right) d\mathbf{x}' \tag{2.33}$$

In this expression, G is a filtering function over 3D space with a characteristic scale, Δ, the filter size. Similarly to the RANS formulation for variable-density flows, the filtered solution vector, at the exception of the filtered density, is based on density-weighted filtering, such that:

$$\widetilde{\Xi}\left(\psi;\mathbf{x},t\right) \equiv \frac{\overline{\rho\Xi}}{\overline{\rho}} \tag{2.34}$$

where

$$\overline{\rho\Xi} = \int_{\Delta} G\left(\mathbf{x},\mathbf{x}'\right) \rho\left(\psi;\mathbf{x}',\mathbf{t}\right) \Xi\left(\psi;\mathbf{x}',\mathbf{t}\right) d\mathbf{x}' \tag{2.35}$$

The Favre-filtered momentum and scalar equations are expressed as follows:

- Continuity

$$\frac{\partial \overline{\rho}}{\partial t} + \nabla \cdot \overline{\rho} \widetilde{\mathbf{u}} = 0, \tag{2.36}$$

- Momentum

$$\overline{\rho} \frac{\partial \widetilde{\mathbf{u}}}{\partial t} + \overline{\rho} \widetilde{\mathbf{u}} \cdot \nabla \widetilde{\mathbf{u}} = -\nabla \overline{p} + \nabla \cdot \overline{\tau} + \overline{\rho} \sum_{k=1}^{N} + \overline{\rho} \sum_{k=1}^{N} \widetilde{Y_k \mathbf{f}_k} + \nabla \cdot [\overline{\rho}(\widetilde{\mathbf{u}}\widetilde{\mathbf{u}} - \widetilde{\mathbf{u}\mathbf{u}})], \tag{2.37}$$

- Species continuity ($k = 1, \cdots, N$)

$$\overline{\rho} \frac{\partial \widetilde{Y}_k}{\partial t} + \overline{\rho} \widetilde{\mathbf{u}} \cdot \nabla \widetilde{Y}_k = \nabla \cdot (-\overline{\rho} \widetilde{\mathbf{V}_k Y_k} + \overline{\omega}_k + \nabla \cdot \left[\overline{\rho} \left(\widetilde{\mathbf{u}} \widetilde{Y}_k - \widetilde{\mathbf{u} Y_k} \right) \right], \tag{2.38}$$

- Energy

$$\overline{\rho} \frac{\partial \widetilde{e}}{\partial t} + \overline{\rho} \widetilde{\mathbf{u}} \cdot \nabla \widetilde{e} = -\nabla \cdot \widetilde{\mathbf{q}} - \overline{p \nabla \cdot \mathbf{u}} + \overline{\tau : \nabla \mathbf{u}} + \overline{\rho} \sum_{k=1}^{N} \widetilde{Y_k \mathbf{f}_k \cdot \mathbf{V}_k} + \nabla \cdot [\overline{\rho} (\widetilde{\mathbf{u}} \widetilde{e} - \widetilde{\mathbf{u} e})]. \tag{2.39}$$

We have kept the same symbols for operations as above, although they correspond to spatial filtering operations instead of ensemble or time-averaging as in RANS. Here the overbar corresponds to a process of unweighted spatial filtering and the '$\sim$' corresponds to a density-weighted spatial filtering. Again, considering the revised solution vector, $\overline{\Xi} = (\overline{\rho}, \widetilde{\mathbf{u}}, \widetilde{\mathbf{Y}}, \widetilde{e})$, additional terms are present in the governing equation, which correspond to subgrid scale stresses $\overline{\rho}(\widetilde{\mathbf{u}}\widetilde{\mathbf{u}} - \widetilde{\mathbf{u}\mathbf{u}})$ and scalar fluxes, $\overline{\rho}(\widetilde{\mathbf{u}}\widetilde{Y}_k - \widetilde{\mathbf{u}Y_k})$ and $\overline{\rho}(\widetilde{\mathbf{u}}\widetilde{e} - \widetilde{\mathbf{u}e})$. These terms also are unclosed and require modeling. The molecular diffusion term, $\nabla \cdot (-\overline{\rho} \widetilde{\mathbf{V}_k Y_k})$, may be insignificant in the governing equation relative to the scalar flux on the LES resolution, it acts at scales that are fundamentally different from those of the scalar fluxes; and therefore, its contribution may be closely tied to the reaction source term and its closure. The process of averaging or filtering of the governing equations invariably leaves the contributions of the unresolved physics unclosed, and similar challenges of closure are found.

Similarly to the RANS formulation, an important closure term corresponds to the reaction source terms, $\overline{\omega}_k$. Here, a similar concept to the PDF may be used based on the filtered-density function (FDF) [15]:

$$\overline{\omega}_k = \int_{\psi} \omega_k(\psi) F(\psi) d\psi \tag{2.40}$$

and

$$\overline{\Omega(\Xi)} = \int_{\psi} \Omega(\Xi(\psi)) F(\psi) d\psi \tag{2.41}$$

Here $F(\Xi)$ is the filtered-density function.

At this point, it is important to contrast the concepts of PDF, which discussed within the context of RANS, and FDF, which is discussed within the context of LES. We introduce the concept of a fine-grained PDF [30], which represents the time- and spatially-resolved joint PDF. This fine-grained PDF may be expressed as follows:

$$\zeta\left(\psi,\phi\left(\mathbf{x},t\right)\right)=\delta\left(\psi-\phi\left(\mathbf{x},t\right)\right)=\prod_{\alpha=1}^{N_\psi}\delta\left(\psi_\alpha-\phi_\alpha\left(\mathbf{x},t\right)\right) \tag{2.42}$$

The fine-grained PDF, $\zeta\left(\psi,\phi\left(\mathbf{x},t\right)\right)d\psi$ represents probability that at $\mathbf{x}$ and t, that $\psi_\alpha \leq \phi_\alpha \leq \psi_\alpha + d\psi_\alpha$ for all values of α. Within the context of RANS, the joint scalar PDF may be expressed as follows:

$$f\left(\phi;\mathbf{x},t\right)=\overline{\zeta\left(\psi,\phi\left(\mathbf{x},t\right)\right)} \tag{2.43}$$

The FDF is expressed in terms of the fine-grained PDF as follows:

$$F=\int_{-\infty}^{+\infty}\zeta\left(\psi,\phi\left(\mathbf{x},t\right)\right)G\left(\mathbf{x},\mathbf{x}'\right)d\mathbf{x}' \tag{2.44}$$

Therefore, the PDF represents a distribution built over time or ensembles of realizations of the scalar values at one single position, $\mathbf{x}$. In contrast, the FDF represents an instantaneous subfilter distribution of the same scalars over a prescribed filter volume.

The closure of the reaction source term is a principal modeling challenge in combustion LES; and often, strategies implemented for RANS have been extended to LES as well, as discussed in various chapters in this book.

2.4 Addressing the Closure Problem

The scope of turbulent combustion modeling is related to the representation of reactive scalar statistics. The traditional strategy is based on the RANS averaging framework. However, LES is becoming a viable framework for turbulent combustion models. The challenges are fundamentally similar. Averaging or filtering results in the closure problem for key terms in the conservation equations, including primarily the chemical source terms. The bulk of chapters in this book (Chapters 3–14) attempt to address the different approaches to turbulent combustion closure.

In a recent review, Bilger et al. [4] discussed traditional paradigms that defined turbulent combustion modeling over the last 40 or so years. Principal strategies resulting from these paradigms are based on either a 1) separation of scales and/or 2) separation of model elements that address the model description of moments of reactive scalars in terms of scalar description in state-space and model for the distribution (PDF or FDF) function. Examples of models based on the separation of scales include the assumption of fast chemistry (e.g. the eddy dissipation model (EDM) [31], the eddy break-up model (EBU) [51]) and the laminar flamelet model [39] where the flames thicknesses are below the energetic turbulence scales.

We illustrate the second strategy by revisiting the Eqs. (2.29) and (2.40). The mean or filtered reactions is constructed through a weighted average of the instantaneous reaction rate $\omega_k(\psi)$ and the distribution, $f(\psi)$ or $F(\psi)$. For the instantaneous reaction rate term, a 'reactor' model is needed that is representative of the state-space conditions encountered. For example, a flamelet library or CMC solutions may represent such reactor models. For the distribution description different strategies may be adopted depending on whether a reduced description of the state-space variables is available. For example, in the standard laminar non-premixed flamelet model and in CMC, models for the mean mixture fraction and its variance may be used to construct presumed shape PDF functions for reactive scalars. In the Bray-Moss-Libby (BML) model, a simple PDF function is adopted in terms of the reaction progress variable for premixed flames, although knowledge of the PDF shape is not always guaranteed. The more general form for the determination of the joint PDF involves the solution for a transport equation for the PDF and the FDF. However, intermediate strategies may be adopted as well. These include 1) the construction of PDF/FDF's using independent stochastic simulations, or 2) optimally build PDF's, such as the ones based on the statistically-most likely distribution [45]. The mapping closure approach (MMC; see Chapter 7) illustrates a strategy where a PDF transport equation is adopted for the construction of a statistical distribution and the CMC approach is used for the state-space relations.

Given the scope of the Bilger et al. [4] paper as related to paradigms in turbulent combustion, other modeling approaches have not been discussed; they will be presented here and the remaining chapters in the book will address them in more detail.

2.5 Outline of Upcoming Chapters

In this chapter, we have attempted to provide a brief outline of the challenges associated with turbulent combustion modeling. These challenges may be addressed by improved physical models of turbulent combustion processes; great strides have been made in the last two decades since the later contribution of Libby and Williams [28] and the more recent combustion literature. Moreover, rapid advances in computational sciences (hardware and algorithms) have fueled important advances in high-fidelity simulations of turbulent combustion flows that provide direct solutions of unresolved physics.

This book attempts to highlight recent progress in the modeling and simulation of turbulent combustion flows. It is divided into four parts, which include 1) two introductory chapters (Chapters 1 and 2) and 2) that motivate the growth of the disciplines associated with turbulent combustion flows from a societal and technological perspectives, 2) progress and trends in turbulent combustion models, 3) progress and trends in a new class of models based on multiscale simulation strategies, and 4) cross-cutting science that may be needed to move the subject forward. In Part II, emphasis is placed on recent progress in advanced combustion models, including the flamelet approach for non-premixed systems (Chapter 3), approaches for premixed combustion(Chapter 4), CMC (Chapter 5), MMC (Chapter 7) and the

PDF approach (Chapter 6). In Part III, emphasis is placed on multiscale strategies that seek to directly or indirectly compute subgrid scale physics. This part is preceeded by an introductory chapter highlighting the driving motivation behind multiscale strategies in turbulent combustion. Topics covered in this part include the role of chemistry reduction and acceleration (Chapter 9), the linear-eddy model (LEM) (Chapter 10), the one-dimensional turbulence (ODT) model (Chapter 11), flame-embedding (Chapter 12), adaptive-mesh refinement (AMR) (Chapter 13), wavelet-based methods (Chapter 14). Our coverage of existing models in Parts I and II is admittedly incomplete; but, it provides a flavor of current state-of-the-art and trends in turbulent combustion models. This state-of-the-art can be contrasted with the general strategies adopted during the last three decades to gauge recent progress. Part IV addresses cross-cutting science, which include the basic tools to advance the discipline of turbulent combustion modeling. Experiment has played, and will continue to play, a central role in the development of new and the refinement of old strategies. The role of experiment is discussed in Chapter 15. From the computational side, two principal drivers for improving turbulent combustion modeling and simulation are addressed. The first chapter (Chapter 16) deals with the subject of uncertainty quantification as an emerging requirement to improve the ability of turbulent combustion modeling and simulation tools to predict practical flows. The second chapter (Chapter 17) addresses the need to develop effective strategies to build optimized software tools to predict turbulent combustion flows. Chapter 18 presents the homogeneous multiscale method (HMM) as a mathematical multiscale framework for turbulent combustion. Finally, Chapter 19 reviews the lattice-Boltzmann method (LBM), which represents a viable alternative to the standard Navier-Stokes equations for a large class of flows.

References

1. Bell, J.B., Day, M.S., Shepherd, I.G., Johnson, M.R., Cheng, R.K., Grcar, J.F., Beckner, V.E., Lijewski, M.J.: Numerical simulation of a laboratory-scale turbulent V-flame. Proc. Nat. Acad. Sci. **102**, 10006–10011 (2005)
2. Bell, J.B., Day, M.S., Grcar, J.F., Lijewski, M.J., Driscoll, J.F., Filatyev, S.A.: Numerical simulation of laboratory-scale turbulent slot flame. Proc. Combust. Inst. **31**, 1299–1307 (2009)
3. Bilger, R.W.: The structure of turbulent nonpremixed flames. Proc. Combust. Inst. **22**, 475–488 (1988)
4. Bilger, R.W., Pope, S.B., Bray, K.N.C., Driscoll, J.F.: Paradigms in turbulent combustion research, Proc. Combust. Inst. **30**, 21–42 (2005)
5. Brown, P.N., Byrne, G.D., Hindmarsh, A.C.: VODE: A variable coefficient ODE solver. SIAM J. Sci. Stat. Comput. **10**, 1038–1051 (1989)
6. Chen, J.-Y.: A General procedure for constructing reduced reaction mechanisms with given independent relations, Combust. Sci. Technol. **57**, 89–94 (1988)
7. Chen, J.H., Choudhary A., de Supinski, B., DeVries, M., Hawkes, E.R., Klasky, S., Liao, W.K., Ma, K.L., Mellor-Crummey, J., Podhrszki, N., Sankaran, R., Shende, S., and Yoo, C.S.: Terascale direct numerical simulations of turbulent combustion using S3D. Comput. Sci. Discovery **2**, 015001 (2009)
8. Dixon-Lewis, G.: A FORTRAN computer code for the evaluation of gas-phase multicomponent transport properties. Proc. Royal Soc. **A304**, 111–134 (1968)

9. Dixon-Lewis, G.: Structure of laminar flames Proc. Combust. Inst. **23**, 305–324 (1990)
10. Dubois, T., Jauberteau, F., Temam, R.: Dynamic multilevel methods and the numerical simulation of turbulence, Cambridge University Press (1999)
11. Echekki, T., Chen, J.H.: Unsteady strain rate and curvature effects in turbulent premixed methane-air flames. Combust. Flame **106**, 184–202 (1996)
12. Eggenspieler, G., Menon, S.: Combustion and emission modelling near lean blow-out in a gas turbine engine. Prog. Comput. Fluid Dyn. **5** 281–297 (2005)
13. Ern, A., Giovangigli, V.: EGLIB: A general purpose FORTRAN library for multicomponent transport property evaluations, Software Manual (1986)
14. Favre, A.: Equations des gas turbulents compressible. J. Mec. **4**, 361–390 (1965)
15. Gao, F., O'Brien, E.E.: A large-eddy simulation scheme for turbulent reacting flows. Phys. Fluids **5**, 1282–1284 (1993)
16. Germano, M., Piomelli, U., Moin, P., Cabot, W.H.: A dynamic subgrid-scale eddy viscosity model. Phys. Fluids A **3**, 1760–1765 (1991)
17. Hawkes, E.R., Sankaran, R., Sutherland, J.C. and Chen, J.H.: Structure of a spatially developing turbulent lean methane-air Bunsen flame. Proc. Combust. Inst. **31**,1291–1298 (2007)
18. Jones, W.P.: Models for turbulent flows with variable density and combustion. In Prediction Methods for Turbulent Flows, Kollman, W. ed., pp. 379–421. Hemisphere (1980)
19. Kee, R.J., Dixon-Lewis, G., Warnatz, J., Coltrin, M.E., Miller, J.A.: A FORTRAN computer code package for the evaluation of gas-phase multicomponent transport properties, SAND86-8246, Sandia National Laboratories (1986)
20. Kee, R.J., Rupley, F.M., and Miller, J.A.: Chemkin-II: A FORTRAN Chemical Kinetics Package for the Analysis of Gas-Phase Chemical Kinetics. Sandia National Laboratories Report No. SAND 89–8009 (1989)
21. Kee, R.J., Coltrin, M.E., Glarborg, P.: Chemically Reacting Flow: Theory and Practice, Wiley-Interscience, New Jersey (2003)
22. Kim, S.H., Pitsch, H.: Mixing characteristics and structure of a turbulent jet diffusion flame stabilized on a bluff-body. Phys. Fluids **18**, 075103 (2006)
23. Lam, S.H., Goussis, D.A.: Understanding complex chemical kinetics with the computational singular perturbations. Proc. Combust. Inst. **22**, 931–941 (1988)
24. Law, C.K.: On the applicability of direct relation graph to the reduction of of reaction mechanisms. Combust. Flame **146**, 472–483 (2006)
25. Law, C.K.: Combustion Physics, Cambridge University Press, New York (2006)
26. Law, C.K.: Combustion at a crossroads: Status and prospects. Proc. Combust. Inst. **31**, 1–29 (2006)
27. Libby, P.A., Williams, F.A.: Turbulent Reacting Flows, Springer-Verlag, Heidelberg (1980)
28. Libby, P.A., Williams, F.A.: Turbulent Reacting Flows, Academic Press, London (1994)
29. Lu, T.F., Law, C.K.: Toward accommmodating realistic chemistry in large-scale computations. Prog. Energy Combust. Sci. **35**, 192–215 (2009)
30. Lundgren, T.S.: Distribution of functions in the statistical theory of turbulence. Phys. Fluids **10**, 969 (1967)
31. Magnussen, B.F., Hjertager, B.H.: On mathematical modeling of turbulent combustion with special emphasis on soot formation and combustion. Proc. Combust. Inst. **16**, 719–729 (1976)
32. Mizobuchi, Y., Sinjo, J., Ogawa, S., Takeno, T.: A numerical study of the formation of diffusion flame islands in a turbulent hydrogen jet lifted flame. Proc. Combust. Inst. **30**, 611–619 (2005)
33. Mizobuchi, Y., Tachibana, S., Shinjo, J., Ogawa, S., Takeno, T.: A numerical analysis of the structure of a turbulent hydrogen jet lifted flame. Proc. Combust. Inst. **29**, 2009–2015 (2002)
34. Moin, P., Squires, K., Cabot, W., Lee, S.: A dynamic subgrid-scale model for compressible turbulence and scalar transport. Phys. Fluids A **3**, 2746–2757 (1991)
35. Navarrao-Martinez, S., Kronenburg, A., Di Mare, F., Conditional moment closure for large-eddy simulations. Flow Turbul. Combust. **75**, 245–274 (2005)
36. McIlroy, A., McRae, G., Sick, V., Siebers, D.L., Westbrook, C.K., Smith, P.J., Taatjes, C., Trouve, A., Wagner, A.E., Rohlfing, E., Manley, D., Tully, F., Hilderbrandt, R., Green, W.,

Marceau, D., O'Neal, J., Lyday, M., Cebulski, F., Garcia, T.R., Strong, D., Basic research needs for clean and efficient combustion of 21st century transportation fuels. Department of Energy Office of Science Report (2006)

37. Patel, N., Kirtas, M., Sankaran, V., Menon, S.: Simulation of spray combustion in a lean-direct injection combustor. Proc. Combust. Inst. **31**, 2327–2334 (2007)
38. Paul, P.H.: DFRM: A new package for the evaluation of gas-phase transport properties, SAND98-8203, Sandia National Laboratories (1997)
39. Peters, N.: Local quenching due to flame stretch and non-premixed turbulent combustion. Combust. Sci. Techol. **30**, 1–17 (1983)
40. Peters, N.: Turbulent Combustion, Cambridge University Press, UK (2000)
41. Petzold, L.R., A description of dassl: A differential/algebraic system solver, SAND82-8637, Sandia National Laboratories (1982)
42. Pitsch, H., Large-eddy simulation of turbulent combustion. Ann. Rev. Fluid Mech. **38**, 453–482 (2006)
43. Poinsot, T., Candel, S., Trouve A.: Applications of direct numerical simulation to premixed turbulent combustion. Prog. Energy Combust. Sci. **21**, 531–576 (1995)
44. Poinsot, T., Veynante, D., Theoretical and Numerical Combustion, Second Ed., RT Edwards (2005)
45. Pope, S.B.: The statistical theory of turbulent flames. Philos. Trans., Roy. Soc. London Ser. A **291**, 529–568 (1979)
46. Pope, S.B.: Computations of turbulent combustion: Progress and challenges. Proc. Combust. Inst. **23**, 591–612 (1990)
47. Pope, S.B., Maas, U.: Simplifying chemical kinetics: Intrinsic low-dimensional manifolds in composition space. Combust. Flame **88**, 239–264 (1992)
48. Pope, S.B., Ren, Z.: Efficient implementation of chemistry in computational combustion. Flow Turbul. Combust. **82**, 437–453 (2009)
49. Selle, L, Lartigue, G., Poinsot, T., Koch, R., Schildmacher, K.: Compressible large eddy simulation of turbulent combustion in complex geometry on unstructured meshes. Combust. Flame **137**, 489–505 (2004)
50. Smooke, M.D., Giovangigli, V.: Reduced Kinetic Mechanisms and Asymptotic Approximations for Methane-Air Flames. Lecture Notes in Physics, Springer-Verlag, Berlin **384** (1991)
51. Spalding, D.B.: Mixing and chemical reaction in steady confined turbulent flames Proc. Combust. Inst. **13**, 649–657 (1971)
52. Triantafyllidis, A., Mastorakos, E., Eggels, R.L.G.M.: Large eddy simulations of forced ignition of a non-premixed bluff-body methane flame with conditional moment closure. Combust. Flame **156**, 2328–2345 (2009)
53. Valorani, M., Paolucci, S.: The G-scheme: A framework for multi-scale adaptive model reduction. J. Comput. Phys. **228**, 4665–4701 (2009)
54. Vervisch, L., Poinsot, T.: Direct numerical simulation of non-premixed turbulent flames. Ann. Rev. Fluid Mech. **30**, 655–691 (1998)
55. Veynante, D., Vervisch, L.: Turbulent combustion modeling. Prog. Energy Combust. Sci. **28**, 193–266 (2002)
56. Warnatz, J.: Calculation of structure of laminar flat flames. 1. Flame velocity of freely propagating ozone decomposition flames. Ber. Bunsenges. Phys. Chem. Phys. **82**, 193–200 (1978)
57. Williams, F.A., Combustion Theory: The Fundamental Theory of Chemically Reacting Flow Systems, Second Ed., Benjamin-Cummings (1985)

Part II
Recent Advances and Trends in Turbulent Combustion Models

Chapter 3
The Flamelet Model for Non-Premixed Combustion

Bénédicte Cuenot

Abstract The flamelet approach for non-premixed combustion is based on the description of the turbulent flame as a collection of laminar flame elements embedded in a turbulent flow and interacting with it. The local structure of the flame at each point of the flame front is supposed to be similar to a laminar flamelet, while the interaction with turbulence is reduced to the front evolution. This view is supported by the introduction of the mixture fraction, which allows to decouple the turbulent transport and the flame structure. One key parameter of the flamelet structure is the scalar dissipation rate, which controls the reactant fluxes to the reaction zone and is related to the flow velocity gradients. Probability density functions or flame surface density are then used to describe the turbulent flame and relate the flamelet description to the turbulent flame front. As unsteady effects may become significant, various transient flamelet approaches also exist to take into account the flame history. The flamelet approach may be used either in the RANS or LES context and is still being developed to account for additional complexities such as heat losses and sprays.

3.1 Introduction

Non-premixed combustion occurs when the fuel and the oxidizer are injected separately into the combustor, and experience simultaneous mixing and burning. This is the case in many industrial systems, mainly for safety reasons. Another advantage is that there is no need for a specific premixing device. Non-premixed flames are found for example in furnaces, diesel engines or gas turbines. They are also called diffusion flames as diffusion is usually the controlling mechanism. As combustion is a fast process compared to convective and diffusive transport mechanisms, a common assumption is then to consider the flame chemistry infinitely fast. This leads to

B. Cuenot
CERFACS, 31057 Toulouse, France, e-mail: Benedicte.Cuenot@cerfacs.fr

T. Echekki, E. Mastorakos (eds.), *Turbulent Combustion Modeling*,
Fluid Mechanics and Its Applications 95, DOI 10.1007/978-94-007-0412-1_3,

important simplifications in the flame structure description. There are however cases where finite-rate chemistry plays a role, for example when the flame is submitted to a high stretch rate. Then non-equilibrium effects must be taken into account.

Diffusion flames do not propagate and cannot be characterized, as premixed flames, with well determined time and length scales (such as laminar flame speed and laminar flame thickness). This explains why most models for non-premixed combustion use the "flamelet approach", which describes the basic features of the flame and introduces the required scales. Although the interaction of diffusion flames with turbulence is similar to turbulent premixed flames in many aspects (e.g. front wrinkling, effect of stretch), the absence of own dynamics makes them more sensitive to turbulence. The critical extinction stretch rate is one order of magnitude smaller than for premixed flames, and the increased probability of quenching make the use of the flamelet description less justified. Extensions to transient flamelets allow to partly overcome this difficulty but introduce additional complexity in the computations and are more delicate to apply in complex flows.

The various flamelet models differ by their use of the flame structure in the mixture fraction space, and therefore have some similarities with the Conditional Moment Closure methods (see Chapter 5). For premixed flames, the flamelet concept is described in detail in Chapter 4 and will not be discussed further here. This chapter reviews the fundamentals of the method for non-premixed combustion, applications in the context of RANS, developments in the context of LES, and discusses some future directions. It is largely inspired by the books of Peters [32] and Poinsot and Veynante [38].

3.2 Fundamental Concepts

Following the earlier ideas of [27, 47], Peters [31] introduced a turbulent combustion model based on the flamelet concept. This term now designates a variety of physical models in which the turbulent flame is viewed as a collection of laminar flame elements embedded in a turbulent flow and interacting with it. The main advantage of the flamelet concept is that it decouples the complex chemical structure of the flame from the flow dynamics, which can be modeled independently. The flamelet structure is either described with a simplified analytical formulation or with a flamelet library involving the detailed chemical structure of the flame. For non-premixed combustion, it corresponds to the laminar counterflow flame, which determines a steady diffusion flame structure. The flamelet is therefore submitted to strain only and curvature is ignored, although it may have non negligible effects [13, 23].

3.2.1 The Mixture Fraction

The decoupling of the flame structure from the flow dynamics is obtained via the introduction of the mixture fraction variable Z. The temperature and species mass fractions equations differ only by their chemical source terms, and one can combine them so as to cancel the chemical source term. For a single-step reaction described as:

$$\nu_F \mathrm{F} + \nu_O \mathrm{O} \rightarrow \nu_P \mathrm{P} \tag{3.1}$$

the combination $sY_F - Y_O$, where $s = \nu_O W_O / \nu_F W_F$ and W_k is the molecular weight of species k, is a passive scalar; its transport equation carries no source terms. This passive scalar can then be used to describe the flame dynamics, while the flame structure is still described by the mass fractions and the temperature. If normalized so as to vary from 1 in the fuel side to 0 in the oxidizer side, this passive scalar corresponds to the mixture fraction, defined as:

$$Z = \frac{sY_F - Y_O + Y_O^0}{sY_F^0 + Y_O^0} \tag{3.2}$$

where Y_F^0 and Y_O^0 are the fuel and oxidizer mass fractions in the unburnt fuel and oxidizer streams. The mixture fraction Z measures the degree of premixing of both reactants. The stoichiometric value Z_{st} is calculated by setting $sY_F - Y_O = 0$ in Eq. 3.2.

There is no unique definition of the passive scalar, as multiple combinations of the temperature and species conservation equations can be done. For example the following quantities can also be used for the flamelet analysis:

$$Z_1 = \frac{T^* - T_O^{0*} + Y_F}{T_F^{0*} - T_O^{0*} + Y_F^0} \tag{3.3}$$

or

$$Z_2 = \frac{s(T^* - T_O^{0*}) + Y_O - Y_O^0}{s(T_F^{0*} - T_O^{0*}) - Y_O^0} \tag{3.4}$$

where T^* is a normalized temperature. In fact $Z = Z_1 = Z_2$ as they are all solutions of the same equation with the same boundary conditions.

A more general way to define the mixture fraction is related to the conservation of chemical elements, as the mass of elements is conserved throughout the reaction. A common definition is for example the one by Bilger [3].

Note that in case of multiple fuel injections, the problem cannot be described with one single mixture fraction anymore but requires multiple passive scalars [20].

3.2.2 The Flamelet Solution

The basic principle of the flamelet approach relies on the variable change $(x,t) \to Z$. If the dynamics and flame structure are decoupled, then the space and time evolutions of the temperature and species profiles are fully described by the space and time evolution of Z.

There are different ways of expressing flame variables as functions of Z. The most simple, called the Burke-Schumann solution, makes use of the infinitely-fast chemistry assumption. In this case, the reaction zone reduces to an infinitely thin layer and the two reactants never coexist. Then, the flame solution is written as:

$Z \geq Z_{st}$:	$Z \leq Z_{st}$:
$Y_F = Y_F^0 \frac{Z-Z_{st}}{1-Z_{st}}$	$Y_F = 0$
$Y_O = 0$	$Y_O = Y_O^0 \left(1 - \frac{Z}{Z_{st}}\right)$
$T^* = ZT_F^{0*} + (1-Z)T_O^{0*} + Y_F^0 Z_{st} \frac{1-Z}{1-Z_{st}}$	$T^* = ZT_F^{0*} + (1-Z)T_O^{0*} + Y_F^0 Z$

Outside the reaction zone, the temperature and species profiles are linear functions of Z. If the infinitely-fast chemistry involves reversible reactions, the flame is at equilibrium at each value of the mixture fraction.

However, real flames have finite-rate chemistry and non-equilibrium effects are present. The variable change $(x,t) \to (Z,t)$ applied to the conservation equation of any flame variable, for example the mass fraction of species k, gives:

$$\frac{\partial Y_k}{\partial t} = \rho D (\nabla Z)^2 \frac{\partial^2 Y_k}{\partial Z^2} + \dot{\omega}_k, \qquad (3.5)$$

where equal diffusivities have been assumed for all species. If chemistry is sufficiently fast, the unsteady term of Eq. 3.5 can be skipped as the time dependence of the flame is mainly due to transport, and therefore already described by the time evolution of Z. Fast chemistry also means a thin reaction zone around the value Z_{st}, and the Z-gradient can be evaluated at stoichiometry. The steady flamelet equation is then obtained:

$$\rho D (\nabla Z)^2_{st} \frac{\partial^2 Y_k}{\partial Z^2} + \dot{\omega}_k = 0 \qquad (3.6)$$

This equation applies to all species and the temperature and allows a full description of the flame structure as a function of Z. The scalar dissipation rate $\chi_{st} = 2D(\nabla Z)^2_{st}$ plays a key role in diffusion flame structure and is a very important parameter in non-premixed turbulent flame modelling. Equation 3.6 may be solved analytically for simplified expressions of the reaction rate. In the case of sufficiently fast chemistry (high Damköhler numbers), asymptotic analysis allows to solve the complete problem for single-step chemistry [25]. To obtain detailed chemical solutions, flamelets can be computed with dedicated software, and stored in 'flamelet libraries' which can then be used in complex turbulent flame calculations. One difficulty here is to determine the passive scalar which guarantees the uniqueness of the solution.

3.2.3 The Counterflow Diffusion Flame

The counterflow diffusion flame is very often used in experiments and numerical studies as it is the simplest configuration of a steady one-dimensional diffusion flame. It is then used as the generic flamelet in the steady flamelet approach and flamelet libraries are built using this configuration.

In the counterflow configuration, a planar diffusion flame stabilises at the stoichiometric plane between the opposed fuel and air flows. The flame is subjected to strain, which imposes the scalar dissipation rate. The flow is described by the potential solution:

$$u = ax \tag{3.7}$$

$$v = -ay \tag{3.8}$$

where a is the strain rate. The mixture fraction is the solution of the one-dimensional stationary transport equation:

$$u\frac{\partial Z}{\partial x} = D\frac{\partial^2 Z}{\partial x^2} \tag{3.9}$$

Replacing u with ax in the above equation, and using the boundary conditions $Z = 1$ on the fuel side ($x \longrightarrow -\infty$) and $Z = 0$ on the oxidizer side ($x \longrightarrow +\infty$), one easily finds the solution for Z:

$$Z(x) = \frac{1}{2}\left(1 - \text{erf}\left(\frac{x}{\sqrt{2D/a}}\right)\right) \tag{3.10}$$

Variable-density effects can be included by replacing x with $\eta = \int_0^x (\rho/\rho_0)\,dx$ where ρ_0 is a reference value. The scalar dissipation rate then is written as:

$$\chi_{st} = \frac{a}{\pi}\exp\left(-\frac{a}{D}x_{st}^2\right) \tag{3.11}$$

where x_{st} is the position of the stoichiometric plane, and can be found from Z_{st} and Eq. 3.10. The scalar dissipation is proportional to the strain rate, i.e. the velocity gradient along the flame front, which is useful to evaluate in turbulent flows.

This counterflow flame structure is equivalent to the unsteady unstrained flamelet, when Z is now the solution of:

$$\frac{\partial Z}{\partial t} = D\frac{\partial^2 Z}{\partial x^2} \tag{3.12}$$

Introducing the similarity variable $\zeta = x/\sqrt{2Dt}$ (or $\zeta = \eta/\sqrt{2Dt}$), the solution for Z is written:

$$Z(x,t) = \frac{1}{2}\left(1 - \text{erf}\left(\frac{\zeta}{\sqrt{2}}\right)\right) \tag{3.13}$$

Here the scalar dissipation rate becomes:

$$\chi_{st} = \frac{1}{2\pi t}\exp\left(-\frac{a}{D}x_{st}^2\right) \tag{3.14}$$

This means that a steady flamelet with a strain rate a has the same structure in the mixture fraction space as in the unsteady flamelet at time $t = 1/2a$.

Both flamelet descriptions (steady or unsteady) introduce space and time scales, which are useful to characterize turbulent diffusion flames. The diffusive flame thickness can be defined as $\delta_f = (1/\nabla z)_{st} = \sqrt{2D/\chi_{st}}$, while a chemical time scale can be evaluated from the reaction constants, for example as $\tau_c = A\ \exp\left(-T_a/T_f\right)$ where A is the pre-exponential factor and T_a the activation temperature. Asymptotic analysis [13] leads to more precise evaluations of the chemical time scale. The reaction layer thickness is much smaller than the diffusive thickness, and is also related to the chemical parameters.

3.2.4 Validity of the Flamelet Approach

It is generally believed that the flamelet concept is valid in the range of high Damköhler number, which compares the chemical time scale τ_c to the flow time scale τ_t:

$$\mathrm{Da} = \frac{\tau_t}{\tau_c} \tag{3.15}$$

High Da corresponds to fast chemistry compared to the turbulent time scale. In this case, the chemistry is fast enough to adapt instantaneously to the flow changes and unsteady effects can be neglected. Another condition is that the flame thickness is sufficiently small compared to the turbulent length scales, so that vortices cannot disturb the inner structure of the flame. Even if chemistry is fast, i.e. the reaction layer is thin, the diffusive flame thickness is larger and may be perturbed by the flow. Both space and time-scale ratios are usually recast in two non-dimensional numbers: Da and the turbulent Reynolds number Re_t. It can be shown that the time ratio condition corresponds to $\mathrm{Da}=\mathrm{Da}_0\sqrt{\mathrm{Re}_t}$ where Da_0 is a critical value under which unsteady effects appear. This value can be determined either with asymptotic analysis or Direct Numerical Simulation of turbulent diffusion flames. A typical regime diagram is shown in Fig. 3.1, using a log-log scale for both Da and Re_t. In this diagram, the different regimes are separated by lines corresponding to Da $\propto \sqrt{\mathrm{Re}_t}$ [38].

There are diverse approaches to turbulent flamelet models, but they all include the same ingredients:

- A laminar flamelet description, as presented above, describing the local flame structure,
- A turbulence model, describing the unclosed terms of the averaged or filtered equations,
- A coupling model to describe the interaction of the flame with the turbulent flow.

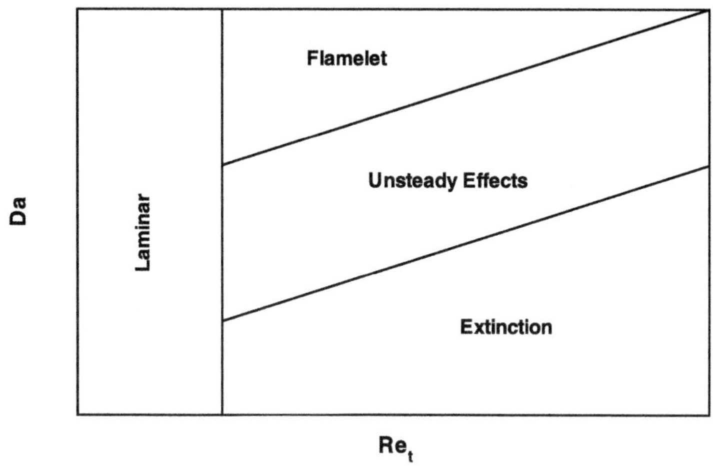

Fig. 3.1: Regimes for turbulent diffusion flames. Adapted from [38].

In the following, we will discuss the last item, which is the main differencing aspect among the various flamelet models. These models have been all derived in the RANS context. Their extension to LES already have yielded good results but also have raised new questions related to the subgrid scale motion.

3.3 RANS Flamelet Modeling

In the flamelet approach, flame dynamics are described by the mixture fraction Z and the solution of a classical Favre-averaged transport equation for Z:

$$\overline{\rho}\frac{\partial \widetilde{Z}}{\partial t} + \overline{\rho}\widetilde{v} \cdot \nabla \widetilde{Z} = \nabla \cdot \left(\overline{\rho D \nabla Z} - \overline{\rho} \widetilde{v''Z''} \right) \tag{3.16}$$

where the turbulent flux $\widetilde{v''Z''}$ is usually modeled via the gradient transport assumption:

$$\widetilde{v''Z''} = -D_t \nabla \widetilde{Z} \tag{3.17}$$

The molecular diffusivity is often neglected as it is much smaller than the turbulent diffusivity D_t. To fully describe the mixing problem, higher moments of Z are usually required, such as the variance $\widetilde{Z''^2}$, for which a conservation equation is also written:

$$\overline{\rho}\frac{\partial \widetilde{Z''^2}}{\partial t} + \overline{\rho}\widetilde{v} \cdot \nabla \widetilde{Z''^2} = \nabla \cdot (\rho D_t \nabla \widetilde{Z''^2}) + 2\overline{\rho} D_t (\nabla \widetilde{Z})^2 - \overline{\rho}\widetilde{\chi} \tag{3.18}$$

where the mean scalar dissipation rate is defined as:

$$\widetilde{\chi} = 2D\widetilde{(\nabla Z)^2} \tag{3.19}$$

This scalar dissipation rate measures the decay of the mixture fraction fluctuations and plays for Z the equivalent role of the dissipation rate of the kinetic energy ε for the velocity. The associated time scale $\widetilde{Z''^2}/\widetilde{\chi}$ is then often set proportional to the flow time $\widetilde{k}/\varepsilon$, leading to the following expression for $\widetilde{\chi}$:

$$\widetilde{\chi} = c_\chi \frac{\varepsilon}{\widetilde{k}} \widetilde{Z''^2} \tag{3.20}$$

where the constant c_χ is of order unity.

Both equations 3.16 and 3.18 can be solved in the RANS solver, and combined to the flamelet structure to describe the turbulent flame [32]. The most common approach uses a PDF approach, either of presumed shape or computed with a PDF transport equation. Examples of presumed PDF shapes include the δ Dirac, the β and the Gaussian distributions. These distributions are constructed from a minimum set of moments (e.g. two for the β or Gaussian distributions) and may be deficient in describing important non-equilibrium related distributions, which may require higher moments or when the parameters that characterize these distributions are not independent. However, intermediate strategies between presumed PDFs and transported PDFs have been developed. A statistically most likely distribution (SMLD) [22] represents a systematic strategy for the construction of statistical distributions based on prescribed moments. The SMLD approach is based on the maximum entropy principle. Another alternative approach is based on constructed PDFs from simulations, such as the ones based on the linear-eddy model (LEM) [18].

An alternative approach is based on the flame surface density, defined as the flame surface area per unit volume. Several approaches exist to calculate this quantity, based on algebraic models [6, 29], fractal theory [19] or on an additional transport equation with sink and source terms [27, 39, 43, 44].

Although the PDF and flame surface density models are conceptually different, they are closely linked through relationships derived by Veynante and Vervisch [45].

3.3.1 Steady Flamelets

In the PDF approach, any averaged flame variable $\widetilde{\psi_i}$ (being either species mass fraction, temperature or chemical source term) can be expressed as:

$$\overline{\psi_i} = \int_0^1 (\overline{\psi_i|Z})p(Z)dZ \tag{3.21}$$

where the quantity $(\overline{\psi_i|Z})$ designates the conditional average of ψ_i for a given value of the mixture fraction Z, and $p(Z)$ is the probability density function of Z. As

Eq. 3.21 can be written either for conserved or primitive variables, or for reaction rates, there are two possible methods to use the PDF approach. In the first method, the mean mass fractions and temperature are calculated with Eq. 3.21, using the flamelet formulation to express $(\overline{\rho Y_k|Z})$ and $(\overline{\rho T|Z})$ as functions of Z. In this case the conservation equations for these variables can be skipped, and knowledge of $p(Z)$ is sufficient to describe the turbulent flame. In the second method, only the mean reaction rates are calculated with Eq. 3.21, and are introduced in the averaged conservation equations of species and energy. The flamelet is used here for reaction rates only. The first method is clearly less time-consuming, as the number of equations to be solved is considerably reduced. On the other hand, it does not allow to include additional flow effects on the flame structure, such as heat losses, curvature or reactants inhomogeneity in the fresh gases. Both approaches are not strictly equivalent, mainly because the impact of the turbulent species and heat fluxes on the mean reaction rate is not taken into account in the first method.

If the infinitely-fast chemistry assumption is used, the conditional average of the flame variables is simply:

$$(\overline{\rho Y_k|Z}) = \rho(Z)Y_k(Z), \quad (\overline{\rho T|Z}) = \rho(Z)\,T(Z). \tag{3.22}$$

This assumption is often combined with a presumed-shape PDF. Since the mixture fraction varies between 0 and 1, the β-function is often used:

$$p(Z) = \frac{\Gamma(a+b)}{\Gamma(a)+\Gamma(b)} Z^{a-1}(1-Z)^{b-1}, \tag{3.23}$$

where the Γ function is defined as:

$$\Gamma(a) = \int_0^{+\infty} e^{-x} x^{a-1} dx. \tag{3.24}$$

The parameters a and b are calculated from the mean and variance of Z:

$$a = \widetilde{Z}\gamma; \quad b = (1-\widetilde{Z})\gamma \quad \text{with} \quad \gamma = \frac{\widetilde{Z}\left(1-\widetilde{Z}\right)}{\widetilde{Z''^2}} - 1. \tag{3.25}$$

The β-pdf takes various forms depending on the parameters a and b, and can be either (almost) bimodal or unimodal for higher values of γ. It is however not able to represent all situations, and other forms have been proposed in the literature.

The simplicity of the infinitely-fast chemistry flamelet model makes it very attractive, but it fails to predict situations with finite-rate chemistry. The steady flamelet analysis gives:

$$\dot{\omega}_k = -\frac{1}{2}\rho\chi\frac{\partial^2 Y_k}{\partial Z^2} \tag{3.26}$$

This reaction rate is then a function of Z and χ, and a joint PDF $p(Z,\chi)$ must be used to evaluate the mean value:

$$\overline{\dot{\omega}_k} = -\frac{1}{2}\int_0^1 \int_0^\infty \rho\chi \frac{\partial^2 Y_k}{\partial Z^2} p(Z,\chi) d\chi dZ \tag{3.27}$$

From Eq. 3.5, the second derivative of Y_k is a Dirac-function at $Z = Z_{st}$, and, for example, one finds for the fuel:

$$\overline{\dot{\omega}_F} = -\frac{Y_F^0}{2(1-Z_{st})}\overline{\rho}\widetilde{\chi}_{st}\, p(Z_{st}) \tag{3.28}$$

A similar result has been derived for infinitely-fast chemistry (see the review in Ref. [3]). Introducing finite-rate chemistry, the flamelet approach assumes that, as long as the reaction zone is thinner than the smallest size eddies, the flame locally keeps the structure of a laminar flame. This corresponds to the high Damköhler number introduced in Sec. 3.2.4. Therefore, fluctuations of flame variables inside the reaction zone thickness can be neglected, whereas fluctuations related to turbulent mixing in the diffusive layer must be described with the mixture fraction and its scalar dissipation rate. The mean variables are then expressed as:

$$\overline{\rho}\widetilde{Y}_k = \int_0^\infty \int_0^1 \rho Y_k(Z,\chi_{st})\, p(Z,\chi_{st})\, dZ\, d\chi_{st} \tag{3.29}$$

$$\overline{\rho}\widetilde{T} = \int_0^\infty \int_0^1 \rho T(Z,\chi_{st})\, p(Z,\chi_{st})\, dZ\, d\chi_{st} \tag{3.30}$$

where the functions $\rho Y_k(Z,\chi_{st})$ and $\rho T(Z,\chi_{st})$ are taken from the steady laminar flamelet library. The difficulty again is in the determination of the joint PDF $p(Z,\chi_{st})$, and to simplify the problem statistical independence is often assumed:

$$p(Z,\chi_{st}) = p(Z)\, p(\chi_{st}) \tag{3.31}$$

Assumed shapes for the PDF of χ_{st} can be Dirac functions: $p(\chi_{st}) = \delta(\chi_{st} - \widetilde{\chi_{st}})$, corresponding to a constant scalar dissipation at the mean value $\widetilde{\chi}_{st}$. However, it has been shown that this PDF is more like a log-normal distribution [15]:

$$p(\chi_{st}) = \frac{1}{\chi_{st}\sigma\sqrt{2\pi}} \exp\left(-\frac{(\ln(\chi_{st})-\mu)^2}{2\sigma^2}\right) \tag{3.32}$$

where $\mu = \ln(\widetilde{\chi}_{st}) - \sigma^2/2$ and σ is the variance of $ln(\chi_{st})$: $\widetilde{\chi_{st}''^2} = \widetilde{\chi}_{st}^2(\exp(\sigma^2)-1)$, often taken constant of the order unity. In the above expressions, the mean stoichiometric scalar dissipation $\widetilde{\chi}_{st}$ is still unknown. However, the mean scalar dissipation $\widetilde{\chi}$ is known from Eq. 3.20, and may be used to evaluate $\widetilde{\chi}_{st}$ through their analytical relation given by flamelet analysis: $\chi/F(Z) = \chi_{st}/F(Z_{st})$ where the function $F(Z)$ is:

$$F(Z) = \exp\left[-2\left(\text{erf}^{-1}(2Z-1)\right)^2\right] \tag{3.33}$$

The mean reaction rate can be calculated with the same methodology, taking $\dot{\omega}_k(Z,\chi_{st})$ from the flamelet library:

$$\overline{\dot{\omega}_k} = \int_0^{\infty}\int_0^1 \dot{\omega}_k(Z,\chi_{st})p(Z,\chi_{st})dZ\,d\chi_{st} \tag{3.34}$$

Alternatively, the flame surface density concept may be used [27, 39, 41–44]. In this approach, the flamelet library contains only the total (i.e. integrated across the flame front) reaction rate, and the mean reaction rate is written:

$$\overline{\dot{\omega}_k} = \overline{\dot{\Omega}_k(\chi_{st})}\;\Sigma \tag{3.35}$$

where $\overline{\dot{\Omega}_k(\chi_{st})}$ is the integrated reaction rate averaged along the flame front, and depends only on the scalar dissipation and Σ is the mean flame surface density defined as:

$$\Sigma = \overline{|\nabla Z|}_{Z=Z_{st}}\,p(Z_{st}). \tag{3.36}$$

In the case of fast chemistry (thin flame front), the following relation can be derived [45]:

$$\overline{\dot{\omega}_k} = \int_0^{\infty}\int_0^1 \dot{\omega}_k(Z,\chi_{st})p(Z,\chi_{st})dZd\chi_{st} = \left(\int_0^{\infty} \Omega_k(\chi_{st})p(\chi_{st})d\chi_{st}\right)\Sigma \tag{3.37}$$

showing the link between PDF and flame surface density approaches. Assuming $p(\chi_{st})$ as a Dirac function centered on $\widetilde{\chi}_{st}$, one finds that:

$$\overline{\dot{\Omega}_k\left(\chi_{st}\right)} = \dot{\Omega}_k(\widetilde{\chi}_{st}). \tag{3.38}$$

The quantity $\dot{\Omega}_k(\widetilde{\chi}_{st})$ is then taken from the flamelet library. The main issue in this approach is the determination of Σ. Marble and Broadwell [27] were the first to propose a conservation equation, including convective and turbulent transport as well as sink and source terms:

$$\frac{\partial \Sigma}{\partial t} + \nabla\cdot(\widetilde{v}\Sigma) + \nabla\cdot(\widetilde{v''\Sigma}) = P - S \tag{3.39}$$

where P is a production term, proportional to $\kappa\Sigma$, where κ is a mean strain rate and S is a destruction term corresponding to mutual annihilation of flame surface.

In the above formulations, one single parameter (χ_{st}) is used to describe the flamelet library. It allows to correctly describe flames in the high Damköhler number limit, but is not sufficient to capture additional effects such as curvature, ignition or extinction, flame history and transient effects.

3.3.2 Transient Flamelets

If the scalar dissipation rate changes rapidly compared to the flame chemical time scale, the unsteady term in the flamelet equations must be retained as in Eq. 3.5, where the scalar dissipation is now a function of time. Haworth et al. [21] made an *ad hoc* modification of the flamelet model, introducing an 'equivalent strain rate model', to limit the effect of rapidly changing χ. This idea is consistent with other studies of flames subjected to periodic oscillations of the strain rate, where it was found that, depending on the flame response time compared to the frequency, unsteady effects on the local flame structure may stay moderate. However, it was also shown that, at high frequencies, the strain rate may exceed the laminar quenching limit without flame extinction [14]. This is due to the delayed flame response, which implies a certain time necessary to fully quench the flame. Such unsteady effects were analysed in detail by Cuenot et al. [12].

In turbulent jet diffusion flames, unsteady effects are described by a Lagrangian residence time of the flamelet, which characterizes the velocity at which the flamelet is transported along the flame front, and is based on the flow velocity:

$$\bar{t} = \int_0^x \frac{1}{\widetilde{u}(x)|Z = Z_{st}} dx \tag{3.40}$$

where x is the distance from the nozzle and $\widetilde{u}$ is evaluated along the stoichiometric line. Alternatively an averaged velocity may be used at each flame height, defined along the radial direction in the zone where the flame exists (determined from a given range of the mixture fraction):

$$< u(x) >= \frac{1}{A_Z} \int_{A_Z} \widetilde{u} dA'_Z \tag{3.41}$$

The importance of transient effects in turbulent diffusion flames was also discussed by Pitsch et al. [36]. They introduced a diffusion time, that is the time needed to transport mass and energy over a distance (ΔZ) in the mixture fraction space:

$$t_\chi = \frac{(\Delta Z)^2}{\chi_{st}} \tag{3.42}$$

where (ΔZ) is typically the flame thickness. If this time is short compared to the flamelet residence time, the flame is able to adapt fast enough to the flow variations, and unsteady effects can be neglected. Generally, this is true in the first part of the flame jet, up to 30 diameters [36], where turbulence is strong enough to maintain high values of χ_{st}. Further downstream the scalar dissipation decreases, the flame thickens and unsteady effects appear.

Mauss et al. [28] introduced unsteady flamelets to simulate flamelet extinction and re-ignition in a statistically-steady turbulent jet diffusion flame. They used the residence time $\bar{t}$ as an additional parameter in the flamelet library, also viewed as a stochastic variable with a probability density function having a mean and a variance. Assuming a Gaussian distribution, they found the probability of the residence time as:

$$p(t) = \frac{1}{2} \left(1 + \mathrm{erf} \left(\frac{t - \bar{t}}{\sqrt{2}\sigma} \right) \right) \tag{3.43}$$

where the variance σ is unity. A stochastic approach was also used by Blanquart and Pitsch [4] to introduce a random scalar dissipation in the transient flamelet equations.

Other and more complex approaches exist to account for unsteady effects and are described below.

3.3.3 Representative Interactive Flamelets (RIF) Model

The RIF model was proposed by Pitsch et al. [35] and is based on the interactive coupling of two different solvers: a CFD code computing all necessary flow and flame-averaged variables, as well as the averaged mixture fraction, its variance and scalar dissipation rate, and an unsteady flamelet code calculating the unsteady flamelet in the mixture fraction space. The averaged species mass fractions are calculated as:

$$\widetilde{Y}_k(x) = \int_0^1 Y_k(Z,\bar{t},\widetilde{\chi}_{st})p(Z)dZ \tag{3.44}$$

where $p(Z)$ is a presumed-shape PDF of Z, based on the mean $\widetilde{Z}(x)$ and variance $\widetilde{Z''^2}(x)$ calculated by the CFD code. In the above expression, a constant scalar dissipation is assumed, but the model can be easily extended to account for a scalar dissipation distribution. The flame residence time $\bar{t}$ is calculated with Eq. 3.40. The model is illustrated in Fig. 3.2. Calculations are performed iteratively, starting with a steady flamelet solution. Convergence is reached when the scalar dissipation converges within a prescribed tolerance.

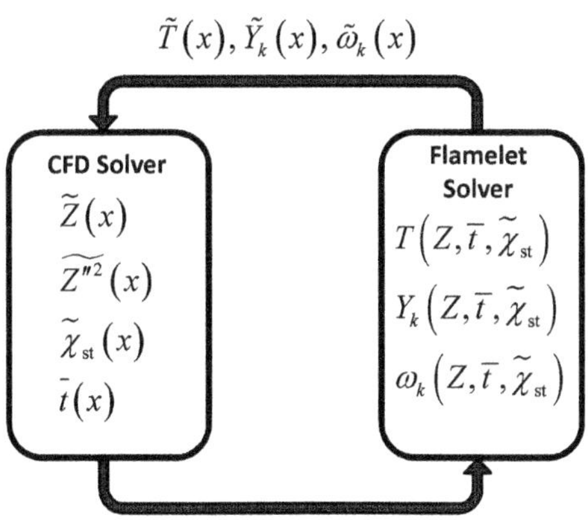

Fig. 3.2: Iterative loop in the RIF model. Adapted from [32].

3.3.4 Eulerian Particle Flamelet Model (EPFM)

The use of a single unsteady flamelet for the whole turbulent flame limits the validity of the approach and may lead to unphysical behaviors, in particular in the flame stabilisation zone. Multiple unsteady flamelets may be used, corresponding to different values of the scalar dissipation describing its evolution along the flame front [46]. The Eulerian Particle Flamelet Model proposed by Barths et al. [1, 2] introduces fictitious marker particles to follow the unsteady flamelets evolution. The mass-weighted fraction of particles corresponding to the flamelet l follows the equation:

$$\frac{\partial \overline{\rho} \widetilde{I}_l}{\partial t} + \nabla \cdot (\widetilde{v} \overline{\rho} \widetilde{I}_l) - \nabla \cdot \left(\overline{\rho} \frac{\nu_t}{\mathrm{Sc}_l} \nabla \widetilde{I}_l \right) = 0 \tag{3.45}$$

where Sc_l is the Schmidt number characterizing turbulent diffusion. The above equation allows the determination of the probability of finding the flamelet l at each location. The flamelet time is started from the time at which the particles are released. A time-dependent scalar dissipation is obtained from a weighted average over the region where the particles are dispersed. Integrating the unsteady flamelet equations over the residence time needed to reach a particular location along the flame front, and using the presumed-shape pdf at this location, all mean flame variables can be calculated.

3.3.5 Flamelet–Progress Variable (FPV) Models

To fully describe the flamelet structure, and in particular the reaction zone, the flamelet-progress variable approach combines both the mixture fraction Z and progress variable C [24, 26]. The progress variable describes the reaction progress and evolves through the flame from 0 in the fresh gases to 1 in the burnt gases. It is mainly used to describe premixed flames, where mixing is already achieved before burning. In diffusion flames, the progress variable may be useful to describe the reaction zone as the mixture fraction only describes mixing and ignores the flame. Using the assumption that the reaction rate is the same for diffusion flames and premixed flames at the same mixture fraction, Bradley et al. [5] built premixed flamelet libraries to calculate the reaction rate $\dot{\omega}_k(Z,C)$, where Z is interpreted as the premixed flame equivalence ratio, and used these libraries to compute turbulent diffusion flames at the same value of Z.

The progress variable is constructed from the flame variables, with the constraint of being monotonic from fresh to burnt gases, and non-dimensionalized to vary in the range [0;1]. In simple (or reduced) chemistry, the temperature, reactants or products may be used but not the intermediate species:

$$C = \frac{T - T_u}{T_b - T_u} \quad \text{or} \quad C = \frac{Y_k - Y_{k,u}}{Y_{k,b} - Y_{k,u}} \tag{3.46}$$

where the subscripts b and u stand for 'burnt' and 'unburnt'. In complex chemistry the definition of C is more difficult. First the temperature does not always stay monotonic. Second the reaction zone is a superposition of multiple reactions, most of them between intermediate species, and reactants or products do not fully represent the reacting zone. In this case combinations of species, involving for example CO and CO_2, are built to verify monotonicity.

Introduced in LES by Pierce and Moin [33], the FPV approach may also be viewed as an unsteady flamelet model. The unsteady evolution of the flamelet equation is then embedded in the progress variable transport equation, where the reaction rate is kept:

$$\frac{\partial \overline{\rho} \widetilde{C}}{\partial t} + \nabla \cdot (\widetilde{v} \overline{\rho} \widetilde{C}) = \nabla \cdot \left(\overline{\rho} D_t \nabla \widetilde{C} \right) + \overline{\rho} \widetilde{\dot{\omega}_C} \tag{3.47}$$

The mean flame variables are then reconstructed from the expression:

$$\widetilde{\psi}_i = \int_0^1 \int_0^1 \psi_i(Z,C) p(Z,C) dZ dC \tag{3.48}$$

for a constant scalar dissipation. The above expression introduces a joint PDF, modelled by first writing:

$$p(Z,C) = p(C|Z)p(Z) \tag{3.49}$$

where $p(Z)$ is classically assumed to be a β-pdf, and the conditional $p(C|Z)$ is taken from a single flamelet relation:

$$p(C|Z) = \delta(C - C(Z, \chi_{st})) \tag{3.50}$$

where $C(Z, \chi_{st})$ is the progress variable as a function of Z for a flamelet at scalar dissipation χ_{st}. The scalar dissipation χ_{st} is chosen in order to satisfy:

$$\widetilde{C} = \int_0^1 C(Z, \chi_{st}) p(Z) dZ \tag{3.51}$$

The implementation of the model is similar to that of the SLFM, where the parameter χ_{st} is replaced by the progress variable $\widetilde{C}$. This is an important improvement, as the modeling of the scalar dissipation (Eq. 3.20) is the weak point of the flamelet approach. One issue is the definition of the progress variable. In single-step chemistry, any species, or temperature, may be used as the system has only one degree of freedom in addition to the mixture fraction. For large complex chemistry descriptions, the definition of C is not so clear and bad choices may lead to important errors. In particular, the progress variable must guarantee a unique mapping of the flame state, i.e. have a monotonic behavior across the flame.

The FPV model was further extended by Pitsch and Ihme [37] to account for unsteady effects.

3.4 LES Flamelet Modeling

Although all modelling issues have not been yet addressed, LES is now currently applied to combustion applications. Most of the flamelet models developed in the RANS context have been adapted to LES [34]. Filtered conservation equations replace averaged equations, and mixing subgrid-scale models use the turbulent viscosity concept. PDF approaches can be directly applied to LES, where the PDFs represent only the subgrid fluctuations of the variables and may also be described with β-functions, using the filtered value and subgrid fluctuations. The filtered flame surface density may also be calculated via a filtered transport equation and adapted source terms.

In the case of infinitely-fast chemistry, flame variables depend only on the mixture fraction. Introducing the LES filter $F(x)$, the filtered flame variables can be written as [38]:

$$\begin{aligned}\widetilde{\psi_i} &= \int_V \int_0^1 \psi_i(Z^*)\delta(Z(x)-Z^*)\, F(x-x')\, dZ^*\, dx' \\ &= \int_0^1 \psi_i(Z^*) \int_V \delta(Z(x)-Z^*)\, F(x-x')\, dx'\, dZ^* \end{aligned} \tag{3.52}$$

where the quantity $\int_V \delta(Z(x)-Z^*)F(x-x')dx'$ represents the filtered PDF of Z.

3.4.1 Subgrid Scale Modelling

In addition to the filtered mixture fraction, and possibly filtered progress variable or filtered flame surface density, flamelet models require the knowledge of the subgrid scale variance $\widetilde{Z''^2}$ and, in the case of finite-rate chemistry, the filtered scalar dissipation $\widetilde{\chi}$.

The mixture fraction subgrid scale variance may be obtained with a transport equation, which in turn requires modeling of unclosed subgrid terms. In LES, the scale similarity assumption is often used, leading to:

$$\widetilde{Z''^2} \propto \left(\widehat{\widetilde{Z}^2} - \hat{\widetilde{Z}}^2 \right) \tag{3.53}$$

where $\hat{Z}$ denotes a test filter of larger size than the LES filter.

Using assumptions of local homogeneity and local equilibrium for the subgrid scales, Pierce and Moin [33] derived algebraic models for the subgrid variance and dissipation rate of Z:

$$\overline{\rho}\widetilde{Z''^2} = C_Z \overline{\rho} \Delta^2 |\nabla \widetilde{Z}|^2 \tag{3.54}$$

$$\overline{\rho}\widetilde{\chi} = \overline{\rho}(D+D_t)|\nabla \widetilde{Z}|^2 \tag{3.55}$$

where $\widetilde{Z}$ is now the filtered mixture fraction, Δ is the cutoff length of the LES filter and D_t is the subgrid diffusivity. The constant C_Z may be evaluated through a dynamic procedure.

3.5 Conclusion

A large variety of flamelet approaches exist and have been successfully applied to various configurations. Extensions of the flamelet equation have found use even for the study autoignition of completely premixed problems [10]. Flamelet models are particularly adapted to LES, which provides good prediction of the filtered and subgrid variables, as well as their time evolution, to be used in the flamelet library. One important feature of flamelet methods is that they give access to the fully detailed chemical structure of the flame, including all minor species. Optimized and efficient tabulation techniques (e.g. ILDM, ISAT, FPI) [11, 16, 40, 48] enable the use of flamelet libraries in complex flow simulations. This is of major importance to predict pollutant emissions and soot. Flamelet models are still being developed, in particular to account for heat losses or liquid fuel sprays [17, 26] or to predict diesel engine auto-ignition for which early examples may be found in [7, 8, 49].

Although a large majority of industrial systems are fed with separate fuel and oxidizer flows, the flame in the chamber is usually partially-premixed, with premixed (with variable equivalence ratio) and non-premixed flame elements. In this case unified models adapted to both premixed and diffusion flames must be developed. Flamelet methods are good candidates, as they are able to involve both the mixture fraction and the progress variable. For example, a turbulent flame speed method that gives $S_T(Z)$ from premixed laminar flamelets at various Z (i.e. various equivalence ratios) and a PDF for Z was applied to turbulent jet diffusion flame lift-off [9, 30].

References

1. Barths, H., Hasse, C., Bikas, G., Peters, N.: Simulation of combustion in direct injection diesel engines using a Eulerian particle flamelet model. Proc. Combust. Inst. **28**, 1161–1168 (2000)
2. Barths, H., Peters, N., Brehm, N., Mack, A., Pfitzner, M., Smiljanovski, V.: Simulation of pollutant formation in a gas turbine combustor using unsteady flamelets. Proc. Combust. Inst. **27**, 1841–1847 (1998)
3. Bilger, R.: The structure of turbulent non-premixed flames. Proc. Combust. Inst. **22**, 475–488 (1988)
4. Blanquart, G., Pitsch, H.: Modeling autoignition in non-premixed turbulent combustion using a stochastic flamelet approach. Proc. Combust. Inst. **30**, 2745–2753 (2005)
5. Bradley, D., Gaskell, P.H., Gu, X.J.: The mathematical modeling of liftoff and blowoff of turbulent non-premixed methane jet flames at high strain rates. Proc. Combust. Inst. **27**, 1199–1206 (1998)

6. Bray, K.N.C., Champion, M., Libby, P.: The interaction between turbulence and chemistry in premixed turbulent flames. Turbulent Reactive Flows, Lecture notes in engineering, Springer-Verlag pp. 541–563 (1989)
7. Bruel, P., Rogg, B., Bray, K.N.C.: On auto-ignition in laminar and turbulent non-premixed systems. Proc. Combust. Inst. **23**, 759–766 (1990)
8. Chang, C., Zhang, Y., Bray, K.N.C., Rogg, B.: Modelling and simulation of autoignition under simulated diesel-engine conditions. Combust. Sci. Technol. **113–114**, 205–219 (1996)
9. Chen, M., Herrmann, M., Peters, N.: Flamelet modeling of lifted turbulent methane/air and propane/air jet diffusion flames. Proc. Combust. Inst. **28**, 167–174 (2000)
10. Cook, D.J., Pitsch, H., Chen, J.H., Hawkes, E.R.: Flamelet-based modelling of auto-ignition with thermal inhomogeneities for application to HCCI engines. Proc. Combust. Inst. **31**, 2903–2911 (2007)
11. Correa, C., Niemann, H., Schramm, B., Warnatz, J.: Reaction mechanism reduction for higher hydrocarbons by the ILDM method. Proc. Combust. Inst. **28**, 1607–1614 (2000)
12. Cuenot, B., Egolfopoulos, F., Poinsot, T.: An unsteady laminar flamelet model for non-premixed combustion. Combust. Theory Model. **4**, 77–97 (2000)
13. Cuenot, B., Poinsot, T.: Effects of curvature and unsteadiness in diffusion flames. Implications for turbulent diffusion flames. Proc. Combust. Inst. **25**, 1383–1390 (1994)
14. Darabiha, N.: Transient behaviour of laminar counterflow hydrogen-air diffusion flames with complex chemistry. Combust. Sci. Technol. **86**, 163–181 (1992)
15. Effelsberg, E., Peters, N.: Scalar dissipation rates in turbulent jets and jet diffusion flames. Proc. Combust. Inst. **22**, 693–700 (1988)
16. Fiorina, B., Gicquel, O., Vervisch, L., Carpentier, S., Darabiha, N.: Approximating the chemical structure of partially premixed and diffusion counterflow flames using FPI flamelet tabulation. Combust. Flame **140**, 174–160 (2005)
17. Ge, H., Gutheil, E.: Simulation of a turbulent spray flame using coupled PDF gas phase and spray flamelet modeling. Combust. Flame **153**, 173–185 (2008)
18. Goldin, G., Menon, S.: A scalar pdf construction model for turbulent non-premixed combustion. Combust. Flame **155**, 70–89 (2008)
19. Gouldin, F., Bray, K., Chen, J.: Chemical closure model for fractal flamelets. Combust. Flame **77**, 241 (1989)
20. Hasse, C., Peters, N.: A two mixture fraction flamelet model applied to split injections in a DI diesel engine. Proc. Combust. Inst. **30**, 2755–2762 (2005)
21. Haworth, D., Drake, M., Pope, S., Blint, R.: The importance of time-dependent flame structures in stretched laminar flamelet models for turbulent jet diffusion flames. Proc. Combust. Inst. **22**, 589–597 (1988)
22. Ihme, M., Pitsch, H.: Prediction of extinction and re-ignition in non-premixed turbulent flames using a flamelet/progress variable model 1. A priori study and presumed pdf closure. Combust. Flame **155**, 70–89 (2008)
23. Kortschik, C., Honnet, S., Peters, N.: Influence of curvature on the onset of autoignition in a corrugated counterflow mixing field. Combust. Flame **142**, 140–152 (2005)
24. Lehtiniemi, H., Mauss, F., Balthasar, M., Magnusson, I.: Modeling diesel spray ignition using detailed chemistry with a progress variable approach. Combust. Sci. Technol. **178**, 1977–1997 (2006)
25. Liñan, A.: The asymptotic structure of counterflow diffusion flames for large activation energies. Acta Astronautica **1**, 1007–1039 (1974)
26. Ma, C.Y., Mahmud, T., Fairweather, M., Hampartsoumian, E., Gaskell, P.H.: Prediction of lifted, non-premixed turbulent flames using a mixedness-reactedness flamelet model with radiation heat loss. Combust. Flame **128**, 60–73 (2002)
27. Marble, F.E., Broadwell, J.: The coherent flame model of non-premixed turbulent combustion. Report TRW-9-PU (1977)
28. Mauss, F., Keller, D., Peters, N.: A Lagrangian simulation of flamelet extinction and re-ignition in turbulent jet diffusion flames. Proc. Combust. Inst. **23**, 693–698 (1990)

29. Meneveau, C., Poinsot, T.: Stretching and quenching of flamelets in premixed turbulent combustion. Combust. Flame **86**, 311–332 (1991)
30. Müller, C.M., Breitbach, H., Peters, N.: Partially premixed turbulent flame propagation in jet flames. Proc. Combust. Inst. **25**, 1099–1106 (1994)
31. Peters, N.: Laminar diffusion flamelets in non-premixed turbulent combustion. Prog. Energy Combust. Sci. **3**, 319–339 (1984)
32. Peters, N.: Turbulent Combustion. Cambridge University Press (2000)
33. Pierce, C.D., Moin, P.: Progress-variable approach for large-eddy simulation of non-premixed turbulent combustion. J. Fluid Mech. **504**, 73–97 (2004)
34. Pitsch, H.: Large-eddy simulation of turbulent combustion. Ann. Rev. Fluid Mech. **38**, 453–482 (2006)
35. Pitsch, H., Barths, H., Peters, N.: Three-dimensional modelling of NOx and soot formation in DI-diesel engines using detailed chemistry based on the interactive flamelet approach. SAE Paper 962057, 103–117 (1996)
36. Pitsch, H., Chen, M., Peters, N.: Unsteady flamelet modelling of turbulent hydrogen-air diffusion flames. Proc. Combust. Inst. **27**, 1057–1064 (1998)
37. Pitsch, H., Ihme, M.: An unsteady / flamelet progress variable method for LES of nonpremixed turbulent combustion. In: 43rd AIAA Aerospace Sciences Meeting and Exhibit, pp. 1–14 (2005)
38. Poinsot, T., Veynante, D.: Theoretical and Numerical Combustion. R.T. Edwards, Inc., Philadelphia (2005)
39. Pope, S.: The evolution of surfaces in turbulence. Int. J. Engng. Sci. **26**, 445–469 (1988)
40. Pope, S.B.: Computationally efficient implementation of combustion chemistry using *in situ* adaptive tabulation. Combustion Theory and Modelling **1**, 41–63 (1997)
41. Tap, F.A., Hilbert, R., Thevenin, D., Veynante, D.: A generalized flame surface density modelling approach for the auto-ignition of a turbulent non-premixed system. Combust. Theory Model. **8**, 165–193 (2004)
42. Tap, F.A., Veynante, D.: Simulation of flame lift-off on a diesel jet using a generalized flame surface density modeling approach. Proc. Combust. Inst. **30**, 919–926 (2005)
43. Trouve, A., Poinsot, T.: The evolution equation for the flame surface density. J. Fluid Mech. **278**, 1–31 (1994)
44. Vervisch, L., Bidaux, E., Bray, K., Kollmann, W.: Surface density function in premixed turbulent combustion modeling, similarities between probability density function and flame surface approaches. Phys. Fluids **7**, 2496–2503 (1995)
45. Veynante, D., Vervisch, L.: Turbulent combustion modeling. Prog. Energy Combust. Sci. **28**, 193–266 (2002)
46. Wan, Y., Pitsch, H., Peters, N.: Simulation of autoignition delay and location of fuel sprays under diesel-engine relevant conditions. Journal of Engines **106**, 1611–1621 (1997)
47. Williams, F.A.: A review of some theoretical combustions of turbulent flame structure. AGARD Conference Proceedings **II.1**, 1–25 (1975)
48. Woelfert, A., Nau, M., Maas, U., Warnatz, J.: Application of automatically simplified chemical kinetics in PDF calculations of turbulent methane-air diffusion flames. IWR-SFB-359–94-69 (1994)
49. Zhang, Y., Rogg, B., Bray, K.N.C.: 2-D simulation of turbulent autoignition with transient laminar flamelet source term closure. Combust. Sci. Technol. **105**, 211–227 (1995)

Chapter 4
RANS and LES Modelling of Premixed Turbulent Combustion

Stewart Cant

Abstract Premixed combustion is becoming more common in practical combustion systems in response to increasing regulatory pressure to reduce unwanted emissions. The inherent ability of premixed flames to propagate into the unburned mixture leads to a more active response to turbulence by comparison with non-premixed flames. It is clear that the sheet-like reaction zone within the premixed flame is highly resistant to disruption by turbulent eddies. Hence flamelet structure is prevalent in premixed combustion over a broad range of turbulent velocity fluctuation magnitudes and turbulent eddy length scales. Modelling of turbulent transport in premixed flames is rendered more difficult by the occurence of countergradient transport in the presence of strong heat release. Modelling of the mean turbulent reaction rate has involved a variety of approaches involving either algebraic expressions or additional transport equations. A brief review of current modelling practice is presented, covering some simple models together with the flame surface density approach, the G equation and the more recent scalar dissipation rate model. The emphasis is on models that are applicable in the context of both Reynolds-averaged Navier Stokes (RANS) and Large Eddy Simulation (LES).

4.1 Introduction to Premixed Flames

Mixing of the fuel and air streams prior to combustion is becoming more common in practical energy-conversion devices due to greater legislative emphasis on the reduction of unwanted emissions. A high degree of premixing provides for effective control of the stoichiometry of the flame. This brings significant benefits by ensuring that the combustion reaction takes place under benign thermochemical conditions. As an example, it becomes possible using premixing to specify lean

Stewart Cant
Department of Engineering, University of Cambridge, Trumpington Street, Cambridge CB2 1PZ, United Kingdom, e-mail: rsc10@cam.ac.uk

T. Echekki, E. Mastorakos (eds.), *Turbulent Combustion Modeling*,
Fluid Mechanics and Its Applications 95, DOI 10.1007/978-94-007-0412-1_4,

fuel-air mixtures which avoid the emission of significant amounts of unburned fuel, CO or particulates by virtue of the chemistry, and which minimise the formation of thermal NO_X by limiting the post-flame temperature. The use of well-designed premixed burners can also help to reduce emissions of CO_2 by contributing towards greater energy efficiency of the plant.

Premixed combustion has been used in spark-ignition automotive engines for many years. Often the fuel-air mixture strength is controlled by the engine management system to ensure that combustion takes place under stoichiometric conditions for compatibility with exhaust clean-up using a catalytic converter. In recent years, premixed combustion under very lean conditions has been widely adopted for industrial gas turbines mainly in order to minimise emissions of NO_X. Premixed burners are also used for industrial furnaces, for example in the glass-making industry, where highly-efficient low-emission combustion is a necessity. In the field of aerospace, premixed combustion is used in reheat systems for supersonic aircraft but has yet to be adopted in the main combustors of gas turbine aero-engines.

The outstanding technological issues in premixed combustion centre mainly on flame stability. Premixed burners must be carefully designed to avoid unwanted phenomena such as blow-off and flashback, while premixed flames are notoriously prone to convectively- and acoustically-coupled instabilities. Relatively small variations in the fuel-air mixture strength can have a major effect on the dynamics of the flame, and the distribution of heat release must be designed wherever possible to avoid acoustic coupling with the natural resonant modes of the combustion system. It is rarely possible in practice to ensure that the mixing between fuel and air is entirely complete before combustion takes place, resulting in partially-premixed flames. Under some circumstances these are found to be more stable than a fully-premixed flame under the same global stoichiometry, and the consequent slight loss of emissions performance is accepted as an engineering compromise. Greater understanding of premixed flames would help to avoid this issue.

4.2 Modelling Framework for RANS and LES

4.2.1 Introduction

Premixed turbulent flames are often found to contain thin, highly-wrinkled reaction surfaces in which most of the combustion chemistry takes place. Experimental observations of premixed flames under a broad range of conditions of temperature, pressure, mixture strength and turbulence intensity, and using a wide variety of measurement techniques, have shown that such structures are remarkably robust. This is the basis of flamelet modelling which has become well accepted for premixed flames and which has proved successful in many different applications. This is in contrast to the modelling of non-premixed combustion, where the flamelet concept has proved to be less universally applicable.

4.2.2 Regimes of Premixed Turbulent Combustion

Two measurable properties of laminar premixed flames, namely the laminar burning velocity and the laminar flame thickness, have proved to be invaluable in parameterising regimes of behaviour for the turbulent case. The laminar burning velocity S_L is the speed at which a laminar flame will propagate normal to itself into the reactant mixture. It is a well-defined but non-simple function of the pressure, the reactant temperature and reactant composition. It emerges from mathematical analysis as an eigenvalue of the two-point boundary value problem that yields the flame structure as a solution, and it can be measured accurately using modern diagnostic techniques and also computed reliably given a suitably detailed treatment of the chemical reaction mechanism and the molecular transport properties. It provides a natural velocity scale for premixed flames and has no equivalent in non-premixed flames. Typical values for stoichiometric hydrocarbon-air flames at ambient conditions are in the region of 40-50 cm/s. The laminar flame thickness δ_L is harder to define, given that the flame structure merges smoothly with the surrounding volumes of reactants and products. Standard measures of δ_L include the normalized inverse of the maximum temperature gradient or the distance between 5% and 95% of the temperature rise through the flame. Dimensional reasoning also provides an estimate of δ_L as α_T/S_L where α_T is the thermal diffusivity. The reaction zone thickness δ_R is known to be somewhat thinner than the overall flame structure which also includes the preheat and equilibration zones found respectively ahead of and behind the flame. For hydrocarbon-air flames, most estimates of δ_R are on the order of 1 mm.

The regimes of premixed turbulent combustion can be plotted on the well-known Borghi diagram [7] as modified by Peters [54] and shown in Fig. 15.1. The turbulence is represented by the velocity fluctuation magnitude u' together with the integral length scale l_T, while the thermochemistry is represented by S_L and δ_L. This allows any given premixed turbulent flame to be plotted on the diagram, and its location to be characterised by a set of dimensionless numbers such as the Reynolds number Re, the Damköhler number Da and the Karlovitz number Ka, respectively defined for the flame as

$$\mathrm{Re} = \frac{u' l_T}{S_L \delta_L}; \qquad \mathrm{Da} = \frac{S_L l_T}{u' \delta_L}; \qquad \mathrm{Ka} = \frac{\delta_L u'}{S_L \lambda} \tag{4.1}$$

where λ is the Taylor microscale of the turbulence.

Flamelet behaviour is seen to be prevalent except for high values of the Karlovitz number Ka where distributed combustion occurs. The major subdivisions into the corrugated flamelets regime and the thin reaction zones regime reflect the observation that the reaction zone can survive even in circumstances where the preheat and/or equilibration zones have been disrupted by small-scale turbulence. An additional Karlovitz number $\mathrm{Ka}_\delta = \delta_R u'/S_L \lambda$ is required [54] to account for this.

Clearly the Borghi diagram is an oversimplification, since the turbulence in practical cases is rarely well-represented by single values of u' and of l_T. Moreover, the

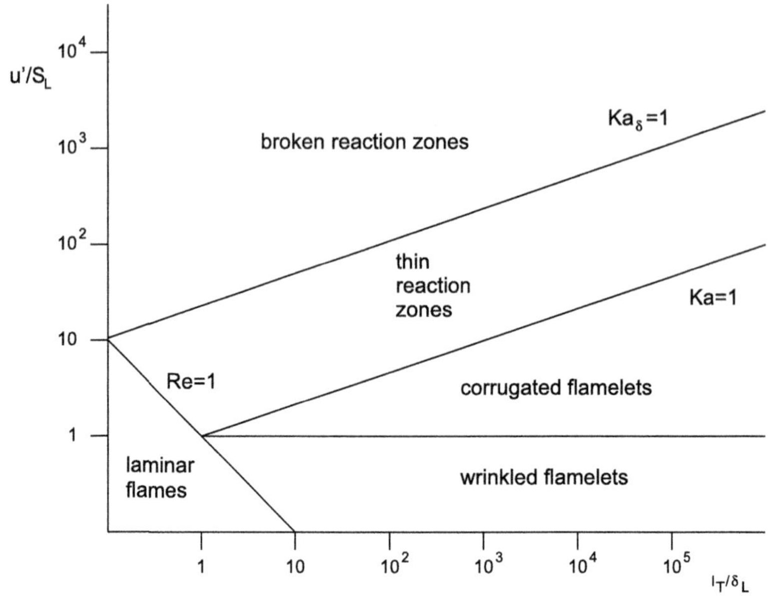

Fig. 4.1: Regime diagram for premixed turbulent combustion. Adapted from Peters [54].

transition between regimes is not sharply defined. Nevertheless the concept is useful and can be used to guide the most appropriate modelling approach. There are extensions of the diagram to account for the effects of filtering in LES [58]. Extension to partially-premixed turbulent flames may be as straightforward as drawing an ellipse around each point in order to represent the ranges of S_L and δ_L encompassed by the variations in the fuel-air mixture composition, provided that the mixture remains within the flammability limits.

4.2.3 Averaging and Filtering

The large range of length scales present in practical turbulent combustion problems means that full resolution is unattainable at reasonable computational cost. Hence it is necessary to treat at least the smallest scales on a statistical basis. Inevitably this involves some form of averaging in which information is lost. The purpose of modelling is to restore the averaged effect of the missing scales.

In the classical Reynolds Averaged Navier Stokes (RANS) approach, standard Reynolds time averaging of any quantity Q results in the mean $\bar{Q}$ according to

$$\bar{Q}(\underline{x}) = \frac{1}{T}\int_0^T Q(\underline{x},t)dt \tag{4.2}$$

where T is the timescale for averaging, assumed to be long relative to any flow time of interest. This approach assumes that the problem is statistically stationary, and is often too restrictive for many combustion applications. Instead, Reynolds ensemble averaging may be employed, using the definition

$$\bar{Q}(\underline{x},t) = \frac{1}{N}\sum_{n=1}^{N} Q(\underline{x},t;n) \tag{4.3}$$

where the average is taken over a notional (and large) number of realisations N. This approach allows for some level of unsteadiness, at least on a time scale that is slow relative to the turbulent fluctuations. In either case the instantaneous fluctuation from the mean is denoted by Q'. Note that the averaging operation for RANS is applied equally to all length scales. In principle no length scale is favoured over any other. In practice, the nature of turbulence ensures that flow features that are uncorrelated with each other are averaged out, and only features that are well-correlated over the averaging timescale are retained. Hence, turbulent eddies of all scales are removed leaving only the mean flow field.

In Large Eddy Simulation (LES) a spatial filtering operator is employed in order to remove flow features that are smaller than a specified filter size Δ. The filtering operation involves the convolution integral

$$\bar{Q}(\underline{x},t) = \int_{-\infty}^{\infty} Q(\underline{x}',t)F(\underline{x}-\underline{x}';\Delta)d\underline{x}' \tag{4.4}$$

where the integral is taken over the entire flow domain and F is a filter function which must satisfy the normalisation condition

$$\int_{-\infty}^{\infty} F(\underline{x}';\Delta)d\underline{x}' = 1 \tag{4.5}$$

The function F is often chosen to have a simple top-hat form in either physical space or wavenumber space, although there are advantages in choosing a Gaussian form instead [61]. Note that the filtering operation is applied in space only. In contradistinction to RANS averaging, the nature of turbulence then ensures that there is an equivalent filtering operation in time, with an implied temporal filter width that is proportional to the spatial filter width Δ divided by the velocity scale of the turbulent eddy at that length scale. This filtering timescale can be interpreted as the eddy turnover time of the filter-scale eddy.

In turbulent combustion, density fluctuations arise due to the combined effects of the turbulence and the thermal expansion due to heat release. It has become standard to use Favre density-weighted averaging or filtering in order to avoid the occurrence of density correlation terms in the governing equations. The Favre decomposition for a quantity Q is written as $Q = \tilde{Q} + Q''$ where the Favre mean is given by

$$\tilde{Q} = \frac{\overline{\rho Q}}{\bar{\rho}} \tag{4.6}$$

and the Favre fluctuation Q'' has the properties $\overline{Q''} \neq 0$ but $\overline{\rho Q''} = 0$. The operation denoted by the overbar may correspond to any of the averaging or filtering techniques outlined above.

Given that RANS and LES both have implied filtering length and time scales that are related through the nature of turbulence, it should be possible to use similar modelling in both approaches in order to restore the lost information at small length and time scales. Indeed this turns out to be the case, although care must be taken in order to interpret the different meaning of apparently similar quantities in each approach.

4.2.4 Modelling Principles

In both RANS and LES the requirement for modelling arises from the need to provide closure of the averaged or filtered governing equations. Assuming that a suitable closure model already exists for the momentum equations, the remaining task is to provide a consistent closure model for the transport equations describing the thermochemistry. In premixed turbulent combustion, it is common practice to define a reaction progress variable c which rises monotonically from zero in the fresh reactants to unity in the fully-burned products. If the Lewis number is close to unity, the Mach number is low and the problem is adiabatic, then the reaction progress variable alone is sufficient to describe the complete thermochemical state of the system. The various quantities of interest, such as the species mass fractions and temperature, can be obtained by choosing a suitable definition for the reaction progress variable. In the simplest case, c is treated as a scaled mass fraction of one or more product species Y_P:

$$c = \frac{Y_P}{Y_{P\infty}} \tag{4.7}$$

where $Y_{P\infty}$ is the value of Y_P in the fully-burned or equilibrium products. Alternatively, c can be defined in terms of the fuel mass fraction Y_F according to:

$$c = \frac{Y_F - Y_{F0}}{Y_{F\infty} - Y_{F0}} \tag{4.8}$$

where Y_{F0} is the fuel mass fraction in the fresh reactants and $Y_{F\infty}$ is the (notional) value of Y_F in the products. Clearly for lean combustion $Y_{F\infty} = 0$. More complicated definitions for c in terms of combinations of species mass fractions can be devised to suit particular cases. The principal criteria are that c must be monotonic, bounded between zero and unity, and provide a reasonable representation of the flame structure in terms of its thickness and location. In the rather specific case of low Mach number adiabatic flames with unity Lewis number, it is possible to define a reaction progress variable based on temperature according to

$$c = \frac{T - T_0}{T_\infty - T_0} \tag{4.9}$$

where T_0 and T_∞ are the limiting temperatures in the reactants and products respectively. This definition becomes invalid whenever the temperature is affected by any factor external to the flame, including acoustic activity, heat losses or differential transport of heat and mass. Thus, a definition of c in terms of temperature lacks flexibility and is not recommended for more general use.

With a suitable definition of c, a transport equation for reaction progress variable can be derived from the relevant transport equations for species mass fraction. The exact transport equation for c may be stated as

$$\frac{\partial}{\partial t}\rho c + \frac{\partial}{\partial x_k}\rho u_k c = w + \frac{\partial}{\partial x_k}\left(\rho D_c \frac{\partial c}{\partial x_k}\right) \tag{4.10}$$

in which w is the chemical production rate of c and ρD_c is the molecular diffusivity of c.

For RANS, the Favre-averaged reaction progress variable transport equation is

$$\frac{\partial}{\partial t}\bar{\rho}\tilde{c} + \frac{\partial}{\partial x_k}\bar{\rho}\tilde{u}_k\tilde{c} = \bar{w} + \frac{\partial}{\partial x_k}\left(\overline{\rho D_c \frac{\partial c}{\partial x_k}}\right) - \frac{\partial}{\partial x_k}\overline{\rho u_k'' c''} \tag{4.11}$$

The terms on the right-hand side of this equation represent the mean chemical reaction rate, the mean rate of molecular transport and the Reynolds flux or mean rate of turbulent transport. All three terms are unclosed and require modelling, although for high Reynolds number flows the molecular transport term is often simply neglected.

For LES, the Favre-filtered transport equation for c has the form

$$\frac{\partial}{\partial t}\bar{\rho}\tilde{c} + \frac{\partial}{\partial x_k}\bar{\rho}\tilde{u}_k\tilde{c} = \bar{w} + \frac{\partial}{\partial x_k}\left(\overline{\rho D_c \frac{\partial c}{\partial x_k}}\right) - \frac{\partial}{\partial x_k}\left(\overline{\rho u_k c} - \bar{\rho}\tilde{u}_k\tilde{c}\right) \tag{4.12}$$

where now the terms on the right-hand side represent the filtered chemical reaction rate, the filtered rate of molecular transport and the sub-grid scale contribution to the rate of turbulent transport. Again, all of these terms are unclosed and require modelling. Note that the form of the transport equation for c is similar for both RANS and LES. This reflects the fact that in both approaches the combustion is taking place at unresolved scales. This gives some hope that modelling devised for RANS may well remain applicable in LES, and *vice versa*.

The molecular transport term is usually retained in LES, since the grid-scale Reynolds number is not sufficiently large to justify its neglect relative to the turbulent transport term. The simplest closure model for this term is based on the assumption that the mass diffusivity ρD_c is uncorrelated with reaction progress variable:

$$\left(\overline{\rho D_c \frac{\partial c}{\partial x_k}}\right) = \overline{\rho D_c}\frac{\partial \tilde{c}}{\partial x_k} \tag{4.13}$$

in which an additional assumption has been made to allow for the replacement of the filtered value $\bar{c}$ with the Favre-filtered value $\tilde{c}$ inside the spatial derivative. In reality, the mass diffusivity ρD_c is often a function of c, but more complicated modelling is rarely justified in view of the more pressing modelling issues associated with the other terms in the transport equation.

4.3 Transport Modelling for Premixed Turbulent Flames

In a turbulent premixed flame the mean reaction progress variable rises from zero to unity across the flame brush. In passing from reactants to products through a statistically one-dimensional flame, the mean density falls due to heat release and the mean velocity normal to the flame rises due to mass continuity. Hence the conditional mean velocity in the products is greater than that in the reactants, i.e $\bar{u}_P > \bar{u}_R$. Fluctuations in the instantaneous velocity and reaction progress variable will occur due to the turbulence, and heat release ensures that an upwards (downwards) fluctuation in c will be associated with an upwards (downwards) fluctuation in u. Thus the physics ensures that u and c are statistically correlated, and moreover the correlation expressed by the Reynolds flux $\overline{\rho u''c''}$ is positive along the direction through the flame in which the gradient of mean reaction progress variable is positive also.

For this situation the standard gradient model for turbulent scalar transport may be written as:

$$\overline{\rho u_k''c''} = -\bar{\rho} D_T \frac{\partial \tilde{c}}{\partial x_k} \tag{4.14}$$

where D_T is a turbulent diffusivity. A difficulty arises at once, since the gradient model predicts a negative value of the Reynolds flux when the mean progress variable gradient is positive and *vice versa*. Thus the standard gradient transport model is not valid in general for turbulent premixed flames, in which the phenomenon of countergradient transport may well occur instead.

In physical terms, countergradient transport is a pressure-driven effect, and is a consequence of thermal expansion due to heat release within the flame. The resulting increase in the velocity through the flame brush is associated with a self-induced favourable pressure gradient, which acts preferentially to accelerate the lower-density product gases in preference to the higher-density reactants. This tends to transport these gases further up the gradient of reaction progress variable, in the opposite sense to that suggested by the gradient hypothesis [46].

In reality, turbulent transport in premixed flames involves a balance between mixing due to the turbulence and acceleration due to heat release. Turbulent mixing acts to promote gradient transport while heat release effects act to promote countergradient transport. Even in turbulent premixed flames where heat release is strong and countergradient transport prevails, there is a small region close to the leading edge of the flame brush in which heat release effects are weak and turbulence ensures that gradient transport still occurs [68]. From a strict interpretation of the flamelet theory the Reynolds flux in the direction normal to the turbulent flame brush may be expressed as [75]

$$\overline{\rho u''c''} = \bar{\rho}\tilde{c}(1-\tilde{c})\left(\tau S_L - 2\alpha u'\right) \tag{4.15}$$

where α is an efficiency function [52] accounting for the effects of different turbulent eddy sizes. Here, countergradient transport corresponds to positive values of the Reynolds flux. Hence this expression gives rise to a dimensionless ratio N_B which has become known as the Bray number:

$$N_B = \frac{\tau S_L}{2\alpha u'} \tag{4.16}$$

A Bray number greater (less) than unity indicates that countergradient (gradient) transport is likely to be prevalent.

Modelling of turbulent transport in premixed flames is made significantly harder by the occurrence of countergradient transport. In RANS, several *ad-hoc* models have been suggested, but the difficulty can be overcome only by resorting to a full second-moment closure approach with separate balance equations for each component of Reynolds stress $\overline{\rho u_i'' u_j''}$ and Reynolds flux $\overline{\rho u_i'' c''}$. This is computationally expensive, and requires additional modelling effort in order to close terms in the second-moment balance equations. The Bray-Moss-Libby (BML) model [12] remains one of the most successful closure approaches in this regard.

In LES, studies of countergradient transport are rare as yet. Countergradient effects in sub-grid scale turbulent scalar transport were explored by Tullis [72] and found to be non-negligible. Some simple models were suggested based on existing formulations including BML.

Recent work has focussed on Lewis number effects on scalar transport in turbulent premixed flames. Contrary to early expectations, the effects of differential molecular transport of heat and mass are not masked by turbulent transport [24]. Instead, the destabilising effects on laminar flame propagation experienced at low Lewis numbers give rise to greatly enhanced countergradient transport in the turbulent flame. Locally-unstable flames become more wrinkled, leading to enhanced production of flame surface area and higher heat release per unit volume. This leads to greater thermal expansion and hence to a stronger self-induced pressure gradient through the flame. A revised Bray number has been obtained taking Lewis number effects into account [24], and new modelling has been developed for terms in the scalar flux transport equation [25].

4.4 Reaction Rate Modelling for Premixed Turbulent Flames

4.4.1 Simple Models

Simple algebraic closure modelling for $\bar{w}$ is well worth pursuing due to the low computational cost. An important class of models is based on the Eddy Break Up (EBU) approach [69]. The underlying concept is that the chemistry is fast and hence the mean reaction rate is governed by the rate at which turbulent mixing processes can bring reactants and products together. In the RANS context, the rate of mixing is taken to be inversely proportional to the turbulence timescale $\tilde{k}/\tilde{\varepsilon}$, leading to a reaction rate model of the form

$$\bar{w} = C_{\mathrm{EBU}} \frac{\tilde{\varepsilon}}{\tilde{k}} \overline{\rho c'' c''} \tag{4.17}$$

in which the progress variable variance is used as a measure of the scalar fluctuation magnitude. The EBU model is very simple, has very low computational cost and is often used to obtain a first estimate of a solution as an initial guess for something more sophisticated. The disadvantages are lack of sensitivity to chemistry, a tendency to produce unphysical solutions close to walls, and the need for the model constant C_{EBU} to be retuned for each individual case. Straightforward extensions of the EBU model include the Magnussen model [49] which brings in some level of sensitivity to mixture composition. A closely-related class of models is known as the Eddy Dissipation Concept (EDC) which has enjoyed a recent resurgence due to its simplicity and flexibility [41].

It is entirely feasible to adapt EBU-type models for LES. The timescale is the turn-over time for a filter-scale eddy, given by Δ/u', and the model may be paraphrased as [6]:

$$\bar{w} = C_{\Delta} \frac{\tilde{c}(1-\tilde{c})}{\Delta} \tag{4.18}$$

where C_{Δ} is a model constant and the local sub-grid variance is now evaluated in the strict flamelet limit as with the BML approach. This form of the EBU model still lacks sensitivity to chemistry, but it is better behaved close to walls and requires rather less *ad-hoc* tuning.

Incorporation of the effects of chemistry within a simple algebraic model is difficult. One interesting approach is called the beta transformation [16]. The concept is based on the observation that the laminar flame speed is determined by the geometric mean of the diffusivity and the reaction rate, i.e. $S_L \propto \sqrt{D_c w}$ where D_c is a representative mass diffusion coefficient (or thermal diffusivity) and w is the corresponding species reaction rate (or heat release rate). If the diffusivity is multiplied by a constant denoted by β, and the reaction rate is divided by the same constant, then the flame will be thickened due to the increased rate of diffusion but the flame speed will remain unchanged. Hence it becomes possible to resolve the flame structure on a coarse mesh without affecting the propagation speed. Furthermore, it is possible in principle to use a fully-detailed chemical reaction mechanism to obtain w. The approach was used for a time in RANS but was superseded by EBU and other models. More recently the concept has been revived for LES as the thickened flame model [30]. The thickened flame structure can be resolved over an affordable number of LES cells, and complex chemistry can be incorporated. A disadvantage of the approach is that it is no longer possible to represent flame-turbulence interactions accurately with eddies close to the filter scale. Current thickened-flame models make use of an efficiency function [2] in an attempt to overcome this difficulty. The thickened-flame model has been used to good effect in LES of highly-complex industrial combustion systems [14, 67].

4.4.2 Flame Surface Density Modelling

4.4.2.1 Algebraic Flame Surface Density Modelling

In its simplest form, the flame surface density (FSD) model provides an algebraic expression for the amount of flame surface area per unit volume Σ at each point within the premixed turbulent flame brush. If the flamelet assumption holds then the local flame structure remains quasi-laminar and the local propagation speed remains close to the unstrained planar laminar burning velocity S_L^0. Hence

$$\bar{w} = \bar{\rho}_R S_L^0 I_0 \Sigma \tag{4.19}$$

where $\bar{\rho}_R$ is the mean density in the reactants and I_0 is a factor accounting for the change in the local burning velocity due to straining and curvature effects [9]. Since S_L^0 and I_0 are quantities that depend essentially on the thermochemistry, the principal modelling problem is how to determine Σ. Following on from earlier work on time-series analysis [10, 11], a model was constructed [18] using spatial information as available from flame imaging techniques. Given an instantaneous two-dimensional image of a flame superimposed on a contour plot of the averaged flame brush, it is possible to count the number of points at which the instantaneous flame crosses the averaged contour, and this can be repeated over a large number of images. Using telegraph-signal analysis, an expression for the mean number of crossings per unit length $\bar{\nu}_y$ is obtained as

$$\bar{\nu}_y = \frac{g\bar{c}(1-\bar{c})}{\hat{L}_y} \tag{4.20}$$

where g is a constant and $\hat{L}_y$ is the integral length scale of the flame crossing process. The mean reaction rate per crossing is found from the mass flux through the flame at the crossing point corrected geometrically for the crossing angle θ:

$$\bar{w}_y = \frac{\bar{\rho}_R S_L^0 I_0}{|\cos\theta|} \tag{4.21}$$

Thus, the mean turbulent reaction rate is given by the product of the mean reaction per crossing and the mean number of crossings per unit length. Assuming statistical isotropy in the plane of the flame, the mean reaction rate per unit volume becomes

$$\bar{w} = \bar{\rho}_R S_L^0 I_0 \frac{g\bar{c}(1-\bar{c})}{\hat{L}_y |\cos\theta|} \tag{4.22}$$

Comparing (4.19) with (4.22) indicates that the flame surface density Σ is given by

$$\Sigma = \frac{g\bar{c}(1-\bar{c})}{\hat{L}_y \overline{|\cos\theta|}} \tag{4.23}$$

The quantities g and $\overline{|\cos\theta|}$ have been determined by experiment [29], while $\hat{L}_y$ has been modelled with reference to the turbulent integral length scale [18]. It is interesting to note that the model predicts a symmetric distribution of FSD through the flame brush conforming to the parabolic form $\bar{c}(1-\bar{c})$. Observations from experiment and from DNS suggest that the distribution of FSD tends to be biased towards either side of the flame brush due to Huygens propagation effects. This can be accounted for in the model by (for example) changing the exponents of $\bar{c}$ and $(1-\bar{c})$ in (4.23) to be different from unity. Further modelling is required in order to parameterise the factor I_0 based on strain rate and curvature statistics obtained using information available from standard turbulence models [63, 78]. Models based on expressions similar to (4.23) have been used extensively in RANS and have proved successful in different application areas [15, 44].

4.4.2.2 FSD transport equation

A transport equation for FSD has been derived using a kinematic approach [60] and also using a more geometrical approach [17]. The underlying concept is based on the earlier Coherent Flame Model developed principally for application to non-premixed flames [51]. In common with the algebraic FSD models, the surface area in question is that of a single isosurface of reaction progress variable at $c = c^*$ where c^* is a chosen constant. The reaction progress variable is a field quantity defined everywhere in three-dimensional space, but the fine-grained flame surface density Σ' is non-zero only on the infinitesimally-thin two-dimensional isosurface at the spatial locus of points corresponding to $c(\underline{x},t) = c^*$. In order to allow the FSD to be treated also as a field variable it is necessary to define an expectation value $\Sigma = \overline{\Sigma'}$ using a suitable local averaging technique. A unit vector based on progress variable gradient may be defined as

$$\underline{N} = -\frac{\nabla c}{|\nabla c|} \tag{4.24}$$

which is valid at all points where the progress variable gradient magnitude is non-zero. At $c = c^*$ the vector $\underline{N}$ becomes a normal vector to the chosen isosurface, and the velocity of the isosurface in laboratory coordinates can be expressed as

$$\dot{\underline{X}} = \underline{u} + S_d \underline{N} \tag{4.25}$$

where u is the velocity vector of the gas mixture at the isosurface location. The quantity S_d is the displacement speed defined as the speed of the isosurface normal to itself relative to the local gas mixture. With these definitions the exact transport equation for FSD may be stated as

$$\frac{\partial \Sigma}{\partial t} + \frac{\partial}{\partial x_k}\Sigma \langle \dot{X}_k \rangle_s = \Sigma \langle \dot{S} \rangle_s \tag{4.26}$$

Here, $\langle Q \rangle_s = \overline{Q\Sigma}/\Sigma$ denotes a surface average for any quantity Q, and $\dot{S}$ is the stretch rate defined as

$$\dot{S} = \frac{1}{A}\frac{dA}{dt} = (\delta_{ij} - N_i N_j)\frac{\partial \dot{X}_i}{\partial x_j} \tag{4.27}$$

Using the definition (4.25) the stretch rate may be decomposed according to

$$\dot{S} = (\delta_{ij} - N_i N_j)\frac{\partial u_i}{\partial x_j} + S_d \frac{\partial N_k}{\partial x_k} = a_T + 2 S_d h_m \tag{4.28}$$

where a_T is the hydrodynamic strain rate in the plane locally tangential to the isosurface and h_m is the geometrical mean curvature of the isosurface. With the definition (4.25) the FSD transport may be written as

$$\frac{\partial \Sigma}{\partial t} + \frac{\partial}{\partial x_k}\langle u_k \rangle_s \Sigma = \langle a_T \rangle_s \Sigma - \frac{\partial}{\partial x_k}\langle S_d N_k \rangle_s \Sigma + \langle S_d \frac{\partial N_k}{\partial x_k} \rangle_s \Sigma \tag{4.29}$$

For RANS applications, it is standard practice to use a Favre decomposition of the velocity for compatibility with the usual form of the reaction progress variable equation. Hence the FSD transport equation becomes

$$\begin{aligned}\frac{\partial \Sigma}{\partial t} + \frac{\partial}{\partial x_k}\tilde{u}_k \Sigma = \\ - \frac{\partial}{\partial x_k}\langle u_k'' \rangle_s \Sigma + \langle \bar{a}_T \rangle_s \Sigma + \langle a_T'' \rangle_s \Sigma - \frac{\partial}{\partial x_k}\langle S_d N_k \rangle_s \Sigma + \langle S_d \frac{\partial N_k}{\partial x_k} \rangle_s \Sigma\end{aligned} \tag{4.30}$$

Note that for further compatibility of numerical implementation it is possible to replace Σ with a flame surface area per unit mass $\sigma = \Sigma/\rho$ [17]. The five terms on the right-hand side of (4.30) are identified as (i) the turbulent transport term; (ii) the mean strain rate term; (iii) the turbulent strain rate term; (iv) the propagation term and (v) the curvature term. All of these terms are unclosed and require modelling.

A standard gradient model for the turbulent transport term may be written as

$$\langle u_k'' \rangle_s \Sigma = -\frac{\nu_T}{\sigma_\Sigma}\frac{\partial \Sigma}{\partial x_k} \tag{4.31}$$

where ν_T is the turbulent kinematic viscosity and σ_Σ is the turbulent Schmidt number for FSD. Realisability arguments, valid in both RANS and LES [38], indicate that σ_Σ should be taken equal to the turbulent Schmidt number for reaction progress variable.

The strain rate terms are usually considered as production terms since flame surface area is generated on average by fluid-mechanical straining. The mean strain rate term may be written as

$$\langle \bar{a}_T \rangle_s = (\delta_{ij} - \langle N_i N_j \rangle_s) \frac{\partial \tilde{u}_i}{\partial x_j} \tag{4.32}$$

where there is assumed to be no correlation between the flame normal vector and the Favre-mean velocity. Here, a model is required for the correlation $\langle N_i N_j \rangle_s$ between the components of the flame normal vector. There has been some DNS [21] and experimental [74] investigation of this quantity. A suitable model may be expressed as [19]

$$\langle N_i N_j \rangle_s = n_{ij} = M_i M_j + \frac{1}{3} \delta_{ij} (1 - M_k M_k) \tag{4.33}$$

where $\underline{M}$ is the surface-averaged normal vector given by the exact result [19]

$$\underline{M} = \langle \underline{N} \rangle_s = -\frac{\nabla \bar{c}}{\Sigma}. \tag{4.34}$$

The fluctuating strain rate term has been modelled [19] as

$$\langle a_T'' \rangle_s = C_A \left(\frac{1 - M_k M_k}{\tau_\eta} \right) \tag{4.35}$$

where τ_η is the Kolmogorov time scale. The quantity $1 - M_k M_k$ is called the orientation factor α and measures the tendency of the local flame surface to align with the direction of mean propagation. An alternative model for the strain rate term is based on integral scaling as [31]

$$\langle a_T'' \rangle_s = C_A \frac{\tilde{\varepsilon}}{\tilde{k}} \tag{4.36}$$

In both cases C_A can be treated as a model constant, or it can be made to depend on the local turbulent velocity scales [78]. Further investigation of the strain rate terms is ongoing [23, 40].

The propagation term is either neglected or it is taken in combination with the curvature term. The combined effect is usually treated as a net sink for FSD due to the role of propagation and curvature in the destruction of surface area by Huygens propagation effects at the trailing edge of the flame brush. A typical model is

$$\langle S_d \frac{\partial N_k}{\partial x_k} \rangle_s \Sigma = -C_H \langle S_d \rangle_s \frac{\Sigma}{1 \quad \tilde{c}} \tag{4.37}$$

where C_H is a model constant. Again, if required, C_H can be made to depend on the local turbulence timescales as well as on the orientation factor α [19].

It is worth noting that there are several outstanding issues concerning the modelling of the FSD transport equation for RANS. As indicated above, the surface average is not equivalent to a standard Reynolds average and some careful interpretation is needed. Moreover, the displacement speed is not equivalent to the laminar burning velocity and further modelling is required.

In LES, it has become standard to use the generalised FSD defined as $\Sigma_{\text{gen}} = \overline{|\nabla c|}$ [6]. In principle, using this quantity removes the need to choose a single isosurface of reaction progress variable, although extra care is required in the modelling. With a Favre decomposition of the filtered velocity, the filtered transport equation for the generalised FSD may be stated as

$$\frac{\partial \Sigma_{\text{gen}}}{\partial t} + \frac{\partial}{\partial x_k}\tilde{u}_k \Sigma_{\text{gen}} =$$

$$-\frac{\partial}{\partial x_k}\left(\langle u_k \rangle_s - \tilde{u}_k\right)\Sigma_{\text{gen}} + \langle a_T \rangle_s \Sigma_{\text{gen}} - \frac{\partial}{\partial x_k}\langle S_d N_k \rangle_s \Sigma_{\text{gen}} + \langle S_d \frac{\partial N_k}{\partial x_k}\rangle_s \Sigma_{\text{gen}} \quad (4.38)$$

This leaves the four terms on the right-hand side to be modelled. In order, they are (i) the sub-grid transport term; (ii) the strain rate term; (iii) the propagation term and (iv) the curvature term. In the turbulent transport term, the velocity difference $\langle u_i \rangle_s - \tilde{u}_i$ gives rise to a sub-grid flux of FSD which is closely related to the turbulent scalar flux of reaction progress variable [19]. In particular, it is clear that countergradient transport of FSD is to be expected under similar conditions to those which promote countergradient transport of reaction progress variable. Hence, models for turbulent transport of FSD are of the form [26, 37]

$$\left(\langle u_i \rangle_s - \tilde{u}_i\right)\Sigma_{\text{gen}} = -\frac{\nu_T}{\sigma_\Sigma}\frac{\partial \Sigma_{\text{gen}}}{\partial x_i} + \tau\left(K - \tilde{c}\right)S'_d \langle N_i \rangle_s \Sigma_{\text{gen}} \quad (4.39)$$

where ν_T is a sub-grid scale turbulent kinematic viscosity which may be evaluated using a standard Smagorinsky or similar model, σ_Σ is a turbulent Schmidt number, τ is the heat release parameter, K is a model constant and S'_d is a modified displacement speed given by $S'_d = \langle \rho S_d \rangle_s / \rho_R$. The first term on the right-hand side is a classical gradient transport model which caters for the occurrence of gradient transport of FSD. The other term embodies the effects of acceleration across the flame due to heat release which tends to promote countergradient transport of FSD [37, 71]. It is important to note that in LES, using the generalised FSD, the surface-averaged normal vector $\langle \underline{N} \rangle_s$ is no longer equal to the resolved normal vector $\underline{M}$:

$$\langle \underline{N} \rangle_s = -\frac{\nabla \bar{c}}{\Sigma_{\text{gen}}}; \qquad \underline{M} = -\frac{\nabla \bar{c}}{|\nabla \bar{c}|}. \quad (4.40)$$

These quantities each depend on the filtered reaction progress variable $\bar{c}$ and it is necessary to relate this quantity to its Favre-filtered value $\tilde{c}$. This can be done using an expression derived from the Bray-Moss Libby formulation in the limit of thin flamelets:

$$\bar{c} = \frac{(1+\tau)\tilde{c}}{1+\tau\tilde{c}} \quad (4.41)$$

In LES there may well be partial resolution of the flame and the sub-grid conditions may not allow the use of the simple BML approach. Instead a slightly more complicated expression is required [26, 37]:

$$\bar{c} = \frac{(1+\tau)\tilde{c}}{1+\tau\tilde{c}}\left[1-\exp\left(-\Theta\Delta/\delta_L\right)\right]+\tilde{c}\exp\left(-\Theta\Delta/\delta_L\right) \tag{4.42}$$

where $\Theta \simeq 0.2$ is a model constant.

The strain rate term may be decomposed according to

$$\langle \bar{a}_T \rangle_s = \left(\delta_{ij} - \langle N_i N_j \rangle_s\right)\frac{\partial \tilde{u}_i}{\partial x_j} + S_{\text{hr}} + S_{\text{sg}} \tag{4.43}$$

in which the three terms on the right-hand side represent the resolved contribution, a contribution due to heat release, and a sub-grid contribution. The correlation between the normal vector components can be treated in a similar manner to that used in RANS:

$$\langle N_i N_j \rangle_s = n_{ij} = \langle N_i \rangle_s \langle N_j \rangle_s + \frac{1}{3}\delta_{ij}\left(1 - \langle N_k \rangle_s \langle N_k \rangle_s\right) \tag{4.44}$$

The heat release contribution may be expressed in terms of the flame normal acceleration in a manner very similar to that used to derive the countergradient transport model in (4.39) above. The resulting model is

$$S_{\text{hr}} = -\tau\left(K - \tilde{c}\right)S_d' \frac{\partial \langle N_k \rangle_s}{\partial x_k}\Sigma_{\text{gen}} \tag{4.45}$$

The sub-grid contribution is commonly modelled using a scaling based on an estimate of the strain rate at the LES filter scale, resulting in the form

$$S_{\text{sg}} = \Phi \frac{u'}{\Delta}\Phi\Gamma_K\left(\frac{u'}{S_L}, \frac{\Delta}{\delta_L}\right) \tag{4.46}$$

where Φ is a model constant and Γ_K is an efficiency function [2, 52] that accounts for the variation in the effectiveness of different eddy sizes in straining the flame. Several different forms have been proposed for the efficiency function [2, 28] and assessment of these is ongoing [26].

For LES modelling it has been found convenient [37, 39] to combine the propagation and curvature terms and to consider their decomposition into resolved and sub-grid scale contributions as

$$\langle S_d \frac{\partial N_k}{\partial x_k} \rangle_s \Sigma_{\text{gen}} - \frac{\partial}{\partial x_k}\langle S_d N_k \rangle_s \Sigma_{\text{gen}} = P_{\text{mean}} + C_{\text{mean}} + C_{\text{sg}} \tag{4.47}$$

The mean propagation term may be stated as

$$P_{\text{mean}} = -\frac{\partial}{\partial x_k}\langle S_d \rangle_s \langle N_k \rangle_s \Sigma_{\text{gen}} \tag{4.48}$$

where the correlation between the displacement speed and the surface-averaged normal vector has been assumed to be negligible [37]. Note that further modelling is required for the surface-averaged displacement speed. Similarly, the mean curvature term may be stated as

$$C_{\text{mean}} = \langle S_d \rangle_s \frac{\partial \langle N_k \rangle_s}{\partial x_k} \Sigma_{\text{gen}} \tag{4.49}$$

This leaves only the sub-grid curvature term which is modelled in the same manner as in RANS, resulting in the expression

$$C_{\text{sg}} = -\alpha_N \beta_1 S_L \frac{\Sigma_{\text{gen}}^2}{(1-\tilde{c})} \tag{4.50}$$

in which β_1 is a model constant and $\alpha_N = 1 - \langle N_k \rangle_s \langle N_k \rangle_s$ is an orientation factor, also known in LES as the resolution factor [37]. It serves to quantify the extent to which the unresolved flame surface is aligned with the mean direction of propagation, and vanishes either when the surface-averaged normal vector magnitude is unity, or when the flame is fully resolved, i.e. when the filter size tends towards zero. A complete modelled transport equation for FSD has been suggested recently [26] and similar approaches are proving successful in applications to real devices [65].

4.4.2.3 Displacement speed modelling

In LES modelling for the FSD transport equation it is necessary to account for the response of the displacement speed S_d to the effects of straining and curvature. Careful treatment of the displacement speed enables the extension of FSD modelling from the corrugated flamelets regime to the thin reaction zones regime [22]. Unfortunately, there is no simple relationship between the displacement speed S_d and the laminar burning velocity S_L, also known as the consumption speed.

For a one-dimensional turbulent flame brush, mass conservation indicates that the surface-averaged displacement speed must vary through the flame according to

$$\langle S_d \rangle_s = \frac{\langle \rho S_d \rangle_s}{\langle \rho \rangle_s} \tag{4.51}$$

while the density-weighted displacement speed remains essentially constant through the flame. This observation has been confirmed for using DNS data [20] and experiment [36], and has been found to break down only when the local flame curvature is large.

An expression for the density-weighted displacement speed can be obtained using terms from the right-hand side of the filtered reaction progress variable equation (4.12) in the form

$$\overline{w} + \overline{\nabla.(\rho D_c \nabla c)} = \overline{\rho S_d |\nabla c|} \tag{4.52}$$

Decomposing the filtered molecular diffusion term into normal and tangential components allows for the decomposition of the density-weighted displacement speed into contributions arising from reaction, normal diffusion and tangential diffusion [55]:

$$\overline{\rho S_d} = \overline{\rho S_r} + \overline{\rho S_n} + \overline{\rho S_t} \tag{4.53}$$

where

$$\overline{\rho S_r} = \frac{\bar{w}}{|\nabla c|} \tag{4.54}$$

$$\overline{\rho S_n} = \frac{\overline{\underline{N}.\nabla(\rho D_c \underline{N}.\nabla c)}}{\overline{|\nabla c|}} \tag{4.55}$$

$$\overline{\rho S_t} = -\overline{\rho D_c \nabla.\underline{N}} = -2\overline{\rho D_c} \overline{h}_m \tag{4.56}$$

Data from DNS [20, 54] has indicated that the contributions from reaction and normal diffusion may be combined as $\bar{\rho} S_{LS} = \overline{(\rho S_r + \rho S_n)}_s$ and that this combination is insensitive to local curvature. Hence S_{LS} can be modelled using planar laminar flame data. The tangential contribution S_t is deterministically dependent on curvature, and a model is required for the sub-grid mean curvature $\overline{h}_m$. A scale-similarity approach has been suggested [22] and has been tested using DNS data.

4.4.3 G-equation Modelling

The G-equation approach is based on the use of a scalar field variable G whose definition in principle is completely arbitrary. An isosurface of G is fixed at some chosen value G_0, and the level set of the scalar at this value is taken to represent the spatial location of the flame surface [76]. The location of the level set $\underline{x}_F$ at any time t is the solution of the equation $G(\underline{x}_F, t) = G_0$. Differentiating this equation with respect to time yields

$$\frac{\partial G}{\partial t} + \nabla G.\frac{\partial \underline{x}_F}{\partial t} = 0 \tag{4.57}$$

Noting that

$$\frac{\partial \underline{x}_F}{\partial t} = \underline{u} + S_d \underline{N} \tag{4.58}$$

where $\underline{N}$ is the normal vector in the direction of propagation, given by $\underline{N} = -\nabla G/|\nabla G|$, the exact G equation can be written as

$$\frac{\partial G}{\partial t} + \underline{u}.\nabla G = S_d |\nabla G| \tag{4.59}$$

This is an eikonal equation which describes the advection and propagation of the level set. It contains no explicit reaction or diffusion terms, but it can be used to provide a marker for the location of an isosurface within a flame. In this sense the G-equation approach is not a model *per se*, but instead provides a mathematical framework for modelling.

Since the scalar variable G is arbitrary, it can be interpreted as a signed distance from the level set at $G = G_0$. The incremental distance from the level set is denoted by

$$dx_n = -\underline{N}.d\underline{x} = \frac{\nabla G}{|\nabla G|}.d\underline{x} \tag{4.60}$$

If the G field is assumed to be frozen for the purposes of evaluating the distance, the increment of the scalar G is given by $dG = \nabla G.d\underline{x}$ and the incremental distance becomes

$$dx_n = \frac{dG}{|\nabla G|} \tag{4.61}$$

In practice, identity between the distance field and the G field is ensured by enforcing the reinitialisation condition $\nabla G = 1$ together with the condition $x_n = 0$ at $G = G_0$. The distance field interpretation allows for the development of models based on flame structures that are located around the level set.

An important feature of the G-equation approach is to incorporate a decomposition of the displacement speed as

$$S_d = S_L^0 - \mathscr{L}_a a_T - S_L^0 \mathscr{L}_h \kappa \tag{4.62}$$

where $\kappa = 2h_m = \nabla.\underline{N}$ is used to represent the curvature of the level set. The three terms of the right-hand side of (4.62) represent the unstrained planar laminar burning velocity, the effects of straining and the effects of curvature. Linear models are used for both straining and curvature and are parameterised by the respective Markstein lengths $\mathscr{L}_a$ and $\mathscr{L}_h$. Since the curvature κ can be expressed as

$$\kappa = \nabla.\underline{N} = -\nabla.\frac{\nabla G}{|\nabla G|} \tag{4.63}$$

the resulting G-equation becomes

$$\frac{\partial G}{\partial t} + \underline{u}.\nabla G = \left(S_L^0 - \mathscr{L}_a a_T\right)|\nabla G| - D_h \underline{N}.\nabla\left(\underline{N}.\nabla G\right) + D_h \nabla^2 G \tag{4.64}$$

where the Markstein diffusivity $D_h = S_L^0 \mathscr{L}_h$. The three terms on the right-hand side represent the effects of (i) laminar flame propagation including a dependence on strain rate, (ii) normal diffusion and (iii) curvature. It may be observed that the curvature term involves a second derivative of G. This has an important practical effect in providing numerical stability for the G equation and in helping to prevent the occurence of self-intersections of the surface at $G = G_0$.

Applications of the G equation in RANS of practical combustion systems requires the use of a Favre-averaged form. Starting with the standard G equation (4.59) and using the Favre decomposition $G = \tilde{G} + G''$ yields a transport equation for the Favre average $\tilde{G}$:

$$\bar{\rho}\frac{\partial \tilde{G}}{\partial t} + \bar{\rho}\underline{\tilde{u}}.\nabla\tilde{G} = \overline{(\rho S_d)|\nabla G|} - \nabla.\left(\overline{\rho \underline{u}'' G''}\right) \tag{4.65}$$

This equation is in a more conventional form and the right-hand side consists of somewhat familiar unclosed terms, the first representing the mean turbulent reaction rate and the second representing the turbulent flux of G. These can be modelled using a slight variation on the usual arguments. For a statistically stationary planar turbulent flame, the averaged G equation reduces to

$$(\bar{\rho} S_T)\frac{\partial \tilde{G}}{\partial x} = \overline{(\rho S_d)\,|\nabla G|} \tag{4.66}$$

where, by conservation of mass, $\bar{\rho} S_T$ is the constant mass flow rate through the flame and S_T is the turbulent burning velocity. In essence, this relation can be viewed as a statement of Damköhler's hypothesis, since $|\nabla G|$ is a generalised surface density for the level set. The increment of distance from the turbulent flame brush may be expressed in the same manner as (4.61) as

$$dx_T = \frac{\nabla \tilde{G}}{|\nabla \tilde{G}|} \tag{4.67}$$

and this expression can be used to generalise back to the three-dimensional case yielding a model for the mean turbulent reaction rate in (4.65) as

$$\overline{(\rho S_d)\,|\nabla G|} = (\bar{\rho} S_T)\,|\nabla \tilde{G}| \tag{4.68}$$

It is possible to treat the turbulent flux term in (4.65) using a classical gradient transport model in the form

$$\nabla.\left(\overline{\rho \underline{u}'' G''}\right) = -\nabla.\left(\bar{\rho} D_T \nabla \tilde{G}\right) \tag{4.69}$$

where D_T is the turbulent diffusivity for G. This is deemed to be undesirable in the context of the G-equation approach [55] since it is not consistent with the mathematical character of the original eikonal equation. Instead, the gradient transport model is split into its normal and tangential components according to

$$\nabla.\left(\bar{\rho} D_T \nabla \tilde{G}\right) = \underline{\tilde{N}}.\nabla\left(\bar{\rho} D_T \underline{\tilde{N}}.\nabla \tilde{G}\right) - \bar{\rho} D_T \tilde{\kappa} |\nabla \tilde{G}| \tag{4.70}$$

where $\underline{\tilde{N}}$ is the Favre-averaged normal vector and $\tilde{\kappa}$ is the Favre-averaged curvature, both valid on the surface at $\tilde{G}(\underline{x},t) = G_0$. Since normal diffusion has been accounted for already by the inclusion of a turbulent burning velocity in the reaction rate term, only the tangential diffusion component needs to be retained explicitly in the model [55]. This results in the final modelled equation for the Favre average $\tilde{G}$:

$$\bar{\rho}\frac{\partial \tilde{G}}{\partial t} + \bar{\rho}\underline{\tilde{u}}.\nabla \tilde{G} = (\bar{\rho} S_T)\,|\nabla \tilde{G}| - \bar{\rho} D_T \tilde{\kappa} |\nabla \tilde{G}| \tag{4.71}$$

A full closure of this equation requires additional information about the turbulent burning velocity. Typically this is provided using an empirical correlation of the form

$$\frac{S_T}{S_L} = 1 + C\left(\frac{u'}{S_L}\right)^n \tag{4.72}$$

where C and n are constants. This marks a key difference between the G-equation approach and the FSD approach, in that the turbulent burning velocity must be specified as an input to the former, but emerges as a result from the latter. A balance equation for the variance of G may also be derived and modelled [54].

The G-equation approach has proved popular in RANS applications to various problems of practical interest [55]. The extension to LES appears to be straightforward and several formulations have been proposed [43, 77] based on essentially the same modelling concepts as described above. Further developments have made greater use of the special mathematical symmetries of the G-equation [53]. A recent formulation of the G-equation for LES [56, 57] makes use of the notion that the level set $G = G_0$ already represents the filtered flame position, and hence G does not have to be filtered. This results in the G equation

$$\frac{\partial G}{\partial t} + \underline{\hat{u}}.\nabla G = -(S_L + S_\kappa)\underline{N}.\nabla G \tag{4.73}$$

in which $\underline{N}$ is the normal vector to the unfiltered flame surface, $\hat{u}$ is the filtered velocity conditioned on $G = G_0$, S_L is the laminar flame propagation speed and S_κ is the curvature contribution to the propagation speed. Closure modelling for the conditionally-filtered velocity is provided in terms of the familiar Favre-filtered velocity $\tilde{u}$ [56]. Modelling for the propagation term serves to illustrate the need to consider partial resolution of the flame. The model may be stated in the form [56]

$$(S_L + S_\kappa)\underline{N} = \left(S_T - D_\kappa \hat{\kappa} + \hat{S}_L - D\hat{\kappa}\right)\underline{\hat{N}} \tag{4.74}$$

where $\underline{\hat{N}}$ is the normal vector to the surface at $G = G_0$ and $\hat{\kappa}$ is the curvature of that surface. The contributions from normal propagation and curvature effects at the unresolved length scales of the flame surface are denoted by S_T and $D_\kappa \hat{\kappa}$. In addition, there are contributions from the resolved scales as denoted by $\hat{S}_L$ and $D\hat{\kappa}$. The sub-grid scale turbulent burning velocity S_T is modelled based on previous arguments, while D_κ is related to the sub-grid scale turbulent diffusivity. Both of these quantities must vanish as the filter width is reduced, leaving only the resolved contributions which must be formulated to represent the true behaviour of the flame.

4.4.4 Scalar Dissipation Rate Modelling

The scalar dissipation rate has proved to be a very important quantity for the modelling of non-premixed flames [4]. In that context the scalar dissipation rate appears as a sink term in the balance equation for the variance of the mixture fraction. In premixed flames the Favre-averaged scalar dissipation rate is defined in terms of the reaction progress variable as

$$\bar{\rho}\tilde{\varepsilon}_c = \overline{\rho D_c \frac{\partial c''}{\partial x_k}\frac{\partial c''}{\partial x_k}} \tag{4.75}$$

It is interesting to note that the scalar dissipation rate is defined in terms of molecular diffusion only, i.e. its definition does not include any reaction rate information. Nevertheless, the strong coupling between reaction and diffusion in both non-premixed and premixed flames ensures that it is possible to use the scalar dissipation rate as a measure of reaction rate. The scalar dissipation rate may be interpreted as a reciprocal time scale for molecular mixing. If the chemistry is fast and hence the combustion is mixing-controlled, then the scalar dissipation rate may be interpreted also as a reciprocal time scale for reaction. In the fast-chemistry limit, a relation between the scalar dissipation rate and the mean turbulent reaction rate may be stated as [8]

$$\bar{w} = \frac{2\bar{\rho}\tilde{\varepsilon}_c}{2C_m - 1} \tag{4.76}$$

where $C_m = \overline{cw}/\bar{w}$ and may be treated as a model constant. The scalar dissipation rate may be modelled by analogy with the dissipation rate of turbulent kinetic energy ε, using an expression of the form

$$\bar{\rho}\tilde{\varepsilon}_c = C_D \frac{\tilde{\varepsilon}}{\tilde{k}} \overline{\rho c'' c''} \tag{4.77}$$

where C_D is a model constant which is normally assigned a value close to unity. This approach leads to an algebraic closure model for the mean turbulent reaction rate that is closely related to the EBU model [8, 45].

An alternative approach is to derive a balance equation for the scalar disspation rate [50]. This provides greater insight into the physics and allows for a proper order-of-magnitude analysis of contributing terms [70], and for modelling of individual terms using information from experiment [36] and from DNS [27]. A balance between dominant terms yields the algebraic expression

$$\bar{\rho}\tilde{\varepsilon}_c = \left(1 + \frac{2}{3} C_{\varepsilon_c} \frac{S_L^0}{\tilde{k}^{\frac{1}{2}}}\right) \left(C_{D_c} \frac{S_L^0}{\delta_L^0} + C_D \frac{\tilde{\varepsilon}}{\tilde{k}}\right) \overline{\rho c'' c''} \tag{4.78}$$

in which C_{ε_c}, C_{D_c} and C_D are model constants. The first set of parentheses includes a timescale associated with flame stretch while the two terms in the second set of parentheses represent a chemical timescale and molecular dissipation timescale respectively. Hence the model offers a significant advance over the classical EBU approach, in that additional physics is captured and can be justified on a theoretical basis. Note that a classical EBU model is recovered in the limit of intense turbulence ($\tilde{k}^{\frac{1}{2}} \gg S_L^0$) and fast chemistry ($\delta_L/S_L^0 \to \infty$). The scalar dissipation rate approach can be shown to have a direct equivalence to the FSD approach [13] and the two quantities can be related according to the expression

$$\Sigma = \frac{\bar{\rho}\tilde{\varepsilon}_c}{K_\Sigma S_L} \tag{4.79}$$

where $K_\Sigma = \rho_R(2C_m - 1)/2$. This relation allows for the derivation of improved algebraic models for Σ with guidance from the transport equation for the scalar dissipation rate.

4.4.5 Other Approaches

Modelling for turbulent premixed flames has been dominated by approaches based to a large extent on the phenomenology of flamelets. There is abundant evidence to support their effectiveness, and the Borghi diagram supports their applicability. Nevertheless, it is clear that flamelet structure cannot prevail everywhere. In flames close to extinction due to turbulent straining, or close to the flammability limits, the balance between reaction and diffusion is weakened and the identification of a representative flame structure becomes difficult.

In such situations a more general modelling approach is desirable. The transported pdf approach [42, 59] makes no assumptions about the structure of the flame and has proved successful in its application to turbulent non-premixed flames (see Chapter 6). There have been applications of pdf transport modelling to premixed flames [47, 62] in which convincing results have been obtained. The advantages of the approach lie in its potential for generality especially in cases where direct chemical effects may be important, but the computational cost is high and there are technical issues concerned with the modelling of mixing processes in the presence of the high scalar gradients typical of premixed flames.

Another approach that has proved successful in non-premixed combustion is the Conditional Moment Closure (CMC) approach, in which the fluctuations of all variables about the conditional mean mixture fraction are assumed to be small (for more details, see Chapter 5). In principle the CMC approach could be extended to treat premixed flames, possibly by using the reaction progress variable as a conditioning variable [34]. The use of a marker field variable has been suggested [5] but it is questionable whether this would bring any advantage over the existing G-equation approach. One promising avenue is the Conditional Source Term Estimation (CSTE) approach [35] which is closely related to CMC and which may provide a more realistic way to model chemical effects in premixed flames.

In general, turbulent premixed flames well away from extinction can be modelled without explicit consideration of the chemical reaction rate. Chemical information is encapsulated in quantities such as the laminar burning velocity and its response to straining. The computational costs of detailed chemistry are high, and it is preferable to treat the chemistry separately from any turbulent flame calculation. This is true especially in LES where computational costs are already high due to the demand for fine spatial resolution and the need for unsteady solutions. Chemistry can be precomputed and stored in tabulated form using a number of different approaches. Perhaps the simplest is a premixed flamelet library, in which the laminar burning velocity can be stored as a function of mixture strength, pressure and reactant temperature. Other independent variables can be included, such as strain rate and mean curvature. In order to avoid the computational costs of searching in tables, it is is possible to make use of correlations for laminar burning velocity which have been derived from either experimental or one-dimensional computations [1]. A more advanced approach is to use the concepts behind Intrinsic Low Dimensional Manifolds (ILDM) [48], suitably extended to include molecular transport effects in a manner representative of flamelet structure. An approach such as Flamelet Generated Manifolds (FGM) [73] or Flamelet Prolongation of ILDM (FPI) [32] allows for a reasonable level of chemical detail that is computationally inexpensive - at least within the turbulent flame calculation - and which is largely free from the restrictions of the original flamelet concept.

4.5 Future

There is an increasing technological requirement to operate combustion systems in a manner that does not correspond to either premixed or non-premixed burning. For example, many lean premixed gas turbine combustors involve the use of a much richer or even non-premixed pilot flame which acts to stabilise the main premixed flame closer to the lean limit than would be possible otherwise. Modern gas turbine aero-engines are being designed to pass a much greater flow of air through the fuel injector in order to minimise rich burning and hence reduce unwanted emissions. This results in flames which burn in non-uniform fuel-air mixtures and which exhibit features such as edge flames or triple flames. Modelling efforts for such partially-premixed flames are in their infancy and generally involve straightforward extensions of existing models. A common assumption is that the variables used to characterise the strength of the mixture (e.g, mixture fraction) and the progress of the reaction (e.g, reaction progress variable) are uncorrelated, and this is unlikely to be true in general. Improved modelling is under development using fundamental data from DNS [33] and from experiment [3]. A more sophisticated approach is the Libby-Williams-Poiters (LW-P) model [64] in which the conceptual basis behind the Bray-Moss-Libby formulation is extended to capture the probabilistic behaviour of the mixture fraction. This approach has been validated against experimental data [66] and has been used also to derive improved models for the turbulent stresses and scalar fluxes in RANS.

Many aspects of turbulent premixed flames have been modelled more-or-less successfully in both RANS and LES using the flamelet concept in its different guises. It is now clear that the FSD, G-equation and scalar dissipation rate approaches are all essentially similar, and a great deal can be learned by comparing and contrasting these approaches in their various forms in order to explore the various assumptions and to elucidate additional physics. There is less certainty around the modelling of turbulent premixed flames under intense turbulence, or close to extinction. More work is necessary also in the immediate aftermath of ignition, where the mixture is within the flammability limits but a flame structure is not established. Here, the flame may become premixed or non-premixed depending on the circumstances, or it may simply fail to burn. More general models are required in order to handle such situations, and also fully to bridge the gap between premixed and non-premixed combustion. The inherently low computational cost of flamelet-type models may prove to be a thing of the past as more and more chemistry is required for an accurate representation. At the same time, greater computer power, for example in the form of the much greater parallelism now becoming available from multi-core processors, may make this irrelevant as the affordable resolution for LES is steadily increased.

Acknowledgements Thanks are due to many people including Dr Evatt Hawkes and Dr Nilanjan Chakraborty, and to Professor Ken Bray for all his help and for many inspirational discussions. This chapter is dedicated to the memory of Dr Arthur McNaughtan 1956-2009.

References

1. Abu-Orf, G.M., Cant, R.S.: A turbulent reaction rate model for premixed turbulent combustion in spark-ignition engines. Combust. Flame **122**, 233–252 (2000)
2. Angelberger, C., Veynante, D., Egolfopoulos, F., Poinsot, T.: LES of combustion instabilities. Proc Summer Program 1998, Center for Turbulence Research, Stanford/NASA Ames (1998)
3. Anselmo-Filho, P., Hochgreb, S., Barlow, R.S., Cant, R.S.: Experimental measurements of geometric properties of turbulent stratified flames. Proc. Combust. Inst. **32**, 1763–1770 (2009)
4. Bilger, R.W.: Structure of diffusion flames. Combust. Sci. Technol. **13**, 155–170 (1976)
5. Bilger, R.W.: Marker fields for turbulent premixed combustion. Combust. Flame **138**, 188–194 (2004)
6. Boger, M., Veynante, D., Boughanem, H., Trouvé, A.: Direct numerical simulation analysis of flame surface density concept for large eddy simulation of turbulent premixed combustion. Proc. Combust. Inst. **27**, 917–925 (1998)
7. Borghi, R.: On the structure and morphology of turbulent premixed flames. In: Casci, C. (ed.) Recent advances in the aerospace sciences. Plenum, New York (1985)
8. Bray, K.N.C.: The interaction between turbulence and combustion. Proc. Combust. Inst. **17**, 223–233 (1978)
9. Bray, K.N.C., Cant, R.S.: Some applications of Kolmogorov's turbulence research in the field of combustion. Proc. Roy. Soc. Lond. **434**, 217–240 (1991)
10. Bray, K.N.C., Libby P.A.: Passage times and flamelet crossing frequencies in prelixed turbulent combustion. Combust. Sci. Technol. **47**, 253–274 (1986)
11. Bray, K.N.C., Libby P.A., Moss J.B.: Flamelet crossing frequencies and mean reaction rates in premixed turbulent combustion. Combust. Sci. Technol. **41**, 143–172 (1984)
12. Bray, K.N.C., Libby P.A., Moss J.B.: Unified modelling approach for premixed turbulent combustion part 1 - general formulation. Combust. Flame **61**, 87–102 (1985)
13. Bray, K.N.C., Swaminathan, N.: Scalar dissipation and flame surface density in premixed turbulent combustion. C. R. Mecanique **334**, 466–473 (2006)
14. Boileau, M., Staffelbach, G., Cuenot, B., Poinsot, T., Bérat, C.: LES of an ignition sequence in a gas turbine engine. Combust. Flame **54**, 2–22 (2008)
15. Brookes, S.J., Cant, R.S., Dupere, I.D.J., Dowling A.P.: Computational modelling of self-excited combustion instabilities. Trans. ASME **123** 322–326 (2001)
16. Butler, T.D., O'Rourke, P.J.: A numerical method for two-dimensional unsteady reacting flows. Proc. Combust. Inst. **16**, 1503–1515 (1977)
17. Candel, S.M., Poinsot, T.J.: Flame stretch and the balance equation for the flame area. Combust. Sci. Technol. **70**, 1–15 (1990)

18. Cant, R.S., Bray, K.N.C.: Strained laminar flamelet calculations of premixed turbulent combustion in a closed vessel. Proc. Combust. Inst. **22**, 791–799 (1988)
19. Cant, R.S., Pope, S.B., Bray, K.N.C.: Modelling of the flamelet surface-to-volume ratio in turbulent premixed combustion. Proc. Combust. Inst. **23**, 809–815 (1990)
20. Charkraborty, N.: Comparison of displacement speed statistics of turbulent premixed flames in regimes representing combustion in corrguated flamelets and thin reaction zones. Phys. Fluids **19**, 105109 (2007)
21. Charkraborty, N., Cant, R.S.: Statistical behavior and modeling of the flame normal vector in turbulent premixed flames. Num. Heat Transfer A **50**, 623–643 (2006)
22. Charkraborty, N., Cant, R.S.: A priori analysis of the curvature and propagation terms of the flame surface density transport equation for large eddy simulation. Phys. Fluids **19**, 105101 (2007)
23. Charkraborty, N., Cant, R.S.: Influence of Lewis number on strain rate effects in turbulent premixed flame propagation. Int. J. Heat Mass Transfer **49**, 2158–2172 (2007)
24. Charkraborty, N., Cant, R.S.: Effects of Lewis number on scalar transport in turbulent premixed flames. Phys. Fluids **21**, 035110 (2009)
25. Charkraborty, N., Cant, R.S.: Effects of Lewis number on turbulent scalar transport and its modelling in turbulent premixed flames. Combust. Flame **156**, 1427–1444 (2009)
26. Charkraborty, N., Cant, R.S.: Direct numerical simulation analysis of the flame surface density transport equation in the context of large eddy simulation. Proc. Combust. Inst. **32**, 1445–1453 (2009)
27. Charkraborty, N., Rogerson, J.W., Swaminathan, N.: A priori assessment of closures for scalar dissipation rate transport in turbulent premixed flames using direct numerical simulation. Phys. Fluids **20**, 045106 (2008)
28. Charlette, F., Trouvé, A., Boger, M., Veynante, D.: A flame surface density model for large eddy simulations of turbulent premixed flames. Proc. Joint Meeting of the British, French and German Sections of The Combustion Institute, Nancy (1999)
29. Chew T.C., Britter R.E., Bray, K.N.C.: Laser tomography of turbulent premixed bunsen flames. Combust. Flame **75**, 165–174 (1989)
30. Colin, O., Ducros, F., Veynante, D., Poinsot, T.: A thickened flame model for large eddy simulations of turbulent combustion. Phys. Fluids **12**, 1843–1863 (2000)
31. Duclos, J.M., Veynante, D., Poinsot, T.: A comparison of flamelet models for premixed turbulent combustion. Combust. Flame **95**, 101–117 (1993)
32. Gicquel, O., Darabiha, N., Thévenin, D.: Laminar premixed hydrogen/air counterflow flame simulations using flame prolongation of ILDM with preferential diffusion. Proc. Combust. Inst. **28**, 1901–1908 (2000)
33. Grout, R.W.: Interaction of turbulence and scalar fields in premixed flames. Phys. Fluids **18**, 045102 (2006)
34. Grout, R.W.: An age extended progress variable for conditioning reaction rates. Phys. Fluids **19**, 105107 (2007)
35. Grout, R.W., Bushe W.K., Blair C.: Predicting the ignition delay of turbulent methane jets using conditional source term estimation. Combust. Theory Model. **11**, 1009–1028 (2007)
36. Hartung, G., Hult, J., Kaminski, C.F., Rogerson, J.W., Swaminathan, N.: Effect of heat release on turbulence and scalar-turbulence interaction in premixed combustion. Phys. Fluids **20**, 035110 (2008)
37. Hawkes E.R., Cant, R.S.: A flame surface density approach to large eddy simulation of premixed turbulent combustion. Proc. Combust. Inst. **28**, 51–58 (2000)
38. Hawkes, E.R., Cant, R.S.: Physical and numerical realizability requirements for flame surface density approaches to large eddy and Reynolds averaged simulation of premixed turbulent combustion. Combust. Theory Model. **5**, 699–720 (2001)
39. Hawkes, E.R., Cant, R.S.: Implications of a flame surface density approach to large eddy simulation of turbulent premixed combustion. Combust. Flame **126**, 1617–1629 (2001)
40. Hawkes, E.R., Chen, J.H.: Evaluation of models for flame stretch due to curvature in the thin reaction zones regime. Proc. Combust. Inst. **30**, 647–655 (2004)

41. Hong, S., Assanis, D.N., Wooldridge, M.S., Im, H.G., Kurtz, E., Pitsch, H.: Modelling of diesel combustion and NO emissions based on a modified eddy dissipation comcept. SAE Paper 2004-01-0107 (2004)
42. Jones, W.P.: The joint scalar probability density function. In: Launder, B.E., Sandham, N.D. (Eds.) Closure Strategies for Turbulent and Transitional Flows, Cambridge University Press, Cambridge, UK (2002)
43. Kim, W.W., Menon, S., Mongia, H.C.: Large-eddy simulation of a gas turbine combustor flow. Combust. Sci. Technol. **143**, 25–62 (1999)
44. Knikker, R., Veynante, D., Meneveau, C.: A priori testing of a similarity model for large eddy simulations of turbulent premixed combustion. Proc. Combust. Inst. **29**, 2105–2111 (2002)
45. Kolla, H., Rogerson, J.W., Charkraborty, N., Swaminathan, N.: Scalar dissipation rate modelling and its validation. Combust. Sci. Technol. **181**, 518–535 (2009)
46. Libby, P.A., Bray, K.N.C.: Counter gradient diffusion in premixed turbulent flames. AIAA J. **19**, 205–213 (2001)
47. Lindstedt, R.P., Vaos, E.M.: Transported PDF modelling of high-Reynolds number premixed turbulent flames. Combust. Flame **145**, 495–511 (2006)
48. Maas, U., Pope, S.B.: Simplifying chemical kinetics: intrinsic low-dimensional manifolds in composition space. Combust. Flame **88**, 239–264 (1992)
49. Magnussen B., Hjertager B.: On mathematical modelling of turbulent combustion with special emphasis on soot formation and combustion. Proc. Combust. Inst. **16**, 719–727 (1977)
50. Mantel, T., Borghi, M.: A new model of premixed wrinkled flame propagation based on a scalar dissipation equation. Combust. Flame **96**, 443–457 (1984)
51. Marble, F.E., Broadwell, J.E.: The coherent flame model for turbulent chemical reactions. Project Squid Technical Report TRW-9-PV, Purdue University (1977)
52. Meneveau, C., Poinsot, T.: Stretching and quenching of flamelets in premixed turbulent combustion. Combust. Flame **86**, 311–332 (1991)
53. Oberlack, M., Wenzel, H., Peters, N.: On symmetries and averaging of the G equation for premixed combustion. Combust. Theory Model. **5**, 1–20 (2001)
54. Peters, N. The turbulent burning velocity for large and small scale turbulence. J. Fluid Mech. **384**, 107–132 (1999)
55. Peters, N.: Turbulent Combustion. Cambridge University Press, Cambridge, UK (2000)
56. Pitsch, H.: A consistent level set formulation for large-eddy simulation of premixed turbulent combustion. Combust. Flame **143**, 587–598 (2005)
57. Pitsch, H.: Large-eddy simulation of turbulent combustion. Ann. Rev. Fluid Mech. **38**, 453–482 (2006)
58. Pitsch, H., Duchamp de Lageneste, L.: Large-eddy simulation of premixed turbulent combustion using a level set approach. Proc. Combust. Inst. **29**, 2001–2008 (2002)
59. Pope, S.B.: PDF methods for turbulent reacting flows. Prog. Energy Combust. Sci. **11**, 119–192 (1985)
60. Pope, S.B.: The evolution of surfaces in turbulence. Int. J. Eng. Sci. **26**, 445–469 (1988)
61. Pope S.B.: Turbulent Flows. Cambridge University Press, Cambridge, UK (2000)
62. Pope, S.B., Cheng, W.K.: The stochastic flamelet model of turbulent premixed combustion. Proc. Combust. Inst. **22**, 781–815 (1989)
63. Pope, S.B., Yeung, S.S., Girimaji, S.S.: The curvature of surfaces in isotropic turbulence. Phys. Fluids A **1**, 2010–2018 (1989)
64. Ribert, G., Champion, M., Plion, P.: Modelling turbulent reactive flows with variable equivalence ratio: Application to the calculation of a reactive shear layer. Combust. Sci. Technol. **176**, 907–923 (2004)
65. Riched, S., Colin, O., Vermorel, O., Benkenida, A., Angelberger, C., Veynante, D.: Towards large eddy simulation of combustion in spark ignition engines. Proc. Combust. Inst. **31**, 3059–3066 (2007)
66. Robin, V., Mura, A., Champion, M., Plion, P.: A second-order model for turbulent reactive flows with variable equivalence ratio. Combust. Sci. Technol. **180**, 1707–1732 (2008)

67. Selle, L., Lartigue, G., Poinsot T., Koch R., Schildmacher K.-U., Krebs, W., Prade, B., Kaufmann, P., Veynante, D.: Compressible large eddy simulation of turbulent combustion on unstructured meshes. Combust. Flame **137**, 489–505 (2004)
68. Shepherd, I.G., Moss, J.B., Bray, K.N.C.: Turbulent transport in a confined premixed flame. Proc. Combust. Inst. **19**, 423–431 (1982)
69. Spalding D.B.: Mixing and chemical reaction in steady confined turbulent flames. Proc. Combust. Inst. **13**, 649–657 (1971)
70. Swaminathan, N., Bray, K.N.C.: Effects of dilatation on scalar dissipation in turbulent premixed flames. Combust. Flame **143**, 549–565 (2005)
71. Tullis, S., Cant, R.S.: Counter-gradient scalar transport in large eddy simulation of turbulent premixed flames. In: Pollard, A., Candel, S. (eds) Proc. IUTAM Symposium on Turbulent Mixing and Combustion, Kluwer, Dordrecht (2001)
72. Tullis S., Cant, R.S.: Scalar transport modelling in large eddy simulation of turbulent premixed flames. Proc. Combust. Inst. **29**, 2097–2104 (2002)
73. van Oijen, J.A., de Goey, L.P.H.: Modelling of premixed laminar flames using flamelet-generated manifolds. Combust. Sci. Technol. **161**, 113–138 (2000)
74. Veynante, D., Piana, J., Duclos, J.M., Martel, C.: Experimental analysis of flame surface density models for premixed turbulent combustion. Proc. Combust. Inst. **26**, 413–420 (1996)
75. Veynante D., Trouvé A., Bray K.N.C., Mantel T.: Gradient and countergradient turbulent scalar transport in turbulent premixed flames. J. Fluid Mech. **332**, 263–293 (1997)
76. Williams, F.A.: Turbulent combustion. In: Buckmaster, J. (ed.) The Mathematics of Combustion, SIAM, Philadelphia (1985)
77. Pitsch, H., Duchamp de Lageneste, L.: Large-eddy simulation of premixed turbulent combustion using a level-set approach. Proc. Combust. Inst. **29**, 2001–2008 (2002)
78. Yeung, P.K., Girimaji, S.S., Pope, S.B.: Straining and scalar dissipation on material surfaces in turbulence: implications for flamelets. Combust. Flame **79**, 340–365 (1990)

Chapter 5
The Conditional Moment Closure Model

A. Kronenburg and E. Mastorakos

Abstract The relatively recent Conditional Moment Closure methods for turbulent reacting flows have advanced from application to relatively well behaved, simple laboratory flames to complex flow geometries and flame conditions with intense turbulence-chemistry interactions. The progress on second order closures, double conditioning approaches, two-phase and premixed CMC is reviewed in the first part of this chapter, while the second part is largely dedicated to numerical methods to solve the CMC equations and to the model's capability to address questions of direct engineering interest such as the modelling of diesel engine combustion and the analysis of flame stabilization mechanisms.

5.1 Introduction

More than ten years ago, the first and - to date - last comprehensive review of the Conditional Moment Closure (CMC) method was published by Klimenko and Bilger [35]. Yet, great strides have been made in advancing the method from its rather typical application to relatively simple and "well behaved" diffusion flames to more complex flow geometries and flame conditions.

The development of the Conditional Moment Closure (CMC) method was motivated by the need to provide accurate closures for the average of the non-linear turbulent reactive source term. It was conceptually derived as a mixture fraction based approach for non-premixed turbulent combustion and has as such some similarities with laminar flamelet methods. The basic idea is to exploit a strong correlation be-

A. Kronenburg
Institut für Technische Verbrennung, University of Stuttgart, 70569 Stuttgart, Germany, e-mail: kronenburg@itv.uni-stuttgart.de

E. Mastorakos
Department of Engineering, University of Cambridge, Cambridge, CB2 1PZ, UK, e-mail: em257@eng.cam.ac.uk

T. Echekki, E. Mastorakos (eds.), *Turbulent Combustion Modeling*,
Fluid Mechanics and Its Applications 95, DOI 10.1007/978-94-007-0412-1_5,

tween reactive scalar species and the mixture fraction and hence fluctuations in reactive scalar space can be associated with fluctuations in mixture fraction space. Conditioning of the reactive species on mixture fraction then leads to relatively small fluctuations around the conditional mean, and a simple first order closure for the chemical source can be found. Hence, transport equations for the reactive species mass fractions conditioned on mixture fraction have been derived, and some unclosed terms such as the conditional velocity and the conditionally averaged scalar dissipation need to be modelled. Relatively simple models for these terms suffice and the solution of the temporal and spatial evolution of the conditional moments gives relatively good predictions of finite-rate chemistry effects for a wide range of turbulent diffusion flames [25, 31, 66, 67, 69].

During the last decade, the majority of CMC-related theoretical development has focused on the modelling of flame regimes where local correlations between reactive scalars and mixture fraction are weakened. This decorrelation will for example occur in flames with local extinction, in lifted flames where fuel and oxidizer mix without combustion before the stabilization point and more generally, in any flow with regions of partial or complete premixing prior to ignition. Here, fluctuations around the conditional mean will start to become significant, but it is very important to realize that CMC is not rendered invalid if conditioned and conditioning quantities are not well correlated. The CMC transport equations can be derived without making assumptions on the degree of correlation. However, closures of unclosed terms, and in particular of the conditionally averaged chemical source term, need improvements to account for the effects of fluctuations on the evolution of the conditioned moments.

The second major focus of CMC research has been on the application of the method to more complex flow geometries and flame conditions with technical relevance such as engine and gas turbine related environments. The two foci of work determine the structure of the remainder of this chapter which is largely split into two major sections. The next section is called "methodological developments" where we will briefly review the standard formulation of the conditional moment closure method and then elaborate on techniques that account for flame conditions where reactive scalars do not correlate well with mixture fraction, namely the second order closures and double conditioning. Further, specific subsections are dedicated to the discussion of the first steps in CMC related premixed flame modelling and advances in two-phase flow CMC. The second major section (Section 5.3) is dedicated to applications of CMC to problems of engineering interest. Several research groups have worked on CMC for diesel engine combustion, auto-ignition studies, flame stabilization, CMC as combustion sub-model for Large-Eddy Simulations (LES) for improved flow field modelling in complex geometries, and more generally, on pollutant predictions where finite rate effects dominate. While the theoretical framework for the application of CMC is well established, some implementation and modelling issues arise. Section 5.3 will provide an up-to-date assessment of CMC and its capabilities, but improvements to the technique are certainly needed to allow application to all flow and flame regimes. We will attempt to outline future steps in CMC development in the final part of Section 5.3 that should serve as encouragement and guideline to new researchers in the field of CMC modelling.

5.2 Methodological Developments in CMC

The CMC transport equations for turbulent reacting flows were derived by Klimenko [33] and Bilger [4] using somewhat different methodologies and primary closure assumptions. Details on the differences are discussed in [35], and we limit ourselves here to present one approach only that is called "the decomposition method". Its derivation requires little knowledge of combustion, it is based on the well known instantaneous transport equations for reactive and passive scalars and only standard mathematical techniques need to be used. The derivation in the next subsection will be quite general and standard closures will be introduced. The further subsections discuss more advanced closures such as second order and double conditioning approaches, but also outline CMC's potential for the modelling of premixed flames. Modifications that need to be made for multiphase combustion are described in Section 5.2.5.

5.2.1 The CMC Equations

We will first define the conditional mean, $Q_k(\mathbf{Z}^c,\mathbf{x},t)$, of a scalar $Y_k(\mathbf{x},t)$ as

$$Q_k(\mathbf{Z},\mathbf{x},t) = \langle Y_k(\mathbf{x},t) \mid \mathbf{Y}^c = \mathbf{Z}^c \rangle. \tag{5.1}$$

The angular brackets denote ensemble averages of Y_k, conditioned on $\mathbf{Y}^c = \mathbf{Z}^c$, where $\mathbf{Y}$ is a multidimensional scalar space, and $\mathbf{Z}$ is its sample space. The choice whether a specific scalar is a conditioned or a conditioning quantity is problem dependent, however, the sets of conditioned and conditioning quantities are mutually exclusive and no scalar should be part of both sets. The arrays to the right of the vertical bar in Eq. 5.1 should be understood to represent a (usually small) subset of the entire scalar space, $\mathbf{Y}^c = (Y_1, Y_2, \ldots, Y_{n_c})$ and $\mathbf{Z}^c = (Z_1, Z_2, \ldots, Z_{n_c})$, where the number of conditioning scalars, n_c, should be much smaller than the number of scalars, n_{sc}. The scalar space of the conditioned quantities is then given by $\mathbf{Y}^q = (Y_{n_c+1}, \ldots, Y_k, \ldots, Y_{n_{sc}})$, and the superscripts 'c' and 'q' are used here to indicate scalar arrays with conditioning and conditioned quantities, respectively. Note that the scalar array may include mixture fraction, scalar dissipation and an energy related scalar such as enthalpy or temperature. The conditional mean - or conditional expectation - is related to the joint probability density function through

$$\langle Y_k \mid \mathbf{Y}^c = \mathbf{Z}^c \rangle = \int_{-\infty}^{+\infty} Z_k P(Z_k \mid \mathbf{Y}^c = \mathbf{Z}^c) dZ_k = \frac{\int_{-\infty}^{+\infty} Z_k P(Z_k, \mathbf{Z}^c) dZ_k}{P(\mathbf{Z}^c)}, \tag{5.2}$$

where spatial and temporal dependencies have been omitted for clarity of presentation, and the unconditional mean can be obtained from the conditional mean by integration across the entire sample space of the conditioning variables,

$$\langle Y_k \rangle = \int_{-\infty}^{+\infty} \langle Y_k \mid \mathbf{Y}^c = \mathbf{Z}^c \rangle P(\mathbf{Z}^c) d\mathbf{Z}^c. \tag{5.3}$$

The starting point for the derivation of the CMC equations is the well known transport equation for chemically reactive species, discussed in Chapter 2, which is repeated here for the reader's convenience,

$$\rho \frac{\partial Y_k}{\partial t} + \rho u_i \frac{\partial Y_k}{\partial x_i} - \frac{\partial}{\partial x_i}\left(\rho D_k \frac{\partial Y_k}{\partial x_i}\right) = w_k. \tag{5.4}$$

In Eq. 5.4 we have used Fick's law for the diffusive flux. The instantaneous mass fraction Y_k can now be decomposed into the conditional mean and the fluctuation around the conditional mean,

$$Y_k(\mathbf{x},t) = Q_k(\mathbf{Z},\mathbf{x},t) + Y_k''(\mathbf{x},t) \tag{5.5}$$

and be inserted into Eq. 5.4. The chain rule of differentiation is applied to all temporal and spatial derivatives of Q_k. Note, for example, the application of the chain rule to the time derivative,

$$\frac{\partial Y_k}{\partial t} = \frac{\partial Q_k}{\partial t} + \frac{\partial Q_k}{\partial Z_j}\frac{\partial Z_j}{\partial t} + \frac{\partial Y_k''}{\partial t}, \qquad j = 1, \ldots, n_c \tag{5.6}$$

and its application to the spatial derivatives - or repeated applications for the diffusion term - obeys identical rules. The entire resulting equation needs to be conditionally averaged again, and the final CMC equation results in

$$\begin{aligned} \langle \rho \mid \mathbf{Z}^c \rangle \frac{\partial Q_k}{\partial t} = &- \langle \rho u_i \mid \mathbf{Z}^c \rangle \frac{\partial Q_k}{\partial x_i} + \langle w_k \mid \mathbf{Z}^c \rangle - \langle w_j \mid \mathbf{Z}^c \rangle \frac{\partial Q_k}{\partial Z_j} \\ &+ \langle \rho D_k \frac{\partial Y_j}{\partial x_i}\frac{\partial Y_j}{\partial x_i} \mid \mathbf{Z}^c \rangle \frac{\partial^2 Q_k}{\partial Z_j \partial Z_j} + \langle \rho D_k \frac{\partial Y_j}{\partial x_i}\frac{\partial Y_l}{\partial x_i} \mid \mathbf{Z}^c \rangle \frac{\partial^2 Q_k}{\partial Z_j \partial Z_l} \\ &+ e_q + e_y \end{aligned} \tag{5.7}$$

with

$$\begin{aligned} e_q &\equiv \langle \frac{\partial}{\partial x_i}\left(\rho D_k \frac{\partial Q_k}{\partial x_i}\right) + \rho D_k \frac{\partial Y_j}{\partial x_i}\frac{\partial^2 Q_k}{\partial x_i \partial Z_j} + \frac{\partial}{\partial x_i}\left[\rho (D_k - D_j)\frac{\partial Y_j}{\partial x_i}\right]\frac{\partial Q_k}{\partial Z_j} \mid \mathbf{Z}^c \rangle \\ e_y &\equiv -\langle \rho \frac{\partial Y_k''}{\partial t} + \rho u_i \frac{\partial Y_k''}{\partial x_i} - \frac{\partial}{\partial x_i}\left(\rho D_k \frac{\partial Y_k''}{\partial x_i}\right) \mid \mathbf{Z}^c \rangle. \end{aligned}$$

All terms on the RHS need closure. The e_q-term can usually be neglected under high Reynolds number assumptions [35], but needs to be modeled if differential diffusion effects are to be included [37, 38]. Similarly, the e_y-term that involves all the fluctuations around the conditional mean, must not be omitted in the presence of significant differential diffusion effects, but can be set to zero in the absence of differential diffusion and if a "good" set of conditioning variables is selected. The term "good" refers to the correlation of the (unconditional) fluctuations of the conditioned scalars with the (unconditioned) fluctuations of the conditioning scalars. A perfect correlation would lead to $Y_k'' \approx 0$, however, correlations are hardly ever perfect, and based on the primary closure hypothesis employed in the decomposition method [35], $e_y P(\mathbf{Z}^c) = -\nabla \cdot \left[\langle \rho \mid \mathbf{Z}^c \rangle \langle \mathbf{u}'' Y_k'' \mid \mathbf{Z}^c \rangle P(\mathbf{Z}^c)\right]$, we may propose a gradient diffusion approximation for e_y. The validity of the gradient diffusion model has been corroborated by Richardson et al. [63] for relatively high turbulence levels. Deviations between the model and DNS data at low turbulence levels are due to counter-gradient transport present in the studied expanding flames that involved propagating fronts; such phenomena may be absent in attached non-premixed flames. The importance of e_y, and therefore the importance of the accuracy of its closure, strongly depends on flow and flame conditions. It is of paramount importance for flame stabilization in lifted flames where flame propagation is the dominating stabilization mechanism [15, 24].

The best modeling approaches for the conditional velocity, $\langle \rho u_i \mid \mathbf{Z}^c \rangle$, dissipation, $N_{jj} \equiv \langle D \nabla Y_j \nabla Y_j \mid \mathbf{Z}^c \rangle$ and cross-dissipation, $N_{jl} \equiv \langle D \nabla Y_j \nabla Y_l \mid \mathbf{Z}^c \rangle$, have not yet been established for multidimensional conditioning spaces. Ample experience exists, however, for simple, single conditioning ($n_c = 1$), where $Y_1 = \xi$ and $Z_1 = \eta$ with ξ being mixture fraction and η its sample space. Single conditioning on mixture fraction leads to the standard CMC equations,

$$\rho_\eta \frac{\partial Q_k}{\partial t} + \langle \rho u_i \mid \eta \rangle \frac{\partial Q_k}{\partial x_i} = \rho_\eta N_\eta \frac{\partial^2 Q_k}{\partial \eta^2} + \langle w_k \mid \eta \rangle - \frac{1}{P_\eta} \frac{\partial}{\partial x_i} \left[\rho_\eta \langle u_i'' Y_k'' \mid \eta \rangle P_\eta\right] \quad (5.8)$$

with scalar dissipation $N \equiv D \nabla \xi \nabla \xi$ and subscript η indicating dependence on mixture fraction.

Sreedhara et al. [71] assessed different standard closures for the conditional velocity, $\langle \mathbf{u} \mid \eta \rangle$ and the conditional scalar dissipation, $\langle N \mid \eta \rangle$. The different models for $\langle \mathbf{u} \mid \eta \rangle$ were termed (1) "conditional independence", (2) "linear in terms of unconditional flux" and (3) "gradient diffusion in terms of local PDF". Only minor differences between the models could be observed for the jet flames studied; such comparison is not available for more complicated flows.

Standard models for conditionally averaged scalar dissipation have been termed (1) "AMC"-, (2) "Girimaji's"- and (3) "double integration of the PDF transport equation"-model [71]. Theoretically, the AMC-model [57] requires the presence of unmixed fluid, a condition that will not be satisfied in regions downstream of the potential core of a jet. Girimaji's model [20] has been derived for homogeneous flows and - strictly speaking - may not hold in regions with strong mean gradients. In contrast, the integration of the PDF transport equation ensures consistency between conditionally averaged dissipation and the scalar's probability distribution in all regions of the flow [40] and as such, should be the modeller's choice. However, the PDF transport equation requires a "good" model for the conditionally averaged velocity, and accurate numerical integration of the equation is difficult in regions of low probability. Practically, all three models provide qualitatively similar results for conditionally averaged scalar dissipation in Sandia Flame D [2], with the AMC-model being symmetric in η-space and the PDF-model deviating most from symmetry. 3-D measurements [2] show some asymmetry, but it would be difficult to decide on the superiority of one specific model. Notable differences exist in a bluff body flame, but lack of experimental data makes judgment on the quality of the different models impossible.

Further models with great promise have been developed by Devaud et al. [14], Cha et al. [8] and Mortensen and de Bruyn Kops [51], however, none of the models has been validated by comparison with experimental data, and they are therefore not discussed further in the present review.

The key to successful CMC modelling is the accuracy of the closure of the conditionally averaged chemical source term, $\langle w_k \mid \eta \rangle$. The assumption of relatively small fluctuations around the conditional mean allows for simple first order closure of the reaction rate term, i.e.

$$\langle w_k \mid \eta \rangle = w_k(\mathbf{Q}, \eta). \tag{5.9}$$

This closure has proven adequate for many applications, but more complex closures may become necessary if fluctuations become significant. It is repeated here that the CMC equation (Eq. 5.8) remains valid even if fluctuations are large, however, the accuracy of above closures will suffer and different modelling strategies will need to be employed. Rather unexpectedly, improved closures of the e_y-term may not be necessary, however, accurate modelling of the chemical source term is crucial [36, 43]. The latter can be achieved by multiple conditioning (see Section 5.2.3) or by second order closures that are discussed next.

5.2.2 Advances in Second Order Closures

The reasons for fluctuations around the conditional mean are diverse. Mixture fraction is a good conditioning scalar for all non-premixed flames, however, relatively strong fluctuations in scalar dissipation due to turbulence directly affect the flame structure, this leads locally to deviations from the conditional mean and induces fluctuations. If scalar dissipation fluctuations are large, local extinction can occur and conditional fluctuations are significant as shown in Fig. 5.1. Similar conditional fluctuations can be observed in turbulent flows where fuel and oxidizer partially premix before ignition occurs. In addition, relatively large variations of ignition location can exist even in statistically stationary cases, e.g. Gordon et al. [21] reported fluctuations of lift-off height of around 10 jet diameters in experiments with jets issued into hot vitiated air. Mastorakos [47] reviewed the spatial fluctuations of autoignition spots in detail and found that the fluctuations of the scalar dissipation rate are crucial for generating conditional reaction rate fluctuations.

It seems obvious that the influence of these fluctuations cannot be ignored. However, advanced closures do not necessarily aim at improved modelling of the term that involves the fluctuations, but seek an improvement to the existing closures for the chemical source term. The second order closures presented in this Subsection therefore focus on a second order correction of the conditionally averaged source term only.

The forward reaction rate of any bi-molecular elementary reaction, m, $[C_k] + [C_l] \rightleftharpoons [C_r] + [C_s]$ can be expressed in Arrhenius form,

$$w^m(\rho, \mathbf{Y}, T) = A_m \rho^2 \frac{Y_k Y_l}{M_k M_l} T^{b_m} exp(-T_{a,m}/T) \tag{5.10}$$

and a Taylor expansion around the conditionally averaged first order approximation yields

$$\begin{aligned} \langle w^m \mid \eta \rangle &= w^m(\rho_\eta, \mathbf{Q}, \eta) \cdot \left(1 + \frac{\langle Y_k'' Y_l'' \mid \eta \rangle}{Q_k Q_l} + T_1 + T_2\right) \\ T_1 &\equiv \left(b_m + \frac{T_{a,m}}{Q_T}\right)\left(\frac{\langle Y_k'' T'' \mid \eta \rangle}{Q_k Q_T} + \frac{\langle Y_l'' T'' \mid \eta \rangle}{Q_l Q_T}\right) \\ T_2 &\equiv \frac{1}{2}\left(b_m(b_m - 1) + \frac{2(b_m - 1)T_{a,m}}{Q_T} + \frac{T_{a,m}^2}{Q_T^2}\right) \cdot \frac{\langle T''^2 \mid \eta \rangle}{Q_T^2}. \end{aligned} \tag{5.11}$$

The second term in the brackets in Eq. 5.11 accounts for the correlation between species k and l, T_1 approximates the correlations between species and temperature and T_2 results from higher moments of temperature. These higher order corrections can be significant and easily exceed the first order approximation, in particular for reactions with high activation energies (due to the dependence of T_2 on squared $T_{a,m}$) as can be found in the first Zel'dovich step of NO formation. Closure of Eq. 5.11 requires knowledge of the conditionally averaged correlations between all chemically reactive species, $G_{kl} = \langle Y_k'' Y_l'' \mid \eta \rangle$ and between species and temperature

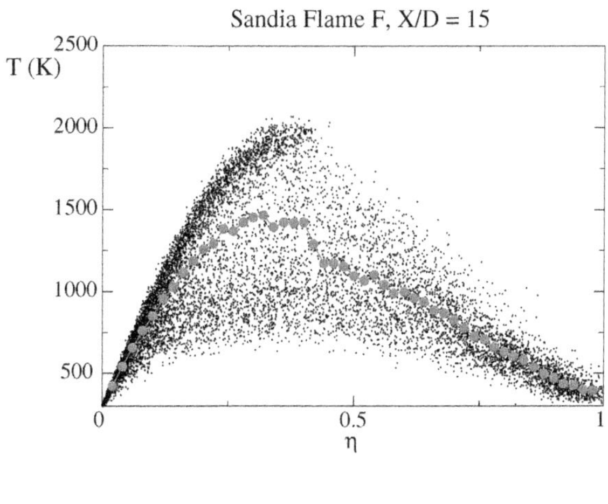

Fig. 5.1: Measurements of temperature as function of mixture fraction in a piloted methane-air jet diffusion flame with significant local extinction and re-ignition (Sandia Flame F). The filled symbols indicate the conditional mean of temperature, $Q_T = \langle T \mid \eta \rangle$. Reprinted from [19] with permission from the Combustion Institute.

$G_{kT} = \langle Y_k'' T'' \mid \eta \rangle$. A major issue is apparent: second order closure requires the solution of $n_{sc}(n_{sc}+1)/2$ additional transport equations for the conditional variance and co-variance, a task hardly feasible when using detailed chemical kinetics for hydrocarbon combustion. Early attempts by Mastorakos and Bilger [48] and Kronenburg et al. [39] used global mechanisms that could be parameterized by two variables, however, extension was not straightforward, and these methods therefore lacked generality. Kim and Huh [29] introduced the idea that second order corrections need to be applied to rate limiting steps only. They identified four rate limiting elementary reactions in a detailed methane-air mechanism (GRI 2.11 with 49 species and 277 reactions) that control the radical pool, the CO-CO_2 conversion, initialization and chain termination. These reactions involve 9 species (and temperature), and the resultant reduction from 1250 to 45 additional equations renders the problem trackable. The conditionally averaged variance and co-variance equations as such can easily be derived using the PDF or decomposition method [26],

$$\begin{aligned}
&\frac{\partial G_{kl}}{\partial t} + \langle u_i \mid \eta \rangle \frac{\partial G_{kl}}{\partial x_i} - N_\eta \frac{\partial^2 G_{kl}}{\partial \eta^2} + \frac{1}{\rho_\eta P_\eta} \frac{\partial}{\partial x_i} \left[\rho_\eta \langle u_i'' Y_k'' Y_l'' \mid \eta \rangle P_\eta \right] \\
&\quad + \frac{\partial Q_k}{\partial x_i} \cdot \langle Y_l'' u_i'' \mid \eta \rangle - \frac{\partial Q_l}{\partial x_i} \cdot \langle Y_k'' u_i'' \mid \eta \rangle \\
&= \quad \langle w_k'' Y_l'' + w_l'' Y_k'' \mid \eta \rangle - 2 \langle D \left(\frac{\partial Y_k''}{\partial x_i} \frac{\partial Y_l''}{\partial x_i} \right) \mid \eta \rangle \\
&\quad + \langle N'' Y_k'' \mid \eta \rangle \frac{\partial^2 Q_l}{\partial \eta^2} + \langle N'' Y_l'' \mid \eta \rangle \frac{\partial^2 Q_k}{\partial \eta^2} + \frac{1}{\rho_\eta P_\eta} \frac{\partial J_{gkl}}{\partial \eta}
\end{aligned} \tag{5.12}$$

with the transport equation for G_{kT} being identical in form. Primary closure assumptions have been invoked (i.e. e_q-terms and differential diffusion effects are neglected) and all terms on the LHS are closed or can be closed with standard closures. However, all terms on the RHS require additional models. Closures have been mainly suggested by the Sydney [45, 72] and Pohang [26, 27, 29] research groups. A set of relatively simple closures is given by

$$\langle w_k'' Y_l'' \mid \eta \rangle = \frac{\partial w_k}{\partial Y_\alpha} |_{Y=Q} \, G_{l\alpha} \tag{5.13}$$

$$2 \langle D \left(\frac{\partial Y_k''}{\partial x_i} \frac{\partial Y_l''}{\partial x_i} \right) \mid \eta \rangle = 2 \sqrt{C_k C_l} \frac{\varepsilon}{k} G_{kl} \tag{5.14}$$

$$\langle N'' Y_k'' \mid \eta \rangle = R_k N_\eta \langle G_{kk} \mid \eta \rangle^{1/2} \tag{5.15}$$

$$J_{gkl} = C_g \rho_\eta N_\eta P_\eta \frac{\partial G_{kl}}{\partial \eta} \tag{5.16}$$

Equation 5.13 is a first order approximation of the correlation between the chemical source term and the species and can be easily obtained from a Taylor expansion of w_k around $w_k(\mathbf{Q})$, multiplication with Y_l'' and subsequent conditional averaging [17]. The dissipation terms (Eq. 5.14) are modelled by an equivalence of the scalar fluctuation and turbulence time scales, C_k and C_l are constants to be discussed below, and successful modelling of the dissipation-scalar correlations (Eq. 5.15) relies on a "good" choice of $R_k = R_{N''Y''_k} N_{rms}/N_\eta$ with $R_{N''Y''_k}$ being the correlation coefficient between N'' and Y''_k. The turbulent transport in conserved scalar space, J_{gkl}, is modelled assuming gradient transport [26].

A second set of alternative closures can be based on laminar flamelet assumptions [29, 72] where scalar fluctuations around their conditional means can be associated with fluctuations in scalar dissipation. The laminar flamelet solution can be parameterized by the value of scalar dissipation at stoichiometric and further assumptions of independence of $P(\eta)$ and $P(N_{st})$ and log-normality of $P(N_{st})$ lead to a closed form of the equations

$$\langle w_k'' Y_l'' \mid \eta \rangle = \frac{G_{ll}^{1/2}}{\sigma_l} \times A$$

$$A = \left(\int w_k^S Y_l^S P_N(N_{st}) dN_{st} - \int w_k^S P_N(N_{st}) dN_{st} \int Y_j^S P_N(N_{st}) dN_{st} \right) \quad (5.17)$$

$$2\langle D \left(\frac{\partial Y_k''}{\partial x_i} \frac{\partial Y_l''}{\partial x_i} \right) \mid \eta \rangle = 2D \int \frac{\partial lnY_k''^S}{\partial \eta} \frac{\partial lnY_l''^S}{\partial \eta} P_N(N_{st}) dN_{st} \quad (5.18)$$

$$\langle N'' Y_k'' \mid \eta \rangle = \frac{N_\eta G_{kk}^{1/2}}{N_{\eta,st} \sigma_i} \times B$$

$$B = \left(\int N_{st} Y_k^S P_N(N_{st}) dN_{st} - \int N_{st} P_N(N_{st}) dN_{st} \int Y_k^S P_N(N_{st}) dN_{st} \right) \quad (5.19)$$

where superscript 'S' indicates laminar flamelet solutions and σ_k is the RMS of Y_k^S. No alternative closure is suggested for J_{gkl}.

Data for *a priori* closure validation is scarce, the accuracy of some approximations has not yet been established (e.g. Eq. 5.13), and all existing studies rely on DNS of reactive mixtures in homogeneous, isotropic turbulence for the assessment of Eqs. 5.13-5.19. Sreedhara et al. [71] reported good qualitative and quantitative agreement between DNS and models based on the flamelet assumption with one notable exception: the flamelet model performed poorly for the closure of the dissipation term (Eq. 5.18). This is in contrast to the findings of Swaminathan and Bilger [72], but -in general- good performance of Eq. 5.18 cannot be expected in flames with local extinction since the flamelet approximation imposes certain gradients in mixture fraction space. This can best be illustrated by analysing conditionally averaged temperature or mass fractions of product species such as CO_2 and H_2O: their maxima around stoichiometric impose zero gradients while extinction and re-ignition events lead to a decorrelation of these scalars and mixture fraction resulting in the disappearance of the local minimum of dissipation around stoichiometry. It further needs to be emphasized that good quantitative agreement could only be established with additional scaling factors in Eqs. 5.17 and 5.19 that account for the "mismatch between mean scalar dissipation and mean reaction rate" [71]. Dissipation and temperature (and therefore reaction rate) are well correlated during extinction, however, they are rather uncorrelated during re-ignition leading to the need of the additional parameter. These scaling factors need to be taken from DNS. Omitting these additional factors, Eqs. 5.17 and 5.19 introduce modelling errors of up to 100% and 30%, respectively. Furthermore, it should be noted that flamelet assumptions for the closure of the conditional variance equation lead to inconsistencies with the transport equation for the conditional mean: they could also be used to model $\langle w_k \mid \eta \rangle$.

Equations 5.15-5.16 provide decent agreement between models and DNS, but errors in peak values of up to 50% may occur. Sreedhara et al. [71] assessed different constants for C_k in Eq. 5.14. Standard values of $C_k = 1$ for major species and temperature and $C_k = 2$ for radicals tend to underpredict dissipation rates while order of magnitude estimates, exploiting proportionality of the gradient of radicals with $C_k = 1/\xi_{st}$ and of major species with $C_k = 1/\sqrt{\xi_{st}(1-\xi_{st})}$, lead to overpredictions. In terms of qualitative behaviour, this model represents the counterpart to the flamelet based model: it always predicts local maxima around stoichiometric and fails to predict local minima that occur while scalar and mixture fraction are well correlated. Last, Eq. 5.15 allows for similar freedom in choosing modelling constants. The correlation coefficient is unknown, and DNS data show large variations between -1 and +1 [48, 72]. Fortunately, $R_{N''Y''_k}$ is approximately constant in mixture fraction regions where the second derivative of Q_l is large, and the modelling of the third and fourth RHS terms in Eq. 5.12 with constant $R_{N''Y''_k} = O(1)$ is adequate.

Even though *a priori* analysis of all suggested closures is not quite satisfactory, predictions of the conditional variance and co-variances are promising. "Good" agreement has been achieved for conditionally averaged temperature and species RMS in several jet diffusion flames [29, 39]. However, second order closures have had limited impact on predictions of major scalars and temperature. Peak temperatures and O_2 mass fractions could markedly be improved in flames with local extinction and re-ignition (Sandia Flames D, E and F) [17, 29], however, leaving room for improvement. Temperature predictions in bluff body flames (flames HM1 and HM3) were hardly affected by changes to the modelling of the chemical source term [70], but effects on minor species are significant. All studies show clear improvements of predicted OH, NO and even CO levels and examples are presented in Fig. 5.2.

In contrast to the above where the second-order correction has been applied to only one or just a few reacting scalars, De Paola et al. [58] used second-order closure without any reduction in the dimensionality of the second-order correlation matrix, i.e. they solved transport equations for all $\langle Y''_k Y''_l | \eta \rangle$ and $\langle Y''_k T'' | \eta \rangle$. The application to autoignition problems with a 32-species chemical mechanism that included low temperature autoignition was successful. The validity of some closures in the second-order CMC equation has also been explored by comparison with DNS of autoigniting jets [64].

It seems that some effects cannot be captured by second order closures, it is unclear why temperature predictions in bluff body flames are rather unimpressed by improved source term modelling and corrections to the chemical source term may not suffice here. It is unlikely that any - yet to be demonstrated - inaccuracy of closures for the variance equations can be blamed; variance and co-variances are reasonably well predicted but no sign of better temperature predictions in HM1 and

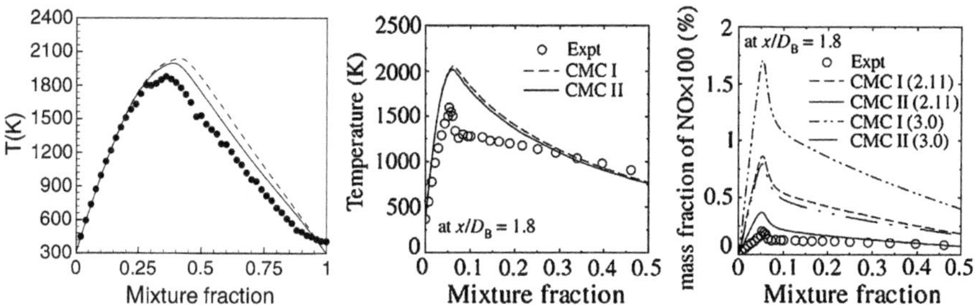

Fig. 5.2: CMC computations of temperature in Sandia Flame E (left), of temperature in the Sydney bluff body flame HM3 (centre) and of NO in HM3 (right). Computations (solid lines denote second order closure and dashed lines are results from first order closure) are from [29, 70] and measurements (symbols) are from [13, 19]. The numbers 2.11 and 3.0 in the right figure refer to GRI mechanisms used for the computations. Reprinted from [29, 70] with permission from the Combustion Institute.

HM3 can be detected. It is possible that the fluctuation terms themselves should not be neglected - after all Kronenburg and co-workers [42, 43] based their statement on the importance of e_y on DNS of reacting flows in homogeneous, isotropic turbulence. The strategy that is pursued in the next Section therefore aims at a reduction of the fluctuations themselves and problems may be avoided.

5.2.3 Advances in Doubly Conditioned Moment Closures

Large fluctuations around the conditional mean indicate that the conditioning variable does not sufficiently parameterize the combustion process. The introduction of further conditioning variables should reduce the fluctuations if a strong dependence of the reactive species on the new conditioning variables can be established, and suitable choices for these variables, their implementation and modelling issues are discussed next.

5.2.3.1 Basics of Double Conditioning

Traditionally, the flamelet equations suggest two quantities that parameterize the composition field: mixture fraction and scalar dissipation. We have seen above in Eqs. 5.17-5.19 that laminar flamelet assumptions can be invoked and instantaneous composition fields can be parameterized by mixture fraction and scalar dissipation if mixture fraction alone ceases to suffice as conditioning scalar. The introduction of scalar dissipation as second conditioning variable seems a particularly adequate choice in flames with flame extinction since large values of scalar dissipation are the cause of extinction and lead to flame quenching. Cha et al. [9] therefore introduced scalar dissipation as second conditioning variable and formulated a closed set of equations for the computation of the evolution of the reactive scalars doubly conditioned on mixture fraction and scalar dissipation at $\eta = 0.5$. Their DNS of reactive scalars in homogeneous turbulence used 1-step chemistry and showed varying degrees of extinction and re-ignition. The results show that doubly conditioned CMC predicts extinction well, but unfortunately, the onset of re-ignition is predicted much too early. Scalar dissipation may be a good indicator of extinction events, however, scalar dissipation does not correlate well with species composition after flame extinction occurred. In addition, conditioning on instantaneous values of scalar dissipation neglect the importance of relevant chemical time scales on the flame. Very short events of very large dissipation values will not necessarily lead to extinction due to a finite response time needed by the flame. Similarly, events of low dissipation values (let them be long or short) will not automatically lead to re-ignition since re-ignition requires heat flux to the mixture.

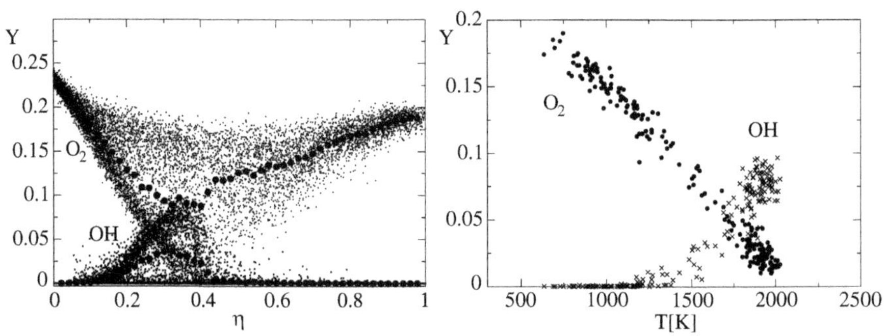

Fig. 5.3: Measurements of O_2 and OH mass fractions in a piloted jet diffusion flame (Sandia Flame F) at X/D=15. (a) O_2 and OH as function of mixture fraction, where filled symbols indicate the conditional means, and (b) O_2 and OH as function of mixture fraction ($\xi \approx \eta_{st}$) and temperature. Measurements are from [1]. Reprinted from [42] with permission from the Combustion Institute.

The key issue here is the lack of correlation of the chemical source term with scalar dissipation, and this results in the flamelet assumptions losing their validity. Now, we may want to remember our original goal of finding an accurate closure for the chemical source term. The chemical source term is largely dependent on the provision of fuel and oxidizer and on temperature. Therefore, Bilger [3] proposed conditioning on mixture fraction and sensible enthalpy and experimental data from a piloted jet diffusion flame with significant extinction provide support. Figure 5.3 (left) shows instantaneous mass fractions of O_2 and OH as function of mixture fraction at a downstream distance of X/D=15. Apparently, mixture fraction alone does not parameterize the flame well, and large scatter around the conditional mean can be observed. Figure 5.3 (right) shows the same data with mixture fraction values around $\xi = 0.351$ (the stoichiometric mixture fraction) as function of temperature.

Now the fluctuations have been greatly reduced, and we may be able to find fits through the data that correspond to the doubly conditioned means.

The governing equations for the doubly conditioned scalars result directly from Eq. 5.7 with $\mathbf{Z}^c = (\eta, \zeta)^T$, where ζ represents the sample space of normalized sensible enthalpy, $\hat{h}_s = \int_{T_0}^{T} c_p dT / \int_{T_0}^{T_{ad}} c_p dT$, and can then be written as

$$\frac{\partial Q_k}{\partial t} + \langle u_i \mid \eta, \zeta \rangle \frac{\partial Q_k}{\partial x_i} = \langle w_k \mid \eta, \zeta \rangle - \langle w_{h_s} \mid \eta, \zeta \rangle \frac{\partial Q_k}{\partial \zeta} + N_{11} \frac{\partial^2 Q_k}{\partial \eta^2} + N_{22} \frac{\partial^2 Q_k}{\partial \zeta^2}$$
$$+ 2N_{12} \frac{\partial^2 Q_k}{\partial \eta \partial \zeta} - \frac{1}{\rho_{\eta,\zeta} P(\eta,\zeta)} \frac{\partial \left[\rho_{\eta,\zeta} \langle u_i'' Y_k'' \mid \eta, \zeta \rangle P(\eta,\zeta)\right]}{\partial x_i} \tag{5.20}$$

As above, high Reynolds numbers have been assumed, differential diffusion effects are neglected and the doubly conditioned dissipation terms are defined as $N_{11} \equiv \langle D \nabla \xi \nabla \xi \mid \eta, \zeta \rangle$, $N_{22} \equiv \langle D \nabla \hat{h}_s \nabla \hat{h}_s \mid \eta, \zeta \rangle$ and $N_{12} \equiv \langle D \nabla \xi \nabla \hat{h}_s \mid \eta, \zeta \rangle$. The last RHS term may approximate zero due to the expected reduction in fluctuations around the mean. In principle, the number of conditioning scalars can be increased arbitrarily up to $n_c = n_{sc}$, but every addition of a conditioning variable increases the dimension of $\mathbf{Q}$, and more than two conditioning variables may not be feasible for the computation of combustion in 2- or 3-dimensional geometries. Two conditioning variables should suffice to characterize a wide range of flame regimes: mixture fraction is the key quantity for non-premixed combustion while sensible enthalpy constitutes a kind of a progress variable that characterizes species compositions in premixed flames. Realizations of species compositions should therefore be close to a two dimensional space parameterized by mixture fraction and sensible enthalpy.

5.2.3.2 Modelling Partially Premixed Flames

An approach that employs conditioning on mixture fraction and sensible enthalpy should be ideally suited for the modelling of partially premixed flames where both the degree of mixing and the reaction progress determine fuel conversion and flame structure. A number of studies have dealt with reactive flows where significant local extinction leads to unburnt, partially premixed regions that later re-ignite due to turbulent and/or molecular diffusion of heat towards the premixed mixture. As indicated above Cha et al. [9] used DNS of flames with extinction and reignition in homogeneous turbulence to establish the lack of correlation between scalar dissipation and a reactive mixture during re-ignition, and in similar studies, Kronenburg [36, 43] assessed the suitability of mixture fraction and sensible enthalpy as conditioning variables. The latter studies demonstrate the potential of the doubly conditioned moment closure approach: the reactive species concentrations correlate well with ξ and $\hat{h}_s$ at all times, fluctuations around the doubly conditioned mean are very small, and CMC predictions agree very well with DNS data. The timing of extinction and the onset of re-ignition are captured accurately and the closure of chemical conversion rates can be based on doubly conditioned values. However, it shall not be

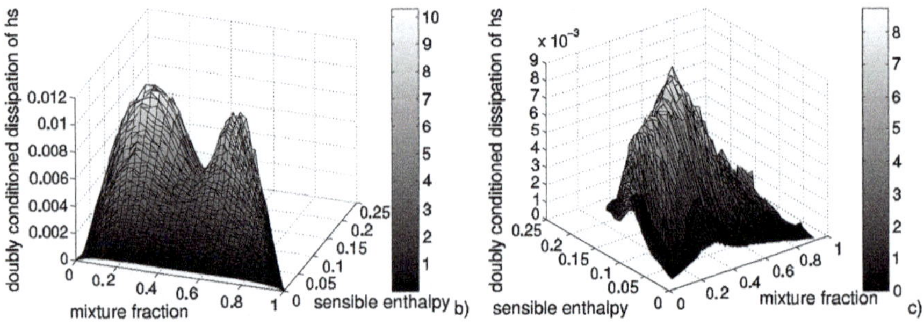

Fig. 5.4: DNS data of doubly conditioned dissipation of normalized sensible enthalpy at two different times: $t^* = 0.8$ (left) and $t^* = 2.5$ (right) where t^* denotes time normalized by the initial eddy turnover time. Reprinted from [36] with permission. Copyright © 2004, American Institute of Physics.

forgotten that double conditioning brings a whole host of new issues that need to be addressed. The doubly conditioned dissipation terms, N_{11} and N_{22}, and their cross-correlation, N_{12}, are not closed and need modelling. Hasse and Peters [22] suggest independence of the scalar dissipation of the respective second conditioning scalar. This may hold if - as in their case - both conditioning scalars are passive, mixture fraction-like. In a similar context, Nguyen et al. [56] parameterize the composition space by mixture fraction and a progress variable. They suggest independence of N_{11} of the progress variable and impose a functional dependence of N_{22} on mixture fraction such that local maxima occur at stoichiometric. This may hold when progress variable and mixture fraction are not well correlated, but should lead to gross overpredictions of the dissipation rate in regions where flamelet solutions exist. This is demonstrated in Fig. 5.4 for N_{22} with the help of DNS data [36]. Initially, temperature - and therefore sensible enthalpy - is a strong function of mixture fraction, temperature maxima exist where $\xi = \eta_{st}$, and N_{22} therefore tends to zero at locations with stoichiometric mixture. Local extinction destroys this correlation and the minimum of N_{22} at η_{st} disappears.

Accurate closures of the dissipation terms thus seem difficult and so far only been attempted as part of DNS related studies where the mean dissipation rates of mixture fraction and sensible enthalpy could be extracted from the DNS. In addition, the joint PDF of mixture fraction and sensible enthalpy and the conditionally averaged velocity do not result from the CMC equations and must be separately modelled. The modelling of the doubly conditioned dissipation rates can be avoided, if we solely focus on the modelling of the chemical source term. Bradley et al. [6] based their closure of the chemical source term on tabulated scalar fields that were obtained from flamelet solutions and parameterized by mixture fraction and temperature. Kronenburg and Kostka [42] improved the method slightly, and the assumption of a β-distribution of the conditional probability density function $P_{\xi,\hat{h}_s}(\zeta \mid \eta)$

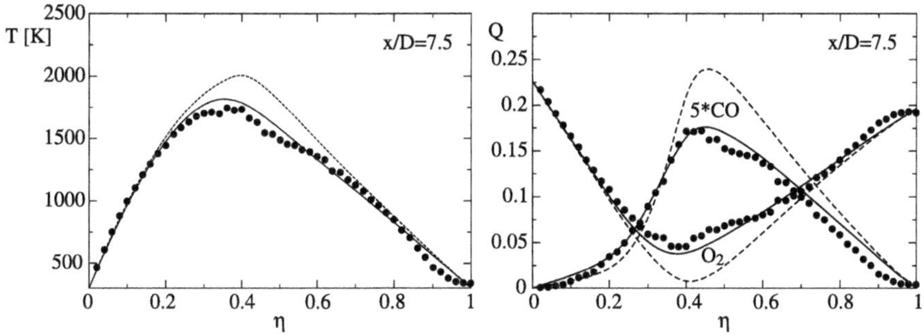

Fig. 5.5: Conditionally averaged temperature (left) and CO and O_2 mass fractions (right) at X/D=7.5 in Sandia Flame E. Symbols represent experimental data from [19], solid lines are CMC predictions with source term closure based on doubly conditioned moments and dashed lines are CMC predictions using a standard first order closure for $\langle w_k \mid \eta \rangle$. Reprinted from [42] with permission from the Combustion Institute.

proved accurate enough to achieve very good agreement of predicted temperature, major species and even CO and NO with measurements of the Sandia Flame series with progressive levels of extinction (Sandia Flames D, E and F). Examples of the results are shown in Fig. 5.5.

Here, we would like to remind the reader that source term modelling by second order closure has not quite led to the expected success with respect to temperature predictions in particular in bluff body flames, but also in flames with significant extinction. This may indicate the importance of the fluctuation terms and/or diffusion in progress variable space that is omitted in singly conditioned approaches. The very good results achieved in [42] may therefore be rather fortunate and accurate modelling of flame regimes with partial premixing may require the solution of the full doubly conditioned moment equations.

We should also mention that such approaches may be used for problems involving autoignition, where locally an autoignition spot is developed and flames propagate around it. The speed of such flames is larger than the conventional flame propagation in unburnt reactants due to the fact that the region away from the autoignition has also been reacting, albeit slowly, and is hence easier to jump to fully-fledged combustion through the action of diffusion of species and heat from the burning region; see Mastorakos [47] for a discussion and Wright et al. [80] for a demonstration of the importance of this quick propagation phase in diesel engines. Double-conditioning may also be used in flame expansion problems in multiple-injection diesel engines or following spark ignition in non-premixed systems; at present such phenomena are captured in single-conditioned CMC (often with acceptable accuracy) through the e_y-term [63, 76].

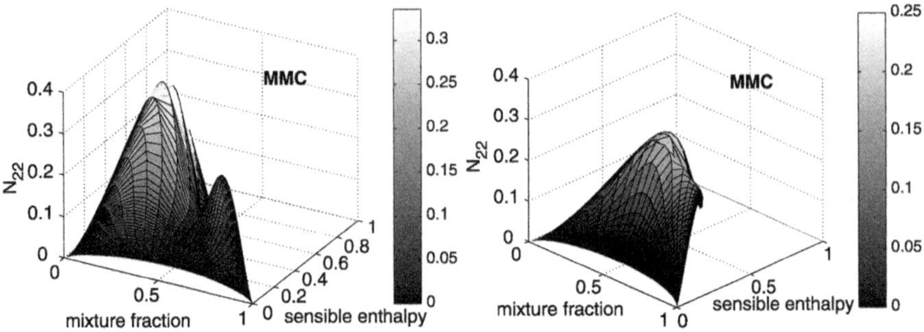

Fig. 5.6: Dissipation of sensible enthalpy predicted by MMC at times $t^* = 1.0$ (left) and $t^* = 2.0$ (right). Reprinted from [41] with permission from the Combustion Institute.

5.2.3.3 Possible Closures Using Multiple Mapping Conditioning

At this point we may want to anticipate Chapter 7 on Multiple Mapping Conditioning (MMC) without giving much detail on the method. We limit ourselves here to state that MMC is *PDF-* and *CMC-consistent*, i.e. the solution of the MMC equation for the conditioning scalars provides their joint probability distribution, and the MMC equations for the conditioned scalars are simple transformations of the CMC equations and therefore provide identical solutions. In this subsection it shall suffice to say that MMC implicitly provides a mapping closure for the conditionally averaged dissipation terms, it succeeds in capturing the evolution of the local minimum of N_{22} as shown in Fig. 5.6. In addition, the evolution of the joint probability of mixture fraction and sensible enthalpy is quite well approximated. MMC correctly predicts a conditional PDF of sensible enthalpy conditioned on mixture fraction with one peak at fully burning conditions and a second peak around $\hat{h}_s = 0.2$. A conventional (conditional) β-PDF would never predict the location of the second peak away from the bounds, and MMC is clearly superior to approaches using presumed PDFs, especially around stoichiometric.

MMC may appear as the solution to the closure issues addressed in the previous sections, but it shall not be forgotten that all the MMC studies presented in this subsection refer to a comparison of MMC with DNS data of reacting flows with reduced chemistry in homogeneous, isotropic turbulence. The feasibility of the method for laboratory flames needs to be assessed, and we may anticipate that extension to complex flow geometries may not be as simple as it first seems. Alternative stochastic implementations of MMC may be more favourable, and this is discussed in much more detail in Chapter 7.

5.2.4 Premixed Combustion

The conditional moment closure method has been derived for non-premixed combustion where mixture fraction makes for an ideal quantity that describes flame structure. We have seen above that the CMC method can be extended to partially premixed flames and further application to premixed flames is straightforward. Chemistry will be strongly linked to a progress variable, and simple first order closure of the chemical source term conditioned on this progress variable can be expected to yield accurate source term closures and pollutant predictions. Martin et al. [46] applied premixed CMC to a flame stabilized on a backward-facing step. This configuration shall mimic the flame stabilization mechanisms in a typical lean premixed gas turbine combustor where high mixing intensities justify the assumption of distributed reaction regimes and therefore spatial homogeneity of the conditional moments. The qualitative agreement between CMC predictions and experiments is good, and CMC with a detailed methane-air mechanism gives a much better match with the measurements than the reduced global mechanisms that are conventionally used in industrial CFD work.

Martin et al. [46] have demonstrated that CMC provides a suitable source term closure, however, CMC itself does not quite address some of the major issues associated with modelling premixed turbulent combustion: the evolution of the progress variable, in particular the temporal and spatial evolution of its probability distribution is one of the major research areas in turbulent combustion modelling (Chapters 4 and 6). Presumed probability distributions may suffice, but the solution of the transport equations for mean and variance require modelling of the mean scalar dissipation and the turbulent scalar flux, two quantities that involve some uncertainty in their modelling. Since the probability distribution $P(\zeta)$ is needed for full closure and the conditionally averaged scalar dissipation of the progress variable and the conditionally averaged turbulent scalar flux appear explicitly in the CMC equations, a series of CMC related studies were dedicated to the analysis of these terms [73, 74]. One of the key issues is the effect of reaction on the scalar dissipation of a reactive scalar. Often, these effects are neglected and dissipations of reactive and passive scalars are modelled equally. Swaminathan and Bray [74] however showed, that the dilatation effects need to be included in the modelling of N_ζ, and they suggested a simple algebraic closure for this quantity. The derivation of conditionally averaged dissipation rates is currently under way, but accurate modelling of the conditionally averaged turbulent flux seems a bit further off.

The two extremes of very fast and very slow chemistry shall be mentioned here, but will not be discussed further. Lee and Huh [44] proposed a zone conditional modelling of premixed flames which corresponds to a CMC approach where the PDF is approximated by two δ-functions at the extremes, and Bilger [5] introduced markers fields for the mapping of the progress variable that may be limited to low Damköhler numbers only. The reader is referred to the above references for more detail on these methods.

5.2.5 Liquid Fuel Combustion

A further area of CMC research that has been enjoying increasing attention recently is the modelling of reacting two-phase flows. It is postulated that liquid fuel combustion is largely determined by the evaporation rate and the mixing between fuel and oxidizer. Conditioning on mixture fraction appears as a promising concept for the accurate closure of the chemical source term. It is well established that in the presence of fuel evaporation, mixture fraction is no longer conserved and additional source terms appear in its transport equations of mean and variance [65]. Mortensen and Bilger [50] derived the fully consistent conditional moment closure equations for spray combustion. The derivation is based on the instantaneous single phase transport equations and a level set/indicator function technique is used to account for the interfaces. The final form of the singly conditioned moment closure equation for two phase flow can be written as

$$\begin{aligned}\frac{\partial Q_k}{\partial t}+\langle u_i\mid\eta\rangle\frac{\partial Q_k}{\partial x_i} &= N_\eta\frac{\partial^2 Q_k}{\partial\eta^2}+\langle w_k\mid\eta\rangle-\frac{1}{\langle\theta\rangle\rho_\eta P_\eta}\frac{\partial}{\partial x_i}\left[\langle\theta\rangle\rho_\eta\langle u_i''Y_k''\mid\eta\rangle P_\eta\right]\\ &+\left[Q_{1,k}-Q_k-(1-\eta)\frac{\partial Q_k}{\partial\eta}\right]\frac{\langle\Pi\mid\eta\rangle}{\langle\theta\rangle}\\ &-\frac{1}{\langle\theta\rangle\rho_\eta P_\eta}\frac{\partial}{\partial\eta}\left[(1-\eta)\rho_\eta\langle\Pi''Y_k''\mid\eta\rangle P_\eta\right]\end{aligned}\tag{5.21}$$

where $Q_{1,k}$ denotes the conditionally averaged mass fraction of species k in the liquid droplet. Several new terms that involve the evaporation rate, Π, and its fluctuation appear in the final CMC formulation, and they require closure. The potential of these new equations has not yet been evaluated and closures have not yet been developed.

A simplified form of the CMC equations that neglected spray interactions was applied to studies of spray autoignition under engine-like conditions with some success. Kim and Huh [32] reported no influence of different models for the conditionally averaged evaporation rate on ignition delay times, and even the effects of different models for the closure of the mixture fraction variance equation are within 5%. The latter finding must be contrasted with the DNS studies by Réveillon and Vervisch [62] who analysed the mixing field and showed that droplet evaporation needs to be considered for the computation of the mixture fraction and its variance; it seems that the extra terms in the mixture fraction variance equation made a small contribution in the flow studied by Kim and Huh [32]. Similarly, Wright et al. [80] confirmed good agreement with measurements of ignition delay times and spray penetration lengths in high pressure chambers using CMC and neglecting the evaporation terms in the CMC and mixture fraction variance transport equations. However, it seems premature to arrive at any conclusions from the above engine-related studies on the importance of the modelling of the evaporation terms. The modelling by Kim and Huh [32] was based on a simplistic model for the conditionally averaged evaporation and compared with no closure. Equally, Wright et al. [80]

used zero closures for the variance and CMC equations. These studies simply show that auto-ignition delay times are not extremely sensitive to the correlations between evaporation and mixture fraction, but sufficient experimental data simply does not exist to allow any assessment of their importance on species predictions and flame structure.

Schroll et al. [77] have corroborated the adequacy of CMC for the modelling of droplet combustion. DNS of droplet evaporation and ignition with three different initial droplet diameters shows the lowest conditionally averaged dissipation at the most reactive mixture composition for the smallest droplets and CMC would therefore correctly predict the shortest ignition times for these smallest droplets. However, multiplication with the mixture fraction PDF and integration across mixture fraction space yields the highest unconditionally averaged dissipation for the smallest droplets that would lead - incorrectly - to the longest predicted ignition delay times for these droplets if estimates were based on unconditional values. The CMC concept is therefore ideally suited for the prediction of droplet combustion but some key issues, such as the modelling of the conditionally averaged evaporation rate and its correlation with species mass fractions and mixture fraction need to be addressed. The mixture fraction PDF and the modelling of the scalar dissipation in the presence of spray evaporation are also important to consider. It is fair to say that CMC of two phase flows is in its infancy, and thorough investigations will be needed for a better assessment of the viability of the method in real applications.

5.3 Application to Flows of Engineering Interest

As time progressed from the mid-90's when the first papers on CMC begun to appear, CMC simulations have developed from spatially-integrated formulations to fully three-dimensional applications with strong temporal and spatial variations of the conditional averages, thereby revealing the true strengths of the method. In addition, a shift from RANS to LES has begun to materialise, as well as the application to large furnaces with important radiation effects and to new fuels such as syngas. Some aspects of these simulations and numerical developments are discussed in this Section.

5.3.1 Dimensionality of the CMC Equation

When using the CMC method for realistic problems, first-order CMC, with single conditioning, is the virtually ubiquitous choice today. The key decision to be taken by the user is the spatial dimensionality of the CMC equation, which should reflect the physical nature of the problem, but also may be influenced by the available computational resources. The spatial resolution needed in the CMC equation should not be confused with the resolution needed in the calculation for the velocity or mixture fraction fields. For statistically-steady attached flames in jets, for example, experiment has suggested that the conditional averages are weak functions of the radial coordinate. Therefore, ignoring streamwise turbulent diffusion and integrating the full CMC equation (Eq. 5.8) across the jet results in:

$$\langle \rho u_i \mid \eta \rangle^* \frac{\partial Q_k}{\partial x_i} = \rho_\eta N_\eta^* \frac{\partial^2 Q_k}{\partial \eta^2} + \langle w_k \mid \eta \rangle \tag{5.22}$$

with the starred quantity being a PDF-weighted integral:

$$\langle \phi \mid \eta \rangle^* = \frac{\int_V \langle \phi \mid \eta \rangle P_\eta dV}{\int_V P_\eta dV} \tag{5.23}$$

with V being the region of integration (line, plane or even volume). For transient problems, such as in diesel engines, volume integration may be performed to result in the time derivative $\partial Q_k / \partial t$ replacing the convective term on the LHS of Eq. 5.22.

The computational requirements for the numerical solution of the resulting parabolic equations is modest, even with quite detailed chemical mechanisms. This approach has provided results of very good accuracy for many problems, especially for jets. The similarity with transient flamelet modelling is evident, although the modelling of $\langle N|\eta \rangle$ and the exact way the volume integration is done tends to differ. Let us denote this simplification to the full multi-dimensional CMC equation as "0DCMC", implying the lack of spatial diffusion.

For some problems, there is little or no direct experimental evidence that the conditional averages are weak functions of space. In lifted jet flames, significant variations of Q_k have been measured [10]. In compression-ignition engines, experimental evidence points to a quite localised first emergence of ignition. Heat losses to engine walls and multiple injections also lead to variations in Q_k [59]. In spark ignition of non-premixed combustors [47], the conditional distributions switch from unburnt to burnt in different regions in space at different times. Such problems necessitate a multi-dimensional CMC formulation, which we will denote as "3DCMC". Numerical solution of the 3DCMC model has been attempted by various research groups and some details are given next.

5.3.2 Numerical Methods

The three-dimensional CMC equation (Eq. 5.8), after modelling the spatial diffusion term, can be written in more generic form as

$$\frac{\partial Q_k}{\partial t} + C(\mathbf{x},t,\eta)\nabla Q_k = N(\mathbf{x},t,\eta)\frac{\partial^2 Q_k}{\partial \eta^2} + \nabla \cdot (D(\mathbf{x},t,\eta)\nabla Q_k) + w_k(\eta, Q_2, \ldots, Q_{n_{sc}}) \tag{5.24}$$

where the coefficients C, N, D are, in general, functions of time, space $\mathbf{x}$, and mixture fraction η and contain information from the fluid mechanical field. Equation 5.24 is a 5-dimensional partial differential equation with a stiff chemical source term, making its solution at least as challenging as the numerical simulation of a multi-dimensional laminar flame. The user of the elliptic CMC method must decide: (i) whether to solve this equation in its full 3D form or whether to perform some averaging across one of the spatial directions; (ii) the resolution to be used (i.e. the number of grid nodes in each spatial direction); (iii) the size of the chemical mechanism; (iv) the numerical method. Some compromises between the conflicting requirements of having very fine resolution in physical space with the wish to employ a very detailed chemical scheme may have to be made. Typically, central differences are used for discretizing the diffusion in mixture fraction space and for the second-order derivative in physical space, while upwind schemes have been used for the convective term. An alternative strategy to solve the CMC equation before dividing by the mixture fraction PDF, i.e. to solve for $P_\eta Q_k$, has also been proposed [11], but the comments below and the discretization issues apply to that formulation as well.

Assuming a detailed chemical mechanism with, say, $N_s = 50$ species, a discretization in mixture fraction space with $N_\eta = 100$ points, and a physical-space grid of, say, $N_x = N_y = N_z = 25$, and assuming that such a system were to be solved by the Method of Lines, which transforms the partial differential equation into a system of $N_s \times N_\eta \times N_x \times N_y \times N_z$ ordinary differential equations, we would arrive at a system of 75×10^6 o.d.e's. Considering the stiffness involved, which necessitates implicit schemes, the size of this system is just too large for present day solvers. Despite this, in two physical-space coordinates ($N_z = 0, N_x = N_y = 20$), this solution method has been used successfully with the GRI3.0 chemical mechanism and the VODPK solver [24, 25].

An alternative is to solve the CMC p.d.e. by a fractional step (operator splitting) method [16, 18, 23, 31, 59, 60, 80]. In such a scheme, one would solve the sequence

$$\frac{\partial Q_k}{\partial t} + C(\mathbf{x},t,\eta)\nabla Q_k = \nabla \cdot (D(\mathbf{x},t,\eta)\nabla Q_k) \tag{5.25}$$

$$\frac{\partial Q_k}{\partial t} = N(\mathbf{x},t,\eta)\frac{\partial^2 Q_k}{\partial \eta^2} \tag{5.26}$$

$$\frac{\partial Q_k}{\partial t} = w_k(\eta, Q_2, \ldots, Q_{n_{sc}}) \tag{5.27}$$

with each fractional step picking-up the solution from where the previous step left it and advancing it for a given timestep. The chemistry fractional step is the one that takes the most computational time; it will require separate solution of $N_\eta \times N_x \times N_y \times N_z$ stiff ordinary differential equations of the type $dQ_k/dt = w_k$. This is now more affordable, since the stiff solver will have to deal with a smaller system, but the solver used will need to be able to deal effectively with the "restart costs" when solving the chemical step while sweeping across the physical and mixture fraction grid nodes. Splitting the various phenomena incurs an error; this has been assessed [59, 80] and it is not negligible, but it can be controlled by a small enough timestep. The Method of Lines is of course more accurate, but as the size of the chemical mechanism increases, the operator splitting approach is the only one that is practical.

It is evident that the numerical cost of solving the three-dimensional CMC equation is formidable, and experience shows that it is typically 50-90% of the total computational cost of the whole simulation. The computational burden of a well-resolved CMC simulation is not far from that of a low Reynolds number DNS.

For established flames, the discretization in mixture fraction space must follow the usual practice of clustering the grid nodes around stoichiometry; 50-100 grid nodes are the usual practice. For autoignition problems, the grid must be fine enough across the whole mixture fraction range to resolve the reaction fronts that will propagate across mixture fraction space following ignition [80]. For attached flames, the physical-space gradients are small and hence relatively coarse physical-space grids may be sufficient. Typical values are 10-20 grid nodes. For lifted jet flames, especially in the axial direction, a higher resolution is necessary. Note also the dangers of numerical diffusion associated with upwind differencing and coarse grids when having to resolve sharp transitions from Q_k corresponding to unburnt fluid to Q_k corresponding to burnt fluid.

Recently, Large-Eddy Simulations have begun to appear with CMC as the combustion sub-model. The CMC equation in LES is virtually identical to the RANS formulation discussed above [55]. Various specific choices concerning the sub-models for the conditional velocity and scalar dissipation must be made, especially since it must be recognized from the outset that the physical-space CMC grid will be coarser than the LES grid; this necessitates to develop a strategy concerning how can the fine-grid fluid mechanical quantities be provided to the coarse grid used to solve the CMC equation. Various options exist and are discussed in detail in [75]. The LES/CMC model has been applied to attached jet flames [55], bluff-body flames [52], autoigniting jets [53] and spark ignition problems [76]. The numerical cost here is very high. From a model development point of view, many of the sub-models that are being used have not been properly validated yet for LES.

5.3.3 Applications and Outlook

The CMC model has been used for gaseous statistically-steady flames. Some examples include non-premixed attached jet flames [66, 67, 69], attached bluff-body flames [28, 52, 70, 76], lifted jet flames [15, 24], autoigniting jets [53, 60, 61], and soot formation [78, 79]. It has also been used for transient methane jets [30, 68], heptane sprays [32, 80], and spark-ignition problems [76]. All these simulations have focused on simple geometries and have, in general, produced good results compared to experimental data. In addition to the DNS-based validation, these simulations have also provided more detailed assessment of specific sub-models. The CMC model has also been examined for reacting flows in porous media [34], chemical engineering [49] and atmospheric flows [7], demonstrating thus the wide applicability and usefulness of the method.

CMC is an attractive tool to model complicated geometries and problems. Multi-dimensional CMC has been used for diesel engines [59] and compartment fires [11, 12] and furnaces [23, 65] that have important radiation effects or low-oxygen combustion. The analysis of the results provides very useful insights on the flame structure and brings CMC to the point of practical calculations for design. First-order CMC with single conditioning and proper chemistry seems sufficient for many engineering problems (except for pollutant production, which may necessitate second-order corrections). For problems with significant variations of the conditional averages, such as extinctions, ignitions or quenching due to radiation or convective heat transfer to walls, the current practice is to use refined elliptic CMC.

However, future applications should not be limited to first-order CMC. Second order CMC provides a useful extension to first-order closures. Second-order CMC has been applied to various laboratory flames with decent results; the theoretical framework is well known, but a wider range of applicability would need to be established, and future efforts should be directed towards improved modelling of the conditional variance and co-variance equations, Eq. 5.12. Future progress in the application of double-conditioning approaches may be more challenging. Double conditioning is certainly attractive for (1) flames with partial premixing, for (2) flames with local extinction in the vicinity of the walls where the second conditioning variable could be a wall distance parameter or for (3) engine calculations with multiple injections. The latter has been attempted in a somewhat different framework by Hasse and Peters [22] with multiple conditioning on two different mixture fractions, however, all applications would require the closure of the doubly (or multiply) conditioned dissipation terms which is far from being trivial. Simplification as suggested by Hasse and Peters [22] are certainly not applicable to configurations with a reactive species as second conditioning scalar, and other alternative methods for closure would need to be applied. MMC might offer one possible solution, but modelling suggestions as brought forward by Nguyen et al. [56] should equally be pursued.

Double conditioning could also be applied to liquid fuel combustion where the second conditioning scalar describes the inter-droplet space. However, some more fundamental issues should be assessed in CMC of multiphase flows first, such as the effects of droplet evaporation and of the correlations between evaporation rate and reactive scalar field on conditional moments and on mixture fraction variance. Existing DNS studies and RANS calculations do not give a clear picture on the importance of these terms, and we would need to know under which conditions they become important and how they could be modelled.

But all these exciting new developments with respect to improvements of the chemical source term closure and two-phase flow modelling should not distract from implementation issues as addressed in Sections 5.3.1 and 5.3.2. In particular the dimensionality and also the CMC cell size need to be chosen carefully. Small CMC cells lead to cells with zero probability for some of the mixture fraction bins. In case of non-zero probability of these CMC bins in the neighbouring cells, it is not clear yet how to model convective and diffusive flux into a cell where a certain mixture fraction bin has zero probability. Standard practice is to neglect fluxes below a certain threshold value of the PDF, however, this may be inaccurate as new simulations for the determination of blow-out limits of lifted flames show [54]. This example simply illustrates that implementation can strongly affect the quality of the predictions, and much more work is needed to understand the correct treatment of zero- (or low) probability moments, in particular in LES-CMC.

5.4 Conclusion

It is of course not possible to give a complete and detailed review on all research activities of the last ten years related to the development and application of novel CMC techniques. The majority of research efforts had been directed towards improvements to the modelling of the conditioned chemical source term and towards CMC applications to flows of practical interest. We have therefore focused in this chapter on rather detailed summaries of second order closures, double conditioning and issues related to CMC in more complex flow geometries. The CMC methodology has advanced within the last ten years, and this chapter should be therefore viewed as complementary to Klimenko and Bilger's review [35]. We hope that it offers some stimuli to new researchers in the field for continued work on the conditional moment closure method.

References

1. Barlow, R.S., Frank, J.H.: Effect of turbulence on species mass fractions in methane/air jet flames. Proc. Combust. Inst. **27**, 1087–1095 (1998)
2. Barlow, R.S., Frank, J.H.: Measurements of scalar variance, scalar dissipation, and length scales in turbulent piloted methane/air jet flames. Flow Turbul. Combust. **72**, 427–448 (2004)
3. Bilger, R.W.: Advanced laser diagnostics: implications of recent results for advanced combustor models. In: R.S.L. Lee, J.H. Whitelaw, T.S. Wung (eds.) Aerothermodynamics in Combustors. Springer Verlag, Berlin (1993)
4. Bilger, R.W.: Conditional moment closure for turblent reacting flow. Phys. Fluids A **5**, 436–444 (1993)
5. Bilger, R.W.: Marker fields for turbulent premixed combustion. Combust. Flame **138**, 188–194 (2004)
6. Bradley, D., Emerson, D.R., Gaskell, P.H., Gu, X.J.: Mathematical modeling of turbulent non-premixed piloted-jet flames with local extinctions. Proc. Combust. Inst. **29**, 2155–2162 (2002)
7. Brown, R.J., Bilger, R.W.: Experiments on a reacting plume - 2. Conditional concentration statistics. Atmos. Environ. **3**, 629–646 (1998)
8. Cha, C.M., de Bruyn Kops, S.M., Mortensen, M.: Direct numerical simulations of the double scalar mixing layer. Part I: Passive scalar mixing and dissipation. Phys. Fluids **18**, 067,106-1–067,106-12 (2006)
9. Cha, C.M., Kosály, G., Pitsch, H.: Modeling extinction and reignition in turbulent non-premixed combustion using a doubly-conditional moment closure approach. Phys. Fluids **13**, 3824–3834 (2001)
10. Cheng, T.S., Wehrmeyer, J.A., Pitz, R.W.: Conditional analysis of lifted hydrogen jet diffusion flame experimental data and comparison to laminar flame solutions. Combust. Flame **150**, 340–354 (2007)

11. Cleary, M.J., Kent, J.H.: Modelling of species in hood fires by conditional moment closure. Combust. Flame **143**, 357–368 (2005)
12. Cleary, M.J., Kent, J.H., Bilger, R.W.: Prediction of carbon monoxide in fires by conditional moment closure. Proc. Combust. Inst. **29**, 273–279 (2002)
13. Dally, B.B., Masri, A.R., Barlow, R.S., Fiechtner, G.J.: Instantaneous and mean compositional structure of bluff-body stabilized nonpremixed flames. Combust. Flame **114**, 119–148 (1998)
14. Devaud, C.B., Bilger, R.W., Liu, T.: A new method of modeling the conditional scalar dissipation rate. Phys. Fluids **16**, 2004–2011 (2004)
15. Devaud, C.B., Bray, K.N.C.: Assessment of the applicability of conditional moment closure to a lifted turbulent flame: first order model. Combust. Flame **132**, 102–114 (2003)
16. El Sayed, A., Devaud, C.B.: Conditional moment closure (cmc) applied to autoignition of high pressure methane jets in a shock tube. Combust. Theor. Modeling **12**, 943–972 (2008)
17. Fairweather, M., Woolley, R.: First- and second-order elliptic conditional moment closure calculations of piloted methane diffusion flames. Combust. Flame **150**, 92–107 (2007)
18. Fairweather, M., Woolley, R.M.: Conditional moment closure calculations of a swirl-stabilized, turbulent nonpremixed methane flame. Combust. Flame **151**, 397–411 (2007)
19. Frank, J.H., Barlow, R.S., Lundquist, C.: Radiation and nitric oxide formation in turbulent non-premixed jet flames. Proc. Combust. Inst. **28**, 447–454 (2000)
20. Girimaji, S.S.: On the modelling of scalar diffusion in isotropic decaying turbulence. Phys. Fluids A **4**, 2529–2537 (1992)
21. Gordon, R.L., Masri, A.R., Mastorakos, E.: Simultaneous Rayleigh temperature, OH- and CH_2O-LIF imaging of methane jets in a vitiated coflow. Combust. Flame **155**, 181–195 (2008)
22. Hasse, C., Peters, N.: A two mixture fraction flamelet model applied to split injections in a DI Diesel engine. Proc. Combust. Inst. **30**, 2755–2762 (2005)
23. Kim, G., Kang, S., Kim, Y., Lee, K.S.: Conditional moment closure modeling for a three-dimensional turbulent non-premixed syngas flame with a cooling wall. Energ. Fuels **22**, 3639–3648 (2008)
24. Kim, I.S., Mastorakos, E.: Simulations of turbulent lifted jet flames with two dimensional conditional moment closure. Proc. Combust. Inst. **30**, 911–918 (2005)
25. Kim, I.S., Mastorakos, E.: Simulations of turbulent non-premixed counterflow flames with first-order conditional moment closure. Flow Turbul. Combust. **76**, 133–162 (2006)
26. Kim, S.H.: On the conditional variance and covariance equations for second-order conditional moment closure. Phys. Fluids **14**, 2011–2014 (2002)
27. Kim, S.H., Choi, C.H., Huh, K.Y.: Second-order conditional moment closure modeling of a turbulent $CH_4/H_2/N_2$ jet diffusion flame. Proc. Combust. Inst. **30**, 735–742 (2004)
28. Kim, S.H., Huh, K.Y.: Use of the conditional moment closure model to predict NO formation in a turbulent CH_4/H_2 flame over a bluff-body. Combust. Flame **130**, 94–111 (2002)
29. Kim, S.H., Huh, K.Y.: Second-order conditional moment closure modeling of turbulent piloted jet diffusion flames. Combust. Flame **138**, 336–352 (2004)
30. Kim, S.H., Huh, K.Y., Fraser, R.A.: Modeling autoignition of a turbulent methane jet by the conditional moment closure model. Proc. Combust. Inst. **28**, 185–191 (2000)
31. Kim, S.H., Huh, K.Y., Tao, L.: Application of the elliptic conditional moment closure model to a two-dimensional nonpremixed methanol bluff-body flame. Combust. Flame **120**, 75–90 (2000)
32. Kim, W.T., Huh, K.Y.: Numerical simulation of spray autoignition by the first order conditional moment closure model. Proc. Combust. Inst. **29**, 569–576 (2002)
33. Klimenko, A.Y.: Multicomponent diffusion of various admixtures in turbulent flow. Fluid Dyn. **25**, 327–331 (1990)
34. Klimenko, A.Y., Abdel-Jawad, M.M.: Conditional methods for continuum reacting flows in porous media. Proc. Combust. Inst. **31**, 2107–2115 (2007)
35. Klimenko, A.Y., Bilger, R.W.: Conditional moment closure for turbulent combustion. Prog. Energy Combust. Sci. **25**, 595–688 (1999)
36. Kronenburg, A.: Double conditioning of reactive scalar transport equations in turbulent non-premixed flames. Phys. Fluids **16, 7**, 2640–2648 (2004)

37. Kronenburg, A., Bilger, R.W.: Modelling differential diffusion in non-premixed reacting flows: application to jet flames. Combust. Sci. Technol. **166**, 195–227 (2001)
38. Kronenburg, A., Bilger, R.W.: Modelling differential diffusion in non-premixed reacting flows: model development. Combust. Sci. Technol. **166**, 175–194 (2001)
39. Kronenburg, A., Bilger, R.W., Kent, J.H.: Second order conditional moment closure for turbulent jet diffusion flames. Proc. Combust. Inst. **27**, 1097–1104 (1998)
40. Kronenburg, A., Bilger, R.W., Kent, J.H.: Computation of conditional average scalar dissipation in turbulent jet diffusion flames. Flow Turbul. Combust. **64**, 145–159 (2000)
41. Kronenburg, A., Cleary, M.J.: Multiple mapping conditioning for flames with partial premixing. Combust. Flame **155**, 215–231 (2008)
42. Kronenburg, A., Kostka, M.: Modeling extinction and reignition in turbulent flames. Combust. Flame **143**, 342–356 (2005)
43. Kronenburg, A., Papoutsakis, A.: Conditional moment closure modeling of extinction and re-ignition in turbulent non-premixed flames. Proc. Combust. Inst. **30**, 759–766 (2005)
44. Lee, E., Huh, K.Y.: Zone conditional modeling of premixed turbulent flames at a high Damköhler number. Combust. Flame **138**, 211–224 (2004)
45. Li, J.D., Bilger, R.W.: Measurement and prediction of the conditional variance in a turbulent reactive-scalar mixing layer. Phys. Fluids **5**, 3255–3264 (1993)
46. Martin, S.M., Kramlich, J.C., Kosály, G., Riley, J.J.: The premixed conditional moment closure method applied to idealized lean premixed gas turbine combustors. J. Eng. Gas Turb. Power **125**, 895–900 (2003)
47. Mastorakos, E.: Ignition of turbulent non-premixed flames. Prog. Energ. Combust. Sci. **35**, 57–97 (2009)
48. Mastorakos, E., Bilger, R.W.: Second-order conditional moment closure for the autoignition of turbulent flows. Phys. Fluids **10**, 1246–1248 (1998)
49. Mortensen, M.: Implementation of a conditional moment closure for mixing sensitive reactions. Chem. Eng. Sci. **59**, 5709–5723 (2004)
50. Mortensen, M., Bilger, R.W.: Derivation of the conditional moment closure equations for spray combustion. Combust. Flame **156**, 62–72 (2009)
51. Mortensen, M., de Bruyn Kops, S.M.: Conditional velocity statistics in the double scalar mixing layer - a mapping closure approach. Combust. Theor. Model. **12**, 929–941 (2008)
52. Navarro-Martinez, S., Kronenburg, A.: LES-CMC simulations of a turbulent bluff-body flame. Proc. Combust. Inst. **31**, 1721–1728 (2007)
53. Navarro-Martinez, S., Kronenburg, A.: LES-CMC simulations of a lifted methane flame. Proc. Combust. Inst. **32**, 1509–1516 (2009)
54. Navarro-Martinez, S., Kronenburg, A.: Numerical study of turbulent non-premixed jet flame stability. In: CD-ROM Proceedings. European Combustion Meeting, Vienna (2009)
55. Navarro-Martinez, S., Kronenburg, A., di Mare, F.: Conditional moment closure for large eddy simulations. Flow Turbul. Combust. **75**, 245–274 (2005)
56. Nguyen, P.D., Vervisch, L., Subramaniam, V., Domingo, P.: Multi-dimensional flamelet-generated manifolds for partially premixed combustion. Combust. Flame (2009)
57. O'Brien, E.E., Jiang, T.L.: The conditional dissipation rate of an initially binary scalar in homogeneous turbulence. Phys. Fluids A **3**, 3121–3123 (1991)
58. Paola, G.D., Kim, I.S., Mastorakos, E.: Second-order conditional moment closure simulations of autoignition of an n-heptane plume in a turbulent coflow of heated air. Flow Turbul. Combust. **82**, 455–475 (2009)
59. Paola, G.D., Mastorakos, E., Wright, Y.M., Boulouchos, K.: Diesel engine simulations with multi-dimensional conditional moment closure. Combust. Sci. Technol. **180**, 883–899 (2008)
60. Patwardhan, S.S., De, S., Lakshmisha, K.N., Raghunandan, B.: CMC simulations of lifted turbulent jet flame in a vitiated coflow. Proc. Combust. Inst. **32**, 1705–1712 (2009)
61. Patwardhan, S.S., Lakshmisha, K.: Autoignition of turbulent hydrogen jet in a coflow of heated air. Int. J. Hydrogen Energ. **33**, 7265–7273 (2008)
62. Reveillon, J., Vervisch, L.: Spray vaporization in nonpremixed turbulent combustion modeling: A single droplet model. Combust. Flame **121**, 75–90 (2000)

63. Richardson, E.S., Chakraborty, N., Mastorakos, E.: Analysis of direct numerical simulations of ignition fronts in turbulent non-premixed flames in the context of conditional moment closure. Proc. Combust. Inst. **31**, 1683–1690 (2007)
64. Richardson, E.S., Yoo, C.S., Chen, J.H.: Analysis of second-order conditional moment closure applied to an autoignitive lifted hydrogen jet flame. Proc. Combust. Inst. **32**, 1695–1703 (2009)
65. Rogerson, J.W., Kent, J.H., Bilger, R.W.: Conditional moment closure in a bagasse-fired boiler. Proc. Combust. Inst. **31**, 2805–2811 (2007)
66. Roomina, M.R., Bilger, R.W.: Conditional moment closure modelling of turbulent methanol jet flames. Combust. Theor. Model. **3**, 689–708 (1999)
67. Roomina, M.R., Bilger, R.W.: Conditional moment closure (CMC) predictions of a turbulent methane-air jet flame. Combust. Flame **125**, 1176–1195 (2001)
68. Sayed, A.E., Milford, A., Devaud, C.B.: Modelling of autoignition for methane-based fuel blends using conditional moment closure. Proc. Combust. Inst. **32**, 1621–1628 (2009)
69. Smith, N.S.A., Bilger, R.W., Carter, C.D., Barlow, R.S.: A comparison of CMC and PDF modelling predictions with experimental nitric oxide LIF/Raman measurements in a turbulent H_2 jet flame. Combust. Sci. Technol. **105**, 357–375 (1995)
70. Sreedhara, S., Huh, K.Y.: Modeling of turbulent, two-dimensional nonpremixed CH_4/H_2 flame over a bluff-body using first- and second-order elliptic conditional moment closures. Combust. Flame **143**, 119–134 (2005)
71. Sreedhara, S., Huh, K.Y., Ahn, D.H.: Comparison of submodels for conditional velocity and scalar dissipation in CMC simulation of piloted jet and bluff body flames. Combust. Flame **152**, 282–286 (2008)
72. Swaminathan, N., Bilger, R.W.: Study of the conditional covariance and variance equations for second order conditional moment closure. Phys. Fluids **11**, 2679–2695 (1999)
73. Swaminathan, N., Bilger, R.W., Cuenot, B.: Relationship between turbulent scalar flux and conditional dilatation in premixed flames with complex chemistry. Combust. Flame **126**, 1764–1779 (2001)
74. Swaminathan, N., Bray, K.N.C.: Effect of dilatation on scalar dissipation in turbulent premixed flames. Combust. Flame **143**, 549–565 (2005)
75. Triantafyllidis, A., Mastorakos, E.: Implementation issues of the conditional moment closure in large eddy simulations. Flow Turbul. Combust. **84**, 481–512 (2010)
76. Triantafyllidis, A., Mastorakos, E., Eggels, R.L.G.M.: Large eddy simulation of forced ignition of a non-premixed bluff-body methane flame with conditional moment closure. Combust. Flame **156**, 2328–2345 (2009)
77. Schroll, P., Wandel, A.P., Cant, R.S., Mastorakos, E.: Direct numerical simulations of autoignition in turbulent two-phase flows. Proc. Combust. Inst. **32**, 2275–2282 (2009)
78. Woolley, R.M., Fairweather, M., Yunardi: Conditional moment closure modelling of soot formation in turbulent, non-premixed methane and propane flames. Fuel **88**, 393–407 (2009)
79. Woolley, R.M., Fairweather, M., Yunardi: Conditional moment closure prediction of soot formation in turbulent, nonpremixed ethylene flames. Combust. Flame **152**, 360–376 (2008)
80. Wright, Y.M., De Paola, G., Boulouchos, K., Mastorakos, E.: Simulations of spray autoignition and flame establishment with two-dimensional CMC. Combust. Flame **143**, 402–419 (2005)

Chapter 6
Transported Probability Density Function Methods for Reynolds-Averaged and Large-Eddy Simulations

D.C. Haworth and S.B. Pope

Abstract Probability density function (PDF) methods provide an elegant and effective resolution to the closure problems that arise from averaging or filtering chemical source terms and other nonlinear terms in the equations that govern chemically reacting turbulent flows. PDF methods traditionally have been associated with studies of turbulence-chemistry interactions in laboratory-scale, atmospheric pressure, nonluminous, statistically-stationary nonpremixed turbulent flames; and Lagrangian particle-based Monte Carlo numerical algorithms have become the predominant method for solving modeled PDF transport equations. The emphasis in this chapter is on recent advances, new trends and perspectives in PDF methods. These include advances in particle-based algorithms, alternatives to particle-based algorithms (e.g., Eulerian field methods), treatment of combustion regimes beyond low-to-moderate-Damköhler-number nonpremixed systems (e.g., premixed flamelets), extensions to include radiation heat transfer and multiphase systems (e.g., soot and fuel sprays), and the use of PDF-based methods as the basis for modeling in large-eddy simulation.

6.1 Introduction

Probability density function (PDF) methods have emerged as one of the most promising and powerful approaches for accommodating the effects of turbulent fluctuations in velocity and/or chemical composition in computational fluid dynamics (CFD)-based modeling of turbulent reacting flows. Here the term "PDF method" refers to an approach based on solving a modeled transport equation for the one-point, one-time Eulerian joint PDF of a set of variables that describe the hydro-

D.C. Haworth
The Pennsylvania State University, University Park, PA, USA, e-mail: dch12@psu.edu

S.B. Pope
Cornell University, Ithaca, NY, USA, e-mail: s.b.pope@cornell.edu

T. Echekki, E. Mastorakos (eds.), *Turbulent Combustion Modeling*,
Fluid Mechanics and Its Applications 95, DOI 10.1007/978-94-007-0412-1_6,

dynamic and/or thermochemical state of a reacting medium: that is, a transported (versus presumed) PDF method.

Dopazo and O'Brien [14, 15] were the first to consider a modeled equation for the PDF of a set of scalars that describes the thermochemical state of a reacting medium (a composition joint PDF) to model mixing and chemical reaction in turbulent reacting flows. The relationship between particle models and PDF methods was established by Pope [53], and particle methods have become the dominant approach for modeling and solving PDF transport equations. The point of departure for PDF methods in their modern form, including the associated Lagrangian-particle-based solution algorithms, often is taken to be Pope's 1985 paper [55]. Detailed treatments of PDF methods are provided in the books by Pope [60] and Fox [22]. A comprehensive, up-to-date (2010) review can be found in [25].

In keeping with the theme of the book, this chapter emphasizes recent advances in PDF methods. A baseline approach that is representative of the current state-of-the-art is established in Sec. 6.2. Several recent developments with respect to this baseline are highlighted in Sec. 6.3. In Sec. 6.4, one particularly important area of development is discussed: the extension of PDF-based methods to model the effects of unresolved fluctuations in large-eddy simulation. The chapter concludes with a summary of key recent advances, outstanding issues, and promising directions for future development. The reader who seeks more information is referred to [25].

6.2 A Baseline PDF Formulation

The velocity-composition-turbulent frequency PDF (VCF-PDF) method introduced by Pope and coworkers represents the current state-of-the-art [7, 51, 74]. In that case, the PDF considered is the one-point, one-time joint PDF of the three components of velocity $\mathbf{u}$, N_ϕ composition variables ϕ, and a turbulent frequency ω. The N_ϕ composition variables are a set of scalars from which one can determine the local thermochemical properties (mixture mass density, molecular weight, and specific heats), species chemical production rates, and where required, the molecular transport properties and radiation properties. For a low-Mach-number, ideal-gas, single-phase, multicomponent reacting mixture, the species mass fractions $\mathbf{Y}$ and mixture specific enthalpy h usually are an appropriate set of composition variables. The turbulent frequency ω is the inverse of a local turbulence time scale.

Standard notation is used to distinguish conventionally-averaged mean quantities ($\langle\ \rangle$) from Favre-averaged or density-weighted mean quantities ($\widetilde{\ }$), and to distinguish a fluctuation about a conventionally-averaged mean value ($'$) from a fluctuation about a Favre-averaged mean value ($''$). The density-weighted VCF-PDF is denoted as $\widetilde{f}_{\mathbf{u}\phi\omega}(\mathbf{V},\psi,\theta;\mathbf{x},t)$. Here $\mathbf{V}$, ψ, and θ are the sample-space variables corresponding to $\mathbf{u}$, ϕ, and ω, respectively. This PDF contains complete one-point statistical information about the turbulent velocity, composition, and frequency at all locations $\mathbf{x}$ and times t. An important property of the PDF is that one-point joint statistics can be expressed as integrals of the PDF over its sample space. For example, the turbulent scalar fluxes $\widetilde{u_i''\phi_\alpha''}$ are given by:

$$\widetilde{u_i''\phi_\alpha''} = \int\int\int (V_i - \widetilde{u}_i)(\psi_\alpha - \widetilde{\phi}_\alpha)\widetilde{f}_{\mathbf{u}\phi\omega} d\mathbf{V} d\psi d\theta \ , \tag{6.1}$$

and integration is over the entire sample space. Analogous expressions can be written for the Reynolds stresses $\widetilde{u_i''u_j''}$, the scalar covariances $\widetilde{\phi_\alpha''\phi_\beta''}$, and any other one-point statistics of interest. The velocity-composition PDF (VC-PDF), $\widetilde{f}_{\mathbf{u}\phi}(\mathbf{V},\psi;\mathbf{x},t)$, and the composition PDF (C-PDF), $\widetilde{f}_\phi(\psi;\mathbf{x},t)$, also are widely used to model turbulent reacting flows. These can be obtained from the VCF-PDF by integrating over frequency space or velocity-frequency space, respectively.

Equations for $\widetilde{f}_{\mathbf{u}\phi\omega}$, $\widetilde{f}_{\mathbf{u}\phi}$, and $\widetilde{f}_\phi$ can be derived from the equations expressing conservation of mass, momentum, species composition, and energy for a reacting system [25, 55]. The resulting PDF equations contain terms that must be modeled. The modeling proceeds by considering a system of *notional particles* that are, in some sense, analogous to fluid particles that obey the underlying governing conservation equations. The idea is to devise a system of notional particles whose one-point, one-time joint PDF evolves in the same manner as the one-point, one-time joint PDF for the real reacting fluid system, while circumventing a key difficulty that arises in the real system: the wide dynamic range of scales that characterizes turbulent reacting flow at high Reynolds and Damköhler number. Here a superscript asterisk * denotes a notional particle value. For a high-Reynolds-number, low-Mach-number, open system, neglecting body forces and radiation heat transfer, and adopting standard physical models, the equations governing notional particle properties (position $\mathbf{x}^*(t)$, velocity $\mathbf{u}^*(t)$, composition $\phi^*(t)$, and turbulent frequency $\omega^*(t)$) can be written as:

$$dx_i^* = u_i^* dt \ , \tag{6.2}$$

$$du_i^* = -\frac{1}{\rho(\phi^*)}\frac{\partial\langle p\rangle}{\partial x_i}dt - \left(\frac{1}{2}+\frac{3}{4}C_0\right)\left(u_i^* - \widetilde{u}_i\right)\Omega dt + \left(C_0 k\Omega\right)^{1/2} dW_i \ , \tag{6.3}$$

$$d\phi_\alpha^* = S_\alpha(\phi^*)dt - \frac{1}{2}C_\phi(\phi_\alpha^* - \widetilde{\phi}_\alpha)\Omega dt \ , \tag{6.4}$$

$$d\omega^* = -C_3(\omega^* - \widetilde{\omega})\Omega dt - S_\omega \Omega\omega^* dt + \left(2C_3C_4\widetilde{\omega}\Omega\omega^*\right)^{1/2} dW' \ . \tag{6.5}$$

Here all mean quantities are evaluated at the particle location: $S_\omega = C_{\omega 2} - C_{\omega 1}P/(k\Omega)$ is the sink of frequency, $P = -\widetilde{u_i''u_j''}\partial\widetilde{u}_i/\partial x_j$ is the turbulence production, $k = \widetilde{u_i''u_i''}/2$ is the turbulence kinetic energy, and Ω is a conditionally-averaged turbulent frequency given by $\Omega = C_\Omega\langle\rho^*\omega^*|\omega^* \geq \widetilde{\omega}\rangle/\langle\rho\rangle$.

Particle advection in physical space (Eq. 6.2) is treated exactly; this includes transport by the mean velocity and by turbulent velocity fluctuations. Equation (6.3) is the simplified Langevin model for notional particle velocities [28]. There p is pressure, $\rho(\phi)$ is the density, and $\mathbf{W}(t)$ is a vector of three independent stochastic Wiener processes. The mean pressure gradient is treated exactly; that is, the mean pressure gradient term in the notional particle equations is the same as the corresponding term in the fluid particle equations. The modeled terms (terms involving Ω) represent the effects of the fluctuating pressure gradient and molecular (viscous) transport. Equation (6.4) governs the evolution of notional particle compositions. There $\mathbf{S}$ is the chemical source term in the species mass fraction equations, which is treated exactly. The modeled term (term involving Ω) represents the effects of molecular transport (scalar mixing). Here the interaction by exchange with the mean (IEM) mixing model [75] has been used, for simplicity. Finally, Eq. (6.5) is the stochastic frequency model for particle turbulent frequencies [74]. There $W'(t)$ is a stochastic Wiener process that is independent of the Wiener process in the particle velocity equation. There are seven model constants: C_0, C_ϕ, $C_{\omega 1}$, $C_{\omega 2}$, C_3, C_4, and C_Ω. Standard values are given in [51]. The VCF-PDF equation corresponding to Eqs. (6.2-6.5) is:

$$\begin{aligned}
&\frac{\partial \langle \rho \rangle \widetilde{f}_{\mathbf{u}\phi\omega}}{\partial t} + \frac{\partial V_i \langle \rho \rangle \widetilde{f}_{\mathbf{u}\phi\omega}}{\partial x_i} - \frac{1}{\rho}\frac{\partial \langle p \rangle}{\partial x_i}\frac{\partial \langle \rho \rangle \widetilde{f}_{\mathbf{u}\phi\omega}}{\partial V_i} + \frac{\partial S_\alpha \langle \rho \rangle \widetilde{f}_{\mathbf{u}\phi\omega}}{\partial \psi_\alpha} \\
&= \left(\frac{1}{2} + \frac{3}{4}C_0\right)\frac{\partial}{\partial V_i}[(V_i - \widetilde{u}_i)\langle \rho \rangle \widetilde{f}_{\mathbf{u}\phi\omega}] + \frac{1}{2}C_0 k \Omega \frac{\partial^2 \langle \rho \rangle \widetilde{f}_{\mathbf{u}\phi\omega}}{\partial V_i \partial V_i} \\
&+ \frac{1}{2}\frac{\partial}{\partial \psi_\alpha}[C_\phi \Omega(\psi_\alpha - \widetilde{\phi}_\alpha)\langle \rho \rangle \widetilde{f}_{\mathbf{u}\phi\omega}] + \Omega \frac{\partial \theta S_\omega \langle \rho \rangle \widetilde{f}_{\mathbf{u}\phi\omega}}{\partial \theta} \\
&+ C_3 \Omega \frac{\partial (\theta - \widetilde{\omega})\langle \rho \rangle \widetilde{f}_{\mathbf{u}\phi\omega}}{\partial \theta} + C_3 C_4 \Omega \widetilde{\omega} \frac{\partial^2 \theta \langle \rho \rangle \widetilde{f}_{\mathbf{u}\phi\omega}}{\partial \theta^2} . \qquad (6.6)
\end{aligned}$$

Here terms that are treated exactly are on the left-hand side, and modeled terms are on the right-hand side. The equation is *closed*; all quantities that appear in Eq. (6.6) can be deduced from $\widetilde{f}_{\mathbf{u}\phi\omega}(\mathbf{V}, \psi, \theta; \mathbf{x}, t)$. (The mean pressure field is determined by a Poisson equation, whose source can be expressed in terms of $\widetilde{f}_{\mathbf{u}\phi\omega}$ [25, 55].)

In a VC-PDF method, turbulent frequency is not carried as a particle property. Then turbulence time-scale information must be specified externally. For example, Ω in Eqs. (6.3) and (6.4) can be replaced by ε/k, where ε is the viscous dissipation rate of turbulence kinetic energy, and a separate modeled equation can be solved for ε (e.g., the ε equation from a standard two-equation turbulence model). A separate equation for k is not required, since k can be determined from $\widetilde{f}_{\mathbf{u}\phi}(\mathbf{V}, \psi; \mathbf{x}, t)$.

In a C-PDF method, velocity and turbulent frequency are not carried as particle properties. As for a VC-PDF method, turbulence time-scale information must be specified externally, and this can be done by writing Ω in Eq. (6.4) as ε/k, and solving separate modeled equations for k and ε. In this case, k cannot be determined from $\widetilde{f}_\phi(\psi; \mathbf{x}, t)$. In addition, the mean velocity and mean pressure fields must be specified externally (e.g., from a conventional CFD solver), and transport by turbulent velocity fluctuations must be modeled. This usually is done by invoking a gradient transport approximation, which is equivalent to replacing Eq. (6.2) by a stochastic diffusion equation that corresponds to a random walk in physical space [25, 55].

It is important to recognize that in all three formulations (VCF-PDF, VC-PDF, and C-PDF), the principal modeling issue that needs to be addressed in most alternative approaches is resolved: namely, the closure problem that arises from averaging the highly nonlinear source terms ($\widetilde{\mathbf{S}(\phi)} \neq \mathbf{S}(\widetilde{\phi})$). For an arbitrarily complex chemical mechanism, $\widetilde{S_\alpha} = \int S_\alpha(\psi)\widetilde{f}_\phi d\psi = \int\int S_\alpha(\psi)\widetilde{f}_{\mathbf{u}\phi} d\mathbf{V} d\psi = \int\int\int S_\alpha(\psi)\widetilde{f}_{\mathbf{u}\phi\omega} d\mathbf{V} d\psi d\theta$.

The single most important, and also probably the weakest, aspect of the physical modeling is the scalar mixing model. Mixing models in PDF methods are closely related to the scalar dissipation rate, a central quantity in all turbulent combustion modeling [22, 52]. Currently the dominant scalar mixing models are IEM, variants of Curl's model (or coalescence-dispersion - CD - model) [11, 68], and the Euclidean minimimum spanning tree (EMST) model [69]. Recent trends and developments in mixing models are aimed at extending PDF methods to higher-Damköhler-number systems and flamelet regimes of combustion, at accounting explicitly for a spectrum of turbulence scales, and at capturing differential diffusion, among other things. Mixing models are discussed further in Sec. 6.3.1.

Most modern numerical solution algorithms for modeled PDF transport equations employ a Lagrangian particle Monte Carlo algorithm that is coupled with a grid-based Eulerian CFD code: a hybrid Lagrangian particle/Eulerian mesh (LPEM) algorithm. While such algorithms have been available for over 20 years [2, 27], it has been only within the last 10 years that algorithms have matured to the point where PDF methods can be applied to general statistically nonstationary, three-dimensional turbulent reacting flows. Early algorithms failed to consider important mathematical and physical consistency requirements that are implicit in the particle representation, and that were not well understood at the time. Modern consistent hybrid LPEM methods are robust, efficient, and accurate. This important area of development is discussed further in Sec. 6.3.2.

The concept of statistically equivalent representations is implicit in the notional particle representation that has been introduced above. In principle, many different systems can give rise to the same one-point, one-time PDF equation. The "trick" is to devise a system whose one-point, one-time PDF equation is as close as possible to the PDF equation that corresponds to the real turbulent reacting flow, while reducing the dynamic range of scales that must be resolved numerically. The stochastic particle representation and physical models of Eqs. (6.2-6.5) have been designed for this purpose, and equivalent particle systems are the current mainstream approach for modeling and solving PDF transport equations. Recently alternative approaches based on equivalent stochastic or deterministic Eulerian fields have been explored, and this line of development is discussed in Sec. 6.3.3.

PDF methods traditionally have been associated with studies of turbulence-chemistry interactions in laboratory-scale, atmospheric, statistically-stationary, non-luminous nonpremixed turbulent flames burning simple gaseous fuels at low-to-moderate Damköhler numbers. These are not inherent limitations. PDF methods can be, and have been, applied to high-Damköhler-number systems, to premixed systems, to multiphase systems (e.g., liquid fuels and sooting luminous flames), and to practical combustion devices. The stochastic modeling and solution strategies provide a natural and powerful framework for accommodating the important nonlinear couplings across widely disparate scales that characterize turbulent combustion and other multiscale, multiphysics systems. For example, the VCF-PDF method is inherently a multiscale approach that accounts for a distribution of turbulence scales at each spatial location and time. Developments in multiscale and multiphysics modeling are discussed in Sec. 6.3.4.

6.3 Recent Advances in PDF Methods

Trends in mixing models, advances in particle-based methods, alternatives to particle-based methods, and extensions to multiscale, multiphysics systems are discussed in the following four subsections; examples are provided in Sec. 6.3.5. A fifth important area of development, the use of PDF-based methods to model the effects of unresolved fluctuations in large-eddy simulation, is discussed in Sec. 6.4.

6.3.1 Mixing Models

The most essential characteristics of scalar mixing models for PDF methods are [22, 60, 69]: 1) mean scalar quantities should not change as a result of mixing; 2) scalar variances should decay at the correct rate; and 3) scalar quantities should remain bounded (e.g., mass fractions should remain between zero and unity, and should sum to unity). Additional criteria have been proposed for mixing models used in LES [43]. In LES, it is important to treat the limit where the flow is locally fully resolved (the DNS limit) properly. This is discussed further in Sec. 6.4.

The earliest mixing models (IEM and Curl) possess the three most essential characteristics, but they do not perform well at high Damköhler numbers and/or in flamelet regimes, for example. Models that enforce locality in composition space (e.g., EMST [69]) generally perform better. A different approach for high Damköhler numbers was proposed in [40] for premixed systems. There a modified Curl's model was used, and the mixing time scale for reactive scalars was modified to include an explicit dependence on the ratio of the laminar burning velocity to the Kolmogorov velocity scale. The "parameterized scalar profile" (PSP) model proposed by Meyer and Jenny [47, 48] considers statistics of scalar profiles. Flamelet regimes can be accommodated, and the model provides joint statistics of scalar and scalar dissipation rate.

Mapping closures [57] provide another route to modeling scalar mixing in PDF methods. In a mapping closure, the stochastic physical field of interest is mapped to a Gaussian reference field. The advantage is that for a Gaussian field, all *multipoint* statistics are known in terms of the mean, the variance, and the two-point correlation function. A particular strength of mapping closures is that they cause the PDF of conserved scalars to relax to a Gaussian distribution in statistically-homogeneous systems. Klimenko and Pope [36] proposed a "multiple mapping conditioning" (MMC) approach for mixing and reaction. MMC is based on a generalization of mapping closures, and combines elements of conditional moment closure (CMC) and PDF methods. MMC has been developed subsequently by Klimenko and others [34, 76]. In some of the most recent work [8, 9], a novel sparse-Lagrangian numerical algorithm has been developed for solving PDF transport equations that have been closed using the MMC approach.

Other approaches that have been proposed include models based on Langevin equations [73] or the closely related Fokker-Planck equation [19, 20], and models that explicitly incorporate scale (spectral) information [21, 71]. Other mixing models are discussed in Chapter 6 of [22]. Some models have been designed to accommodate differential molecular diffusion [43].

It has been argued that relatively simple mixing models should suffice for LES, because the models only need to deal with the unresolved fluctuations about the local resolved-scale composition. This has been borne out in numerical studies, and is consistent with the arguments presented in [35]; there it has been shown that many mixing models are essentially the same when the physical distance between particles becomes infinitesimally small (a DNS limit). For these reasons, IEM and variants have been used in most LES modeling studies reported to date.

6.3.2 Hybrid Lagrangian Particle/Eulerian Mesh Methods

Three areas where important advances are being made in methods to solve the PDF equations are: mathematical/physical consistency; numerical accuracy; and computational efficiency. Each of these is discussed in turn in the following paragraphs.

An intrinsic requirement for the notional particle representation to be consistent with the PDF equation is that the spatial distribution of particle mass must remain consistent with the spatial distribution of fluid mass in the system. The extent to which this constraint is satisfied depends on the particle-tracking algorithm, among other things. For example, the interpolation scheme that is used to determine the value of the local mean velocity at particle locations should guarantee a discrete velocity-divergence field that is consistent with the mean continuity equation [44, 50, 51, 79]. Even with such an interpolation scheme, a correction algorithm may be required to redistribute particles in physical space [51, 79] to prevent the accumulation of numerical errors. Energy consistency is another important consideration. An approach that simultaneously addresses consistency and robustness of the particle/mesh coupling was introduced in [50, 51]. There different forms of the energy equations are solved on the mesh side and on the particle side. The mean energy equation on the mesh side contains a chemical source term that is estimated using particle values. This smooths the noisy particle data through an elliptic PDE, thus enhancing robustness while still capturing the important influence of turbulent fluctuations.

Characterization and quantification of numerical errors for LPEM methods is another area where advances have been made. Errors can be categorized as statistical error, bias error, spatial truncation error, and temporal truncation error [50]. The first two of these are peculiar to Monte Carlo methods. Statistical error refers to the random error that results from using a finite number of particles; it scales as $N_{pc}^{-1/2}$, where N_{pc} is the number of particles per cell. Bias refers to a deterministic error that results from using a finite number of particles, and is not reduced by averaging; it scales as N_{pc}^{-1} [58]. Systematic studies of numerical errors have yielded recommendations for optimum values of key numerical parameters [50, 58]. Stochastic terms in the notional particle equations (Eqs. 6.2-6.5) complicate the construction of time-accurate methods. Higher-order implementations require extreme care in the sequencing of operations and in the updating of coefficients; formally second-order accurate schemes have been published [6].

Efforts to reduce computational cost are a third area where advancements are being made. Chemical kinetics usually dominate the computational effort when finite-rate chemical mechanisms involving more than a small number (10, say) of chemical species are used. *In situ* adaptive tabulation (ISAT) [41, 59] is a storage-retrieval algorithm that has proved effective in improving computational efficiency, especially for statistically stationary systems. ISAT tabulates the composition at the end of a reaction fractional step (obtained by integrating the ODE's $d\phi_\alpha/dt = S_\alpha(\phi)$). Increasingly effective parallelization strategies are being developed for LPEM methods, both with and without ISAT. The number of particles per cell remains approximately constant in most modern algorithms. In that case, the same spatial domain decomposition that is used to parallelize the underlying Eulerian CFD code is effective for most aspects of the particle calculations. An exception is chemical reaction; the computational time required to integrate the stiff chemical source terms for a particle varies widely, depending on its location in composition space. Achieving uniform load balance across processors requires a composition-space (versus physical space) distribution strategy, and approaches are being developed to accomplish this.

6.3.3 Eulerian Field Methods

Eulerian field representations are alternatives to the notional particle representation that is discussed in Sec. 6.2. Both stochastic and deterministic Eulerian field methods have been proposed.

Two developments of stochastic field methods have appeared in the literature. One originated with Valiño [72] and another was introduced by Sabel'nikov and Soulard [65]. Different physical and mathematical arguments are invoked in the two developments. For example, in the Valiño formulation the stochastic fields must remain spatially smooth, while the Sabel'nikov formulation allows for discontinuities. The statistical equivalence of the two approaches has been established in Appendix B of [65]. An overview of the Valiño formulation for a composition PDF follows. The turbulent reacting flow is represented by a number N_F of notional Eulerian fields. With superscript $^{\#}$ referring to any one of the fields, the stochastic Eulerian field equation is:

$$d\phi_\alpha^\# = -\widetilde{u}_i \frac{\partial \phi_\alpha^\#}{\partial x_i} dt + S_\alpha(\phi^\#) dt - \frac{1}{2} C_\phi (\phi_\alpha^\# - \widetilde{\phi}_\alpha) \omega dt$$
$$+ \frac{1}{\langle \rho \rangle} \frac{\partial}{\partial x_i} \left[\Gamma_T \frac{\partial \phi_\alpha^\#}{\partial x_i} \right] dt + (2\langle \rho \rangle^{-1} \Gamma_T)^{1/2} \frac{\partial \phi_\alpha^\#}{\partial x_i} dW_i^\# . \quad (6.7)$$

Here $\mathbf{W}^\#$ denotes a vector-valued Wiener process that varies in time but is independent of spatial location. The first three terms on the right-hand side of Eq. (6.7) correspond, respectively, to advection by the mean flow, chemical reaction, and IEM mixing. The terms involving the apparent turbulence diffusivity Γ_T correspond to a gradient transport model for turbulent velocity fluctuations. Estimates of density-weighted mean quantities are obtained by ensemble averaging over the fields. Between 10 and 30 fields have been used in modeling studies reported to date.

A deterministic approach is the "multi-environment PDF" (MEPDF) method proposed by Fox [22, 70]. In a composition MEPDF method, a system of equations is solved for the weights (or probabilities) and the scalar values of each of N_E "environments." Mean quantities are estimated as probability-weighted sums over the environments. The system is closed by adding source terms that are prescribed to force a subset of the moment equations from the MEPDF to match the corresponding moment equations from a modeled PDF transport equation: a direct-quadrature-method-of-moments (DQMOM) approach. The equations corresponding to gradient transport and IEM mixing for $N_E = 2$ and $N_E = 3$ are given in [70]. In contrast to stochastic Eulerian field methods, the MEPDF equations are deterministic. However, the source terms become poorly conditioned with increasing N_E, and most results reported to date have been for $N_E = 2$. Such low values of N_E correspond to radical approximations of the PDF equations. Moreover, the approach does not guarantee realizable compositions with multispecies chemistry.

Eulerian-field methods (stochastic or deterministic) are more straightforward to implement into mesh-based Eulerian CFD codes than are particle methods. Deterministic Eulerian-field methods are free of statistical error, and the Valiño stochastic Eulerian field formulation is free of spatially varying stochastic errors; the special techniques that have been developed for particle/mesh coupling through the mean density and/or energy in hybrid LPEM methods may not be necessary. On the other hand, simple mixing models (IEM) have been used in Eulerian-field PDF modeling studies reported to date. Mixing models that enforce locality in composition space (e.g., EMST [69]) probably cannot be implemented effectively in the Eulerian-field frameworks. In cases where chemical mechanisms containing more than 10 species (say) are used, the computational effort is dominated by the chemical source terms. In that case, the computational effort per field in an Eulerian method for a mesh containing N_C cells (the solution of a PDE) is comparable to the computational effort per N_C particles in a Lagrangian particle method (the solution of N_C stochastic ODE's). Unless the number of fields that is required for an Eulerian field method is smaller than the number of computational particles per cell that is required for a Lagrangian particle method, little advantage can be expected in computational time.

6.3.4 Multiscale, Multiphysics Modeling

Extensions of PDF methods to deal with flamelet combustion, walls, high-speed flows, soot, sprays, and radiation are discussed briefly in the following paragraphs.

Flamelets have been accommodated in various ways. One approach is to solve a modeled PDF equation for a reduced set of composition variables (e.g., a mixture fraction and/or a small number of progress variables) and to relate the local thermochemical state to this reduced set using precomputed flamelet libraries [26]. Another approach is to build the flamelet structure explicitly into the formulation and modeling. Early in the development of PDF methods, it was recognized that in laminar premixed flames the molecular transport term in the PDF equation can be closed by introducing a reaction progress variable [55, 63]; this was done in several modeling studies of premixed turbulent flames [1, 63]. A third approach is to modify the scalar mixing model to include an explicit dependence on laminar burning velocity and/or Kolmogorov scales [40] (Sec. 6.3.1).

Modifications to the PDF or particle equations have been introduced to include wall effects, following one of two approaches. Models based on wall functions are designed to reproduce a logarithmic mean velocity profile and other statistical features of equilibrium turbulent boundary layers [17, 49]. In contrast, models based on elliptical relaxation incorporate wall effects without explicitly invoking wall or damping functions [16, 77]. Wall-function approaches are computationally more efficient, while elliptical relaxation is more general.

In high-speed flows, the effects of pressure fluctuations on density generally cannot be neglected. It then is necessary to include pressure and/or other thermodynamic quantities in the composition variables. A joint PDF of species mass fractions and enthalpy was extended to high-speed flows, including flows with shocks, in [29]. A formulation based on the joint PDF of velocity, pressure, an energy variable, and a turbulent frequency was proposed and tested in [12, 13].

PDF-based modeling studies of sooting turbulent flames have been reported in [39, 45, 46]. To capture the effects of turbulent fluctuations on soot processes, additional quantities are transported on each notional particle. For example, in the method-of-moments approaches used in [39, 45, 46], the lower-order concentration moments of the soot particle size distribution function are transported, and a closure model is invoked to truncate the infinite series of moments (e.g., interpolative closure). Source terms in the moment equations are designed to capture the key physical and chemical processes involved in soot formation and oxidation.

A dual-Lagrangian formulation has been adopted in most PDF-based studies of turbulent spray flames. There one particle ensemble represents the dispersed liquid phase (e.g., the fuel spray) and a separate particle ensemble represents the carrier gas phase (the notional PDF particles). The principal liquid-to-gas coupling is through the vaporization model. This is appropriate for external group combustion regimes, where combustion occurs around groups of droplets. Composition-PDF modeling studies of turbulent spray flames include [24, 31, 37, 80]. A dual-Lagrangian velocity-composition PDF method also has been developed [4].

PDF methods are particularly well suited for dealing with radiation heat transfer. Radiative emission is determined by local properties, and therefore appears in closed form in PDF equations. This includes the so-called turbulence-radiation interactions (TRI) associated with emission: the mean radiative emission from a turbulent flame is not the same as the radiative emission that would be determined based on the local mean temperature and composition [38, 42, 46]. Radiative absorption requires further modeling, and approaches that are compatible with the notional particle representation have been developed for that purpose [45, 46, 78].

6.3.5 *Examples*

Three recent examples are discussed to illustrate the range of physical phenomena that can be captured, and the diverse physical models and numerical algorithms that have been developed.

6.3.5.1 A Statistically Stationary 1D System with Imperfect Mixing

A plug-flow-reactor often is used to study chemical kinetics, and/or as an idealization of the combustion process in some devices. An imperfect-mixing variant was introduced in [54] as an early test case for PDF methods, and (as now defined) was used in [30] to study the influence of different numerical implementations of PDF methods. A single reaction progress variable, c, characterizes this statistically stationary, one-dimensional system. Unburned reactants ($c = 0$) enter at $x = 0$, and c approaches unity (fully burned products) at $x = L$. Finite-rate chemistry proceeds at rate that is proportional to $1 - c$. For IEM mixing and constant values for all model parameters, analytic solutions can be obtained for the steady-state mean and rms progress-variable profiles (Fig. 6.1). Results from a stochastic Lagrangian particle method, a stochastic Eulerian field method, and a deterministic Eulerian field (two-environment MEPDF with DQMOM closure) are compared in Fig. 6.1 [30]. For the two stochastic methods, time-accurate algorithms are used, the mean and rms values at each x location are obtained by averaging over the local particle or field values, and these are further averaged in time to reduce statistical error. Each individual field in the stochastic Eulerian field method exhibits random fluctuations in time, but remains spatially smooth. For the MEPDF method, starting from (almost) arbitrary initial conditions, each environment ($c_1(x)$, $c_2(x)$) converges to a steady-state profile, and each weight or probability ($w_1(x)$, $w_2(x)$) converges to a uniform value that is equal to the value that is specified at $x = 0$. No simple physical interpretation can be assigned to the environments here, and there is no unique specification for the environment weights at $x = 0$; these can be set to any values between zero and unity that satisfy $w_1 + w_2 = 1$ (with $c_1 = c_2 = 0$). Here $w_1 = 0.1$, $w_2 = 0.9$ have been specified. The steady-state environment profiles $c_1(x)$ and $c_2(x)$ would be different for different choices of the inlet weights, but the mean and rms profiles would

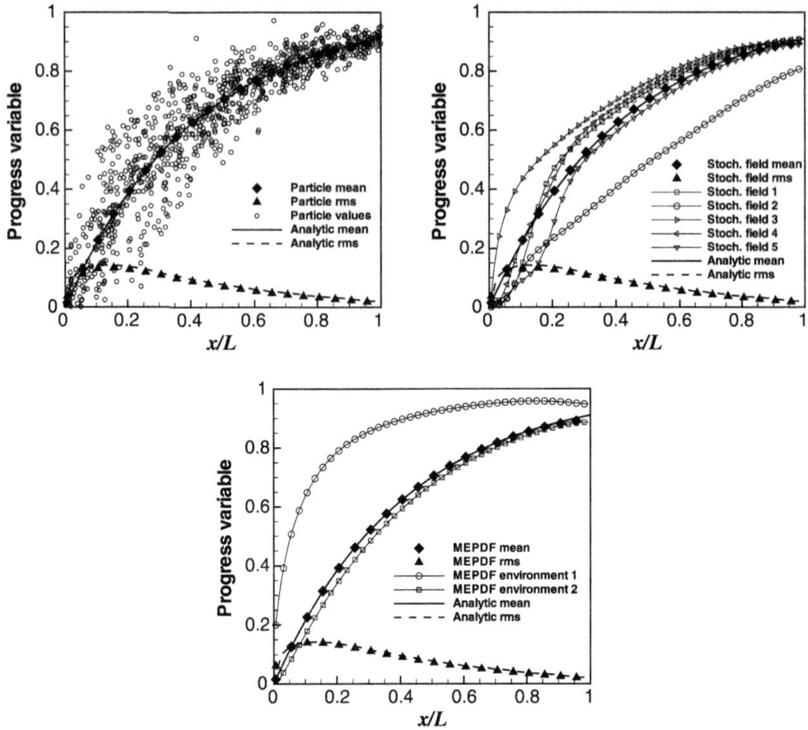

Fig. 6.1: Mean and rms reaction progress variable profiles for a statistically stationary 1D system with imperfect mixing. Analytic mean (bold solid lines) and rms (bold dashed lines) profiles and computed mean (filled diamonds) and rms (filled triangles) profiles for each method are shown [30]. a) 1,000 instantaneous particle values (open symbols) from a Lagrangian particle method. b) Five instantaneous 1D fields (thin solid lines with open symbols) from a stochastic Eulerian field method. c) Steady-state environment profiles (thin solid lines with open symbols) from a two-environment MEPDF method.

remain the same. For all three methods, the mean and rms profiles agree well with the analytic solutions. This configuration is too simple (and nonphysical) to draw conclusions regarding the relative merits of the three methods, but it serves well to illustrate the very different discrete representations of the PDF that are used in each method.

6.3.5.2 Lifted Jet flames in Vitiated Coflow

Lifted flames are useful to study autoignition and stabilization mechanisms. Laboratory-scale, atmospheric-pressure axisymmetric jets of H_2/N_2 mixtures issuing into a hot coflow of lean H_2-air combustion products have been studied experimentally [5]. A VCF-PDF method and a consistent hybrid LPEM solution algorithm were applied to this configuration in [7]. The study featured the simplified Langevin model and a stochastic frequency model for particle velocities and turbulent frequencies, respectively. The effects of coflow temperature, inlet velocity profiles, inlet turbulence intensity, mixing models, and chemical mechanisms were explored. Results are very sensitive to small variations in the temperature that is prescribed for the coflow; a 10 K change in the temperature can change the computed liftoff height by as much as a factor of two, and the experimental uncertainty in the coflow temperature is ± 25 K. Computed and measured liftoff heights versus coflow temperature are shown in Fig. 6.2. There two sets of experimental measurements are shown ("Wu" and "Gordon"), and computed results are shown for two chemical mechanisms ("Mueller" and "Li") and for three mixing models (MC - modified Curl, IEM, and EMST). The models capture the experimental trends in all cases. Differences between model results obtained using the two chemical mechanisms are large, while differences among the mixing models are relatively small; this suggests that these flames are primarily controlled by chemical kinetics.

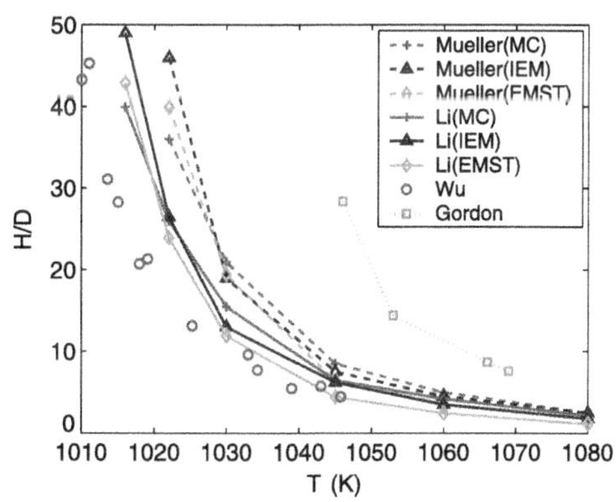

Fig. 6.2: Computed and measured lift-off heights H (normalized by fuel-nozzle diameter D) versus coflow temperature for jets of H_2/N_2 issuing into a coflow of H_2-air combustion products. Results for two sets of experimental measurements, two chemical mechanisms, and three mixing models are shown. Reprinted from [7] with permission from the Combustion Institute.

6.3.5.3 Luminous Flames

The third example illustrates the ability of PDF methods to capture complex interactions of turbulence, gas-phase chemistry, soot and radiation. Here detailed soot and radiation models have been implemented, and the models have been applied to luminous, atmospheric-pressure, nonpremixed, ethylene- and methane/ethylene-fueled turbulent jet flames [45, 46]. A composition PDF method and consistent hybrid LPEM algorithm are used with skeletal gas-phase chemistry (a 33-species mechanism), a detailed soot model (method of moments with interpolative closure), and a spectral photon Monte Carlo radiation solver. Six different turbulent flames were simulated with no changes in the model parameters, and uniformly good agreement was observed between computed and measured mean temperature, mean soot volume fraction, and radiative intensity (where available) profiles. One example is shown in Fig. 6.3, for a 90% methane/10% ethylene-fueled flame with 30% oxygen in the oxidizer stream. There the "equivalent soot volume fraction" was measured using a line-of-sight extinction technique. Results from two sets of simulations are shown: in one case, turbulence-radiation interactions have been ignored by computing radiative emission and absorption using cell-mean values; in the other, turbulence-radiation interactions have been included by computing radiative emission and absorption using particle values. In both cases, turbulence-chemistry interactions are retained by computed chemistry at the particle level. Radiant heat fluxes are up to 50% higher with consideration of TRI (not shown), and soot levels decrease by up to a factor of three with consideration of TRI. Both heat fluxes and soot levels are in better agreement with experimental measurements when TRI are included. These results clearly demonstrate that TRI – often ignored in model calculations – can have a major effect.

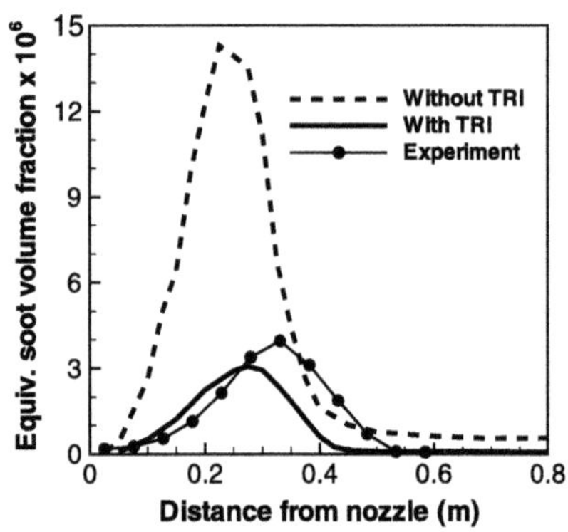

Fig. 6.3: Computed (lines) and measured (symbols) soot volume fraction profiles for a 90% CH_4/10% C_2H_4-fueled flame with 30% O_2 in the oxidizer [46]. Two sets of calculations are shown: one where turbulence-radiation interactions have been neglected (dashed line), and one where turbulence-radiation interactions have been included (solid line).

6.4 PDF-Based Methods for Large-Eddy Simulation

The predominant approach to extending PDF-based modeling to LES is the filtered density function (FDF) method. The filtered density function was introduced by Pope [56], and a transport equation for a composition FDF first was derived and modeled by Gao and O'Brien [23]. FDF methods currently are being developed by several groups [10, 18, 32, 64, 66, 67]. The LES-FDF method is reviewed in Secs. 6.4.1 and 6.4.2. Issues have been raised concerning the appropriateness of the FDF as a basis for modeling in LES. An alternative LES-PDF formulation that addresses these issues has been proposed recently by Pope [62], and this is discussed in Sec. 6.4.3. Examples of LES-FDF methods are provided in Sec. 6.4.4.

6.4.1 Spatial Filtering, FDFs, and FDF Transport Equations

The local spatially-filtered value of a physical quantity Q, denoted $\langle Q\rangle_\Delta$, is,

$$\langle Q\rangle_\Delta = \langle Q(\mathbf{x},t)\rangle_\Delta \equiv \int Q(\mathbf{y},t)G(|\mathbf{x}-\mathbf{y}|)d\mathbf{y} \ , \tag{6.8}$$

where integration is over the entire flow domain. Here the low-pass spatial filter function, $G(|\mathbf{x}-\mathbf{y}|)$, satisfies $\int G(\mathbf{x})d\mathbf{x} = 1$, is non-negative (for FDF methods), and has a characteristic filter width Δ. The instantaneous value of any physical quantity $Q = Q(\mathbf{x},t)$ can be decomposed into a filtered component (the resolved field) and a fluctuation about the filtered component (the residual field): $Q(\mathbf{x},t) = \langle Q(\mathbf{x},t)\rangle_\Delta + Q'_\Delta(\mathbf{x},t)$. Density weighting is useful in variable-density flows; the corresponding fluctuation about a density-weighted spatially-filtered value (denoted $\widehat{Q}_\Delta$) is denoted using a double prime: $Q = \widehat{Q}_\Delta + Q''_\Delta$.

The FDF is the G-weighted spatial average of the fine-grained PDF in a neighborhood of $\mathbf{x}$ [56]. For a composition FDF, for example, $f_{\Delta,\phi}(\psi;\mathbf{x},t)d\psi$ is the G-weighted fraction of the fluid in a neighborhood of $\mathbf{x}$ whose composition is in the range $\psi \le \phi < \psi + d\psi$. Important properties of the FDF are that it is non-negative, it integrates to unity over ψ-space, and the local spatially-filtered value of any function of ϕ can be expressed as an integral over the FDF. These are analogous to the corresponding properties of a PDF. However, there are important differences between FDF's and PDF's. The FDF varies on length scales down to Δ, versus the turbulence integral scale l_T in the case of the PDF; normally Δ is smaller than l_T. The FDF varies in time even for statistically-stationary flows, and in all three spatial coordinates even for flows having one or more directions of statistical homogeneity. The FDF is a random quantity. And the PDF is the expected value of the FDF in the limit as the filter width Δ shrinks to zero.

FDF equations can be derived following procedures that are similar to those used to derive PDF equations, and the resulting equations have essentially the same structure as PDF equations [25]. There are two principal differences: mean quantities in the PDF equations are replaced by spatially-filtered quantities in the FDF equations; and the mean-quantity-based turbulence scales in modeled PDF equations are replaced by sub-filter-scale turbulence scales in modeled FDF equations. As is the case for PDF methods, the single most compelling reason for pursuing FDF methods is that the chemical source terms (and other important one-point processes, including radiative emission - Sec. 6.3.4) remain in closed form even as the dynamics of the small scales are removed by spatial filtering. For example, the filtered chemical source term $\widehat{\mathbf{S}}$ is closed in terms of the VCF-FDF, the VC-FDF, or the C-FDF.

6.4.2 Equivalent Representations, Models, and Algorithms

The Lagrangian notional particle and Eulerian field representations for FDF methods are essentially the same as the those introduced earlier for PDF methods. Local mean values that appear in PDF formulations are replaced by local spatially-filtered values in FDF formulations. And in constructing models, the specification of turbulence scales needs to be modified to account for the fact that, in FDF methods, the models are responsible only for the sub-filter-scale fluctuations about the local spatially-filtered values. The appropriate choice for a turbulence length scale normally is one that is proportional to the local filter width Δ. With a suitable estimate for the sub-filter-scale turbulence kinetic energy k_Δ, or for the apparent sub-filter-scale turbulence viscosity ν_Δ, the local sub-filter-scale turbulent frequency ω_Δ is taken to be proportional to $k_\Delta^{1/2}/\Delta$ or to ν_Δ/Δ^2. In most LES-FDF modeling studies to date, a composition FDF method has been used with a constant-coefficient or dynamic Smagorinsky sub-filter-scale turbulence model. Gradient transport models have been invoked for sub-filter-scale turbulent transport. And simple mixing models (e.g., IEM) have been used, since the models only need to deal with the local sub-filter-scale fluctuations; the latter are expected to small compared to the fluctuations about the local mean. Exceptions are the works of Givi and coworkers, who have been developing VC-FDF [66] and VCF-FDF [67] methods.

In LES, there are likely to be regions of the computational domain where the flow is locally fully resolved. Therefore, it is important to formulate the sub-filter-scale models so that they recover the "DNS limit." This has implications for the modeling of molecular transport. McDermott and Pope [43] present a set of criteria for LES-FDF models that differ somewhat from the criteria that are applied in PDF methods. The resulting models reduce appropriately to DNS in the limit of vanishing filter width, and are able to accommodate differential molecular diffusion.

The fluctuations about local spatially-filtered values in LES can be expected to be smaller than the fluctuations about local mean values in Reynolds-averaged simulations. For that reason, fewer particles per cell have been used in some hybrid LPEM FDF methods, compared to PDF methods. On the other hand, LES is inherently three-dimensional and nonstationary; this limits the extent to which spatial and/or time averaging can used to reduce statistical error. Higher-fidelity Eulerian CFD algorithms normally are used for LES compared to Reynolds-averaged simulations, and commensurately greater care is needed in the construction of the numerical algorithms and coupling strategies for FDF methods compared to PDF methods. For example, the consistency issues that were raised in Sec. 6.3.2 often have been ignored in PDF modeling studies that have been reported to date. It is likely that LES-FDF that does not explicitly deal with these consistency and robustness issues will be unacceptably inefficient and inaccurate, or will fail altogether.

6.4.3 An Alternative Interpretation

Both conceptual and pragmatic issues have been raised concerning the use of spatially-filtered velocities and filtered density functions as the basis for LES [22, 52, 61]. For example, the filtering approach ignores the fact that there is a distribution of physical velocity fields that corresponds to a given filtered velocity field. It is not clear how statistics of filtered fields are related to statistics of the corresponding physical fields, and this gives rise to ambiguities in making quantitative comparisons between LES results and experimental and/or DNS data. And the usual filter-based approaches behave inappropriately in cases where the flow locally is fully resolved (the DNS limit) and near walls. The latter shortcoming is illustrated in Fig. 6.4. In Fig. 6.4a, an error function $\phi(x)$ of width $\sigma = 0.1$ is plotted; this represents a scalar profile in a laminar diffusive layer. The filtered profile $\langle\phi(x)\rangle_\Delta$ (obtained using a top-hat filter of width $\Delta = 1$) also is plotted. In Fig. 6.4b are two corresponding variances: the second central moment of the FDF, $\langle\phi^2(x)\rangle_\Delta - \langle\phi(x)\rangle_\Delta^2$, and the filtered square of the residual, $\langle(\phi(x) - \langle\phi(x)\rangle_\Delta)^2\rangle_\Delta$. In this laminar case, with no physical fluctations, filtering smears the profile and generates a nonphysical variance.

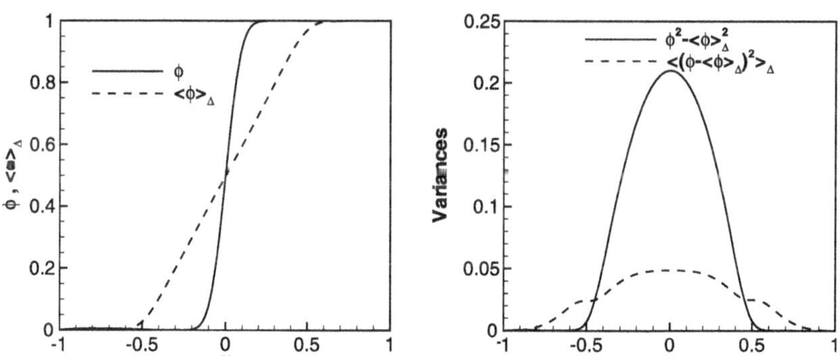

Fig. 6.4: An illustration of the effects of spatial filtering on a laminar diffusive layer. a) $\phi(x)$ and $\langle\phi(x)\rangle_\Delta$. b) $\langle\phi^2(x)\rangle_\Delta - \langle\phi(x)\rangle_\Delta^2$ and $\langle(\phi(x) - \langle\phi(x)\rangle_\Delta)^2\rangle_\Delta$.

The concept of redefining LES fields in terms of conditional means originated with Fox [22]. This idea has been developed and expanded substantially in a recent paper by Pope [62] in terms of "self-conditioned fields." The essence of the self-conditioned-fields (SCF) approach is to introduce a mean velocity field that is conditioned on quantities that can be deduced from the conditioned mean velocity field itself; discrete values of the usual filtered velocity on a grid are appropriate conditioning variables, for example. This bootstrapping approach provides an alternative foundation for LES that resolves many of the key issues that arise with filtering, and is readily extended to a self-conditioned composition PDF method for turbulent reacting flows [62]. Compared to filtering, SCF explicitly account for the distribution of turbulent velocity fields, and behave appropriately in the DNS limit and near walls. Moreover, they resolve the commutation issues that arise with filtering, and the complications that arise from repeated application of spatial filters (e.g., filtering a filtered field does not recover the original filtered field). The SCF corresponding to $\phi(x)$ in Fig. 6.4 is identical to $\phi(x)$, and the SCF variance is zero.

In contrast to LES-FDF methods, self-conditioned fields provide a rigorous PDF-based approach for LES of turbulent reacting flows. The approach is distinct from LES-FDF, and will be denoted as an LES-PDF method. Equations for the self-conditioned mean velocity and for the self-conditioned composition PDF are derived in [62], and some initial proposals for modeling are provided there: the models are essentially conventional sub-filter-scale models that have been modified to recover correct limiting behaviors (e.g., the DNS limit). At the time of this writing, the practical implications for modeling and numerical implementation have not been fully explored. Nevertheless, it is anticipated that this more precise treatment may lead to new insights and to advances in physical modeling for LES of both nonreacting and reacting turbulent flows.

6.4.4 Examples

Two recent examples of FDF modeling studies are discussed to illustrate the important physical phenomena that can be captured, and the different numerical approaches that are being developed.

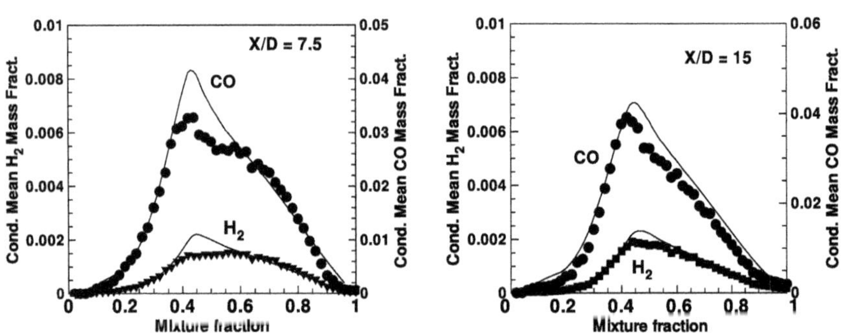

Fig. 6.5: Computed (lines) and measured (symbols) conditional mean CO and H_2 mass fractions at two different axial locations for flame E. This is Fig. 7 of [64]. Reprinted with permission from the Combustion Institute.

A series of piloted methane-air nonpremixed jet flames has been targeted for systematic investigation of turbulence-chemistry interactions by the International Workshop on Measurement and Computation of Turbulent Nonpremixed Flames [3]. Sandia flames D, E, and F exhibit increasing levels of local extinction as the fuel-jet and pilot velocities are increased (that is, as the Damköhler number is decreased); flame D exhibits little local extinction, while flame F is on the verge of global extinction. In [64], a hybrid finite-volume/particle FDF method was applied to flames D and E. The model featured a 16-species chemical mechanism, a dynamic Smagorinsky model for sub-filter stresses, gradient transport for sub-filter-scale turbulent scalar fluxes, and a dynamic formulation to obtain the scalar time scale that is needed in the IEM mixing model. Computational meshes of approximately one million cells were used with 40 particles per cell. Computed radial profiles of mean temperature and mean major species mass fractions were generally in good agreement with experimental measurements (not shown). Computed conditional mean CO and H_2 mass fractions for flame E show some deviations from the measurements for stoichiometric-to-rich mixtures close to the fuel nozzle (Fig. 6.5), but the overall level of agreement is quite encouraging for this flame with strong local extinction and reignition.

In the second example, a C-FDF method is applied to the same vitiated coflow configuration that was considered in Sec. 6.3.5. Here a stochastic Eulerian field method is used with a detailed chemical mechanism, a Smagorinsky sub-filter-scale turbulence model, gradient transport for scalars, and IEM mixing [33]. Sensitivities to the number of stochastic fields and to the coflow temperature were studied, and strong sensitivity of computed lift-off height to coflow temperature was noted, as before. Sixteen stochastic fields and a coflow temperature of 1035 K were used in most of the simulations. Computed and measured scatter plots of temperature versus mixture fraction at an axial location near the flame base are shown in Fig. 6.6. A wide range of states between pure mixing and complete combustion is evident, including locally fuel-rich burning; these features are captured well by the model. When sub-filter-scale fluctuations are neglected, this structure is not captured (not shown).

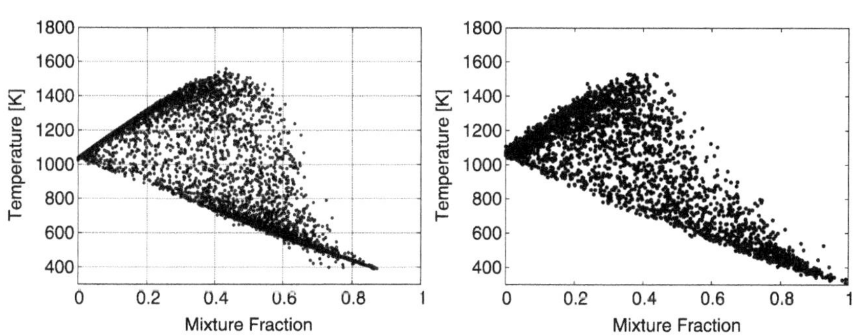

Fig. 6.6: Computed (left) and measured (right) scatter plots of temperature versus mixture fraction near the flame base. This is Fig. 23 of [33]. Reprinted with permission from the Combustion Institute.

6.5 Summary and Conclusions

Significant progress has been made in PDF methods since 1985 [55]. At that time, PDF methods were an academic research tool. Today they remain a powerful tool for science discovery, but they also have become a mainstream modeling approach for laboratory-scale flames, they have been implemented in multiple research and commercial CFD codes, and they are being applied with increasing frequency to practical combustion devices. PDF methods have been applied primarily to reacting ideal-gas mixtures using single-turbulence-scale models. However, multiphysics, multiscale information is readily incorporated. And while most applications to date have been to atmospheric-pressure, laboratory-scale, statistically-stationary, nonluminous, nonpremixed flames, PDF methods can be applied to high-Damköhler-number systems as well as to low-to-moderate-Damköhler-number systems, to premixed systems as well as to nonpremixed and partially premixed systems, and to practical combustion devices as well as to laboratory-scale flames. An up-to-date (2010) review can be found in [25].

PDF methods offer compelling advantages for modeling chemically reacting turbulent flows. They resolve the most important closure problems that arise from averaging or filtering the highly nonlinear chemical source terms, and terms that correspond to other one-point physical processes (e.g., radiative emission). PDF methods provide a rational framework for turbulent combustion modeling, where the best available gas-phase chemical mechanisms, soot models, and radiation models (as established in canonical nonturbulent configurations) can be carried directly into turbulent flames with minimal further approximations or simplifications. Complex interactions among hydrodynamic turbulence, gas-phase chemistry, soot, and thermal radiation then can be captured in a natural and direct manner. For these reasons, PDF methods have been successful where other approaches have not. Examples include their ability to capture strong turbulence-chemistry interactions even in flames with strong local extinction/reignition, and their ability to capture strong turbulence-radiation interactions in luminous flames.

The somewhat unconventional stochastic methods that have been developed to solve modeled PDF transport equations have slowed their widespread adoption. However, stochastic methods continue to gain traction in computational physics and engineering as their benefits are recognized: they are a powerful approach for accommodating the important nonlinear couplings across widely disparate scales that characterize turbulent combustion and other multiscale, multiphysics systems. Significant advances in numerical algorithms have been made over the past decade, and this is an area where substantial further progress is anticipated. This includes hybrid Lagrangian particle/Eulerian mesh methods, and alternatives such as Eulerian field methods.

The trend from Reynolds-averaged turbulent combustion modeling toward LES is expected to continue. PDF-based methods have been extended to LES via the FDF, and alternative LES-PDF formulations are being developed. Essentially similar physical models and numerical algorithms are used in LES as in Reynolds-averaged simulations. It has been argued that relatively simple physical models (e.g., mixing models) may suffice for LES, because the models only need to represent the unresolved (residual) component of the turbulent fluctuations.

Likely directions for next-generation clean and efficient combustion systems include higher pressures, lower temperatures, extremely lean and/or dilute mixtures, and different fuels including reactant mixtures with high levels of H_2, O_2, syngas (CO and H_2), and/or exhaust-gas recirculation. The importance of finite-rate chemistry and turbulence-chemistry interactions and of participating-medium radiation and turbulence-radiation interactions will increase in such systems, compared to current lean-to-stoichiometric conventional hydrocarbon/air systems. PDF methods are uniquely suited to meet these challenges, and it is anticipated that they will be adopted more broadly through the 21st century.

Acknowledgements The authors are grateful for support from DOE, NASA, NSF, CD-adapco, and GM. The contribution of SBP was supported by the Air Force Office of Scientific Research under Grant No. FA-9550-09-1-0047. We thank Prof. W.P. Jones (Imperial College, London, UK) for providing original files for Fig. 6.6.

References

1. Anand, M.S., Pope, S.B.: Calculations of premixed turbulent flames by PDF methods. Combust. Flame **67**, 127–142 (1987)
2. Anand, M.S., Pope, S.B., Mongia, H.C.: A PDF method for turbulent recirculating flows. Turbulent Reactive Flows, Lecture Notes in Engineering **40**, 672–693 (1989)
3. Barlow, R.S.: Intern'l. Workshop on Measurement and Computation of Turbulent Nonpremixed Flames (2008). Combustion Research Facility, Sandia National Laboratories, Livermore, CA; http://www.ca.sandia.gov/TNF/
4. Beishuizen, N.A., Roekaerts, D.: Numerical simulation of a turbulent spray flame using a hybrid Monte Carlo/finite volume method. In: D. Roekaerts, P. Coelho, B.J. Boersma, K. Claramunt (eds.) Computational Combustion 2007, ECCOMAS Thematic Conference (2007). Delft, The Netherlands, 18–20 July
5. Cabra, R., Myhrvold, T., Chen, J.Y., Dibble, R.W., Karpetis, A.N., Barlow, R.S.: Simulataneous laser Raman-Rayleigh-LIF measurements and numerical modeling results of a lifted turbulent H_2/N_2 jet flame in a vitiated coflow. Proc. Combust. Inst. **29**, 1881–1888 (2002)
6. Cao, R., Pope, S.B.: Numerical integration of stochastic differential equations: Weak second-order mid-point scheme for application in the composition PDF method. J. Comput. Phys. **185**, 194–212 (2003)
7. Cao, R.R., Pope, S.B., Masri, A.R.: Turbulent lifted flames in a vitiated coflow investigated using joint PDF calculations. Combust. Flame **142**, 438–453 (2005)
8. Cleary, M.J., Klimenko, A.Y.: A generalized multiple mapping conditioning approach for turbulent combustion. Flow Turbul. Combust. **82**, 477–491 (2009)
9. Cleary, M.J., Klimenko, A.Y., Janicka, J., Pfitzner, M.: A sparse-Lagrangian multiple mapping conditioning model for turbulent diffusion flames. Proc. Combust. Inst. **32**, 1499–1507 (2009)
10. Colucci, P.J., Jaberi, F.A., Givi, P., Pope, S.B.: Filtered density function for large eddy simulation of turbulent reacting flows. Phys. Fluids **10**, 499–515 (1998)
11. Curl, R.L.: Dispersed phase mixing: I. Theory and effects of simple reactors. AIChE J. **9**, 175–181 (1963)
12. Delarue, B.J., Pope, S.B.: Application of PDF methods to compressible turbulent flows. Phys. Fluids **9**, 2704–2715 (1997)

13. Delarue, B.J., Pope, S.B.: Calculations of subsonic and supersonic turbulent reacting mixing layers using probability density function methods. Phys. Fluids **10**, 487–498 (1998)
14. Dopazo, C., O'Brien, E.E.: An approach to the autoignition of a turbulent mixture. Acta Astronautica **1**, 1239–1266 (1974)
15. Dopazo, C., O'Brien, E.E.: Functional formulation of nonisothermal turbulent reactive flows. Phys. Fluids **17**, 1968–1975 (1974)
16. Dreeben, T.D., Pope, S.B.: Probability density function and Reynolds-stress modeling of near-wall turbulent flows. Phys. Fluids **9**, 154–163 (1997)
17. Dreeben, T.D., Pope, S.B.: Wall-function treatment in PDF methods for turbulent flows. Phys. Fluids **9**, 2692–2703 (1997)
18. Drozda, T.G., Sheikhi, M.R.H., Madnia, C.K., Givi, P.: Developments in formulation and application of the filtered density function. Flow Turbul. Combust. **78**, 35–67 (2007)
19. Fox, R.O.: The Fokker-Planck closure for turbulent molecular mixing: Passive scalars. Phys. Fluids A **4**, 1230–1244 (1992)
20. Fox, R.O.: Improved Fokker-Planck model for the joint scalar, scalar gradient PDF. Phys. Fluids **6**, 334–348 (1994)
21. Fox, R.O.: The Lagrangian spectral relaxation model of the scalar dissipation in homogeneous turbulence. Phys. Fluids **9**, 2364–2386 (1997)
22. Fox, R.O.: Computational Models for Turbulent Reacting Flows. Cambridge University Press, Cambridge, UK (2003)
23. Gao, F., O'Brien, E.E.: A large-eddy simulation scheme for turbulent reacting flows. Phys. Fluids A **5**, 1282–1284 (1993)
24. Ge, H.W., Gutheil, E.: Simulation of a turbulent spray flame using coupled PDF gas phase and spray flamelet modeling. Combust. Flame **153**, 173–185 (2008)
25. Haworth, D.C.: Progress in probability density function methods for turbulent reacting flows. Prog. Energy Combust. Sci. **36**, 168–259 (2010)
26. Haworth, D.C., Drake, M.C., Pope, S.B., Blint, R.J.: The importance of time-dependent flame structures in stretched laminar flamelet models for turbulent jet diffusion flames. Proc. Combust. Inst. **22**, 589–597 (1988)
27. Haworth, D.C., El Tahry, S.H.: A PDF approach for multidimensional turbulent flow calculations with application to in-cylinder flows in reciprocating engines. AIAA J. **29**, 208–218 (1991)
28. Haworth, D.C., Pope, S.B.: A PDF modeling study of self-similar turbulent free shear flows. Phys. Fluids **30**, 1026–1044 (1987)
29. Hsu, A.T., Tsai, Y.L.P., Raju, M.S.: Probability density function approach for compressible turbulent reacting flows. AIAA J. **32**, 1407–1415 (1994)
30. Jaishree, J., Haworth, D.C.: Comparisons of Lagrangian and Eulerian PDF methods in simulations of nonpremixed turbulent jet flames with strong turbulence-chemistry interactions. In: Sixth U.S. National Combustion Meeting. Ann Arbor, MI (17–20 May 2009)
31. James, S., Anand, M.S., Pope, S.B.: The Lagrangian PDF transport method for simulations of gas turbine combustor flows. AIAA Paper no. 2002-4017 (2002) (2002)
32. James, S., Zhu, J., Anand, M.S.: Large eddy simulations of turbulent flames using the filtered density function model. Proc. Combust. Inst. **31**, 1737–1745 (2007)
33. Jones, W.P., Navarro-Martinez, S.: Large eddy simulation of autoignition with a subgrid probability density function method. Combust. Flame **150**, 170–187 (2007)
34. Klimenko, A.Y.: Matching the conditional variance as a criterion for selecting parameters in the simplest multiple mapping conditioning models. Phys. Fluids **16**, 4754–4757 (2004)
35. Klimenko, A.Y.: On simulating scalar transport by mixing between Lagrangian particles. Phys. Fluids **19**, 031,702 (2007)
36. Klimenko, A.Y., Pope, S.B.: The modeling of turbulent reactive flows based on multiple mapping conditioning. Phys. Fluids **15**, 1907–1925 (2003)
37. Kung, E.H., Haworth, D.C.: Transported probability density function (tPDF) modeling for direct-injection internal combustion engines. SAE Int. J. Engines **1**, 591–606 (2008)

38. Li, G., Modest, M.F.: Application of composition PDF methods in the investigation of turbulence-radiation interactions. J. Quant. Spec. Rad. Trans. **73**, 461–472 (2002)
39. Lindstedt, R.P., Louloudi, S.A.: Joint-scalar transported PDF modeling of soot formation and oxidation. Proc. Combust. Inst. **30**, 775–783 (2005)
40. Lindstedt, R.P., Vaos, E.M.: Transported PDF modeling of high-Reynolds-number premixed turbulent flames. Combust. Flame **145**, 495–511 (2006)
41. Lu, L., Pope, S.B.: An improved algorithm for *in situ* adaptive tabulation. J. Comput. Phys. **228**, 361–386 (2009)
42. Mazumder, S., Modest, M.F.: A PDF approach to modeling turbulence-radiation interactions in nonluminous flames. Int. J. Heat Mass Trans. **42**, 971–991 (1999)
43. McDermott, R., Pope, S.B.: A particle formulation for treating differential diffusion in filtered density function methods. J. Comput. Phys. **226**, 947–993 (2007)
44. McDermott, R., Pope, S.B.: The parabolic edge reconstruction method (PERM) for Lagrangian particle advection. J. Comput. Phys. **227**, 5447–5491 (2008)
45. Mehta, R.S., Haworth, D.C., Modest, M.F.: Composition PDF/photon Monte Carlo modeling of moderately sooting turbulent jet flames. Combust. Flame **157**, 982–994 (2010)
46. Mehta, R.S., Modest, M.F., Haworth, D.C.: Radiation characteristics and turbulence-radiation interactions in sooting turbulent jet flames. Combust. Theory Model. **14**, 105–124 (2010)
47. Meyer, D.W., Jenny, P.: A mixing model for turbulent flows based on parameterized scalar profiles. Phys. Fluids **18**, 035,105 (2006)
48. Meyer, D.W., Jenny, P.: A mixing model providing joint statistics of scalar and scalar dissipation rate. Proc. Combust. Inst. **32**, 1613–1620 (2009)
49. Minier, J.P., Pozorski, J.: Wall-boundary conditions in probability density function methods and application to a turbulent channel flow. Phys. Fluids **11**, 2632–2644 (1999)
50. Muradoglu, M., Jenny, P., Pope, S.B., Caughey, D.A.: A consistent hybrid finite volume/particle method for the PDF equations of turbulent reactive flows. J. Comput. Phys. **154**, 342–371 (1999)
51. Muradoglu, M., Pope, S.B., Caughey, D.A.: The hybrid method for the PDF equations of turbulent reactive flows: Consistency conditions and correction algorithms. J. Comput. Phys. **172**, 841–878 (2001)
52. Pitsch, H.: Large-eddy simulation of turbulent combustion. Ann. Rev. Fluid Mech. **38**, 453–482 (2006)
53. Pope, S.B.: The relationship between the probability approach and particle models for reaction in homogeneous turbulence. Combust. Flame **35**, 41–45 (1979)
54. Pope, S.B.: A Monte Carlo method for the PDF equations of turbulent reactive flow. Combust. Sci. Technol. **25**, 159–174 (1981)
55. Pope, S.B.: PDF methods for turbulent reactive flows. Prog. Energy Combust. Sci. **11**, 119–192 (1985)
56. Pope, S.B.: Computations of turbulent combustion: Progress and challenges. Proc. Combust. Inst. **23**, 591–612 (1990)
57. Pope, S.B.: Mapping closures for turbulent mixing and reaction. Theoret. Comput. Fluid Dyn. **2**, 255–270 (1991)
58. Pope, S.B.: Particle method for turbulent flows: Integration of stochastic model equations. J. Comput. Phys. **117**, 332–349 (1995)
59. Pope, S.B.: Computationally efficient implementation of combustion chemistry using *in situ* adaptive tabulation. Combust. Theory Model. **1**, 41–63 (1997)
60. Pope, S.B.: Turbulent Flows. Cambridge University Press, Cambridge, UK (2000)
61. Pope, S.B.: Ten questions concerning the large-eddy simulation of turbulent flows. New J. Phys. **6**, 35 (2004). Available at http://www.iop.org/EJ/njp
62. Pope, S.B.: Self-conditioned fields for large-eddy simulations of turbulent flows. J. Fluid Mech. **652**, 139–169 (2010)
63. Pope, S.B., Anand, M.S.: Flamelet and distributed combustion in premixed turbulent flames. Proc. Combust. Inst. **20**, 403–410 (1984)

64. Raman, V., Pitsch, H.: A consistent LES/filtered-density function formulation for the simulation of turbulent flames with detailed chemistry. Proc. Combust. Inst. **31**, 1711–1719 (2007)
65. Sabel'nikov, V., Soulard, O.: Rapidly decorrelating velocity-field model as a tool for solving one-point Fokker-Planck equations for probability density functions of turbulent reactive scalars. Phys. Rev. E **72**, 016,301 (2005)
66. Sheikhi, M.R.H., Givi, P., Pope, S.B.: Velocity-scalar filtered mass density function for large eddy simulation of turbulent reacting flows. Phys. Fluids **19**, 095,106 (2007)
67. Sheikhi, M.R.H., Givi, P., Pope, S.B.: Frequency-velocity-scalar filtered mass density function for large eddy simulation of turbulent reacting flows. Phys. Fluids **21**, 075,102 (2009)
68. Spielman, L.A., Levenspiel, O.: A Monte Carlo treatment for reacting and coalescing dispersed phase systems. Chem. Eng. Sci. **20**, 247–254 (1965)
69. Subramaniam, S., Pope, S.B.: A mixing model for turbulent reactive flows based on Euclidean minimum spanning trees. Combust. Flame **115**, 487–514 (1998)
70. Tang, Q., Zhao, W., Bockelie, M., Fox, R.O.: Multi-environment probability density function method for modelling turbulent combustion using realistic chemical kinetics. Combust. Theory Model. **11**, 889–907 (2007)
71. Vaithianathan, T., Ulitsky, M., Collins, L.R.: Comparison between a spectral and probability density function model for turbulent reacting flows. Proc. Combust. Inst. **29**, 2139–2146 (2002)
72. Valiño, L.: A field Monte Carlo formulation for calculating the probability density function of a single scalar in a turbulent flow. Flow, Turbul. Combust. **60**, 157–172 (1998)
73. Valiño, L., Dopazo, C.: A binomial Langevin model for turbulent mixing. Phys. Fluids A **3**, 3034–3037 (1991)
74. Van Slooten, P.R., Jayesh, Pope, S.B.: Advances in PDF modeling for inhomogeneous turbulent flows. Phys. Fluids **10**, 246–265 (1998)
75. Villermaux, J., Devillon, J.C.: Représentation de la coalescence et de la redispersion des domaines de ségrégation dans un fluide par un modèle d'interaction phénoménologique. In: Proc. Second Intern'l. Symp. on Chemical Reaction Engineering, pp. 1–13. Elsevier, New York (1972)
76. Vogiatzaki, K., Kronenburg, A., Cleary, M.J., Kent, J.H.: Multiple mapping conditioning of turbulent jet diffusion flames. Proc. Combust. Inst. **32**, 1679–1685 (2009)
77. Waclawczyk, M., Pozorski, J., Minier, J.P.: Probability density function computation of turbulent flows with a new near-wall model. Phys. Fluids **16**, 1410–1422 (2004)
78. Wang, A., Modest, M.F.: Photon Monte Carlo simulation for radiative transfer in gaseous media represented by discrete particle fields. J. Heat Trans. **128**, 1041–1049 (2006)
79. Zhang, Y.Z., Haworth, D.C.: A general mass consistency algorithm for hybrid particle/finite-volume pdf methods. J. Comput. Phys. **194**, 156–193 (2004)
80. Zhang, Y.Z., Kung, E.H., Haworth, D.C.: A PDF method for multidimensional modeling of HCCI engine combustion: Effects of turbulence/chemistry interactions on ignition timing and emissions. Proc. Combust. Inst. **30**, 2763–2771 (2005)

Chapter 7
Multiple Mapping Conditioning: A New Modelling Framework for Turbulent Combustion

M.J. Cleary and A.Y. Klimenko

Abstract Multiple mapping conditioning (MMC) is a relatively new addition to the list of models for turbulent combustion that unifies the features of the probability density function, conditional moment closure and mapping closure models. This chapter presents the major concepts and theory of MMC without the detailed derivations which can be found in the cited literature. While the fundamental basis remains the same, MMC ideas have undergone considerable evolution since they were first proposed and the result is a generalised combustion modelling framework which can more transparently and simply incorporate the major turbulence models which have been developed over the past decades including LES. A significant part of this chapter is devoted to a review of the published MMC applications comparing model predictions with DNS and experimental flame databases. Finally, the chapter concludes with a list of some of the advances in MMC methodology that we can expect to see in the coming years.

7.1 Introduction

Multiple Mapping Conditioning (MMC), introduced by Klimenko and Pope in 2003 [21], is a theoretically rigorous combination of the Probability Density Function (PDF) [17, 40] and Conditional Moment Closure (CMC) [20] models incorporating a generalisation of mapping closure [7, 41]. The mapping closure is generalised in the sense that assumptions are not made about the type of flow being modelled, whereas conventional mapping closures for combustion (e.g. amplitude mapping closure [7]) are formally valid in homogeneous turbulence only. Rather than being a specific turbulent combustion model, MMC can be viewed more as a framework for turbulent combustion modelling. This framework contains a gen-

The University of Queensland, School of Mechanical and Mining Engineering, St Lucia, Queensland, 4072, Australia, e-mail: m.cleary@uq.edu.au; a.klimenko@uq.edu.au

T. Echekki, E. Mastorakos (eds.), *Turbulent Combustion Modeling*, Fluid Mechanics and Its Applications 95, DOI 10.1007/978-94-007-0412-1_7, eral set of principles and equations from which specific MMC based models can be formulated to suit a particular turbulent combustion problem.

Of all the models available, the PDF models (reviewed in Chapter 6) provide the most detailed information about the stochastic characteristics of all species involved in a combustion process, and, most importantly, they permit an exact evaluation of the reaction rates. However, realistic chemical processes involve hundreds of species, n_s, and the differential equations which describe those realistic kinetics are always stiff. Therefore a direct evaluation of the joint composition PDF is expensive as it requires the solution of equations in that n_s-dimensional composition space. While the complete composition space in a turbulent flow is indeed highly dimensional, it is not necessary in a practical model to allow all species to fluctuate in all ways. There are constraints due to conservation of elements and other conservation principles, there are fast reactions of some species forcing them to be close to their partial equilibrium states, and (simply stated) fluctuations of some species are unimportant to the major combustion processes [21]. This concept has lead to the notion of an n_m-dimensional reduced manifold ($n_m < n_s$) onto which the full n_s-dimensional composition space is projected. The dimension of the manifold should be commensurate with the effective dimension of the accessed region in composition space for the flow under consideration and this can be expected to increase with flow complexity [45]. From this manifold notion alternative modelling approaches have evolved. One approach, of which the intrinsic low-dimensional manifold (ILDM) method [32] is the prime example, involves dimension reduction by systematically reducing the number of species in the chemical kinetics scheme. This reduced number of species then defines the low-dimensional manifold to which the eliminated species have a functional relationship. In a numerical implementation the source terms for the manifold species are determined from the reduced kinetics and the eliminated species may be tabulated. While a deficiency of ILDM is that it neglects turbulent mixing effects in obtaining the low-dimensional manifold, related kinetics reduction methods such as the reaction-diffusion manifold method (REDIM) [5] explicitly address chemistry-transport coupling. Ren and Pope [46] review ILDM, REDIM and related methods.

A second modelling approach to evolve from the notion of reduced manifolds is to retain arbitrarily detailed chemical kinetics schemes (i.e. with n_s species) but derive transport equations which effectively restrict the compositions to a certain manifold. The primary example of such models in the recent literature is CMC [20] (reviewed in Chapter 5) which is founded on the hypothesis that, in non-premixed combustion, there is a strong correlation between turbulent fluctuations of reactive scalars and the fluctuations of the mixture fraction. Flamelet models [39] (reviewed in Chapter 3) exploit this correlation also and often include parameterisation by the scalar dissipation effectively creating a two-dimensional manifold. In CMC, the mass fractions of the reactive scalars are conditionally averaged on the mixture fraction leaving an equation which has only a single composition dimension (i.e. mixture fraction space) in addition to the dimensions of time and space. Simple first-order closures can be found for the conditional chemical reaction rates by making the assumption that the conditional fluctuations (i.e. the fluctuations of reactive species concentrations relative to mean concentrations conditioned on the mixture fraction) are negligibly small. As a result of its low dimensionality and the simplicity of the source term closure, the cost of first-order CMC computations is much lower than the cost of PDF computations. In partially- or fully-premixed combustion, or in non-premixed combustion with strong local extinction, the first-order CMC closure of chemical source terms is less accurate and CMC models with higher order reaction rate closures [26, 33] and CMC models with an additional temperature related conditioning variable [27] have proven successful in some of these cases but they also introduce additional terms which are difficult to model.

MMC effectively unifies the PDF and CMC approaches and allows for all of the generality of PDF methods while also exploiting some of the advantages of CMC. Both deterministic and stochastic MMC formulations exist. Deterministic MMC is the natural extension of CMC and the stochastic MMC is in fact a complete joint PDF method with MMC playing the role of a mixing model which enforces localness within a defined manifold. Since the basic MMC framework was first proposed [21] a number of specific MMC models have been developed and tested [8, 10, 11, 29, 54–57]. The key feature common to all is the use of reference variables which are related to the physical quantities in turbulent combustion (e.g. mixture fraction, sensible enthalpy, velocity and scalar dissipation). In the original form of MMC [21] the reference variables are used as conditioning variables which form a manifold constraining the computed compositions. Fluctuations around quantities conditionally averaged on that reference space are considered to be small and in the basic form of MMC they are neglected for the purposes of calculating reaction rates. These fluctuations are specific to MMC and are called minor fluctuations. Later a more generalised interpretation of reference variables in stochastic MMC emerged [23], whereby reference variables can take other roles in addition to conditioning such as emulating scalar dissipation fluctuations. This generalised interpretation allows fluctuations relative to the reference manifold to be exploited (rather than minimised) so as to better model the physical conditional fluctuations while keeping computational cost small.

The remainder of this review chapter is organised as follows. Section 7.2 presents the basic MMC as it was first proposed [21]. It includes a description of the context and concepts of MMC, a brief explanation of mapping closures, a presentation of the deterministic and stochastic model equations, a discussion of the qualitative properties of the model and a brief discussion on the replacement of reference variables resulting in equivalent MMC models with alternatively distributed and physically meaningful reference spaces. Section 7.3 deals with the generalised interpretations of MMC. Here we discuss MMC with conditioning and non-conditioning reference variables and in the context of large eddy simulations. Section 7.4 reviews the published MMC applications for a range of homogeneous and inhomogeneous reacting flows. Finally in Section 7.5 we summarise the major features of MMC and the different forms the model can take and suggest areas of research which we believe could dominate the coming decade of research in the field.

7.2 The Basic MMC Framework

7.2.1 Context and Concepts

The multidimensional reacting scalar space $\boldsymbol{Y}(\boldsymbol{x},t) = (Y_1,\ldots,Y_I,\ldots,Y_{n_s})$ is governed by the well known transport equation

$$\frac{\partial \rho Y_I}{\partial t} + \nabla\cdot(\rho \mathbf{v} Y_I) - \nabla\cdot(\rho D \nabla Y_I) = w_I. \tag{7.1}$$

Here $\mathbf{v} = \mathbf{v}(\boldsymbol{x},t)$ is the fluid velocity, D is the diffusivity which is assumed to be the same for all species, ρ is the density, and w_I is the rate of creation of species I due to chemical reactions. In turbulent flows the stochastic distribution of compositions can be represented by the Favre one-time, one-point joint PDF, $\widetilde{P}_Y(\mathbf{y};\boldsymbol{x},t) = \prod_{I=1}^{n_s} \rho(\mathbf{y})\delta(y_I - Y_I)/\overline{\rho}$ where the lower case $\mathbf{y}$ denotes the sample space for $\boldsymbol{Y}$, and the tilde and overline denote Favre and conventional averages, respectively. In high Reynolds number flows the transport equation for $\widetilde{P}_Y$ derived from (7.1) is given by [40]

$$\frac{\partial \overline{\rho}\widetilde{P}_Y}{\partial t} + \nabla\cdot(\overline{\rho}\mathbf{u}\widetilde{P}_Y) + \frac{\partial \overline{\rho} W_I \widetilde{P}_Y}{\partial y_I} + \frac{\partial^2 \overline{\rho} N_{IJ} \widetilde{P}_Y}{\partial y_I \partial y_J} = 0 \tag{7.2}$$

where

$$\mathbf{u}(\mathbf{y};\boldsymbol{x},t) \equiv \langle \rho \mathbf{v} | \boldsymbol{Y} = \mathbf{y}\rangle / \rho_Y \tag{7.3}$$

$$W_I(\mathbf{y};\boldsymbol{x},t) \equiv \langle \rho w_I | \boldsymbol{Y} = \mathbf{y}\rangle / \rho_Y. \tag{7.4}$$

$$N_{IJ}(\mathbf{y};\boldsymbol{x},t) \equiv \langle \rho D \frac{\partial Y_I}{\partial x_k}\frac{\partial Y_J}{\partial x_l} | \boldsymbol{Y} = \mathbf{y}\rangle / \rho_Y \tag{7.5}$$

$$\rho_Y(\mathbf{y};\boldsymbol{x},t) \equiv \langle \rho | \boldsymbol{Y} = \mathbf{y}\rangle. \tag{7.6}$$

The fundamental assumption of MMC is that the compositions which occur in the different realisations of the flow are confined to an n_m-dimensional manifold within the n_s-dimensional composition space where $n_m < n_s$. The n_m species in this manifold are called "major species" and their turbulent fluctuations are called "major fluctuations" while the term "minor species" refers to the remaining set of $n_\alpha = n_s - n_m$ species. The word "species" is interpreted to include chemical species and also other quantities related to the composition such as mixture fraction and enthalpy. Furthermore, our terminology of major and minor species does not imply that those species are present in large and small concentrations. A major species is denoted by Y_i (lower case Roman subscript) and Y_α (lower case Greek subscript) is used to denote a minor species. The set of all major species is denoted as $\boldsymbol{Y}^m$ and the set of minor species is $\boldsymbol{Y}^{\alpha}$. As will be seen when the transport equations are presented, MMC does not discriminate in its treatment of major and minor species and this demarcation has been introduced as a conceptual aid only. Major species are those which are permitted in the model to fluctuate in any physically realisable way whereas minor species are restricted to fluctuate only jointly with the major species and can be fully characterised by mean values conditionally averaged on the major species. Provided that the major species are properly selected then the joint PDF of all species can be effectively replaced by the marginal PDF of major species, $\widetilde{P}_{Y^m}(\mathbf{y}^m;\boldsymbol{x},t)$, supplemented by the conditional means of the minor species $Q_\alpha(\mathbf{y}^m;\boldsymbol{x},t) = \langle Y_\alpha | \boldsymbol{Y}^m = \mathbf{y}^m \rangle$ such that

$$\widetilde{P}_Y = \widetilde{P}_{Y^m} \cdot \delta(\boldsymbol{Q} - \mathbf{y}^\alpha). \tag{7.7}$$

The reduced PDF of major species and conditional expectations of minor species are governed by

$$\frac{\partial \overline{\rho} \widetilde{P}_{Y^m}}{\partial t} + \nabla \cdot (\overline{\rho} \mathbf{u} \widetilde{P}_{Y^m}) + \frac{\partial \overline{\rho} W_i \widetilde{P}_{Y^m}}{\partial y_i} + \frac{\partial^2 \overline{\rho} N_{ij} \widetilde{P}_{Y^m}}{\partial y_i \partial y_j} = 0 \tag{7.8}$$

and

$$\frac{\partial Q_\alpha}{\partial t} + \mathbf{u} \nabla Q_\alpha + W_i \frac{\partial Q_\alpha}{\partial y_i} - N_{ij} \frac{\partial^2 Q_\alpha}{\partial y_i \partial y_j} = W_\alpha. \tag{7.9}$$

From the assumption that the n_m-dimensional major species manifold effectively describes the accessed region in the n_s-dimensional composition space, terms of the form of $\widetilde{P}_{Y^m}^{-1} \nabla \cdot (\langle \mathbf{u}'' Y_\alpha'' | \boldsymbol{Y}^m = \mathbf{y}^m \rangle \widetilde{P}_{Y^m})$, which would normally appear in CMC, have been omitted from (7.9). The double prime denotes conditional fluctuations.

Like any model for the joint scalar PDF, (7.8) and (7.9) contain two unclosed terms – the conditional velocity, $\mathbf{u}$, and the conditional scalar dissipation, N_{ij}. The development of good closure models, particularly models for N_{ij}, is an ongoing area of research in both the joint PDF and CMC communities [17, 50]. As described in the next section MMC employs an indirect approach based on a generalisation of mapping closure to solve the above equations consistently and in a numerically stable manner.

7.2.2 Mapping Functions

The mapping closure concept was first introduced by Chen et al. [7] and a detailed description of mapping closures for turbulent combustion is provided by Pope [41]. Only the main concepts are repeated here. We introduce the n_r-dimensional set of random variables $\boldsymbol{\xi} = (\xi_1, \ldots, \xi_i, \ldots, \xi_{n_r})$ called the reference space whose distribution is prescribed and represented by the joint reference PDF, $\widetilde{P}_\xi(\boldsymbol{\xi};\boldsymbol{x},t)$. The aim is to find a set of mapping functions $\boldsymbol{X}(\boldsymbol{\xi};\boldsymbol{x},t) = (X_1, \ldots, X_I, \ldots, X_{n_s})$ such that $\boldsymbol{X}$ is statistically equivalent to $\boldsymbol{Y}$. In basic MMC a reference variable is assigned to emulate the turbulence of each of the major species and therefore $n_r = n_m$. The word "emulate" implies that reference variables do not model the physical scalars directly, but that there is statistical equivalence between the mapping functions in that reference space and the physical scalar field. Statistical information about $\boldsymbol{Y}$ is then obtained simply from the modelled mapping functions and the prescribed reference space joint PDF:

$$\widetilde{Y}_I(\boldsymbol{x},t) = \int_{\boldsymbol{\xi}} X_I \widetilde{P}_{\xi} d\boldsymbol{\xi}$$

$$\widetilde{Y_I'2}(\boldsymbol{x},t) = \int_{\boldsymbol{\xi}} (X_I - \widetilde{Y}_I)^2 \widetilde{P}_{\xi} d\boldsymbol{\xi} \tag{7.10}$$

$$\vdots$$

$$\widetilde{P}_Y(\mathbf{y};\boldsymbol{x},t) = \int_{\boldsymbol{\xi}} \delta(\boldsymbol{X} - \mathrm{y}) \widetilde{P}_{\xi} d\boldsymbol{\xi}.$$

The concept of mapping functions may be more readily understood by examining the case of a single inert major scalar $Y_Z(t)$ (the mixture fraction) whose mapping $X_Z(\xi,t)$ is the function of a single standard Gaussian distributed reference variable with mean $\langle\xi\rangle = 0$ and variance $\langle\xi'^2\rangle = 1$. Figure 7.1 shows the time evolution of the mixture fraction PDF, $\widetilde{P}_Z(\eta;t)$ (where η is the sample space variable for Y_Z) and the corresponding mapping function X_Z in a homogeneous field with decaying turbulence [8]. The field is initialised at $t^* = 0$ (t^* is a normalised time) so that fuel ($Y_Z = 1$) and air ($Y_Z = 0$) are mostly segregated with only a small amount of smoothing between the two components. This corresponds to $\widetilde{P}_Z$ being close to two delta functions and X_Z is close to a Heaviside function in reference space with $X_Z = 0$ for $\xi < 0$ and $X_Z = 1$ for $= \xi > 0$. As time evolves $\widetilde{P}_Z$ approaches a Gaussian distribution with a mean of $\widetilde{Y}_Z = 0.5$ and with decaying variance. As there is always a linear mapping between any two Gaussian distributed fields X_Z approaches a straight line with a decreasing gradient.

7.2.3 The Deterministic MMC Model

The MMC model equation governing the evolution of mapping functions in time, and physical and reference spaces is [21]

$$\frac{\partial X_I}{\partial t} + \mathbf{U}\cdot\boldsymbol{\nabla} X_I + A_k \frac{\partial X_I}{\partial \xi_k} - B_{kl}\frac{\partial^2 X_I}{\partial \xi_k \partial \xi_l} = W_I. \tag{7.11}$$

This equation is valid for general inhomogeneous flows. From the assumption that minor fluctuations are negligibly small a first-order closure of the conditional reaction rate is applied such that $W_I = W_I(\boldsymbol{X})$. Equation (7.11) introduces the condi-

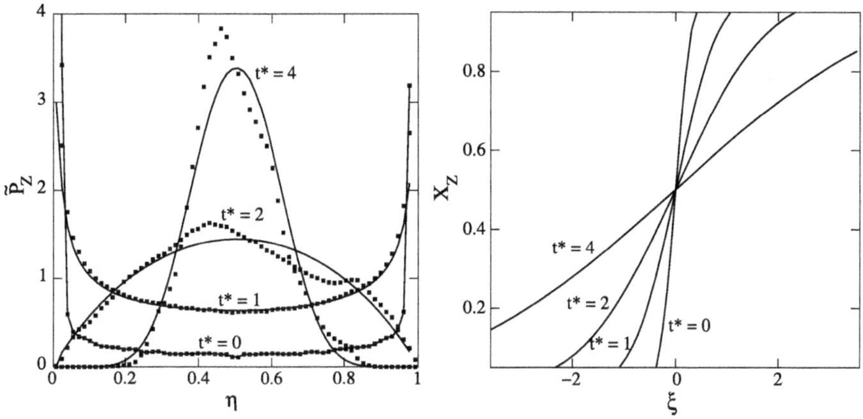

Fig. 7.1: Mixture fraction PDF, $P_Z(\eta)$ (left) and mixture fraction mapping function, $X_Z(\xi)$ (right) in homogeneous decaying turbulence at various normalised times. Symbols are DNS data and lines are MMC model results. Figures adapted from [8].

tional velocity, drift and diffusion coefficients $\mathbf{U}(\boldsymbol{\xi};\boldsymbol{x},t)$, $A_k(\boldsymbol{\xi};\boldsymbol{x},t)$, and $B_{kl}(\boldsymbol{\xi};\boldsymbol{x},t)$, respectively, whose closures are discussed below. We remind readers that the upper case Roman subscript I represents all scalars (both major and minor) while the lower case Roman subscripts k and l are for the major scalars only. An elegant aspect of MMC is that, despite the conceptual division into major and minor species, a single equation governs the transport of all species without discrimination. Mapping functions which satisfy the deterministic model equation (7.11) are themselves deterministic functions and the stochasticity of the modelled scalar field results from the stochasticity of the reference field, $\boldsymbol{\xi}$, whose one-point, one-time joint PDF must satisfy the equation [21]

$$\frac{\partial \overline{\rho}\widetilde{P}_\xi}{\partial t} + \nabla \cdot (\overline{\rho}\mathbf{U}\widetilde{P}_\xi) + \frac{\partial \rho A_k \widetilde{P}_\xi}{\partial \xi_k} + \frac{\partial^2 \rho B_{kl} \widetilde{P}_\xi}{\partial \xi_k \partial \xi_l} = 0. \tag{7.12}$$

There is not space in this review to demonstrate the compliance of the MMC model equations (7.11) and (7.12) with equations (7.8) and (7.9) for $\widetilde{P}_{Y^m}$ and Q_α. However it is explicitly demonstrated by Klimenko and Pope [21]. We note that this compliance is not restrictive on the dimensions involved except the trivial requirement that $n_r \leq n_s$. If $n_r = n_s$ is selected then MMC is a full joint PDF model with a generalised mapping closure for the conditional scalar dissipation. Note that solution of (7.11) for $n_r >> 1$ via finite difference methods is likely to be computationally intractable and the efficient stochastic form of the equations presented in Section 7.2.4 is recommended. MMC reference variables can be selected to emulate fluctuations induced by variations of the mixture fraction and other scalars, dissipation, velocity and, in principle, any other useful quantity [21]. Although a poor selection of reference variables does not render MMC invalid it brings additional complexity without improving the quality of the modelling. From its roots in CMC, practical MMC tends to focus on mixture fraction conditioning and then on conditioning by other quantities. The CMC limit of MMC is achieved when a single reference variable ($n_r = 1$) is chosen to emulate the mixture fraction. Under these conditions MMC is effectively first-order CMC plus a consistent closure of the mixture fraction PDF and conditional mean scalar dissipation.

The MMC velocity, drift and diffusion coefficients are selected so that (7.12) is satisfied for a prescribed (or independently determined) distribution of the joint reference PDF, P_ξ. It is possible to determine the coefficients for any reasonable distribution, however, following the mapping closure convention the reference variables are given standard Gaussian distributions (i.e. zero mean and unit variance):

$$\widetilde{P}_\xi(\boldsymbol{\xi};\boldsymbol{x},t) = \widetilde{P}_\xi(\boldsymbol{\xi}) = G(\xi_1)G(\xi_2)\dots G(\xi_{n_r})$$
$$G(\xi_k) = \frac{1}{\sqrt{2\pi}}\exp\left(-\frac{\xi_k^2}{2}\right). \tag{7.13}$$

For this reference PDF distribution, it can be shown [21] that (7.12) is satisfied when the velocity and drift coefficients are selected to take the following forms:

$$\mathbf{U}(\boldsymbol{\xi};\boldsymbol{x},t) = \mathbf{U}^{(0)} + \mathbf{U}_k^{(1)}\xi_k \tag{7.14}$$

$$A_k(\boldsymbol{\xi};\boldsymbol{x},t) = -\frac{\partial B_{kl}}{\partial \xi_l} + B_{kl}\xi_l + \frac{1}{\overline{\rho}}\nabla\cdot(\overline{\rho}\boldsymbol{U}_k^{(1)}). \tag{7.15}$$

A linear conditional velocity model is commonly used in CMC models [20] and is also suggested for MMC. There is further discussion on this matter below. The last term in (7.15) is particular to MMC and does not appear in conventional mapping closures [41]. Because of that term, MMC as a PDF model is a generalised mapping closure method that makes no assumption about the homogeneity of the flow. In addition, of course, MMC is a CMC model for the minor species.

The velocity terms $\mathbf{U}^{(0)}$ and $\mathbf{U}_k^{(1)}$ depend on the model employed for the drift coefficient, B_{kl}. Selecting B_{kl} to be independent of ξ gives

$$\mathbf{U}^{(0)}(\boldsymbol{x},t) = \widetilde{\mathbf{v}} \tag{7.16}$$

$$\mathbf{U}_k^{(1)}(\boldsymbol{x},t)\langle \xi_k X_i\rangle = \widetilde{\mathbf{v}'Y_i'} \tag{7.17}$$

$$B_{kl}(\boldsymbol{x},t)\left\langle \frac{\partial X_i}{\partial \xi_k}\frac{\partial X_j}{\partial \xi_l}\right\rangle = \widetilde{N}_{ij}. \tag{7.18}$$

In the above the averages in angular brackets are determined by integration weighted by the reference PDF and $\widetilde{N}_{ij}$ is the unconditional Favre-averaged scalar dissipation. Note that these relations are the interactions between the turbulence and reacting scalar fields. The great advantage of employing mapping closure is that the turbulence-chemistry interactions are closed using the unconditional Favre-averaged velocity, turbulent scalar flux and scalar dissipation rather than the more difficult to model conditional averages required to close joint PDF and CMC equations (7.8) and (7.9).

Compliance with the reference PDF does not imply that the MMC coefficients are unique and multiple different forms are possible. The velocity $\boldsymbol{U}$, expressed in (7.14) in terms of $\boldsymbol{\xi}$, $\boldsymbol{x}$ and t, represents a model for conditional velocity $\langle \mathbf{v}|\boldsymbol{Y}\rangle$, which can conventionally be approximated by $\langle \mathbf{v}|Y_Z\rangle$ where Y_Z is the mixture fraction. The simplest, linear approximation of the conditional velocity $\langle \mathbf{v}|Y_Z\rangle = \widetilde{\mathbf{v}} + \beta Y_Z$, where the coefficient β is linked to the turbulent scalar flux as in (7.17), is commonly used in CMC [20]. Kuznetsov and Sabelnikov [30] introduced this approximation and found that the joint Gaussian distribution of the velocity-scalar fluctuations implied by the linear model may be too strong an assumption at the tails of the mixture fraction PDF and may cause convergence difficulties. A clipped version whose shape is an "erf-like" function is more stable and agrees with experiments since larger scalar and velocity fluctuations tend to be less correlated than smaller fluctuations. The same authors [30] note that the linear approximation does not yield the same turbulent diffusivity for the first and the second moment of the mixture fraction ($D_{t2} \neq D_{t1}$). While Kuznetsov and Sabelnikov [30] believed that the different moments should have different turbulent diffusivities, the most common approach in RANS involves assuming the same diffusivities for both moments (i.e. $D_{t2} = D_{t1}$). Mortensen [37] correctly pointed out that using the linear approximation of the conditional velocity in CMC simulations is inconsistent with using the assumption $D_{t2} = D_{t1}$ in the scalar variance equation. This, of course, should not be interpreted as a general inconsistency of the PDF and second moment equations since a consistent equation for any scalar moment is a consequence of, and can be derived from, the PDF equation. The problem of consistency in the common assumption $D_{t2} = D_{t1}$ can be resolved by using Pope's gradient approximation [40] which, as was repeatedly noted [30, 37], yields the same turbulent diffusivities for all moments. This approximation, however, tends to have a shape which is a "tan-like" function and may overestimate velocity-scalar correlations at the tails. On one hand, linear dependence between velocity, whose distribution is close to Gaussian, and the reference variables, which is also Gaussian, can be expected. On the other hand, Gaussian distributions of each of the stochastic variables does not guarantee a joint distribution, which is needed for linear dependence of the conditional expectations. Due to boundedness of the mixture fraction and unboundedness of the Gaussian reference variable, the dependence of $\langle \mathbf{v}|Y_Z\rangle$ on Y_Z determined by (7.14) tends to be "tan"-like. This overestimates dependence at the tails and may cause difficulties in MMC simulations. Recently, Vaishnavi and Kronenburg [53] have suggested a method that can make MMC consistent with any adopted approximation for $\langle \mathbf{v}|Y_Z\rangle$. It seems that this method may become important for inhomogeneous MMC simulations.

7.2.4 The Stochastic MMC Model

Following the methods described by Pope [40] a stochastic form of MMC can be derived which is equivalent to the deterministic model given by Eqs. 7.11 and 7.12. The stochastic formulation is based on the use of Lagrangian particles. In addition to conventional scalar properties, stochastic MMC assigns reference values to those stochastic particles. The equivalent stochastic formulation of MMC is given by [21]

$$d\boldsymbol{x}^{*(p)} = \mathbf{U}(\boldsymbol{\xi}^{*(p)};\boldsymbol{x}^{*(p)},t)dt \tag{7.19}$$

$$d\xi_k^{*(p)} = A_k^0(\boldsymbol{\xi}^{*(p)};\boldsymbol{x}^{*(p)},t)dt + b_{kl}(\boldsymbol{\xi}^{*(p)};\boldsymbol{x}^{*(p)},t)d\omega_l^{*(p)} \tag{7.20}$$

$$dX_I^{*(p)} = (W_I^{*(p)} + S_I^{*(p)})dt \tag{7.21}$$

$$\langle S_I^* | \boldsymbol{\xi}^* = \boldsymbol{\xi}, \boldsymbol{x}^* = \boldsymbol{x} \rangle = 0 \tag{7.22}$$

where

$$A_k^0 = A_k + \frac{2}{\widetilde{P}_\xi}\frac{\partial B_{kl}\widetilde{P}_\xi}{\partial \xi_l} \tag{7.23}$$

$$b_{ki}b_{li} = 2B_{kl}. \tag{7.24}$$

In the above asterisks indicate stochastic quantities, the bracket index (p) indicates a value associated with an individual particle and ω_I^* are Wiener processes. As for the deterministic model, the reference PDF, $\widetilde{P}_\xi$, is prescribed and (7.20) is solved to model the turbulent diffusion of scalars in the reference space.

Equation (7.21) governs transport in scalar space due to chemical reaction, W_I, and a mixing operation, S_I. The latter is not specified beyond the requirement in (7.22) that it not alter the conditional expectations. The treatment of W_I and S_I depends on one's interpretation of stochastic MMC. The first interpretation is that equations (7.19) through (7.22) are an efficient stochastic numerical scheme for solving the deterministic mapping equation (7.11). For $n_r \gg 1$ the stochastic form will be cheaper to compute than a finite difference method applied to the deterministic equations. The goal is to find $\overline{X}_I^* = \langle X_I^* | \boldsymbol{\xi}^* = \boldsymbol{\xi}, \boldsymbol{x}^* = \boldsymbol{x} \rangle$ which can be shown to satisfy (7.11) [21] and we therefore refer to this approach as the conditional interpretation of MMC. We set $W_I^{*(p)} = W_I(\overline{\boldsymbol{X}}^*)$ and the job of the mixing operator is to keep the minor fluctuations $X_I^{\prime\prime\prime(p)} = X_I^{*(p)} - \overline{X}_I^*$ small. Inevitably there will be some scattering around $\overline{X}_I^*$ and this is treated as stochastic error to be minimised by using a large number of particles.

The alternative, probabilistic interpretation of MMC, is to consider the stochastic values $X_I^{*(p)}$ as models for the turbulent realisations of the composition; that is the PDF $\widetilde{P}_X = P(\boldsymbol{X}^*|\boldsymbol{x}^* = \boldsymbol{x})$ is the model for $\tilde{P}_Y$. Practically, stochastic MMC is almost always used as a probabilistic (PDF) model and the recent trend is to imply "probabilistic" when the term "stochastic MMC" is used. In general, the minor fluctuations are still expected to be small when MMC conditioning is effective but deviations from the reference space manifold are now permitted. We set $W_I^{*(p)} = W_I(\boldsymbol{X}^{*(p)})$ and S_I is used to dissipate the usually small but not-negligible minor fluctuations. Note that in conventional joint PDF methods the surrogate mixing models account for the dissipation of all fluctuations, whereas in MMC the mixing operator dissipates only the minor fluctuations and the dissipation of major fluctuations is modelled by diffusion in reference space (see Eq. 7.20). Therefore MMC results are expected to have a lower sensitivity than conventional PDF models to the form of the surrogate mixing model.

In practice, the dissipation of minor fluctuations can occur only if mixing is between particles which are close in ξ-space as demanded by (7.22) and this gives MMC its localness. This localness effectively enforces a CMC-type closure on the mixing model. Here the term CMC is quite general and refers to any method for obtaining conditional means according to (7.9). The probabilistic MMC is a full joint PDF method which, through the mixing model, incorporates the ideas of CMC. Specific surrogate mixing models to dissipate the minor fluctuations can be formulated in a variety of ways but the simplest models are those based on the conventional PDF mixing models such as IEM (interaction by exchange with the mean) [15], IECM (interaction by exchange with the conditional mean) [42] and Curl's model [13] and its modifications [19]. The IECM model can be seen as a special version of MMC-IEM that involves conditioning only on velocity. Traditional MMC pays more attention to conditioning on scalars than to conditioning on velocity and it is not clear whether the true MMC regime can be achieved by IECM [23].

Here we present two alternatives, MMC-IEM and MMC-Curls. MMC-IEM is represented by the mixing operator

$$S_I^{*(p)} = \frac{\overline{X}_I^* - X_I^{*(p)}}{\tau_s}. \tag{7.25}$$

Minor fluctuations are dissipated through adjustments of the relaxation timescale, τ_s, and $\overline{X}_I^*$ is calculated within narrow (i.e. local) ξ bins. This model requires that the number of particles is large so that $\overline{X}_I^*$ can be calculated accurately. In MMC-Curl's model particles p and q are paired on the basis that they are close to each other in ξ-space. During each interaction they have their values reset to the two-particle average $X_I^{*(p)\text{new}} = X_I^{*(q)\text{new}} = (X_I^{*(p)} + X_I^{*(q)})/2$. As particles move randomly in ξ-space new particle pairs are formed as needed to maintain localness in that space. Other two-particle interaction schemes, more sophisticated than the scheme shown above, are possible and practical implementations (e.g. Refs [11, 56] tend to use modified versions of Curl's model [19].

7.2.5 Qualitative Properties of MMC

Being both joint PDF and CMC compliant, MMC inherits the qualitative properties of both. The reaction rates can be modelled by conditional means or by instantaneous stochastic quantities. In the PDF limit convective transport is treated exactly [40] while the convective transport of conditional quantities is modelled by the local properties of the flow (see the ξ dependence of the velocity in (7.14)). An important outcome of MMC is that the PDF of the major scalars and the conditional scalar dissipation are modelled consistently. Lists of desirable properties of conditional scalar dissipation and surrogate mixing models have been suggested and expanded by various authors [16, 44, 51]. MMC adheres to the most essential of these properties [21, 22]: conservation of means, boundedness of scalars and their linear combinations, linearity and independence, localness, equal treatment of all scalars, decay of variances and relaxation to a Gaussian PDF distribution in homogeneous turbulence. Additionally, as MMC in its stochastic form does not specify the form of the surrogate mixing model, the option remains to develop mixing schemes which also include, among other phenomena, the effects of Reynolds number, turbulence length scales and reactions.

A key reason for the observed quality of MMC is the independence of the reference variables and the composition variables ensuring linearity of MMC mixing. This independence does not, of course, imply that $\boldsymbol{\xi}$ and $\boldsymbol{Y}$ are uncorrelated. In fact, correlation is necessary for localisation in reference variables to be a useful constraint. The independence does, however, imply that $\boldsymbol{\xi}$ should be able to fluctuate without taking the local and instantaneous value of $\boldsymbol{Y}$ into account. Practically, a reasonable degree of independence of reference and composition scalar fields is achieved when those fields are modelled by different processes or equations (e.g. $\boldsymbol{\xi}$ can be modelled by the Markov process (7.20) which is independent of the transport of $\boldsymbol{Y}$). Note that this interpretation of independence allows for some quantities, such as density, to be common to both equations.

7.2.6 Replacement of Reference Variables

The velocity, drift and diffusion coefficient closures in Sections 7.2.3 and 7.2.4 are consistent with the reference PDF transport equation when that PDF is a joint standard Gaussian. This is convenient from a mathematical perspective but a better physical understanding of MMC can be gained by replacement of these standard Gaussian reference variables with random variables which more closely resemble the physical major scalars that they emulate.

A reference space transformation from $\boldsymbol{\xi}$ to $\hat{\boldsymbol{\xi}} = \hat{\boldsymbol{\xi}}(\boldsymbol{\xi};\boldsymbol{x},t)$ is achieved by replacing the velocity, drift and diffusion coefficients by [22]

$$\hat{A}_i = \frac{\partial \hat{\xi}_i}{\partial t} + \boldsymbol{U} \cdot \nabla \hat{\xi}_i + A_k \frac{\partial \hat{\xi}_i}{\partial \xi_k} - B_{kl} \frac{\partial^2 \hat{\xi}_i}{\partial \xi_k \partial \xi_l} \tag{7.26}$$

$$\hat{B}_{ij} = B_{kl} \frac{\partial \hat{\xi}_i}{\partial \xi_k} \frac{\partial \hat{\xi}_j}{\partial \xi_l} \tag{7.27}$$

$$\hat{\boldsymbol{U}} = \boldsymbol{U}. \tag{7.28}$$

Although the transformed velocity coefficient in (7.28) is unchanged this does not imply the linear functional form in $\boldsymbol{\xi}$-space corresponds to a linear form in $\hat{\boldsymbol{\xi}}$-space. The new reference space PDF is given by

$$\widetilde{P}_{\hat{\xi}} = \widetilde{P}_{\xi} \det \left(\frac{\partial \hat{\xi}_i}{\partial \xi_k} \right)^{-1}. \tag{7.29}$$

Note that the replacement of variables is simply a mathematically equivalent transformation that does not alter the physical nature of the MMC closures.

An obvious case is the replacement of a single standard Gaussian variable that emulates mixture fraction, ξ, with a new random variable $\hat{\xi} = \eta$ that has the same distribution as the actual mixture fraction (i.e. $\widetilde{P}_\eta = \widetilde{P}_Z$). It is important to remember that the mixture fraction reference variable is not the actual mixture fraction, Y_Z, which is modelled by the mapping X_Z. To preserve the independence of the reference variables, η is a mixture-fraction-like variable with equivalent (or topologically similar) statistics to Y_Z. For this special case the transformed coefficients $\hat{A}_i$ and $\hat{B}_{ij}$ are

$$\hat{A} = 0 \tag{7.30}$$

$$\hat{B} = B \left(\frac{\partial X_Z}{\partial \xi} \right)^2. \tag{7.31}$$

It can be readily seen that replacement of ξ by η and substitution of the new coefficients into the mapping equation (7.11) yields the conditional moment equation (7.9) with conditioning on the mixture fraction. The coefficient $\hat{B}$ appears in the place of, and is therefore a model for, the conditional scalar dissipation

$$\langle N | \eta \rangle = \hat{B} = B \left(\frac{\partial X_Z}{\partial \xi} \right)^2. \tag{7.32}$$

Generally speaking, a higher quality emulation of the mixture fraction by the reference variable η makes modelling of the mixture fraction Y_Z easier.

A detailed application of the replacement of reference variables for a multidimensional reference space emulating mixture fraction and sensible enthalpy is contained in Refs [9, 29]. With such transformations the similarities and differences between the MMC and EMST [51] models are quite obvious. Both models treat conditional scalar dissipation locally in composition space and use mapping closures to achieve this. Where they differ is that EMST uses the stochastic compositions to determine localness, but in so doing violates principles of independence and linearity. MMC, on the other hand, uses reference variables which are formally independent of the stochastic compositions to determine localness and thus it adheres to those principles. However, MMC requires a model for the reference variables and finding a suitable model may not be trivial, especially for reacting quantities.

7.3 Generalised MMC

The basic MMC framework presented in the preceding section is a rather formal model. It assumes that the major species manifold is known and that minor fluctuations are negligibly small. (The probabilistic MMC allows minor fluctuations but until now they have been assumed to be small). The use of standard Gaussian reference variables is conventional and mathematically convenient, but it also removes some physical transparency from the model equations. In this section we present a generalised MMC which, as the name suggests, generalises the concepts of MMC and makes them more amenable to practical implementation. Generalised MMC concepts were first proposed in Ref. [23] to expand the purpose of reference variables beyond conditioning or localisation. A series of subsequent papers [10, 11, 24, 25] developed generalised MMC for the DNS/LES regime and replaced Markov reference variables with Lagrangian variables traced within an Eulerian field. Although the main generalised MMC concepts are presented below, readers interested in a detailed analysis should consult the published articles cited above.

7.3.1 Reference Variables in Generalised MMC

The basic MMC model uses $n_r = n_m$ independent reference variables to emulate each of the major species. In the stochastic form of the model mixing is localised in the reference space effectively linking the modelled composition with the species concentrations conditionally averaged on that space. In the conditional interpretation of MMC minor fluctuations are neglected and therefore the composition is modelled as the conditional mean. In probabilistic MMC, in which minor fluctuations are permitted, the fluctuations are dissipated towards the conditional means by the minor dissipation operator, S_I. The probabilistic MMC interpretation is assumed for the remainder of this section. Reference variables which perform a localisation role (this is the only role we have considered until now) are now called conditioning reference variables to distinguish them from other sorts of reference variables to be discussed below. As before the total number of reference variables is labelled as n_r

and the number of conditioning reference variables is n_c with $n_c \leq n_r$. The set of conditioning variables $\boldsymbol{\eta} = (\eta_1, \eta_2, \ldots, \eta_{n_c})$ forms a subset of $\boldsymbol{\xi}$.

From a practical perspective it may not always be possible or desirable to have a conditioning reference variable to emulate each of the major species. Limiting computational cost is the major reason for this – a greater number of conditioning reference variables requires a larger number of particles to ensure adequate localisation in the space of each of those reference variables. If $n_c < n_m$ then, in general, minor fluctuations are not negligibly small and their variances should be controlled to so that the model predicts the physical conditional variances accurately.

By accepting minor fluctuations in the model we also create the possibility of including reference variables which assist the modelling but which are not used for conditioning purposes. Practically this means that mixing is localised only in the space of the n_c conditioning reference variables, while the non-conditioning reference variables complement the conditioning reference variables and improve the emulation of the physical quantities.

Until now we have only considered reference variables modelled by Markov processes as in (7.20). However, once we allow for the possibility that $n_c < n_r$ or even for $n_c \ll n_r$, any physically relevant process can be used. For example, reference variables can be Lagrangian quantities obtained with the use of DNS or LES. Indeed a non-Markov process can be approximated well by a Markov process of much higher dimension. Motions of Brownian or fluid particles in a turbulent flow are deemed to be non-Markovian while DNS simulations tracing these particles represent a Markov process of a large dimension.

7.3.2 Features of Generalised MMC Models

The generalised MMC model equations were initially proposed in Ref. [23] in the same form as the basic stochastic MMC equations (7.19) through (7.21) and with (7.22) replaced by

$$\langle S_I^* | \boldsymbol{\eta}^* = \boldsymbol{\eta}, \boldsymbol{x}^* = \boldsymbol{x} \rangle = 0. \tag{7.33}$$

while $\langle S_I^* | \boldsymbol{\xi}^* = \boldsymbol{\xi}, \boldsymbol{x}^* = \boldsymbol{x} \rangle$ may be non-zero. It is possible to demonstrate that a failure to satisfy condition (7.22) under the conditional interpretation of MMC will generate a spurious term in the modelled PDF transport equation (the effect of mixing in the direction of non-conditioning reference variables can be interpreted as a mixing-generated diffusion). Hence, using a conditional interpretation of generalized MMC is not recommended. Note that generalized MMC is a stochastic model and it does not generally have a deterministic version. However, with a probabilistic interpretation, generalised MMC remains compliant with the PDF transport equation. Indeed, complying with the PDF equation requires that $\langle S_I^* | \boldsymbol{x}^* = \boldsymbol{x} \rangle = 0$ to allow for representation of the mixing operator in terms of the divergence of dissipation, N_{IJ}°:

$$\widetilde{P}_X \langle S_I^* | \boldsymbol{X}^* = \boldsymbol{X}, \boldsymbol{x}^* = \boldsymbol{x} \rangle = \frac{\partial N_{IJ}^{\circ} \widetilde{P}_X}{\partial X_J}. \tag{7.34}$$

In Ref. [21], this was shown using with the use of (7.22), which represents a sufficient but not necessary condition. A weaker condition, $\langle S_I^* | \boldsymbol{x}^* = \boldsymbol{x} \rangle = 0$, which can be obtained from (7.33), is sufficient for compliance with the PDF equation. Due condition (7.33) generalised MMC mixing does not alter the values of $\overline{X}_I = \langle X_I^* | \boldsymbol{\eta}^* = \boldsymbol{\eta}, \boldsymbol{x}^* = \boldsymbol{x} \rangle$ and hence, in the absence of non-linear reacting terms, $\overline{X}_I^*$ is determined by the properties of the stochastic trajectories of $\boldsymbol{\eta}^*$ and not by the form or quality of the surrogate mixing operator. In other words, MMC enforces the desired conditional properties, through the stochastic properties of the reference variables, on any reasonable surrogate mixing operator. This, however, does not apply to higher conditional moments which are determined by the form and quality of that mixing.

Since stochastic models aim to produce statistically equivalent fields their model equations are not unique and alternative forms can be derived. MMC with Gaussian reference variables is mathematically convenient but some physical transparency of the model is lost. It is, of course, possible to transform the equations for alternatively distributed reference variables according the methods described in Section 7.2.6 and although this improves the physical transparency of the model the transformed drift and diffusion coefficients are complex. An alternative option is to apply the generalised MMC principles within other existing models for turbulent combustion (e.g. any of the various formulations and closures of the joint PDF models). In the broadest sense, then, generalised MMC can be interpreted as the application of the conditioning/localisation condition (7.33) within an existing (or maybe yet to be developed) stochastic combustion model. This interpretation has been taken in practical hybrid binomial Langevin-MMC [57] and MMC-LES applications [10, 11].

The following three points summarise the essential features of a good generalised MMC model [23]:

- The conditioning reference variables should emulate as closely as possible the Lagrangian properties of the key major species to ensure accurate evaluation of conditional species expectations without compromising the independence of the reference space. This can be done with the assistance of non-conditioning reference variables.
- The surrogate mixing operator, S_I, should set the dissipation of minor fluctuations to correspond to the dissipation of physical conditional fluctuations. (Due to the independence of reference and composition scalar fields, minor fluctuations and conditional fluctuations are not the same thing but they are linked).
- The conditioning reference variables should be selected so that minor fluctuations are not too large. This ensures that scalar dissipation is predominantly modelled by diffusion in reference space (e.g. Eq. 7.20) rather than by the surrogate mixing operator, S_I.

7.3.3 MMC with Dissipation-like Reference Variables

We consider an MMC model governed by (7.19) through (7.21) and (7.33) with a single conditioning reference variable η emulating the physical mixture fraction, Y_Z, via the mapping function X_Z. If η does not have a standard Gaussian PDF then the coefficients are modified as described in Section 7.2.6. Conditional fluctuations $Y_I'' = Y_I - \langle Y_I'' | Y_Z \rangle$ are modelled indirectly via the minor fluctuations $X_I'''^{(p)} = X_I^{*(p)} - \overline{X}_I^*$. As it is, the model does not explicitly generate minor fluctuations and they are present only if they appear in the boundary conditions or if generated by the surrogate mixing model [23]. Although generation of minor fluctuations by the mixing model can in principle be used to model the conditional fluctuations [10] it may be difficult to control. An alternative model is to introduce additional non-conditioning reference variables to emulate the scalar dissipation fluctuations which are physically responsible for the appearance of conditional fluctuations. The reference space is defined as $\boldsymbol{\xi} = (\eta, \xi_{d_1}, \xi_{d_2}, \ldots)$ where ξ_{d_i} are called dissipation-like reference variables. The MMC model with dissipation-like variables has the modified diffusion coefficients [23]

$$B_{\eta\eta} = \overline{B}_{\eta\eta}\Phi, \quad B_{\eta d_i} = B_{d_i \eta} = 0, \quad B_{d_i d_j} = \frac{\delta_{d_i d_j}}{\tau_{d_i}} \tag{7.35}$$

$$\Phi(\xi_{d_i}; \boldsymbol{x}, t) = \exp(c_{d_i}\xi_{d_i} - \frac{c_{d_i}c_{d_i}}{2}) \tag{7.36}$$

$$c_{d_i}c_{d_i} = \ln\left(\frac{\langle N_{\eta\eta}'^2 | Y_Z\rangle}{\langle N_{\eta\eta} | Y_Z\rangle} + 1\right) \tag{7.37}$$

In the above, $\overline{B}_{\eta\eta}$ is the value of the diffusion coefficient without inclusion of the dissipation fluctuations. Each dissipation-like reference variable emulates scalar dissipation fluctuations of a certain frequency, $1/\tau_{d_i}$, where τ_{d_i} spans between the Kolmogorov and macro time scales of the flow. Giving each ξ_{d_i} a standard Gaussian distribution ensures that the conditional scalar dissipation $\langle N_{\eta\eta} | \eta \rangle$ has a lognormal distribution. For modelling where the ratio $\tau_{d_{i-1}}/\tau_{d_i}$ is selected to be the same for all dissipation-like variables it can be shown that the constants c_{d_i} are also equal [23]. In basic MMC, all reference variables $\boldsymbol{\xi}$ must be used for conditioning while generalised MMC may involve conditioning only on η, or η and Φ, which seems to be more practical.

7.3.4 DNS/LES Simulated Reference Variables

The computational tractability of Markov processes such as the random walk given by (7.20) have led to them being widely used in stochastic turbulence models [43]. However, the great advances in computing power mean that LES and maybe even DNS are becoming more viable means of modelling non-reacting stochastic diffusion processes and velocity fields. The cost of performing reacting DNS is still prohibitive while LES does not resolve the thin reaction zones. Therefore hybrid methods such as the LES/joint scalar FDF (filtered density function) model [12, 18] have been developed whereby velocity and passive scalar fields are simulated by conventional Eulerian LES and the reactive scalar field is simulated by a stochastic particle scheme.

As noted in Section 7.3.1, the Markov reference variables can be replaced in generalised MMC by traced Lagrangian values within an Eulerian DNS or LES simulated field (i.e. particle reference variable values are the Eulerian values observed at the particle locations). One can note that the highest quality reference variable is the actual physical variable simulated by a fully resolved DNS. The Eulerian reference field is simulated according to

$$\frac{\partial \rho \xi_i}{\partial t} + \nabla \cdot (\rho \mathbf{v} \xi_i) - \nabla \cdot (\rho D \nabla \xi_i) = w_{\xi_i}. \tag{7.38}$$

For passive scalars the source term $w_{\xi_i} = 0$. The mixture fraction is the most obvious reference variable to be modelled in this way for non-premixed combustion but, in principle, other passive or reactive reference variables could also be selected. In the LES version, the filtered form of (7.38) is solved and a closure is required for the source term of any reactive reference variables. If LES subgrid fluctuations are filtered out a Markov process similar to (7.20) can be used to emulate the subgrid distribution such that $\xi_i = \xi_i^{\mathrm{LES}} + \xi_i'^{\mathrm{RW}}$; (RW = random walk) [23]. Practical applications [10, 11] of MMC in LES tend to have far fewer Lagrangian particles for the stochastic reacting species field than there are Eulerian LES grid cells (see Section 7.4.4). Therefore the explicit inclusion of subgrid fluctuations in the formulation of the reference variables is unlikely to have a significant effect on the determination of localness in reference space. Of course the subgrid component of ξ_i may have a significant effect on conditional velocity (according to the linear closure given by (7.14)) or if some of the reference variables represent velocity components. If the random walk component of ξ_i is neglected (as has been done in practical applications) the subgrid conditional velocity can instead be closed by the alternative gradient model [12, 40] which manifests in the stochastic equation for the spatial transport:

$$d\boldsymbol{x}^{*(p)} = \left(\mathbf{U}^{(0)} + \frac{1}{\overline{\rho}} \nabla(\overline{\rho} D_{\mathrm{eff}})\right) dt + \sqrt{2D_{\mathrm{eff}}}\, d\boldsymbol{\omega}^{*(p)} \tag{7.39}$$

where D_{eff} is the sum of subgrid and (if needed) molecular diffusivities.

7.4 Examples

In the past five years an impressive list of publications have proposed and tested specific deterministic and stochastic MMC (including generalised MMC) models in a range of idealised, homogeneous combustion conditions [8, 9, 29, 56] and inhomogeneous, laboratory non-premixed and partially premixed flame conditions [10, 11, 54, 55, 57]. Each of these specific MMC models has a reference variable to emulate mixture fraction while a few of the deterministic models have additional reference variables to emulate scalar dissipation and/or sensible enthalpy [8, 9, 29]. In the inhomogeneous cases MMC has been coupled with RANS based turbulence models [54, 55], with the binomial Langevin model to model the joint velocity-scalar PDF [57] and LES to simulate the joint scalar FDF [10, 11]. The key features of these applications are summarised below.

7.4.1 MMC in Homogeneous Turbulence

7.4.1.1 Stochastic MMC

The first application of MMC to reacting flow conditions was the stochastic MMC reported by Wandel and Klimenko [56]. Results are compared against DNS data [35] in homogeneous turbulence with finite-rate, one-step chemistry and significant local extinction. A single reference variable emulates the mixture fraction and, as the probabilistic MMC interpretation is used, minor fluctuations of the single reactive scalar, normalised temperature, are present. An MMC-Curl's surrogate mixing model dissipates the minor fluctuations and the mixing timescale, τ_{min}, is set proportional to the macro-mixing timescale, denoted by τ_{maj}. Despite the simplicity of the flow and chemistry this modelling demonstrates the ability of stochastic MMC to capture heavy local extinction and subsequent reignition events more accurately than other commonly used models such as CMC, fast chemistry, Curl's, IEM and EMST. The performance of IEM, Curl's and EMST models is investigated in detail in Mitarai et al. [36] for the same test conditions. It is instructive to compare the scatter plots of temperature versus mixture fraction for those models and DNS in Ref. [36] with the MMC scatter plots in Ref. [56]. IEM fails to produce the correct physical behaviour as it cannot change the shape of the joint PDF from its initial conditions and nor can it generate conditional fluctuations of temperature with respect to mixture fraction. Curl's model produces physically plausible compositions for this case but as mixing is not local in composition space it significantly overpredicts conditional fluctuations and the reignition is very slow compared to the DNS. While EMST is local in composition space it violates the principles of independence and linearity and in its basic form EMST can lead to "stranding" [51] or mixing along certain preferential lines leading to non-physical behaviour. For the test case described MMC produces physically realistic and accurate results.

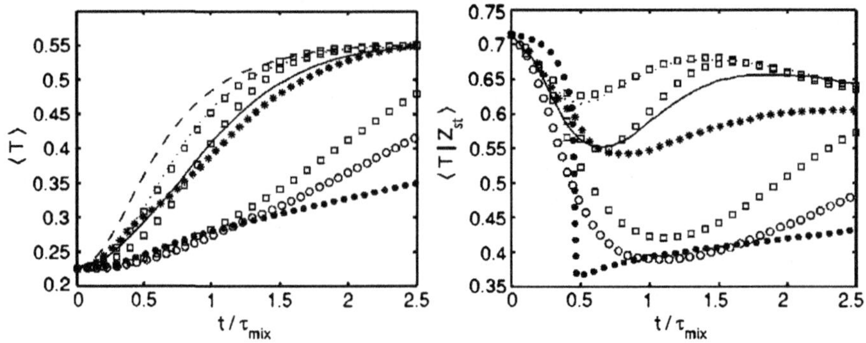

Fig. 7.2: Mean (left) and conditional mean at stoichiometry (right) of temperature. MMC (lowest to highest: $\tau_{\min}/\tau_{\mathrm{maj}} = 1/1.04$, $1/8$ and $1/100$), □; DNS data, –; EMST, ∗; CMC, ⋯; Curl's model, ○; IEM •; fast chemistry, – – –. Figures adapted from [56].

Figure 7.2 below shows the mean temperature (left) and the conditional mean temperature at stoichiometry (right) as a functions of time for the DNS and the various models. It can be seen that both EMST and MMC with $\tau_{\min}/\tau_{\mathrm{maj}} = 1/8$ predict mean temperature very well while the other models either significantly overpredict the rate of temperature rise (CMC and fast-chemistry) or significantly under predict it (IEM and Curl's). For the conditional temperature MMC with $\tau_{\min}/\tau_{\mathrm{maj}} = 1/8$ is the most accurate model. The MMC results are, however, qualitatively and quantitatively sensitive to the parameter $\tau_{\min}/\tau_{\mathrm{maj}}$. By setting $\tau_{\min}/\tau_{\mathrm{maj}} = 1/100$ the model rapidly dissipates any minor fluctuations, and hence conditional fluctuations, of temperature which may be generated. Thus this MMC result closely resembles those for first-order CMC. Alternatively by setting $\tau_{\min}/\tau_{\mathrm{maj}} = 1/1.05$ the model does not dissipate minor fluctuations, and hence conditional fluctuations, fast enough and produces results similar to Curl's model. Although more research is required to determine the best values $\tau_{\min}/\tau_{\mathrm{maj}}$ for a range of practical combustion conditions, the timescale ratio parameter appears to provide a useful mechanism for controlling the level of conditional fluctuations which is not available in many other mixing models. The authors [56] caution that the correct value of $\tau_{\min}/\tau_{\mathrm{maj}}$ is unlikely to be universal or constant with time. The variability with time is illustrated in the results for $\tau_{\min}/\tau_{\mathrm{maj}} = 1/8$ which slightly underpredicts the rate of reignition (see conditional temperature rise in Fig. 7.2). Attempts to reduce that timescale ratio (for all time steps) in order to more rapidly dissipate conditional fluctuations in the later stages of the evolution inadvertently leads to inaccuracy during the initial extinction phase. The creation of a model for $\tau_{\min}/\tau_{\mathrm{maj}}$ would be advantageous but is not trivial. Any such model would need to account for the rate of dissipation and generation of conditional variance by the surrogate mixing model as analysed in Ref. [23].

7.4.1.2 Deterministic MMC

The first application of deterministic MMC is, in fact, contained alongside the original MMC derivation [21]. A three-stream, non-reactive, homogeneous mixing field is modelled with the use of two reference variables emulating two independent mixture fractions. It is shown that the joint PDF of the two mixture fractions is modelled in a very realistic manner and results are in excellent agreement with the analytical solution.

The first applications of deterministic MMC for reactive fields are found in a series of three papers by Cleary and Kronenburg [8, 9, 29]. They propose and test various deterministic MMC models against DNS [28] of homogeneous, decaying turbulence with varying levels of local extinction (up to global extinction). It had previously been established [28] that CMC with conditioning on the mixture fraction alone was inappropriate for these flame conditions due to the importance of conditional fluctuations which are neglected in first-order CMC. A number of previous CMC studies [6, 27, 28] identified scalar dissipation and normalised sensible enthalpy as possible choices for a second conditioning variable for flames with significant local extinction. In fact both quantities have an important role in the physics of local extinction and subsequent reignition. While fluctuating scalar dissipation is the primary instability which causes conditional fluctuations, those conditional fluctuations tend to correlate better with temperature related quantities such as sensible enthalpy, than with scalar dissipation.

The three MMC papers progress incrementally. The first [8] has reference variables emulating mixture fraction and multiple scalar dissipation-like quantities each of which is associated with a certain dissipation timescale (see (7.35) through (7.37)). As expected from the earlier CMC results [6], while conditioning on mixture fraction and a single scalar dissipation variable is able to model the extinction phase well, it cannot accurately predict the subsequent reignition phase which occurs after the turbulence has sufficiently decayed. Any deterministic MMC inevitably forces reference variables to be conditioning variables but at low temperatures there is decorrelation of reactive species and scalar dissipation fluctuations [27] and hence the assumption of negligible conditional/minor fluctuations and first-order reaction rate closures are inappropriate. Although the MMC results improve modestly with additional dissipation-like reference variables the model is illustrated to be unsuitable for a deterministic formulation. Note that dissipation-like reference variables were initially suggested as non-conditioning reference variables in the stochastic formulation of MMC (see Section 7.3).

The second paper [9] proposes an MMC model with reference variables emulating mixture fraction and normalised sensible enthalpy. Like previous CMC calculations with the same conditioning variables [28] results for reactive species are impressive. This is because the manifold comprising of only mixture fraction and sensible enthalpy adequately describes the accessed region in composition space. However, the model does not have a mechanism for introducing the physical instabilities (i.e. fluctuations in scalar dissipation) that cause extinctions to occur in the first place. To overcome this deficiency the fluctuations are imposed on the model using DNS data for the conditional PDF of normalised sensible enthalpy. The result is a hybrid MMC / presumed PDF model which accurately describes the evolution of minor scalars but which does not predict the joint PDF of the major scalars.

The third paper [29] describes an MMC model which is a novel combination of the previous two. There is a reference variable emulating mixture fraction and a second reference variable which emulates normalised sensible enthalpy but that is also a dissipation-like variable which can generate the fluctuations leading to local extinctions. It is explained that any reference variable may adopt the character of a dissipation-like variable and that the physical quantity it emulates is irrelevant. Through the dual-nature of the second reference variable the model exploits the strong negative correlation between sensible enthalpy fluctuations and fluctuations in scalar dissipation during the extinction process. Specifically (7.37) is replaced by

$$c_{d_1} = f_{\mathrm{corr}} \ln \left(\frac{\langle N_{\eta\eta}'^2 | \eta \rangle}{\langle N_{\eta\eta} | \eta \rangle} + 1 \right)^{1/2} \tag{7.40}$$

where the correlation function is simply the conditional normalised sensible enthalpy at stoichiometry $f_{\mathrm{corr}} = -\langle \hat{h}_s | \eta = Y_{Z\mathrm{st}} \rangle$. While this third model is complete and follows the physics of the problem better than the previous two models on which it is based, the quality of the results is mixed. For the flame case with heavy local extinction followed by reignition and another case with global extinction, predictions of major and minor species are in very good agreement with DNS data. A particularly impressive outcome is the model's ability to accurately predict the bimodal distribution of sensible enthalpy in near stoichiometric mixtures as shown in Fig. 7.3. This is in contrast to an assumed β-PDF which cannot give a bimodal distribution between arbitrary minimum and maximum sample space limits. Interestingly, for a third test case exhibiting only moderate local extinction the model performs poorly and noticeably underpredicts the extent of that mild extinction. This is blamed on a realizability constraint which artificially restricts some of the diffusion coefficients to positive values to ensure numerical stability.

7.4.2 MMC with RANS

Two papers by Vogiatzaki et al. document the implementation of deterministic MMC into a RANS computer code and report on model performance for two laboratory jet diffusion flames with complex hydrocarbon chemistry. The first paper [54] reports on modelling of the DLR A and B $CH_4/H_2/N_2$ flames [2, 34] and the second paper [55] presents results and an expanded analysis for the Sandia $CH_4/O_2/N_2$ Flame D [1, 49]. In each case a single Gaussian reference variable emulates the mixture fraction. As the flame cases exhibit low levels of local extinction, conditioning on mixture fraction alone is appropriate as is established by many past accurate CMC and flamelet computations. Figure 7.4 (from

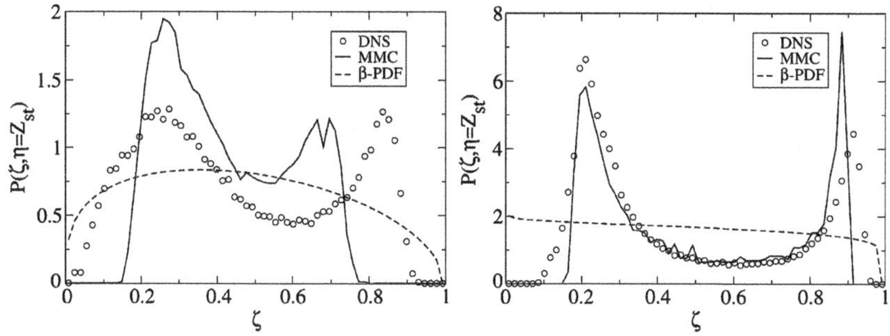

Fig. 7.3: Conditional PDF of normalised sensible enthalpy at stoichiometry at two different times. Figures adapted from [29].

Ref. [55]) shows that MMC predictions of the mixture fraction PDF for Sandia Flame D closely resemble a β-PDF. Close to the nozzle agreement with experimental data is excellent while the downstream discrepancies are linked to the commonly observed underprediction of the mixture fraction variance by the $k-\varepsilon$ turbulence closure. Conditional scalar dissipation does not appear explicitly in MMC but can be determined by (7.32) following a replacement of reference variables from ξ to η. Figure 7.5 shows conditional scalar dissipation profiles for Sandia Flame D conditions by MMC and two alternative closures based on integration of the mixture fraction PDF transport equation [14] and amplitude mapping closure (AMC) [7]. MMC reproduces the profile shapes and the location of the peak value better than the integrated PDF method and is an improvement over AMC which always gives the peak conditional scalar dissipation at mixture fraction equal to 0.5. MMC compares quite well to the 1D experimental data but is unable to capture the slightly bimodal shape which is even more apparent in the more accurate 3D experimental data. Although the obvious qualities of MMC do not make a significant difference to reactive scalar predictions in these simple flames (results are of similar good accuracy to those for CMC with standard PDF and conditional scalar dissipation closures) the MMC computations represent an important first step prior to application to more difficult flame cases which require additional conditioning variables and for which simple PDF shape presumptions and conditional scalar dissipation closures are not available.

7.4.3 MMC with the Binomial Langevin Model

A novel, hybrid model combining the binomial Langevin model [52] and stochastic MMC was introduced by Wandel and Lindstedt [57] to model the joint velocity-

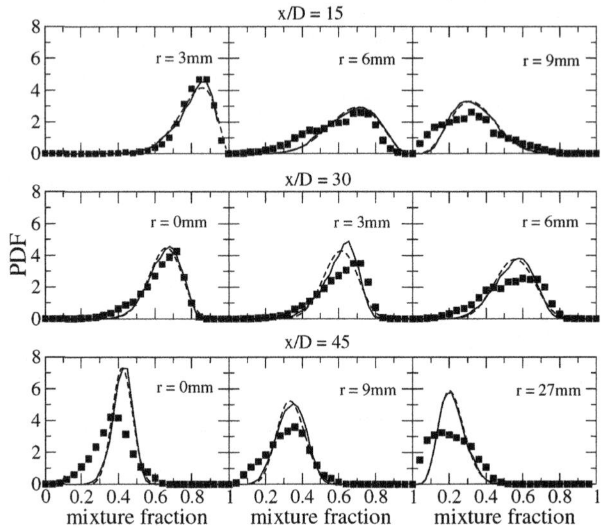

Fig. 7.4: Mixture fraction PDF at various locations for Sandia Flame D. Squares are experimental data, solid lines are MMC predictions, and dashed lines are β-PDFs. Figure adapted from [55].

scalar statistics in an inhomogeneous, reacting scalar mixing layer that was investigated experimentally by Saetran et al. [48] and Bilger and co-workers [3, 31]. The hybrid approach overcomes implementation difficulties associated with producing a bounded scalar field in the binomial Langevin context, while providing a simple and accurate means for obtaining the MMC coefficients (calculation of terms involving the gradient $\partial X/\partial\xi$ can be difficult in stochastic MMC when there is a lot of scatter). The hybrid model employs the principles of a generalised MMC closure. In the MMC part of the hybrid model, rather than solving (7.20), the single conditioning reference variable is instead modelled by inverting (7.14) and some other manipulation to give

$$\eta^{*(p)} = \frac{u_2^{*(p)} - \widetilde{u}_2}{\sqrt{\widetilde{u_2'^2}}}. \tag{7.41}$$

Here u_2 is the dominant velocity component (in this case the transverse component) and it is modelled by the binomial Langevin model. Note that (7.41) has the advantage that it does not contain the diffusion coefficient B and thus it is not necessary to calculate the gradients $\partial X/\partial\xi$. The model also has an additional *pseudo* mixture fraction that is solved according to the binomial Langevin model. The pseudo mixture fraction is a non-conditioning reference variable used only to calculate the extent of mixing between particle pairs while mixing localisation is in η-space

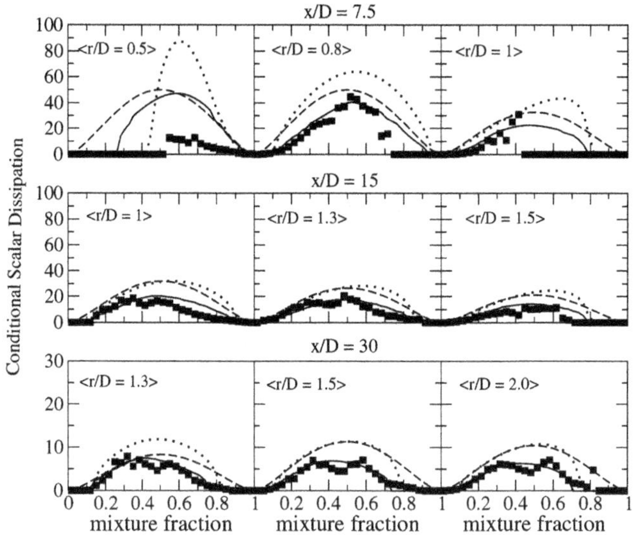

Fig. 7.5: Profiles of local conditional scalar dissipation in mixture fraction space for Sandia Flame D. Squares are 1D experimental data, solid lines are the MMC model, dotted lines are the integrated PDF model, and dashed lines are the AMC model. Figure adapted from [55].

only. (We caution that the notation used in this chapter for conditioning and non-conditioning reference variables is different to that used by Wandel and Lindstedt.)

The paper contains a detailed analysis of the model and makes extensive comparisons with experimental data. These indicate that the model is robust and provides a similar level of accuracy to the binomial Langevin model by itself. While results for mean quantities are in very good agreement with experimental data, the second moments are generally underpredicted signifying a need for future improvements such as better control of the dissipation of minor fluctuations by the surrogate mixing model. Due to the aforementioned elimination of implementation difficulties associated with the binomial Langevin and stochastic MMC models the hybrid model is reported to have a relatively modest computational cost. Application of this model to inhomogeneous flows is underway and preliminary results are encouraging.

7.4.4 MMC with LES

Two recent papers have documented the application of MMC with LES [10, 11] for the Sandia $CH_4/O_2/N_2$ Flame D [1, 49]. The model is a generalised MMC with the reference variable given by the LES filtered mixture fraction. The most compelling aspect is the demonstration of a new very low-cost, sparse-Lagrangian scheme for simulation of the joint scalar FDF, made possible due to the high quality of the MMC mixing closure. Conventional FDF simulations employ an "intensive-Lagrangian" particle scheme with many particles per Eulerian LES grid cell, and the terminology of "sparse-Lagrangian" is introduced to refer to simulations with significantly fewer Lagrangian particles for the joint scalar FDF than there are Eulerian grid cells. The simulations of Flame D are performed for (on average) one particle for every 10 to 12 LES cells culminating in as few as 35,000 particles over the 70 jet-nozzle diameter flow domain. This represents a two or three order of magnitude reduction in particle numbers and computational cost relative to conventional intensive-Lagrangian FDF simulations of the same or similar flame conditions. As a result of the very low cost the sparse-Lagrangian simulations are able to use detailed 219-step chemistry, whereas previous FDF simulations of hydrocarbon flames have required reduced or tabulated chemistry. (A list of recent FDF computations is compiled by Haworth [17].)

The theoretical basis for sparse-Lagrangian simulations is established in two papers [24, 25] while Ref. [10] elaborates on the physical reasoning in support of sparse methods and the reasons for the success of generalised MMC under such conditions. Modelling aside, if it is assumed that a particle within the ensemble representing the one-point, one-time FDF is statistically independent of all other particles, then all those other particles can be removed while the one remaining particle continues to represent that FDF. That probability distribution exists whether we have sufficient numbers of particles to determine it locally or not. From a modelling perspective the problem of using very few particles is that numerical diffusion (i.e. numerical bias in the Lagrangian fields) can become larger than is justified or needed if that diffusion performs a modelling role. Therefore modelling the FDF with very few particles requires a high quality mixing model which specifically minimises numerical diffusion. MMC does this by enforcing localness in the reference space. Whereas non-local mixing models such as IEM are successful in intensive-Lagrangian FDF simulations by virtue of the high spatial resolution of LES, it seems that only models (such as MMC), which enforce localness and adhere to the other desirable qualities of mixing models, are capable when particle numbers are significantly reduced. It should be stressed that the finest details of the fields are not available with so few particles but the published works [10, 11] demonstrate that the major stationary statistics of the reacting scalar fields (i.e. conditional and unconditional means and variances) are in good agreement with experimental data. Figure 7.6 (taken from Ref. [11]) shows radial profiles of unconditional means and variances of temperature and mass fraction of carbon monoxide and the hydroxyl radical. Results are in good agreement with the experimental data and furthermore have a low sensitivity to a five-fold increase in the number of particles.

The model contains two tuneable parameters: a mixing timescale constant and a localness parameter which determines the relative degree of localisation in reference (filtered mixture fraction) and physical spaces. These parameters are adjusted to control the small but not insignificant conditional fluctuations. More work is neces-

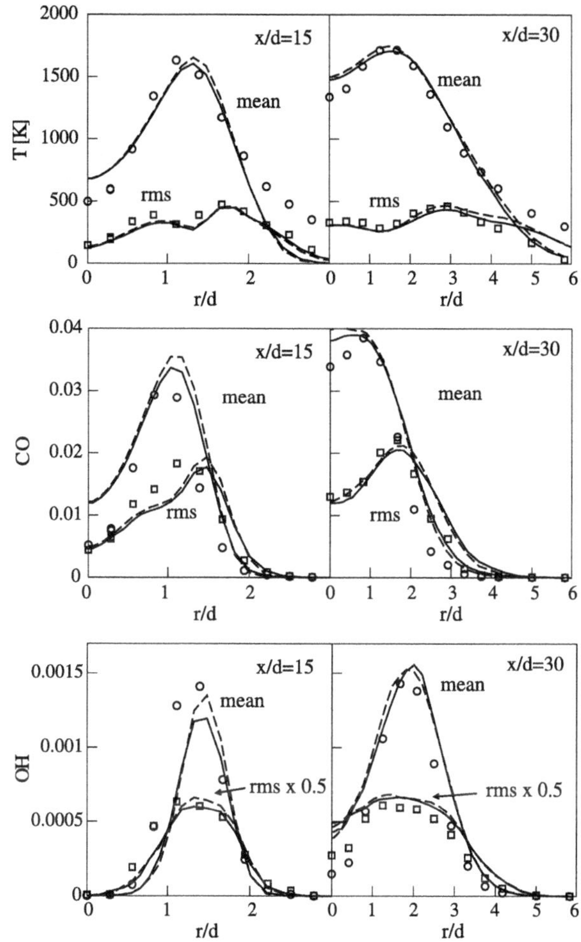

Fig. 7.6: Unconditional mean and rms for temperature, CO and OH at two downstream locations. Open symbols - experimental data, solid lines - MMC with nominally 35,000 particles, broken lines - MMC with nominally 175,000 particles. Figure adapted from [11].

sary to determine the selection of such parameters for a wide range of conditions including those with significant local extinction phenomena. For some complex flame regimes, where conditional fluctuations relative to mean values conditioned on the mixture fraction are very large, additional reference variables (e.g. to emulate sensible enthalpy or other related quantities) may be necessary. However, the need for a filtered source term closure for reacting reference variables could pose a significant challenge if this is attempted.

A principal problem of sparse-Lagrangian methods is related to the evaluation of density which must be obtained from a small number of particles and coupled to the more highly resolved LES flow field. The existing MMC publications do not address this but rather have a tabulated density. Thus there is no coupling or consistency between the Eulerian LES and Lagrangian FDF fields. Ongoing but as yet unpublished research at The University of Queensland has successfully treated the density coupling issue through a conditionally averaged form of the equivalent enthalpy method of Muradoglu et al. [38] which ensures numerical stability and consistency.

7.5 Summary and Future Directions

This chapter has reviewed the MMC concepts and theory, and their evolution, along with the key specific MMC model applications since it was first proposed in 2003. In its most basic deterministic form MMC represents a closed and consistent combination of joint PDF modelling for a set of major species which describe the accessed region in composition space and conditional moment closure for the set of minor species which fluctuate jointly with the major scalars. The closure and consistency is facilitated by mapping closure which is generalised for inhomogeneous flows. MMC has evolved from a deterministic to a stochastic method. Although an equivalent stochastic formulation was introduced initially as a computationally efficient form of the deterministic model it also allows a generalised interpretation where fluctuations of minor species relative to the major species manifold may be exploited and where the reference variables are used to enforce desired properties by conditioning/localisation in mixture fraction space (i.e. a CMC-type mixing model closure) and emulation of scalar dissipation fluctuations. Further evolution of generalised MMC has occurred with the replacement of Markov reference variables by traced Lagrangian quantities in Eulerian DNS or LES. The possibility also exists to obtain reference variables by other simulation or modelling methods.

From the perspective of MMC the use of LES to provide reference variables is expensive but this is more than compensated for through the demonstrated possibility of using a sparse distribution of particles to model the Lagrangian FDF. While a PDF must describe the distributions of all turbulent scales an FDF need only describe the subgrid distributions while the large scale turbulence is resolved by the LES. Therefore from the perspective of LES simulations, generalised MMC allows high-quality, efficient simulations that are dramatically less expensive than conventional intensive-Lagrangian FDF simulations or even LES with chemical source terms modelled using the resolved quantities at the Eulerian grid centres. The concept of sparse-Lagrangian simulations is associated with the FDF method and is not coincident with the concepts of MMC, per se. Sparse simulations with closures other than generalised MMC are certainly possible, however only generalised MMC closures are currently known to work for sparse simulations.

The advances and challenges of MMC research in the coming years will include (but are certainly not restricted to):

- the application of various specific MMC models to a wide range of laboratory combustion conditions and flame regimes to establish the best choice of model parameters;
- the application in inhomogeneous flows of MMC with multiple reference variables emulating more complex accessed composition spaces;
- the expanded testing and development of hybrid MMC methods which borrow ideas from other established models;
- the establishment of criteria for assessing the optimal compromise between quality and economy in current and new sparse-Lagrangian MMC closures;
- the establishment of consistent and stable density coupling methods in sparse-Lagrangian simulations; and
- the application of MMC to conditions of greater engineering and practical interest.

Acknowledgements

The authors would like to thank Dr Andrew Wandel of the University of Southern Queensland and Dr Konstantina Vogiatzaki of Imperial College for their thoughtful, probing and critical comments on the original manuscript. This work was supported by funding from the Australian Research Council.

References

1. Barlow, R.S., Frank, J.: Effects of turbulence on species mass fractions in methane/air jet flames. Proc. Combust. Inst. **27**, 1087–1095 (1998)
2. Bergmann, W., Meier, W., Wolff, D., Stricker, W.: Application of spontaneous Raman and Rayleigh scattering and 2D LIF for the characterization of a turbulent CH4/H2/N2 jet diffusion flame. Appl. Phys. B **66**, 489–502 (1998)
3. Bilger, R.W., Saetran, L.R., Krishnamoorthy, L.V.: Reaction in a scalar mixing layer. J. Fluid Mech. **233**, 211–242 (1991)
4. Bilger, R.W.: Advanced laser diagnostics: implications of recent results for advanced combustor models. In R.S.L. Lee, J.H. Whitelaw, and T.S. Wung, editors, *Aerothermodynamics in Combustors*. Springer-Verlag, Berlin (1993)
5. Bykov, V., Maas, U.: The extension of the ILDM concept to reaction-diffusion manifolds. Combust. Theory Model. **11**, 839–862 (2007)
6. Cha, C.M., Kosaly, G., Pitsch, H.: Modeling extinction and reignition in turbulent nonpremixed combustion using a doubly-conditional moment closure approach. Phys. Fluids **13**, 3824–3834 (2001)
7. Chen, H., Chen, S., Kraichnan, R.H.: Probability distribution of a stochastically advected scalar field. Phys. Rev. Lett. **63**, 2657–2660 (1989)
8. Cleary, M.J., Kronenburg, A.: Multiple mapping conditioning for extinction and reignition in turbulent diffusion flames. Proc. Combust. Inst. **31**, 1497–1505 (2007)

9. Cleary, M.J., Kronenburg, A.: Hybrid multiple mapping conditioning on passive and reactive scalars. Combust. Flame **151**, 623–638 (2007)
10. Cleary, M.J., Klimenko, A.Y.: A generalised multiple mapping conditioning approach for turbulent combustion. Flow Turbul. Combust. **82**, 477–491 (2009)
11. Cleary, M.J., Klimenko, A.Y., Janicka, J., Pfitzner, M.: A sparse-Lagrangian multiple mapping conditioning model for turbulent diffusion flames. Proc. Combust. Inst. **32**, 1499–1507 (2009)
12. Colluci, P.J., Jaberi, F.A., Givi, P., Pope, S.B.: Filtered density function for large eddy simulation of turbulent reacting flows. Phys. Fluids **10**, 499–515 (1998)
13. Curl, R.L.: Dispersed phase mixing: I. Theory and effects of simple reactors. AIChE J. **9**, 175–181 (1963)
14. Devaud, C.B., Bilger, R.W., Liu, T.: A new method for modeling the conditional scalar dissipation rate. Phys. Fluids **16**, 2004–2011 (2004)
15. Dopazo, C., O'Brien, E.E.: An approach to the autoignition of a turbulent mixture. Acta Astronaut. **1**, 1239–1266 (1974).
16. Fox, R.O.: Computational Models for Turbulent Reacting Flows. Cambridge University Press, Cambridge, United Kingdom (2003)
17. Haworth, D.C.: Progress in probability density function methods for turbulent reacting flows. Prog. Energy Combust. Sci. **36**, 168–259 (2009)
18. Jaberi, F.A., Colucci, P.J., James, S., Givi, P., Pope, S.B.: Filtered mass density function for large-eddy simulation of turbulent reacting flows. J. Fluid Mech. **401**, 85–121 (1999)
19. Janicka, J., Kolbe, W., Kollmann, W.: Closure of the transport equation for the probability density function of turbulent scalar field. J. Nonequil. Thermodyn. **4**, 47–66 (1979)
20. Klimenko, A.Y., Bilger, R.W.: Conditional moment closure for turbulent combustion. Prog. Energy Combust. Sci. **25**, 595–687 (1999)
21. Klimenko, A.Y., Pope, S.B.: A model for turbulent reactive flows based on multiple mapping conditioning. Phys. Fluids **15**, 1907–1925 (2003)
22. Klimenko, A.Y.: Matching the conditional variance as a criterion for selecting parameters in the simplest multiple mapping conditioning models. Phys. Fluids **16**, 4754–4757 (2004)
23. Klimenko, A.Y.: Matching conditional moments in PDF modelling of nonpremixed combustion. Combust. Flame **143**, 369–385 (2005)
24. Klimenko, A.Y.: On simulating scalar transport by mixing between Lagrangian particles. Phys. Fluids **19**, 031702 (2007)
25. Klimenko, A.Y.: Lagrangian particles with mixing. Part 2: sparse-Lagrangian methods in application for turbulent reacting flows. Phys. Fluids **21**, 065102 (2009)
26. Kronenburg, A., Bilger, Kent, J.H.: Second order conditional moment closure for turbulent jet diffusion flames. Proc. Combust. Inst. **27**, 1097–1104 (1998)
27. Kronenburg, A.: Double conditioning of reactive scalar transport equations in turbulent non-premixed flames. Phys. Fluids **16**, 2640–2648 (2004)
28. Kronenburg, A., Kostka, M.: Modeling extinction and reignition in turbulent flames. Combust. Flame **143**, 342–356 (2005)
29. Kronenburg, A., Cleary, M.J.: Multiple mapping conditioning for flames with partial premixing. Combust. Flame **155**, 215–231 (2008)
30. Kuznetsov, V.R., Sabelnikov, V.A.: Turbulence and Combustion. Hemisphere, New York (1989)
31. Li, J.D., Bilger, R.W.: Measurement and prediction of the conditional variance in a turbulent reactive-scalar mixing layer. Phys. Fluids A **5**, 3255–3266 (1993)
32. Maas, U., Pope, S.B.: Simplifying chemical-kinetics: intrinsic low-dimensional manifolds in composition space. Combust. Flame **88**, 239–264 (1992)
33. Mastorakos, E., Bilger, R.W.: Second-order conditional moment closure for the autoignition of flows. Phys. Fluids **10**, 1246–1248 (1998)
34. Meier, W., Prucker, S., Cao, M.-H., Stricker, W.: Characterization of turbulent H_2-N_2 air jet diffusion flames by single-pulse spontaneous Raman scattering. Combust. Sci. Technol. **118**, 293–312 (1996)

35. Mitarai, S., Riley, J.J., Kosály, G.: A Lagrangian study of scalar diffusion in isotropic turbulence with chemical reaction. Phys. Fluids **15**, 3856–3866 (2003)
36. Mitarai, S., Riley, J.J., Kosály, G.: Testing of mixing models for Monte Carlo probability density function simulations. Phys. Fluids **17**, 047101 (2005)
37. Mortensen, M.: Consistent modeling of scalar mixing for presumed, multiple parameter probability density functions. Phys. Fluids **17**, 018106 (2005)
38. Muradoglu, M., Pope, S.B., Caughey, D.A.: The hybrid method for the PDF equations of turbulent reactive flows: consistency conditions and correction algorithms. J. Comput. Phys. **172**, 841–878 (2001)
39. Peters, N.: Laminar diffusion flamelet models in non-premixed turbulent combustion. Prog. Energy Combust. Sci. **10**, 319–339 (1984)
40. Pope, S.B.: PDF methods for turbulent reactive flows. Prog. Energy Combust. Sci. **11**, 119–192 (1985)
41. Pope, S.B.: Mapping closures for turbulent mixing and reaction. Theoret. Comput. Fluid Dynamics **2**, 255–270 (1991)
42. Pope, S.B.: On the relationship between stochastic Lagrangian models of turbulence and second-moment closures. Phys. Fluids **6**, 973–985 (1994)
43. Pope, S.B.: Lagrangian PDF methods for turbulent flows. Annu. Rev. Fluid Mech. **26**, 23–63 (1994)
44. Pope, S.B.: Turbulent Flows. Cambridge University Press, Cambridge, United Kingdom (2000)
45. Pope, S.B.: Accessed compositions in turbulent reactive flows. Flow Turbul. Combust. **72**, 219–243 (2004)
46. Pope, S.B., Ren, Z.: Efficient implementation of chemistry in computational combustion. Flow Turbul. Combust. **82**, 437–453 (2009)
47. Repp, S., Sadiki, A., Schneider, C., Hinz, A., Landenfeld, T., Janicka, J.: Prediction of swirling confined diffusion flame with a Monte Carlo and a presumed-PDF model. Int. J. Heat Mass Trans. **45**, 1271–1285 (2002)
48. Saetran, L.R., Honnery, D.R., Stårner, S.H., Bilger, R.W.: Turbulent Shear Flows 6. Springer-Verlag, Berlin (1989)
49. Schneider, Ch., Dreizler, A., Janicka, J., Hassel, E.P.: Flow field measurements of stable and locally extinguishing hydrocarbon-fuelled jet flames. Combust. Flame **135**, 185–190 (2003)
50. Sreedhara, S., Huh, K.Y., Ahn, D.H.: Comparison of submodels for conditional velocity and scalar dissipation in CMC simulation of piloted jet and bluff body flames. Combust. Flame **152**, 282–286 (2008)
51. Subramaniam, S., Pope, S.B.: A mixing model for turbulent reactive flows based on Euclidean minimum spanning trees. Combust. Flame **115**, 487–514 (1998)
52. Valiño, L., Dopazo, C.: A binomial Langevin model for turbulent mixing. Phys. Fluids A **3**, 3034–3037 (1991)
53. Vaishnavi, P., Kronenburg, A.: Multiple mapping conditioning of velocity in turbulent jet flames. Combust. Flame, to be published 2010.
54. Vogiatzaki, K., Kronenburg, A., Cleary, M.J., Kent, J.H.: Multiple mapping conditioning of turbulent jet diffusion flames. Proc. Combust. Inst. **32**, 1679–1685 (2009)
55. Vogiatzaki, K., Cleary, M.J., Kronenburg, A., Kent, J.H.: Modeling of scalar mixing in turbulent jet flames by multiple mapping conditioning. Phys. Fluids, **21**, 025105 (2009)
56. Wandel, A.P., Klimenko, A.Y.: Testing multiple mapping conditioning mixing for Monte Carlo probability density function simulations. Phys. Fluids, **17**, 128105 (2005)
57. Wandel, A.P., Lindstedt, R.P.: Hybrid binomial Langevin-multiple mapping conditioning modeling of a reacting mixing layer. Phys. Fluids, **21**, 015103 (2009)

Part III
Advances and Trends in Multiscale Strategies

Chapter 8
The Emerging Role of Multiscale Methods in Turbulent Combustion

Tarek Echekki

Abstract Turbulent combustion flows are governed by processes that span the range from atomistic scales to device (e.g. engine) scales and beyond (e.g. oil pool fires, thermo-nuclear flames in type Ia supernovae). The multiscale nature of turbulent combustion flows poses both challenges and opportunities. The challenges arise from the need to predict combustion phenomena that are governed by a broad range of scales. The opportunities arise because of the emergence of the multiscale science that permeates many fields, and which for turbulent combustion, has been motivated by the need to predict phenomena in new and evolving combustion technologies, advances in computational and applied mathematics, and the increasing availibility of computational resources. In this chapter, strategies and requirements for the multiscale modeling and simulation of turbulent combustion flows are discussed. The chapter serves as an introductory chapter to Part III of this book.

8.1 Motivation

A principal challenge in predicting turbulent combustion flows arises from their multiscale nature. In a recent review, Peters [36] highlighted some complexities that arise in the prediction of multiscale phenemena associated with chemistry, flame structure, and flame interactions with turbulence. While Peters' review, a more recent review by the author [12] and the present chapter are timely, concrete ideas about addressing the multisale nature of turbulent combustion flows have been implemented for many decades. In physical sciences, constitutive relations (e.g. viscous stresses for Newtonian fluids) are used to represent processes on the molecular scale in a form applicable to macro-scale formulations (e.g. continuum conservation equations for fluid flow). These relations effectively represent multiscale treatments that eliminate the need to resolve atomistic effects. More specifically in com-

Tarek Echekki
North Carolina State University, Raleigh NC 27695-7910, USA, e-mail: `techekk@ncsu.edu`

T. Echekki, E. Mastorakos (eds.), *Turbulent Combustion Modeling*,
Fluid Mechanics and Its Applications 95, DOI 10.1007/978-94-007-0412-1_8,

bustion, turbulent combustion paradigms that rest on the principle of separation of scales have been the cornerstone of turbulent combustion models in computational fluid dynamics (CFD) for more than three decades. The eddy-dissipation model (EDM) [27], the eddy break-up model (EBU) [42] and the flamelet model [34] are based on separating the chemical time scales from turbulent time scales or flame length and time scales from the corresponding turbulence scales, respectively.

However, important advances in applied and computational mathematics and resources combined with evolving requirements in novel combustion technologies are enabling the rapid development of various multiscale strategies for the prediction of turbulent combustion flows. These strategies represent important additions, and quite often extensions, to traditional paradigms in turbulent combustion [3].

The present chapter serves three primary objectives: 1) motivate the need for multiscale strategies for the modeling and simulation of turbulent combustion flows, 2) introduce some basic strategies adopted for these flows, and 3) preview the topics covered in Part III of this book, which deal with multiscale approaches for turbulent combustion. The remaining sections of the chapter are organized as follows. In Sec. 8.2, the multiscale nature of turbulent combustion flows is illustrated. In Sec. 8.3, the case is made for the development of multiscale approaches in turbulent combustion based on progress in multiscale science and the need to model evolving combustion technologies. Fundamental considerations for multiscale approaches in turbulent combustion are discussed in Sec. 8.4. A brief survey of multiscale approaches and a preview of related chapters are presented in Sec. 8.5.

8.2 The Multiscale Nature of Turbulent Combustion Flows

Turbulent combustion flows exhibit a broad range of length and time scales, which encompass phenomena from atomistic to device scales. The most elementary processes at the molecular level, which are associated with the electronic structure of molecules and the breaking of chemical bonds, may occur at time scales as short as 10^{-15} s. However, simpler representations of chemical kinetics, such as through the Arrhenius law for the rate constant and the law of mass action, generally reduce the time-scale requirements to a range from approximately 10^{-10} s for the fastest reactions to a few seconds for the slower reactions. Nonetheless, the presence of a wide range of time scales represents an important challenge to the integration of chemistry and the coupling of this chemistry with transport for combustion problems.

Similar disparities are present for length scales. Molecular scales start with ranges of approximately 10^{-10} m. Beyond this, a broad range of scales represents the mechanisms of soot particle formation at molecular scales to hundreds of nanometers for soot particle aggregates [21]. A large gap separates these processes from continuum scales, which start around the scale of 1 μm. Beyond this range, and within the continuum regime, structures within the reaction zones of flames range in thickness from tens of μm (e.g. within the fuel consumption layer of premixed hydrocarbon flames) to a fraction of a millimeter corresponding to a laminar premixed flame thickness (including the entire reaction zone and a good portion of the preheat zone). Around this upper range of premixed flame thickness, there may be overlap with turbulent scales, starting from the Kolmogorov scale in practical combustion systems. The upper range of the turbulent scales may be comparable to the device scales, which may be of the order of tens of centimeters to meters. Therefore, a broad range of length scales represent the bulk of combustion processes from the atomistic to the device scale.

While the above so-called chemical flames appear untameable, thermonuclear flames, such as the ones encountered in type Ia supernovae exhibit even broader ranges of length and time scales [41, 49]. These flames exhibit much stronger nonlinearity in their nuclear source terms (carbon burning rate during ignition is proportional to the temperature to the 23rd power [49]), turbulence length scales spanning over eleven decades, and flame thicknesses, which are comparable to those of chemical flames [49].

Although, the basic processes of combustion may be described using established physical laws, their direct simulation is beyond today's and near-future computational capabilities. Current estimates of computing resources needed to resolve length scales place an upper limit of three decades of length scales for tera-scale computing and four decades of length scales for peta-scale [29]. Even if computer memory can accommodate large problems, time integration remains an important challenge, and many computations of turbulent reacting flows remain CPU-time-limited. Therefore, multiscale strategies are needed to capture the contribution of processes ranging from atomistic to device scales.

It is possible to carry out computations that are designed for a particular range of scales. For example, quantum dynamics calculations may be used to resolve the necessary electronic structure of molecules to determine kinetic rate parameters and heats of formations of molecules. Molecular dynamic simulations and kinetic Monte-Carlo simulations may be used to model soot processes [21] or to determine molecular transport properties. Fine- and coarse-grained continuum simulations, respectively, may be used to capture particular flame-turbulence interations or to model an actual device. Bootstrapping results from the different computations, which are specialized for different ranges of scales can be done through different strategies depending on the degree of coupling between scales. Here, we must distinguish between two types of coupling [12]: 1) physical coupling, which is associated with the presence of cause-and-effect between two processes occurring at different ranges of scales, and 2) scale coupling, which is associated with the presence of overlap between the length or time scales or both between two processes.

Many important applications in combustion involve a one-way coupling between processes operating at disparate scales. For these applications, various analytical and computational methods are available. Examples of such methods include:

- **Constitutive relations**: In reacting flows, the most common constitutive laws often represent models for atomistic processes in continuum governing equations. An obvious example is the relation between the rate-of-strain and viscous stresses in flows for Newtonian fluids. Another example of relevance to combustion includes the formulation of expressions for species reaction rates based on the Arrhenius law and the law of mass action. Constitutive relations may be based on phenomenological laws (e.g. proportionality between two quantities and scaling relations) or may be derived from more rigorous analysis (e.g. the kinetic theory of gases).

- **Time-scale reduction**: In addition to constitutive relations, various methods of time-scale reduction may be adopted to represent fast processes, which are not coupled with transport, under the assumption of separation of scales and one-way coupling between processes. For example, chemistry reduction using dynamical systems approaches such as computational singular perturbation (CSP) [25] or the intrinsic low-dimensional manifold (ILDM) approach [38] can be used to tabulate the contribution of the 'fast' time scales and integrate primarily the remaining range of 'slow' scales. However, in contrast with traditional constitutive relations for chemistry and molecular transport, these latter approaches (e.g. CSP and ILDM) may depend on the mixture characteristics, the burning modes (e.g. premixed or diffusion flames) or the combustion regime (e.g. flamelet, distributed reaction).

Examples of problems that involve two-way physical coupling and scale separation includes the formation of soot in fires. In this problem, soot is a major contributor to radiative heat loss from the flame zone, and plays a critical role in determining the flame temperature. In turn, the temperature profiles within the flame also contribute to the formation of soot and its transport. Despite the presence of soot-flame coupling, multiscale strategies are enabled by scale separation. Instead of constitutive relations, which address single terms in the governing equations, moment transport equations may be adopted, which features models for transport (e.g. thermophoeresis) and sink and source terms (e.g. condensation, surface growth).

Other examples, which involve separation of scales and strong coupling between the scales, include flame-turbulence interactions in the flamelet regime and thermo-acoustic coupling in a combustion chamber. When an account of the physical coupling is needed to represent physics at all scales, more robust formulations are needed that identify the dynamics of contributions across the scales.

Despite the prevalence of problems in combustion that exhibit scale separation, there is a growing need to address problems with strong physical and scale coupling. This need is addressed in the next section.

8.3 The Case for Multiscale Strategies in Turbulent Combustion

In recent years, important paradigm shifts have occurred in two relevant fronts associated with turbulent combustion flows. The first is associated with the emergence of novel combustion technologies that have pushed the envelope on the types of combustion regimes and modes encountered in practical devices. This shift also entails new requirements for the combustion performance and its prediction associated with combustion efficiency, flexibility of the combustion fuels and operating conditions, the mitigation of pollutants' emissions and safety [29]. The second shift is associated with the emergence of the field of multiscale science as an important discipline in applied and computational mathematics and its increasing implementation in physical sciences. This emergence is fueled by increasing computational resources and technological innovations in various physical sciences (e.g. nanotechnologies, material science).

8.3.1 Emerging Combustion Technologies

New combustion technologies have been proposed in recent decades to address the clean and efficient burning of a growing range of combustion fuels [29]. These new technologies, invariably, push the envelope on the operational regimes of practical combustion devices. Two examples of combustion technologies illustrate these trends.

- In reciprocating internal combustion engines, there is a classical distinction between two competing technologies: diesel and spark-ignited engines, with the first more efficient and the second associated with cleaner burning. Alternative technologies have been proposed in recent years to overcome the limitations of each engine; these technologies include gasoline direct injection (GDI) engines, diesel low-temperature combustion (LTC) and homogeneous charge compression ignition (HCCI) engines [7, 29]. In HCCI engines, combustion is initiated by the compression ignition of nearly homogeneous mixtures of fuel and air, resulting in diesel-like efficiencies and lower emissions of NOx particulate matter through better control of the charge and mixture conditions at combustion. Various control strategies may be adopted, which impact the competition between the rates of mixing and chemistry [7, 45]. When these rates are competitive, a separation of the chemical and transport scales is not feasible, and strategies to couple both effects are needed. Moreover, since the mixture homogeneity may evolve as chemistry evolves, different modes of combustion (with varying degrees of partial premixing) may be encountered during the HCCI combustion process [7].
- In gas turbine engines, strategies to lower emissions and increasing efficiency have been developed as well [5]. One strategy includes burning at lean or ultra-lean fuel-air mixture conditions. However, these conditions also may contribute to the onset of different, and often coupled, combustion instabilities [19]. Important manifestations of these instabilities may include the non-equilibrium effects of extinction and re-ignition events. Capturing non-equilibrium effects provide important challenges to the state-of-the-art turbulent combustion models.

Common among the above-stated problems are requirements of fuel flexibility as a broader range of combustion fuels are considered, including in addition to the standard fuels, heavy hydrocarbons (e.g. oil sands, oil shale) and renewable fuels (e.g. ethanol, biodiesel) [29]. With the above considerations, the traditional paradigms of turbulent combustion based on the assumption of separation of scales may not be applicable when mixing time scales are competitive with chemical time scales or where well-established and stable combustion fronts (i.e. flames) determine the combustion process. More importantly, within the context of moment-based methods in turbulent combustion, three additional challenges emerge:

1. The choice and the number of appropriate transported moments is very critical to capture the desired physics. Generally, to capture more physics, such as extinction and re-ignition, more moments are needed.
2. Closure for non-linear terms in the moment transport equations is very critical as well, especially when different outcomes of the problem may be encountered (e.g. burning and non-burning solutions).
3. The prediction or the assumption of particular statistical distribution (e.g. joint scalars' probability density functions) becomes an additional challenge as well.

Moment-based models have been successful in predicting problems in combustion applications; however, they remain limited by their underlying assumptions, which may be related to the combustion mode (e.g. premixed vs. non-premixed), the combustion regime (e.g. flamelet vs. distributed reaction) or dominant chemistry. These assumptions determine the choice of transported moments, the implementation of closure in the moment equation and the reconstruction of the spatial statistical distributions for these moments. Significant progress has been achieved in all fronts as illustrated by results discussed in the various chapters of this book.

Multiscale strategies within the continuum regime may be used to replace potential empiricism in the closure for unclosed terms in the moment governing equations, construct more robust statistical distributions for the moments, or replace all moments altogether.

8.3.2 Emerging Multiscale Science

Multiscale science permeates many disciplines in the physical (e.g. solid and fluid mechanics, biological systems, chemistry), mathematical and computational sciences. It is concerned with problems and phenomena where an account for a broad range of scales is needed. Renewed interest in multiscale science is motivated in part by growth in computational resources, which in turn fuel interest in computing larger problems in the physical sciences. This growth has enabled scientists and engineers to go beyond scale-specific applications and tools (e.g. quantum dynamics vs. continuum-based calculations) to address the solution of larger problems and couple different scales within the same problem. Invariably, the attempts within the various disciplines are different because of modeling requirements and the extent to which scales are coupled.

Important developments in multiscale science reflect growth in a number of important areas:

- The development of robust mathematical frameworks and scale coupling strategies designed to accurately represent the contribution of all relevant scales and their coupling. A mathematical framework is distinct from the individual model(s) used for the different scales and in different applications. Therefore, they may be applicable to a broad range of problems. The choice of a particular framework depends on a number of factors, which include the extent of coupling (physical and scale) between the different phenomena and the type of inter-dependence between the scales. Therefore, a principal goal of the mathematical frameworks is to reduce the mathematical complexity of multiscale problems associated with their high-dimensionality, high degrees of freedom and the complex inter-dependence between phenomena at different ranges of scales. Two mathematical frameworks appear to be very promising for the study of turbulent combustion flows: the heterogeneous multiscale method (HMM) [9], and variational multiscale method (VMS) [20]. Chapter 18 illustrates the implementation of the HMM for the coupling of multiple scale solutions associated with turbulent flame propagation in turbulence. Although the implementation of VMS for non-reacting turbulent flows has been extensively studied, only recently an extension of VMS for variable-density flows has been formulated [18].

- The development of efficient numerical tools and computational frameworks: Computational tools may include parallel solvers for partial differential equations, linear-algebra suites, multiresolution utilities for the decomposition of the solution vector or data analysis, adaptive mesh refinement tools and other relevant applications. The various computational tools may be common among different disciplines. Therefore, potentially a large pool of users may contribute to their developments. However, certain rules must be put in place to ensure the tools' portability across computational platforms and applications. Chapter 17 illustrates a computational framework that enables a collaborative environment for various tools' development based on the common-component architecture paradigm. The chapter also illustrates the implementation of such a framework for combustion problems.

8.4 Multiscale Considerations for Turbulent Combustion

As stated earlier, mathematical and computational frameworks for the development of multiscale approaches may be shared among different disciplines. However, the choice of which frameworks to adopt and the individual models that represent physics at the different scales is different for each discipline. What drives the choice are the characteristics of the problem, which include: 1) the extent of coupling physical and scale coupling between the different phenomena at different scales, and 2) the choice of the models that are approriate to capture physics at the different scales. This latter consideration is very unique, as it dictates strategies for reduction of order or dimensionality of the problem and the type of closure needed to capture unresolved physics. In the following sections, basic requirements for multiscale approaches and choices of the physical models are discussed. Some of these requirements and associated multiscale approaches may be extended to a broader range of problems beyong turbulent combustion flows [26].

8.4.1 Basic Requirements for Multiscale Approaches in Turbulent Combustion

Important aspects of the multiscale nature of turbulent combustion flows have already dictated the choice of traditional paradigms in turbulent combustion. The first aspect is the prevalence of flames which can propagate and be wrinkled by turbulent flows. These flames represent interfaces between reactants and products, fuels and oxidizers, as well as zones of localized reaction and heat release. In numerous applications, these flames can be thin; and therefore, strategies to couple detailed reactions zones at the interface with fluid dynamics form one potential multiscale approach in turbulent combustion. Indeed, flame-embedding (see Chapter 12) represents one such strategy. Mathematical multiscale frameworks for flame embedding are concerned with the implementation of coupling strategies between the flame solutions and the flow field. Illustrations of such coupling using the heterogeneous multiscale method (HMM) are given in Chapter 18. When flames are present, potential strategies to reduce the physical dimension of the flame solutions are enabled due to the strong gradients along the flame normal.

A second aspect of the multiscale nature of turbulent combustion flows is the clear separation between atomistic scales and continuum scales. Therefore, robust constitutive equations for reaction and molecular transport in the continuum governing equations are a viable substitute to hybrid atomistic-continuum simulations. Nonetheless, there is great effort to be expanded in both scale regimes to develop chemical rate and transport data from atomistic simulations and device-scale simulations from the continuum equations.

Finally, given the non-linear nature of the governing equations, the presence of many coupled variables representing both flow and scalar quantities, and the coupling between fine and coarse scales in combustion, analytical methods (e.g. homogenization, asymptotic methods, renormalization group methods) by themselves are not sufficient to solve or separate scales in turbulent combustion flows. Instead, these tools can be coupled with numerical techniques to develop robust multiscale frameworks for these flows.

At this stage, it is useful to re-iterate the basic desirable attributes for multiscale approaches in turbulent combustion. These approaches are subject to the following constraints:

- All time and length scales are resolved at least within the context of continuum-based transport equations and including a closed-form treatment of chemistry and molecular transport processes using constitutive relations.
- The models must preserve structure (e.g. flames) involving the coupling of reaction and diffusion transport.
- No statistical distribution shapes may be assumed for the reactive scalars; and this distribution must be reconstructed, if needed, as part of the multiscale solution.

8.4.2 General Frameworks for the Governing Equations for Multiscale Models of Turbulent Combustion

The description of the turbulent flow of a multicomponent reacting mixture is prescribed through a set of conservation laws for mass, momentum and composition and auxiliary equations. The most common representation of these laws in the continuum regime are extensions of the Navier-Stokes equations (Eq. 2.2).

In Chapter 2, different mathematical forms of these governing equations in the continuum regime are presented. They include the instantaneous forms used in direct numerical simulations (DNS) and forms involving ensemble/time averaging (i.e. RANS formulations) and spatial averaging (or filtering) (i.e. LES). Averaging, invariably, results in unclosed terms in the governing equations, which must be modeled.

Within the context of multiscale frameworks, only the instantaneous and the LES governing equations enable scale separation. In LES, coarse scales are resolved; while, the smaller scales have to be modeled. In contrast, RANS averaging impacts all scales, and closure compensates for information that is 'lost' across all the scales. Therefore, based on the instantaneous and the LES formulations, two principal classes of multiscale strategies emerge. The first is based on the concept of 'mesh adaptivity' and involve the use of the same set of governing equations and the adoption of a hierarchical grid structure. The level of resolution within this grid structure depends on the spatial and temporal resolution requirements within different zones of the computational domain. The second strategy is based on hybrid combinations of a fine- and a coarse-grained solutions and may adopt a LES implementation for the coarse-grained solution. For the fine-grained solution, different strategies for reducing the model complexity at the small scales may be adopted.

It is important to note that within the context of LES, various extensions of the traditional LES approach may be adopted for turbulent combustion flows. Multilevel strategies may be derived within various contexts, such as homogenization approaches [15] and the variational multiscale method [20].

An alternative strategy to the extended Navier-Stokes formulations is the Boltzmann-Maxwell equation, which models the temporal and spatial evoution of one-particle distribution function. The lattice-Boltzmann method (LBM) is a computationally efficient method to solve the Boltzmann-Maxwell equations. In contrast to the well-established conservations laws based on the Navier-Stokes equations, the LBM approach is more recent. Yet, it is becoming an increasingly popular alternative to the Navier Stokes approach for the solution of complex fluids. LBM can be used to effectively straddle sub-continuum and continuum phenomena; therefore, it may be viewed primarily as a mesoscale approach. Moreover, LBM can naturally accomodate a variety of boundary conditions, including prescribed conditions across fluid phases. The LBM is discussed in detail in Chapter 19.

8.5 Multiscale Approaches in Turbulent Combustion and Preview of Relevant Chapters

The sections below summarizes different strategies to model turbulent combustion flows. Although restricted to only one component of turbulent combustion models, chemistry acceleration is included and an entire chapter (Chapter 9) is dedicated to this topic. The present multiscale strategies in turbulent combustion may be associated closely with either established multiscale numerical/mathematical frameworks or traditional paradigms in turbulent combustion as discussed below. The various approaches can be classified as follows:

1. Time-step acceleration and mesh adaptive methods
2. Flame embedding methods
3. Hybrid LES and low-dimensional models

A brief discussion of the various approaches is provided below:

8.5.1 Time-Step Acceleration

Chemistry integration represents a critical bottle-neck in the time-integration of reactive scalar equations. The disparity of time scales between the fastest and the slowest reactions or time-steps required to resolve transport is at the heart of the stiffness problems in chemical mechanisms. Strategies to accelerate the time-integration of chemistry are based on a combination of different approaches. They include:

- **Chemical mechanism reduction** This step addresses the reduction of the size of the representative species and reactions. Chemistry reduction may involve, as a first step, the development of a skeletal mechanism that contains the most relevant species and elementary reactions. In a second step, a more aggressive reduction strategy may be adopted to develop a global mechanism with even fewer species and global steps. This second step may render the reduction process applicable to a more limited range of problems. More importantly, it does not guarantee that the resulting global mechanism removes the original mechanism stiffness. Important progress has been achieved to address mechanism reduction from the classical strategies of the quasi-steady state (QSS) and the partial equilibrium (PE) approximations, to more systematic model reduction strategies, such as the ones based on the ILDM and the CSP approaches. Chapter 9 provides an extensive review of chemistry reduction strategies.
- **Numerical stiff and mildly-stiff integrators** In the past, stiff chemistry integration have been addressed through implicit integrators, such as VODE [4] and DASSL [37]. However, more recently, explicit as well as implicit-explicit integrators have been developed that are used to efficiently and accurately integrate stiff and mildly stiff ordinary and partial differential equations. There is inherently a trade-off between the computational cost, numerical accuracy and the

coupling requirements with transport operators that must be taken into consideration.

- **Chemistry tabulation, storage and retrieval approaches** Another important strategy for chemistry acceleration is the development of storage and retrival strategies for tabulation results of chemistry integration. Efficient strategies for tabulation, storage and retrieval are available, including the in situ adaptive tabulation (ISAT) algorithm [39], the piece-wise reusable implementation of solution mapping (PRISM) [43] and artificial neural networks (ANN) [6].

8.5.2 Mesh Adaptive Methods

As outlined earlier, one principal consideration for multiscale approaches in turbulent combustion is that combustion and important scalar and velocity gradients may be associated with well-defined regions in the flow (e.g. flames). Mesh adaptivity is based on adapting the spatial resolution (and accordingly, the temporal resolution) to capture fine details of the flow and the scalar field when needed and coarsen when not needed. Mesh adaptive methods can be implemented within the context of both instantaneous and statistical models (e.g. LES).

Chapters 13 and 14 illustrate two adaptive mesh strategies. The first is based on the adaptive mesh refinement (AMR) approach [2] (Chapter 13). In this approach, a recursive refinement of the mesh is implemented until resolution requirements are met within a region of the flow. The second approach is based on the multiresolution wavelet approach [40] (Chapter 14). This approach is based on a decomposition of the solution into spatially- and temporally-varying multiresolution modes (e.g. wavelet modes) to capture the solution at different multiresolution levels.

8.5.3 Flame Embedding Approaches

Flame embedding strategies may represent an extension of the flamelet paradigm [34]; although, some of the ideas of flame-embedding have been proposed earlier [17]. A principal outcome of the flamelet paradigm is the separation of the flame scales from the energetic flow scales, enabling a degree of decoupling of the flamelet and the flow solutions. Therefore, three important implementation elements associated with flame embedding are needed: 1) a solution of the flowfield, 2) a flame solution, and 3) flame tracking.

Through the decoupling of the flow and flamelet solutions, different strategies may be adopted to resolve the flow (e.g. vortex methods) and the reactive scalar within the flame. For the flame, two principal approaches may be identified, which correspond to 1) flame-normal embedding, and 2) flamelet tabulation. Chapter 12 discusses primarily flame-normal strategies where 1D flames are solved along the flame surface normal with prescribed input parameters from the flowfield (e.g. strain rate). Another strategy is based on the transport of a representative scalar for flame tracking and the tabulation of the remaining reactive scalars, such as approaches based on the flamelet-generated manifold (FGM) [44] and the flame-prologation of ILDM (FPI) schemes [13].

Flame embedding strategies can be implemented within the context of the instantaneous equations or statistical models. In these latter models, flamelet solutions with sufficient variability in time and space are used to build statistical models of the flame response in turbulence. Examples of such strategies include the use of the LEMLES approach based on the linear-eddy model (LEM) (see Chapter 10); although, the same strategies may be adopted using the ODTLES approach based on the one-dimensional turbulence (ODT) model (see Chapter 11). More recently, Fureby [15] proposed a homogenization-based LES (hLES) approach, which converts the governing equations into a cascade of equations at different scales. Fureby [15] implemented a flame-embedding strategies using one-dimensional forms of the equations at different scales.

Tabulation strategies represent alternative approaches to *in situ* flame calculations. These approaches represent an extension of the strategies adopted with FGM and FPI. Here, we cite the recent work of Vreman et al. [47] and Fiorina et al. [14]. In both studies, it is proposed that filtered reaction rate solutions of reactive scalars are constructed by filtering of one-dimensional flame solutions.

8.5.4 Hybrid LES-Low-Dimensional Models

Potentially more versatile methods may involve a hybrid formulation involving LES combined with a low-dimensional model for subgrid scale (SGS) physics. Statistical flame-embedding approaches represent an important sub-class of hybrid methods; however, the more general hybrid strategies do not require fine-grained flame solutions to be attached to flame normals or tabulated via transported reactive scalars. Hybrid methods are primarily designed to construct statistical information, such as subgrid scale source terms or filtered scalar quantities. Because a hybrid scheme that spans the range of scales of interest is ultimately costly, strategies for dimensional reduction are needed. The different hybrid strategies adopted for such hybrid LES schemes generally involve either a stochastic or a dynamical-systems formulation incorporated with a reduced physical dimension or a lower order of the solution vector. Among the frameworks adopted for turbulent combustion, we cite:

- Low-dimensional stochastic models, including the linear-eddy model (LEM) [23] (Chapter 10) and the one-dimensional turbulence (ODT) model [24] (Chapter 11).
- Transported probability and filtered density functions with structure-based mixing models (e.g. the PSP approach by Meyer and Jenny [30]).
- The homogenization-based LES (hLES) approach [15].
- The chaotic map approach by McDonough and co-workers [28, 31].

Low-dimensional stochastic models, including the LEM and ODT models, involve the embedding of one-dimensional domains within the 3D computational domain. The 1D transport equations for each LEM or ODT domain feature the usual terms in the governing equations for momentum (in the case of ODT) and scalars, albeit in 1D. The resolution requirements are DNS-like and the computational saving is derived principally from the reduced dimension.

The chaotic map model [28, 31] is based on a fundamentally different strategy from that adopted using 1D stochastic models. The model is based on an estimation procedure for the reactive scalars based on a definition of a chaotic dynamical system. In the chaotic map approach, residual terms of reactive scalars (i.e. instantaneous subtracted filtered values) are represented at each discrete grid point as the product of an amplitude factor, which is derived from classical theories of isotropic turbulence, an anisotropy correction factor, and components of a discrete dynamical system (DDS), whose governing equations are derived from the original reactive scalars transport equations [28, 31]. An additional component of the derivation is the discretization of the solution vector for the reactive scalars into a Fourier series in which only leading coefficients are retained and ordinary differential equations for the Fourier coefficients are solved.

8.6 Concluding Remarks

The direct resolution of all scales in turbulent combustion flows is practically impossible given their multiscale nature. In this chapter, some of the basic challenges and opportunities towards the development of multiscale strategies for turbulent combustion flows are discussed. A principal goal of these strategies is to provide viable alternatives to the traditional closure problem in turbulent combustion, when these latter models fail or are limited in their predictions. The emergence of such strategies is dictated by technological requirements for combustion, advances in computational sciences and the increasing availability of powerful computational resources. A principal challenge to tackle is associated with the representation of chemisty as a critical bottleneck in the temporal integration of solutions of reacting flows. Multiscale approaches for turbulent combustions are based on two general strategies based on either mesh-adaptivity or model-adaptivity. Adaptive mesh refinement (Chapter 13) and wavelet-based methods (Chapter 14) represent promising examples of multiscale approaches based on mesh-adaptivity. Model-adaptive approaches primarily adopt the coupling of fine-scale reduced dimension or order solution and coarse-scale solutions, such as LES. Chapters 10, 11 and 12 illustrate different applications of such model-adaptive methods.

Acknowledgement

Dr. T. Echekki acknowledges support from the Air Force Office of Scientific Research through grants F49620-03-1-0023 monitored by Dr. Julian Tishkoff and FA9550-09-1-0492 monitored by Dr. Fariba Fahroo and the National Science Foundation through grant DMS-0915150 monitored by Dr. Junping Wang.

References

1. Bell, J.B., Day, M.S., Grcar, J.F., Lijewski, M.J., Driscoll, J.F., Filatyev, S.A.: Numerical simulation of laboratory-scale turbulent slot flame. Proc. Combust. Inst. **31**, 1299–1307 (2009)
2. Berger, M., Oliger, J.: Adaptive mesh refinement for hyperbolic partial differential equations. J. Comput. Phys. **53**, 484–512 (1984)
3. Bilger, R.W., Pope, S.B., Bray, K.N.C., Driscoll, J.F.: Paradigms in turbulent combustion, Proc. Combust. Inst. **30**, 21–42 (2005)
4. Brown, P.N., Byrne, G.D., Hindmarsh, A.C.: VODE: A variable coefficient ODE solver. SIAM J. Sci. Stat. Comput. **10**, 1038–1051 (1989)
5. Correa, S.M.: Power generation and aeropropulsion gas turbines: From combustion science to combustion technology. Proc. Combust. Inst. **27**, 1793–1807 (1998)
6. Christo, F.C., Masri, A.R., Nebot, E.M.: An integrated PDF/neural network approach for simulating turbulent reacting systems, Proc. Combust. Inst. **26**, 43–48 (1996)
7. Dec, J.E.: Advanced compression-ignition engines – understanding the in-cylinder processes. Proc. Combust. Inst. **32**, 2727–2742 (2009)
8. Dubois, T., Jauberteau, F., Teman, R.: Dynamic Multilevel Methods and the Numerical Simulation of Turbulence. Cambridge University Press (1999)
9. E, W., Engquist, B.: The heterogeneous multiscale method. Comm. Math. Sci. **1**, 1062–1070 (2003)
10. E, W., Engquist, B.: Multiscale modeling and computation. Notices AMS **50**, 1062–1070 (2003)
11. E, W., Engquist, B., Huang, Z.: Heterogeneous multiscale method: A general methodology for multiscale modelign. Phys. Rev. B **67**, 092101 (2003)
12. Echekki, T.: Multiscale methods in turbulent combustion: strategies and computational challenges. Comput. Sci. Disc. **2**, 013001 (2009)
13. Fiorina, B., Baron, R., Gicquel, O., Thevenin, D., Carpentier, S., Darabiha, N.: Modelling non-adiabatic partially premixed flames using flame-prolongation of ILDM. Combust. Theory Model. **7**, 449–470 (2003)
14. Fiorina, B., Vicquelin, R., Auzillon, P., Darabiha, N., Gicquel, O., Veynante, D.: A filtered tabulated chemistry model for LES of premixed combustion, Combust. Flame. **157**, 465–475 (2010)
15. Fureby, C.: Homogenization based LES for turbulent combustion. Flow Turbul. Combust. **84**, 459–480 (2010)
16. Galpin, J., Naudin, A., Vervisch, L., Angelberger, C., Colin, O., Domingo, P.: Large-eddy simulation of a fuel-lean premixed turbulent swirl-burner. Combust. Flame **155**, 247–266 (2008)
17. Ghoniem, A.F., Chorin, A.J., Oppenheim, A.K.: Numerical modeling of turbulent combustion in premixed gases. Proc. Combust. Inst. **18**, 1375–1385 (1981)
18. Gravemeier, V., Wall, W.A.: Residual-based variational multiscale methods for laminar, transitional and turbulent variable-density flow at low Mach number. Int. J. Num. Meth. Fluids, DOI:10.1002/fld.2242 (2009)
19. Huang, Y., Yang, V.: Dynamics and stability of lean-premixed swirl-stabilized combustion. Prog. Energy Combust. Sci. **35**, 293–364 (2009)

20. Hughes, T.J.R., Mazzei, L., Jansen, K.E.: Large eddy simulation and the variational multiscale method. Comput. Vis. Sci. **3**, 47–59 (2000)
21. Izvekov, S., Violi, A.: A coarse-grained molecular dynamics study of carbon nanoparticle aggregation. J. Chem. Theory Comput. **2**, 504–512 (2006)
22. Keller, J.J.L.: Thermoacoustic oscillations in combustion-chambers of gas-turbines. AIAA J. **33**, 2280–2287 (1995)
23. Kerstein, A.R.: Linear-eddy modeling of turbulent transport. II: Application to shear layer mixing. Combust. Flame **75**, 397–413 (1989)
24. Kerstein, A.R.: One-dimensional turbulence: Model formulation and application to homogeneous turbulence. J. Fluid Mech. **392**, 277–334 (1999)
25. Lam, S.H., Goussis, D.A.: Understanding complex chemical kinetics with the computational singular perturbations. Proc. Combust. Inst. **22**, 931–941 (1988)
26. Law, C.K. Combustion at a crossroads: Status and prospects. Proc. Combust. Inst. **31**, 1–29 (2006)
27. Magnussen, B.F., Hjertager, B.H.: On mathematical modeling of turbulent combustion with special emphasis on soot formation and combustion. Proc. Combust. Inst. **16**, 719–729 (1976)
28. McDonough, J.M., Huang, M.T.: A low-dimensional model of turbulence-chemical kinetics interactions. Proc. Third Int. Symp. Scale Model (10–13 September 2000, Nagoya, Japan) Paper ISSM3-E8 (2000)
29. McIlroy, A., McRae, G., Sick, V., Siebers, D.L., Westbrook, C.K., Smith, P.J., Taatjes, C., Trouvé, A., Wagner, A.E., Rohlfing, E., Manley, D., Tully, F., Hilderbrandt, R., Green, W., Marceau, D., O'Neal, J., Lyday, M., Cebulski, F., Garcia, T.R., Strong, D., Basic research needs for clean and efficient combustion of 21st century transportation fuels. Department of Energy Office of Science Report (2006)
30. Meyer, D.W., Jenny, P.: A mixing model for turbulent flows based on parameterized scalar profiles. Phys. Fluids **18**, 035105 (2006)
31. Mukerji, S., McDonough, J.M., Mengu, M.P., Manickavasagam, S., Chung, S.: Chaotic map models of soot fluctuations in turbulent diffusion flames. Int. J. Heat Mass Transfer **41**, 4095–4112 (1998)
32. Perrier, P., Pironneau, O.: Coupling large with small turbulent structures by theory of homogenization. Comptes Rendus Hebd. Seances Acad. Sci. Ser. A. **286**, 635–638 (1978)
33. Perrier, P., Pironneau, O.: Subgrid turbulence modeling by homogenization. Math. Model. **2**, 295–317 (1981)
34. Peters, N.: Local quenching due to flame stretch and non-premixed turbulent combustion. Combust. Sci. Technol. **30**, 1–17 (1983)
35. Peters, N.: Turbulent Combustion, Cambridge University Press, 2000
36. Peters, N.: Multiscale combustion and turbulence. Proc. Combust. Inst. **32**, 1–25 (2009)
37. Petzold, L.R., A description of dassl: A differential/algebraic system solver, SAND82-8637, Sandia National Laboratories (1982)
38. Pope, S.B., Maas, U.: Implementation of simplified chemical kinetics based on intrinsic low-dimensional manifolds. Proc. Combust. Inst. **24**, 103–112(1992)
39. Pope, S.B.: Computationally efficient implementation of combustion chemistry using in situ adaptive tabulation, Combust. Theory Model. **1**, 41–63 (1997)
40. Prosser, R., Cant, R.S.: On the use of wavelets in computational combustion. J. Comput. Phys. **147**, 337–61 (1998)
41. Röpke, F.K., Bruckschen, R.: Thermonuclear supernovae: a multi-scale astrophysical problem challenging numerical simulation and visualization. New J. Phys. **10**, 125009 (2008)
42. Spalding, D.B.: Mixing and chemical reaction in steady confined turbulent flames. Proc. Combust. Inst. **13**, 649–657 (1971)
43. Tonse, S.R., Moriary, N.W., Brown, N.J., Frenklach, M.: PRISM: piece reusable implementation of solution mapping. An economical strategy for chemical kinetics. Isr. J. Chem. **39**, 97–106 (1999)
44. van Oijen, J.A., de Goey, L.P.H.: Modelling of premixed laminar flames using flamelet-generated manifolds. Combust. Sci. Technol. **161**, 113–138 (2000)

45. Yao, M., Zheng, Z., Liu, H.: Progress and recent trends in homogeneous charge compression ignition (HCCI) engines. Prog. Energy Combust. Sci. **35**, 398–437 (2009)
46. Veynante, D., Fiorina, B., Domingo, P., Vervisch, L.: Using self-similar properties of turbulent premixed flames to downsize chemical tables in high-performance numerical simulations, Combust. Theory Model. **12**, 1055–1088 (2008)
47. Vreman, A.W., van Oijen, J.A., de Goey, L.P.H., Bastiaans, R.J.M.: Subgrid scale modeling in large eddy simulation of turbulent combustion using premixed flamelet chemistry. Flow Turbul. Combust. **82**, 511–535 (2009)
48. Westbrook, C.K., Mizobuchi, Y., Poinsot, T.J., Smith, P.J., Warnatz, J.: Computational combustion. Proc. Combust. Inst. **30**, 125–157 (2005)
49. Woosley, S.E., Almgren, A., Bell, J.B., Glatzmaier, G., Kasen, D., Kerstein, A.R., Ma, H., Nugent, P., Röpke, F., Sankaran, V., Zingale, M.: Type Ia supernovae. J. Phys.: Conf. Ser. **78**, 012081 (2007)

Chapter 9
Model Reduction for Combustion Chemistry

Dimitris A. Goussis and Ulrich Maas

Abstract The description of chemically reacting systems leads very often to reaction mechanisms with far above hundred chemical species (and, therefore, to more than a hundred partial differential equations), which possibly react within more than a thousand of elementary reactions. These kinetic processes cover time scales from nanoseconds to seconds. Due to these scaling problems the detailed simulation of three-dimensional turbulent flows in practical systems using detailed kinetics is beyond the capacity of even today's super-computers. Using simplified sub-models for the chemical kinetics is a way out of this problem. The question arising in mathematical modeling of reactive flows is then: how detailed, or down to which scale has the chemical reaction to be resolved in order to allow a reliable description of the entire process? The aim is the development of models, which should be as simple as possible in the sense of an efficient description, and also as detailed as necessary in the sense of reliability. In particular, an oversimplification of the coupling processes between chemical reaction and turbulent flow should be avoided by all means to allow a predictive character. This chapter describes different methodologies for the reduction of chemical mechanisms for subsequent use in reacting flow calculations.

9.1 Introduction

The interest in the numerical simulation of combustion processes has grown considerably [45, 66, 69, 88, 89] during the last years, and computational fluid dynamics (CFD) has become an important design tool in various industrial applications such as internal combustion engines, gas turbines, power plants, and small

Dimitris A. Goussis
National Technical University of Athens, GR-15780 Athens, e-mail: dagoussi@mail.ntua.gr

Ulrich Maas
Karlsruhe Institute of Technology, D-76131 Karlsruhe, e-mail: Ulrich.Maas@kit.edu

T. Echekki, E. Mastorakos (eds.), *Turbulent Combustion Modeling*,
Fluid Mechanics and Its Applications 95, DOI 10.1007/978-94-007-0412-1_9,

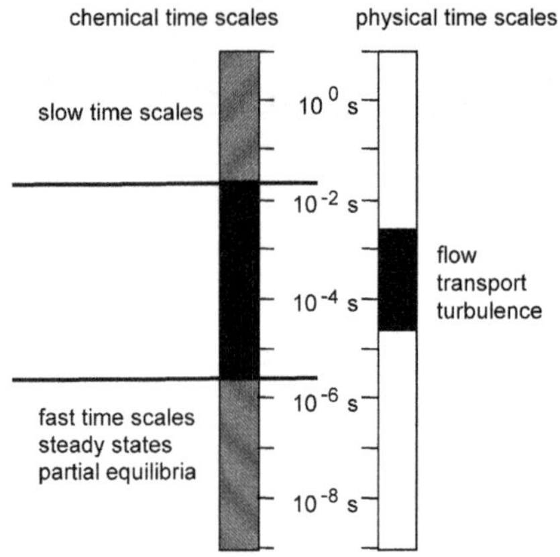

Fig. 9.1: Schematic illustration of the time-scales governing a chemically reacting flow.

scale burners. One problem, however, is that typical applications involve complex three-dimensional geometries, complicated fuel mixtures, and (in most cases) turbulent flow characteristics. At least for practical applications a complete solution of the governing equations (conservation equations for mass, momentum, energy, and species mass fractions) including detailed chemical kinetics is not possible because the chemistry introduces a large number of chemical species and reactions (typically up to 1000) resulting in a large dimension of the underlying system of conservation equations. In order to simplify the numerical treatment of the chemical kinetics, many methods have been devised, such as the assumption of an infinite reaction rate ("mixed equals burnt") or one-step chemistry. On the other hand pollutant formation (NO, soot, unburned hydrocarbons), as well as ignition and extinction are kinetically controlled processes [88], and oversimplified chemical schemes are not able to capture these processes. Thus, there is a strong need for methods which simplify the chemical kinetics while still retaining the essential dynamic features of the reaction system.

As early as one hundred years ago Bodenstein [4, 5] observed that some chemical reactions are so fast that some chemical species in the reaction system are in a quasi-steady state. Comparing the time-scales of the chemical reactions with the typical physical time scales (mixing, molecular transport, residence times) one can observe that the chemical time scales span typically a much larger range (see Fig. 9.1), and because only the time scales of similar order couple, one can decouple the fast ones by assuming them to be in steady state and the slow ones by assuming that they are frozen.

Based on the ideas of Bodenstein, many attempts have been made to develop simplified descriptions of chemical reaction systems, e.g., for the simulation of complex combustion processes, and a variety of different approaches can be found in the literature (see, e.g., [32, 50, 67, 76, 77] for references considering combustion processes). In principle two ways of simplifying the chemical kinetics can be distinguished. One is to use the knowledge about the reaction system, i.e. the information on which species are in quasi-steady state or which reactions are in partial equilibrium. The other is to extract exactly this information from the detailed reaction mechanism based on mathematical methods.

The traditional methodologies for simplifying a chemical kinetics mechanism involve the use of the Quasi Steady-State approximation (QSSA) and/or the Partial Equilibrium approximation (PEA) [44, 88, 90]. The QSSA assumes that the rate of production and the rate of destruction of a number of species are both much larger than their net rate of formation. Setting equal the rates of production and destruction of these species yields a system of algebraic relations among the elementary rates. On the other hand, the PEA assumes that the forward and/or backward rates of some reactions are so large that they form a number of equilibria, which are expressed by an equal number of algebraic relations among the elementary reaction rates. The algebraic system that results from the QSSA or PEA is used:

- in order to compute some of the concentration of species (those in steady state in the case of QSSA or those characterized as being involved in a reaction in partial equilibrium in the case of PEA), and
- to simplify the differential equations of the remaining species in the mechanism (usually by eliminating reactions deemed fast).

It is by now well accepted that the success of the QSSA and PEA methodologies relies on the existence of a number of dissipative time scales, which are much faster than the rest that characterize the long-term evolution of the system [6, 75]. This feature, which is known as *stiffness*, is essential for the development in phase space of low-dimensional surfaces, known as *slow manifolds*, to which all neighboring trajectories are attracted under the influence of the fast time scales [15, 20]. With each additional fast time scale becoming exhausted, the dimensions of the manifolds reduce by one. Therefore, the more fast time scales becoming exhausted, the lower the dimensions of the manifold. Once on the manifold, the fast time scales become exhausted and the solution is characterized by the slow time scales. The mathematical model that describes the evolution on the manifold is usually referred to as the *reduced model*. Clearly, the dimension of the manifold and the size of the reduced model decreases as the number of fast time scales considered exhausted increases.

Starting from the 80's, the QSSA and PEA were extensively employed for the generation of reduced chemical kinetics mechanisms, with great success [64, 65, 67, 68, 76]. When successful, the algebraic system of equations among the elementary rates provides an accurate approximation of the manifold and the reduced mechanism provides an accurate contribution of the kinetics in advancing the solution on the manifold. However, the use of these approximations is hindered by the difficulties in predicting correctly:

- the number of QSSAs and PEAs that are applicable,
- the species to be considered in steady state and, therefore, to be excluded from the stoichiometry of the reduced mechanism,
- the elementary reactions that participate in equilibria, and

- the elementary reactions that are fast and, therefore, must be excluded from the rates of the reduced mechanism.

Several algorithms were developed for the successful application of the QSSA and PEA [13, 24, 56, 79, 81]. However, problems in their use still remain, especially when considering the reduction of large and complex mechanisms of heavy hydrocarbons [61]. There are basically two reasons for these problems. First, a manifold is not always approximated by the algebraic relations that develop when applying the QSSA and/or PEA. The appropriate algebraic relations are usually more complex, involving (i) more reactions than those participating in the assumed QSSAs and/or the PEAs and (ii) the coefficients that multiply the elementary rates are not constant integers, but complex functions of the species concentrations, the temperature, the pressure, etc [28, 29, 85]. The second reason for the limits in the successful application of the QSSAs and PEAs is that the nature and the dimension of the manifold are not fixed, but are both functions of the local chemical and thermophysical properties of the system [29, 58]. The latter issue is demonstrated by the results shown in Fig. 9.2, relating to the analysis of the structure of a stoichiometric, steady, 2D, methane-air edge flame stabilized against an incoming mixing layer [58].

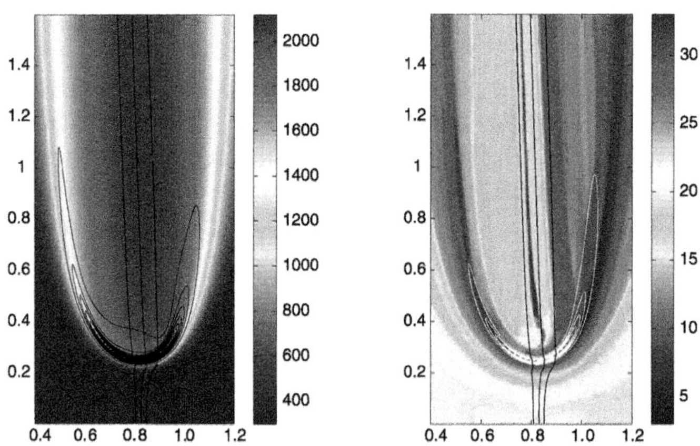

Fig. 9.2: Analysis of methane-air edge flame structure. The temperature field [K] shown with superposed heat release rate (left) and the number of fast dissipative time scales than can be neglected with superposed mass fraction of CH_4 (right). The fuel stream is on the right and the air stream is on the left. Reprinted from [58] with permission from Taylor and Francis.

The chemical kinetics is modeled with the GRImech3.0 mechanism, which involves 53 species. It is shown that the number of fast dissipative time scales that become exhausted, and therefore can be neglected, varies from about 5 to 30, depending on the position in the flame. In particular, the primary reaction zone exhibits the lowest number of fast time scales that can be neglected, while the central diffusion flame region exhibits the highest number. Therefore, optimal reduction within a pre-specified accuracy requires the use of a number of reduced models of varying size.

Such problems in the construction of reduced models lead to the development of algorithmic tools able to provide the appropriate simplified chemistry, without any involvement from the investigator. Such tools can be devised efficiently if one makes use of the facts that:

- not the whole composition space (spanned by the large number of species concentrations) is actually accessed, and
- the dynamics is confined to low-dimensional manifolds in composition space.

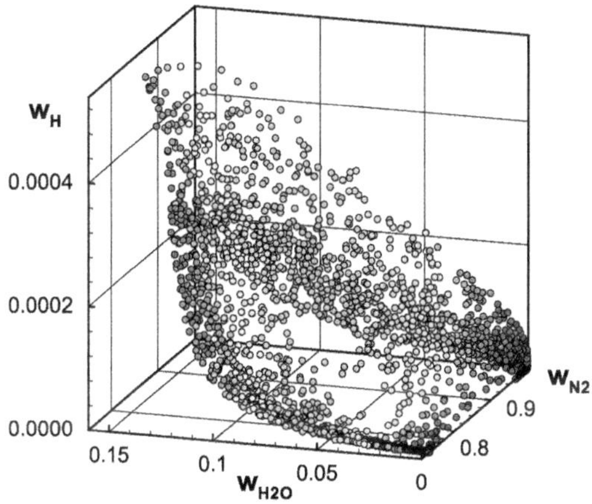

Fig. 9.3: Results from DNS of a hydrogen-oxygen-nitrogen non-premixed flame. Scatter plots in the composition space.

This behavior is typically observed in combustion processes and can be seen nicely from results of direct numerical simulations (DNS). Such DNS are based on a direct solution of the conservation equations without applying any modeling assumption and allow detailed studies of turbulent combustion, even taking into account detailed chemical kinetics. Figure 9.3 shows results of a DNS for a non-premixed hydrogen-air flame (see [55] for details). The results of the DNS have been represented as scatter plots in the state space (here we show only a 3-dimensional plot of the results with the mass fractions of N_2, H_2O, and H as coordinates, and in addition the temperature is coded by the color, where blue is cold and red is hot). The figure shows that not the whole composition space is accessed and furthermore that the accessed states are in the neighborhood of low-dimensional manifolds.

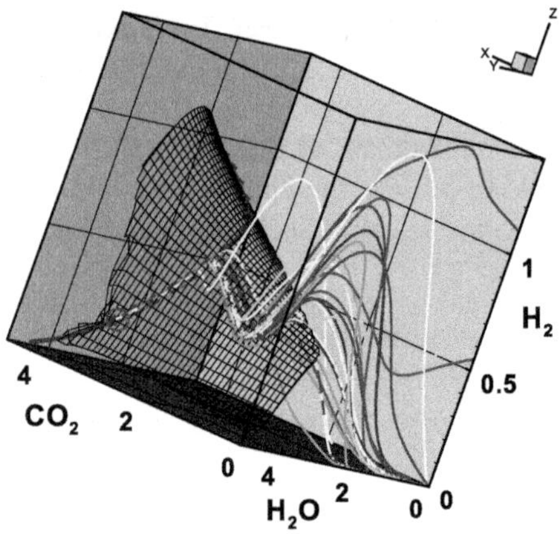

Fig. 9.4: Dynamical behavior of the chemical kinetics in composition space $\left[\frac{w_i}{M_i}\right] = \frac{mol}{kg}$.

This is further illustrated in Fig. 9.4, where trajectories of chemical reactions in composition space have been plotted for different chemical systems (from syngas to n-dodecane) and different initial conditions with the only requirement of having the same enthalpy, pressure and element composition [3]. It can be seen that the systems relax to lower-dimensional manifolds and finally approach equilibrium.

In the following sections of this chapter, the QSSA and PEA will be presented and discussed. A unified view of these two approximations will lead to a reduction strategy that can represent many of the available reduction algorithms currently available. These algorithms will then be displayed. Finally, the strategies followed for the implementation of these algorithms will be presented.

9.2 Traditional Methodologies for Reduction: QSSA and PEA

Relatively large chemical kinetics mechanisms were systematically reduced first with the QSSA and PEA methodologies. These methodologies will be applied here on the *Thermal NO* mechanism for the formation of nitric oxide via the reactions [88]:

$$O + N_2 \xrightarrow{k_1} NO + N \qquad R^1 = k_1[O][N_2] \tag{9.1}$$

$$N + O_2 \xrightarrow{k_2} NO + O \qquad R^2 = k_2[N][O_2] \tag{9.2}$$

$$N + OH \xrightarrow{k_3} NO + H \qquad R^3 = k_3[N][OH] \tag{9.3}$$

where R^i and k_i denote the rate and rate constant of the i-th reaction and $[X]$ denotes the concentration of species X. Considering a homogeneous mixture, when these reactions are part of a larger detailed kinetics mechanism, the rate of formation of NO is given by:

$$\frac{d[\mathrm{NO}]}{dt} = R^1 + R^2 + R^3 + r_{NO} \tag{9.4}$$

while that of the remaining species by:

$$\frac{d[\mathrm{N}]}{dt} = R^1 - R^2 - R^3 + r_N \tag{9.5}$$

$$\frac{d[\mathrm{O}]}{dt} = -R^1 + R^2 + r_O \tag{9.6}$$

$$\frac{d[\mathrm{OH}]}{dt} = -R^3 + r_{OH} \tag{9.7}$$

$$\frac{d[\mathrm{O_2}]}{dt} = -R^2 + r_{O_2} \tag{9.8}$$

$$\frac{d[\mathrm{N_2}]}{dt} = -R^1 + r_{N_2} \tag{9.9}$$

$$\frac{d[\mathrm{H}]}{dt} = R^3 + r_H \tag{9.10}$$

where the functions r_X represent the possible contribution from additional reactions in the detailed mechanism involving the species X.

Let us consider possible ways to reduce the three reactions in the *Thermal NO* mechanism by considering the reaction (9.3) as fast and the species N as the one affected the most by the related fast time scale. In particular, let us examine two alternatives for such a reduction; either to apply the QSSA for [N] or the PEA among the N-producing/consuming reactions (9.1), (9.2) and (9.3). These two alternatives will be examined separately, first for the case of the evolution equation of [NO], and then for the evolution of the concentration of all species in the mechanism.

9.2.1 The QSSA

Assuming the net rate of change of $[N]$ is much smaller than both its rate of production and its rate of destruction, Eq. (9.5) yields:

$$0 \approx R^1 - R^2 - R^3 + r_N \tag{9.11}$$

which, when substituted in Eq. (9.4) in order to eliminate the rate of the fast reaction R_3, yields:

$$\frac{d[\mathrm{NO}]}{dt} \approx 2\,R^1 + (r_{NO} + r_N) \tag{9.12}$$

Instead of the evolution Eq. (9.5), [N] can now be computed from the algebraic expression:

$$[\mathrm{N}] \approx \frac{R^1 + r_N}{k_2[\mathrm{O_2}] + k_3[\mathrm{OH}]} \tag{9.13}$$

9.2.2 The PEA

In its simplest version the PEA assumes that the forward and backward rates of certain reversible reactions are almost equal such that the net rate is approximately zero. Then the law of mass action of these reactions can be used as algebraic relations among the species. The concept of fast reactions can, however, also be applied in a more general context, which allows to analyze the reaction mechanism (9.1)-(9.3), which does not (in this special case) contain reversible reactions. Assuming the fast reaction (9.3) participates in an equilibrium with the reactions (9.1) and (9.2):

$$R^3 \approx R^1 - R^2 \tag{9.14}$$

the PEA is applied by taking the time derivative of Eq. (9.14), substituting from Eqs. (9.5)-(9.10), solving for the rate of the fast reaction R^3:

$$R^3 \approx a_1 R^1 - a_2 R^2 + a_3 \tag{9.15}$$

and then substituting in Eq. (9.4) in order to eliminate R^3:

$$\frac{d[\mathrm{NO}]}{dt} \approx (1+a_1) R^1 + (1-a_2) R^2 + (a_3 + r_{NO}) \tag{9.16}$$

where:

$$a_1 = 1 + \frac{k_1([\mathrm{N_2}]+[\mathrm{O}]) - k_3[\mathrm{N}]}{k_3([\mathrm{N}]+[\mathrm{OH}]) + k_2[\mathrm{O_2}]}$$

$$a_2 = 1 - \frac{(k_3 - k_2)[N] - k_1[N_2]}{k_3([\mathrm{N}]+[\mathrm{OH}]) + k_2[\mathrm{O_2}]}$$

$$a_3 = \frac{[\mathrm{N}](k_2 r_{O_2} + k_3 r_{OH}) + (k_2[\mathrm{O_2}] + k_3[\mathrm{OH}]) r_N - k_1([\mathrm{N_2}] r_O + [\mathrm{O}] r_{N_2})}{k_3([\mathrm{N}]+[\mathrm{OH}]) + k_2[\mathrm{O_2}]}$$

In this case, [N] can be computed either by the algebraic expression (9.14):

$$[\mathrm{N}] \approx \frac{R^1}{k_2[\mathrm{O_2}] + k_3[\mathrm{OH}]} \tag{9.17}$$

by the algebraic expression (9.15):

$$[\mathrm{N}] \approx \frac{a_1 R^1 + a_3}{a_2 k_2[\mathrm{O_2}] + k_3[\mathrm{OH}]} \tag{9.18}$$

or by an evolution equation that can be constructed when substituting the rate of the fast reaction R^3 from Eq. (9.15) to Eq. (9.5):

$$\frac{d[\mathrm{N}]}{dt} \approx (1-a_1)\,R^1 - (1-a_2)\,R^2 - (a_3 - r_N) \tag{9.19}$$

9.2.3 Comments on the QSSA and PEA

The two simplified evolution equations for [NO] constructed with the QSSA and the PEA, Eqs. (9.12) and (9.16), are related by the expression:

$$\left.\frac{d[\mathrm{NO}]}{dt}\right|_{PEA} = \left.\frac{d[\mathrm{NO}]}{dt}\right|_{QSSA} - (1-a_1)\,R^1 + (1-a_2)\,R^2 + (a_3 - r_N) \tag{9.20}$$

Given the relations among the rates, Eqs. (9.11) or (9.14), the QSSA and the PEA provide the same simplified model for [NO] only when the last three terms in the RHS of Eq. (9.20) can be neglected. This is possible only when the quantities:

$$\begin{aligned}
1 - a_1 &= -\frac{k_1([\mathrm{N}_2]+[\mathrm{O}]) - k_3[\mathrm{N}]}{k_3([\mathrm{N}]+[\mathrm{OH}])+k_2[\mathrm{O}_2]}\\
1 - a_2 &= \frac{(k_3-k_2)[N] - k_1[N_2]}{k_3([\mathrm{N}]+[\mathrm{OH}])+k_2[\mathrm{O}_2]}\\
a_3 - r_N &= \frac{[\mathrm{N}](k_2 r_{O_2} + k_3 r_{OH} - k_3 r_N) - k_1([\mathrm{N}_2]r_O + [\mathrm{O}]r_{N_2})}{k_3([\mathrm{N}]+[\mathrm{OH}])+k_2[\mathrm{O}_2]}
\end{aligned}$$

are sufficiently small. Otherwise, the two approximations provide different simplified equations of [NO], the accuracy of which cannot be evaluated *a priori*.

9.2.4 A Common Set-up for the QSSA and PEA

The QSSA and PEA can be viewed from a unified point of view as follows. For simplicity, let us concentrate on the three reactions in the *Thermal NO* mechanism and neglect the possible contribution from other reactions, which are represented by the terms r_X. In that case, Eqs. (9.4)-(9.10) can be cast in vector form as:

$$\frac{d\boldsymbol{\psi}}{dt} = \mathbf{S}_1\,R^1 + \mathbf{S}_2\,R^2 + \mathbf{S}_3\,R^3 = \mathbf{g}(\boldsymbol{\psi}) \tag{9.21}$$

where $\boldsymbol{\psi}$ is the 7-dim. vector of concentrations, $\mathbf{S}_i$ is the 7-dim. stoichiometric vector of the i-th reaction and R^i is the corresponding reaction rate:

$$\boldsymbol{\psi} = \begin{bmatrix} [\mathrm{N}] \\ [\mathrm{NO}] \\ [\mathrm{O}] \\ [\mathrm{OH}] \\ [\mathrm{O_2}] \\ [\mathrm{N_2}] \\ [\mathrm{H}] \end{bmatrix} \qquad \mathbf{S} = \begin{bmatrix} \mathbf{S}_1 \ \mathbf{S}_2 \ \mathbf{S}_3 \end{bmatrix} = \begin{bmatrix} +1 & -1 & -1 \\ +1 & +1 & +1 \\ -1 & +1 & 0 \\ 0 & 0 & -1 \\ 0 & -1 & 0 \\ -1 & 0 & 0 \\ 0 & 0 & +1 \end{bmatrix} \tag{9.22}$$

Both the QSSA and PEA amount to the resolution of the 7-dim. phase space according to a specific set of 7-dim. basis vectors $\mathbf{a}_i$ $(i = 1,7)$:

$$\frac{d\boldsymbol{\psi}}{dt} = a_1 f^1 + \mathbf{a}_2 f^2 + \ldots + a_6 f^6 + \mathbf{a}_7 f^7 \tag{9.23}$$

where f^i are the corresponding amplitudes, defined as $f^i = \mathbf{b}^i \bullet \mathbf{g}$, and the 7-dim. vectors $\mathbf{b}^i$ compose the dual basis ($\mathbf{b}^i \bullet \mathbf{a}_j = \delta^i_j$). Let us denote the set of basis vectors (column) by $\mathbf{a}$ and the set of dual basis vectors (row) by $\mathbf{b}$:

$$\mathbf{a} = \begin{bmatrix} \mathbf{a}_1 \ \mathbf{a}_2 \ \ldots \ \mathbf{a}_6 \ \mathbf{a}_7 \end{bmatrix} \qquad \mathbf{b} = \begin{bmatrix} \mathbf{b}_1 \ \mathbf{b}_2 \ \ldots \ \mathbf{b}_6 \ \mathbf{b}_7 \end{bmatrix}^T \tag{9.24}$$

For the QSSA, appropriate sets of basis and dual vectors are:

$$\mathbf{a} = \begin{bmatrix} -1 & 0 & 0 & 0 & 0 & 0 & 0 \\ +1 & +2 & 0 & +1 & 0 & 0 & -2 \\ 0 & -1 & +1 & -1 & +1 & -1 & +1 \\ -1 & -1 & +1 & 0 & 0 & +1 & -1 \\ 0 & 0 & -1 & 0 & 0 & 0 & +1 \\ 0 & -1 & 0 & 0 & 0 & 0 & +1 \\ +1 & +1 & -1 & 0 & 0 & 0 & +1 \end{bmatrix} \tag{9.25}$$

$$\mathbf{b} = \begin{bmatrix} -1 & 0 & 0 & 0 & 0 & 0 & 0 \\ +1 & 0 & 0 & 0 & +1 & 0 & +1 \\ +1 & 0 & 0 & 0 & -2 & +1 & +1 \\ +1 & +1 & 0 & 0 & 0 & +2 & 0 \\ 0 & +1 & +1 & +1 & +2 & 0 & 0 \\ 0 & 0 & 0 & +1 & 0 & 0 & +1 \\ -1 & 0 & 0 & 0 & +1 & -1 & -1 \end{bmatrix} \tag{9.26}$$

On the basis of these vectors, Eq. (9.23) yields:

$$f^1 = -R^1 + R^2 + R^3 \qquad f^2 = R^1 \qquad f^3 = R^2 \qquad f^4 = \cdots = f^7 \equiv 0 \tag{9.27}$$

$$\frac{d\boldsymbol{\psi}}{dt} = \mathbf{a}_1 f^1 + \mathbf{a}_2 f^2 + \mathbf{a}_3 f^3 \tag{9.28}$$

where the vanishing of the four amplitudes f^4 to f^7 is due to conservation laws; the first three due to the conservation of the N, O and H atoms in the stoichiometric vectors $\mathbf{S}_1$, $\mathbf{S}_2$ and $\mathbf{S}_3$.

Setting $f^1 \approx 0$ is equivalent to invoking the QSSA for [N], Eq. (9.11), so that the resulting simplified system:

$$\frac{d\boldsymbol{\psi}}{dt} \approx \mathbf{a}_2\, f^2 + \mathbf{a}_3\, f^3 \tag{9.29}$$

can be interpreted as the appropriate one for a detailed mechanism consisting of the following reactions and reaction rates:

$$\mathrm{O} + \mathrm{OH} + \mathrm{N_2} \rightarrow 2\mathrm{NO} + \mathrm{H} \qquad f^2 = R^1 \tag{9.30}$$

$$\mathrm{O_2} + \mathrm{H} \rightarrow \mathrm{OH} + \mathrm{O} \qquad f^3 = R^2 \tag{9.31}$$

as can be verified by inspecting the elements of the $\mathbf{a}_2$ and $\mathbf{a}_3$ vectors. Of course, these are not physical reactions; their stoichiometry and rates being a linear combination of the stoichiometry and the rates of the elementary reactions in the original chemical kinetics mechanism (9.1)-(9.3). Note that both the concentration of the fast species [N] and the rate of the fast reaction R^3 are missing in Eqs. (9.30)-(9.31). [N] can be computed from Eq. (9.13), which involves R^3.

For the PEA, appropriate sets of basis and dual vectors are:

$$\mathbf{a} = \begin{bmatrix} A & B & C & D & E & F & F \\ -A & 2-B & -C & -D & -E & -F & -F \\ 0 & -1 & +1 & 0 & 0 & 0 & 0 \\ A & -1+B & 1+C & 1+D & -1+E & 2+F & 2+F \\ 0 & 0 & -1 & -0.5 & +1 & -1 & -1 \\ 0 & -1 & 0 & +0.5 & 0 & 0 & 0 \\ -A & 1-B & -1-C & -1-D & 1-E & -1-F & -2-F \end{bmatrix} \tag{9.32}$$

$$\mathbf{b} = \begin{bmatrix} +1 & 0 & -k_1[\mathrm{N_2}]G & k_3[\mathrm{N}]G & k_2[\mathrm{N}]G & -k_1[\mathrm{O}]G & 0 \\ +0.5 & +0.5 & 0 & 0 & 0 & 0 & 0 \\ +0.5 & +0.5 & +1 & 0 & 0 & 0 & 0 \\ +1 & +1 & 0 & 0 & 0 & +2 & 0 \\ 0 & +1 & +1 & +1 & +2 & 0 & 0 \\ 0 & 0 & 0 & +1 & 0 & 0 & +1 \\ -1 & 0 & 0 & 0 & +1 & -1 & -1 \end{bmatrix} \tag{9.33}$$

where:

$$A = \frac{k_2[\mathrm{O_2}] + k_3[\mathrm{OH}]}{k_2[\mathrm{O_2}] + k_3([\mathrm{N}] + [\mathrm{OH}])} \qquad B = \frac{k_3[\mathrm{N}] - k_1([\mathrm{N_2}] + [\mathrm{O}])}{k_2[\mathrm{O_2}] + k_3([\mathrm{N}] + [\mathrm{OH}])}$$

$$C = \frac{k_1[\mathrm{N_2}] + (k_2 - k_3)[\mathrm{N}]}{k_2[\mathrm{O_2}] + k_3([\mathrm{N}] + [\mathrm{OH}])} \qquad D = \frac{1}{2}\frac{k_1[\mathrm{O}] + (k_2 - 2k_3)[\mathrm{N}]}{k_2[\mathrm{O_2}] + k_3([\mathrm{N}] + [\mathrm{OH}])}$$

$$E = \frac{(-k_2 + k_3)[\mathrm{N}]}{k_2[\mathrm{O_2}] + k_3([\mathrm{N}] + [\mathrm{OH}])} \qquad F = \frac{(k_2 - 2k_3)[\mathrm{N}]}{k_2[\mathrm{O_2}] + k_3([\mathrm{N}] + [\mathrm{OH}])}$$

$$G = (k_2[\mathrm{O_2}] + k_3[\mathrm{OH}])^{-1}$$

Setting $f^1 \approx 0$ is equivalent to invoking the PEA among reactions (9.1)-(9.3), Eq. (9.15), so that the resulting simplified system:

$$\frac{d\boldsymbol{\psi}}{dt} \approx \mathbf{a}_2\, f^2 + \mathbf{a}_3\, f^3 \tag{9.34}$$

can be interpreted as the appropriate one for a detailed mechanism consisting of the following reactions and reaction rates:

$$O + (1-B)OH + N_2 \rightarrow (2-B)NO + (1-B)H + (B)N \qquad f^2 = R^1 \tag{9.35}$$
$$O_2 + (1+C)H + (C)NO \rightarrow (1+C)OH + O + (C)N \qquad f^3 = R^2 \tag{9.36}$$

as can be verified by inspecting the elements of the $\mathbf{a}_2$ and $\mathbf{a}_3$ vectors. Neither of these reactions are physical ones; their stoichiometry and rates being a linear combination of the stoichiometry and the rates of the elementary reactions in the original chemical kinetics mechanism (9.1)-(9.3). Now, the concentration of the fast species [N] is present in Eqs. (9.35)-(9.36), while the rate of the fast reaction R^3 is missing. [N] can be computed either from the algebraic Eqs. (9.17) or (9.18) or from the first component of Eq. (9.34).

The reactions and reaction rates in Eqs. (9.30)-(9.31) and (9.35)-(9.36) represent 2-step reduced models constructed with the QSSA and PEA, respectively. A significant difference between the two reduced models is the fact that the stoichiometry of the QSSA model involves constant coefficients, while that of the PEA model involves variable coefficients. However, the stoichiometry of the two reactions in both reduced models conserves the elements, as in the case of the elementary reactions. Note that the reduced model constructed with the PEA can be reduced to that developed by the QSSA by setting $B = C = 0$. At this point it must be emphasized that the form of the 2-step reduced mechanisms produced by both QSSA and PEA, as stated in Eqs. (9.30)-(9.31) and (9.35)-(9.36), is not unique. This can easily be easily demonstrated by casting Eq. (9.29) or (9.34) as:

$$\frac{d\boldsymbol{\psi}}{dt} \approx (\mathbf{a}_2 + q(\boldsymbol{\psi})\mathbf{a}_3)\, f^2 + \mathbf{a}_3\, (f^3 - q(\boldsymbol{\psi}) f^2) \tag{9.37}$$

where $q(\boldsymbol{\psi})$ is an arbitrary function of $\boldsymbol{\psi}$.

9.2.5 The Need for Algorithmic Methodologies for Reduction

In order to validate the reduced models constructed with the QSSA and PEA in the previous section for the case of the three reactions in the *Thermal NO* mechanism, let us consider a case where the mixture is at constant temperature $T = 1956$ K and initial composition:

$$[NO] = 0.9524\ 10^{-10} \qquad [O] = 0.1984\ 10^{-7} \qquad [OH] = 0.5028\ 10^{-7}$$
$$[O_2] = 0.1165\ 10^{-6} \qquad [N_2] = 0.4273\ 10^{-5} \qquad [H] = 0.4217\ 10^{-7}$$

where the units of concentration are in mol/cm^3 and [N] is computed from the relations $R^1 + R^2 = R^3$, so that Eqs. (9.11) and (9.14) (which are identical in the case considered here), pertinent to the QSSA and PEA, are initially fully satisfied. It can be shown that for this case, $-1 + a_1 \approx 0.5$ and $1 - a_2 \approx 2.5$. Since these quantities are not negligible, according to the discussion in Section 9.2.3 the QSSA and PEA are expected to provide different results.

The evolution in time of [NO] is displayed in Fig. 9.5, as computed from:

- the detailed model; i.e., Eqs. (9.4)-(9.10),
- the QSSA reduced model; i.e., Eq. (9.29) for all concentrations except [N] and Eq. (9.11) for [N],
- a first PEA reduced model; i.e., Eq. (9.34) for all concentrations (including [N]), denoted as PEA (1),
- a second PEA reduced model; i.e., Eq. (9.34) for all concentrations except [N] and Eq. (9.14) for [N], denoted as PEA (2).

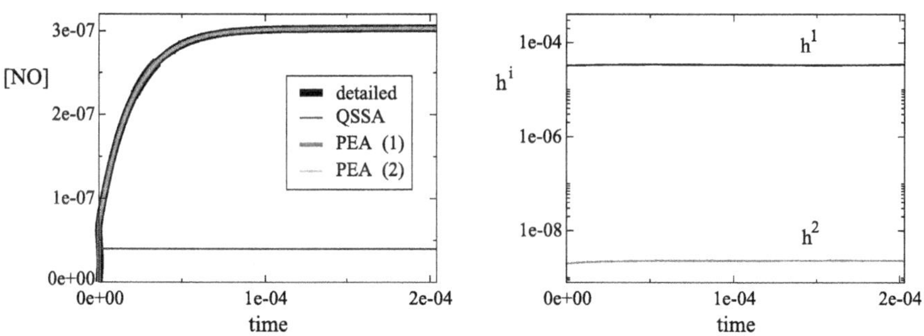

Fig. 9.5: Left: The evolution of [NO] (moles/cm^3) with time (s), determined from the full model, the QSSA model and two PEA models. Right: the evolution of the residuals h^1 and h^2, Eq. (9.38).

It is shown that both PEA models provide excellent accuracy, while the QSSA model is shown to provide poor accuracy. Noting that the construction of both PEA reduced models are based on Eq. (9.15), while the QSSA and second PEA models invoke the same algebraic relation among the elementary rates, Eqs. (9.11) and (9.14), for the computation of [NO], the magnitude of the following ratios is then examined:

$$h^1 = \frac{R^1 - R^2 - R^3}{R^1 + R^2 + R^3} \qquad h^2 = \frac{a_1R^1 - a_2R^2 - R^3}{|a_1R^1| + |a_2R^2| + |R^3|} \tag{9.38}$$

where h^1 measures the cancelations occurring in Eqs. (9.11) or (9.14) and h^2 measures the cancelations occurring in Eq. (9.15). Clearly, the larger the cancelations the greater the validity of the corresponding algebraic relation. Figure 9.5 clearly demonstrates that the reduced model, Eq (9.34), constructed on the basis of the refined Eq. (9.15) is much more accurate than the reduced model, Eq. (9.29), constructed on the basis of Eqs. (9.11) or (9.14), providing thus an explanation for the greater success of the first PEA model.

The treatment of this simple example clearly demonstrated the limitations in the use of the QSSA and PEA approximations. It is not known *a priori* if either of the two approximations are valid, nor the extent of the accuracy they provide. In addition, for the case of the PEA, it was shown how difficult its use is, even for this very simple case. The algorithms to reported next, resolve all these problems.

9.3 Reduction Algorithms

Consider a detailed mechanism, consisting of K reactions, N species and E elements. Considering a homogenous and isothermal mixture, the governing equations for the concentration of the species can be cast as:

$$\frac{d\boldsymbol{\psi}}{dt} = \mathbf{S}\,\mathbf{R}(\boldsymbol{\psi}) = \mathbf{S}_1 R^1 + \mathbf{S}_2 R^2 + \ldots + \mathbf{S}_K R^K = \mathbf{g}(\boldsymbol{\psi}) \tag{9.39}$$

where $\boldsymbol{\psi}$ is the N-dim. vector of concentrations, $\mathbf{S}$ is the $(N \times K)$-dim. matrix of the N-dim. stoichiometric vectors $\mathbf{S}_i$ and $\mathbf{R}$ is the K-dim. vector of the reaction rates R^i:

$$\mathbf{S} = \left[\mathbf{S}_1\ \mathbf{S}_2\ \ldots\ \mathbf{S}_K\right] \qquad \mathbf{R} = \begin{bmatrix} R^1 \\ \vdots \\ R^K \end{bmatrix} \tag{9.40}$$

Let us assume that for a certain period of time there exist M dissipative fast time scales, which are much faster than the remaining time scales that locally characterize the evolution of the system. Let us measure this time scale gap with the ratio of the slowest of the fast time scales to the fastest of the slow ones: $\varepsilon = \tau_M/\tau_{M+1}$. Let us further introduce the N-dim. basis (column) vectors $\mathbf{a}_i$:

$$\mathbf{a}_M = \left[\mathbf{a}_1 \ldots \mathbf{a}_M\right] \qquad \mathbf{a}_S = \left[\mathbf{a}_{M+1} \ldots \mathbf{a}_{N-M-E}\right] \qquad \mathbf{a}_C = \left[\mathbf{a}_{N-M-E+1} \ldots \mathbf{a}_N\right] \tag{9.41}$$

on the basis of which Eq. (9.40) is cast as:

$$\frac{d\boldsymbol{\psi}}{dt} = \mathbf{a}_M\,\mathbf{f}^M + \mathbf{a}_S\,\mathbf{f}^S + \mathbf{a}_C\,\mathbf{f}^C \tag{9.42}$$

where $\mathbf{b}^i \bullet \mathbf{a}_j = \delta^i_j$ are the dual basis (row) vectors and:

$$\mathbf{f}^M = \mathbf{b}^M \mathbf{g}(\boldsymbol{\psi}) \qquad \mathbf{f}^S = \mathbf{b}^S \mathbf{g}(\boldsymbol{\psi}) \qquad \mathbf{f}^C = \mathbf{b}^C \mathbf{g}(\boldsymbol{\psi}) \tag{9.43}$$

$$\mathbf{b}^M = \begin{bmatrix} b^1 \\ \vdots \\ b^M \end{bmatrix} \qquad \mathbf{b}^S = \begin{bmatrix} b^{M+1} \\ \vdots \\ b^{N-M-E} \end{bmatrix} \qquad \mathbf{b}^C = \begin{bmatrix} b^{N-M-E+1} \\ \vdots \\ b^N \end{bmatrix} \tag{9.44}$$

The elements of the vectors $\mathbf{f}^M$, $\mathbf{f}^S$ and $\mathbf{f}^C$ are the amplitudes of the projection of the vector field $\mathbf{g}(\boldsymbol{\psi})$ along the directions of the vectors in $\mathbf{a}_M$, $\mathbf{a}_S$ and $\mathbf{a}_C$, respectively. Let us assume that the vectors in $\mathbf{a}_M$ and their dual vectors $\mathbf{b}^i (i = 1, M)$ define the subspace in which the M fast time scales act. Therefore, when these time scales become exhausted:

$$\mathbf{f}^M = \mathbf{b}^M \, \mathbf{S}\,\mathbf{R} \approx 0 \tag{9.45}$$

where the symbol "$\approx$" denotes the possibility that the basis vectors in $\mathbf{a}_M$ and their corresponding dual vectors provide an approximation only of the fast subspace. Let us further assume that the basis vectors in $\mathbf{a}_C$ define the subspace in which conservation laws act (i.e., the conservation of elements in the elementary reaction), so that $\mathbf{f}^C \equiv 0$. Therefore, when the fast time scales are exhausted the evolution Eq. (9.42) reduces to:

$$\frac{d\boldsymbol{\psi}}{dt} \approx \mathbf{a}_S \, \mathbf{f}^S = \left[\mathbf{I} - \mathbf{a}_M \mathbf{b}^M\right] \mathbf{g}(\boldsymbol{\psi}) \tag{9.46}$$

where $\mathbf{I}$ is the unit matrix. This reduced system is free of the fast time scales and is valid within the constraints imposed by the Eq. (9.45), which defines the slow manifold.

Most of the existing algorithms for the generation of the manifold and the reduced system produce results in the form of Eqs. (9.45) and (9.46); i.e., directly or indirectly they generate a coordinate system for the vector field $\mathbf{g}(\boldsymbol{\psi})$, on the basis of which the system of differential equations reduces to normal form.

In particular, the iterative procedure for defining the slow manifold of Fraser and Roussel through the *invariance equation* [74], can be cast in the form of Eq. (9.43) [30]. This iterative procedure can be extended to produce the reduced model as well [30]. Both the manifold and the reduced model can provide any order of accuracy, since the iterative procedure generates, term by term, the asymptotic expansion of the asymptotic expansion of the slow manifold given by geometric singular perturbation theory [30, 37].

The Computational Singular Perturbation (CSP) algorithm of Lam and Goussis provides two refinement procedures for the computation of the basis vectors $\mathbf{a}_i$ and their dual $\mathbf{b}^i$ [41, 42]. The so called $\mathbf{b}^r$-refinement generates, order by order, the asymptotic expansion of a slow manifold, while the $\mathbf{a}_r$-refinement approximates the tangent spaces to the fast fibers along which solutions relax to the slow manifold [92, 93]. CSP detects the dimensions of the manifold, subject to a pre-specified accuracy, and can also identify the fast variables and fast processes in the detailed model [39, 43].

The Intrinsic Low-Dimensional Manifolds (ILDM) algorithm of Maas and Pope uses the right and left eigenvectors of the Jacobian of the vector field $\mathbf{g}(\boldsymbol{\psi})$ in order to approximate the $\mathbf{a}_i$ and $\mathbf{b}^i$ vectors [52, 53]. With this approach, the manifold and the reduced model provide leading order accuracy, when compared with that provided by geometric singular perturbation theory [37]. ILDM incorporates a tabulation strategy that makes the algorithm very efficient. A variant of the method is the GQL (global quasi-linearization) method, which uses the eigenspaces of a global quasi-linearization matrix instead of those of the local Jacobian [9, 11] or the TILDM method which uses the eigenspaces of JJ^T, where J denotes the local Jacobian ([8].

The Zero Derivative Principle (ZDP) algorithm of Gear and Kevrekidis yields successive approximations to the slow manifold by setting derivatives of successively higher order of the subset of "fast variables" of the state variables equal to zero [21]. This algorithm was later extended to provide approximations of the basis vectors $\mathbf{a}_i$ and their dual $\mathbf{b}^i$, so that it can now provide a reduced model as well [94]. As with the CSP method, the ZDP algorithm in its extended form provides, term by term, the results of the geometric singular perturbation theory. A variant of this algorithm is the one developed by Contou-Carrere and Daoutidis [16].

The Method of Invariant Manifolds (MIM) and the Method of Invariant Grids (MIG) of Gorban and Karlin [14, 26, 27], are based on the introduction of a projector $\mathbf{P} = (\mathbf{I} - \mathbf{a}_M \mathbf{b}^M)$, which is defined on the basis of thermodynamics: the entropy should grow in the fast motion. With this approach, both the manifold and the reduced system can be constructed, in the form of Eqs. (9.45) and (9.46). The slow manifold is considered as the stable fixed point of a relaxation equation or of an algebraic equation solved iteratively with the Newton method with incomplete linearization.

The Rate-Controlled Constrained Equilibrium (RCCE) methodology, originally proposed by Keck [38] and developed further by Jones and Rigopoulos [35, 36], is also used for the construction of reduced models. The basis of RCCE is that, in a system that exhibits substantial time-scale separation, the species governed by the fast time scales will be in a constrained equilibrium state, led by the concentrations of the leading species. Chemical kinetics can then be employed for computing the evolution of the leading species only, and the "fast" species can be calculated via a constrained minimisation of the Gibbs free energy, such that the leading species retain their values computed by kinetics. A system of differential-algebraic equations (DAEs) results, which is readily parametrised by the selection of leading species.

9.4 Interaction of Chemistry with Diffusion

In the presence of diffusion, the governing Eq. (9.39) is modified as:

$$\frac{d\boldsymbol{\psi}}{dt} = \mathbf{L}(\boldsymbol{\psi}) + \mathbf{S}\,\mathbf{R}(\boldsymbol{\psi}) \qquad (9.47)$$

where the N-dim. operator $\mathbf{L}$ accounts for the diffusion terms. In this case, the manifold and the reduced model are stated as:

$$\mathbf{f}^M = \mathbf{b}^M\,(\mathbf{L} + \mathbf{S}\,\mathbf{R}) \approx 0 \tag{9.48}$$

$$\frac{d\boldsymbol{\psi}}{dt} \approx \mathbf{a}_S\,\mathbf{f}^S = \left[\mathbf{I} - \mathbf{a}_M\mathbf{b}^M\right]\,(\mathbf{L} + \mathbf{S}\,\mathbf{R}) \tag{9.49}$$

where $\mathbf{f}^S = \mathbf{b}^S \bullet (\mathbf{L} + \mathbf{S}\;\mathbf{R})$ [33, 39, 53, 73]. In this case, Eqs. (9.48)-(9.49) can provide leading order accuracy only, provided the characteristic time scale due to transport is of the order of the fastest of the slow chemical time scales [33, 40]. Under certain conditions Eq. (9.48) can be simplified to Eq. (9.45), however the projector $\left[\mathbf{I} - \mathbf{a}_M\mathbf{b}^M\right]$ must always operate on both the chemistry and the diffusion terms [33, 40, 52, 73].

In general, there are two alternatives to proceed in the case where transport is present:

- obtain the reduced model, as in Eqs. (9.48)-(9.49) directly for the PDE;
- discretize in space the original PDE, Eq. (9.47), cast the system in the form of Eq. (9.39), and construct the reduced system in the form of Eqs. (9.45)-(9.46) for the resulting ordinary differential equation system.

With the first approach, the number M of fast dissipative time scales that can be considered exhausted might be a function of space, making the time integration not as optimal as the largest simplification possible. The first approach allows for a simpler construction of the simplified system, while the second allows for a simpler time marching scheme [31].

The coupling of the chemical kinetics with diffusion can be handled very efficiently in the context of manifold methods by taking into account the coupling directly in the construction of the manifolds (see next section).

9.5 Manifold Methods and Tabulation Strategies

9.5.1 Principles of Manifold Methods

Manifold methods use consequently the fact that the actual thermokinetic states are confined to or at least in the vicinity of low-dimensional attractors in composition space (cf. Section 9.3). Therefore the actual $n = n_s + 2$- dimensional state space $\boldsymbol{\psi}$ (which is, e.g. given as $\boldsymbol{\psi} = (h, p, w_1/M_1, \ldots, w_{n_s}/M_{n_s})^T$, where n_s is the number of species, h is the specific enthalpy, p the pressure, w_i and M_i the mass fraction and the molar mass of species i, respectively) can be described by a much smaller number m of so-called reduced coordinates or progress variables $\boldsymbol{\theta}$:

$$\boldsymbol{\psi} = \boldsymbol{\psi}(\boldsymbol{\theta}) \tag{9.50}$$

This function can be obtained e.g. using steady-state assumptions, ILDM equations, GQL equations, etc., which has been described above. For an implementation of the developed method of system reduction into a reactive flow calculation it is necessary to derive reduced set of conservation equations. Let us start from a detailed equation system of a typical reacting flow process in symbolic vector form:

$$\frac{\partial \boldsymbol{\psi}}{\partial t} = \mathbf{F}(\boldsymbol{\psi}) - \mathbf{v} \bullet \mathrm{grad}(\boldsymbol{\psi}) - \frac{1}{\rho}\,\mathrm{div}(\mathbf{D} \bullet \mathrm{grad}(\boldsymbol{\psi})) \qquad (9.51)$$

where v represents the velocity field, ρ the density and $\mathbf{D}$ is the (n by n)-dimensional matrix of the transport coefficients [19].

According to the basic assumption of the manifold methods, the state $\boldsymbol{\psi}$ is, at any point of the flow and at any time, close to the manifold (i.e. the system dynamics in the state space are completely described as a movement within the manifold). If all thermochemical states everywhere in chemical reacting system are elements of the manifold, neither the convective term causes movements in the state space perpendicular to the manifold [52], nor the chemical source term causes movements off the manifold for invariant manifolds and marginal movements off the manifold for manifolds that are not invariant. However, a projection of the transport term is required and can be done by a transformation of the transport term into the local coordinates of the invariant slow subspace [52]:

$$\frac{\partial \boldsymbol{\psi}}{\partial t} = \mathbf{F}(\boldsymbol{\psi}) - \mathbf{v} \bullet \mathrm{grad}(\boldsymbol{\psi}) - \frac{1}{\rho}\mathbf{P} \bullet \mathrm{div}(\mathbf{D} \bullet \mathrm{grad}(\boldsymbol{\psi})), \qquad (9.52)$$

where $\mathbf{P}$ denotes the (n by n)-dimensional projection operator for the diffusion term. It should be noted that this projection can be omitted as an approximation if the reduced coordinates $\boldsymbol{\theta}$ are chosen such that the fast relaxing chemical processes cause only a small marginal change of these progress variables. Accordingly, the system (9.52) calculates the full n-dimensional state $\boldsymbol{\psi}$, but it restricts the evolution to a movement tangent to the manifold. This equation system contains now redundant information, because $\boldsymbol{\psi}$ is a function of only m variables $\boldsymbol{\theta}$ and the evolution of $\boldsymbol{\psi}$ is confined to the low-dimensional manifold. This can be attributed by changing the equation system into an equation system for $\boldsymbol{\theta}$. In principle there are two possible choices. The first one is based on a representation of the reduced coordinate in terms of original species in the mechanism, i.e.:

$$\boldsymbol{\theta} = \mathbf{C}\boldsymbol{\psi}, \qquad (9.53)$$

where $\mathbf{C}$ is an (m by n)-dimensional parameterization matrix. If e.g. the species in the detailed mechanism are $CH_4, O_2, CO_2, H_2O, OH, O, \ldots$, and the progress variables are chosen to be CO_2 and H_2O, then the matrix is simply given by:

$$\mathbf{C} = \begin{pmatrix} 0\,0\,1\,0\,0\,0\cdots \\ 0\,0\,0\,1\,0\,0\cdots \end{pmatrix} \qquad (9.54)$$

Pre-multiplying the equation system with $\mathbf{C}$ then yields:

$$\frac{\partial \boldsymbol{\theta}}{\partial t} = \mathbf{CF}(\boldsymbol{\psi}) - \mathbf{v} \bullet \mathrm{grad}(\boldsymbol{\theta}) - \frac{1}{\rho}\mathbf{CP} \bullet \mathrm{div}(\mathbf{D} \bullet \mathrm{grad}(\boldsymbol{\psi}(\boldsymbol{\theta}))), \qquad (9.55)$$

Details can be found e.g. in [54]. A more elegant way is to allow an arbitrary parameterization. This can avoid problems with uniqueness or condition of the parameterization. In this case the n-dimensional governing equation system (9.51) is transformed by premultiplying with the Moore-Penrose-inverse (see e.g [25]). It is given for a regular matrix $\boldsymbol{\psi}_{\boldsymbol{\theta}}^T \bullet \boldsymbol{\psi}_{\boldsymbol{\theta}}$ by

$$\boldsymbol{\psi}_{\boldsymbol{\theta}}^+ = \left(\boldsymbol{\psi}_{\boldsymbol{\theta}}^T \bullet \boldsymbol{\psi}_{\boldsymbol{\theta}}\right)^{-1} \bullet \boldsymbol{\psi}_{\boldsymbol{\theta}}^T, \tag{9.56}$$

yielding

$$\frac{\partial \boldsymbol{\theta}}{\partial t} = \boldsymbol{\psi}_{\boldsymbol{\theta}}^+ \bullet \mathbf{F}(\boldsymbol{\psi}) - \mathbf{v} \bullet \mathrm{grad}(\boldsymbol{\theta}) - \frac{1}{\rho} \boldsymbol{\psi}_{\boldsymbol{\theta}}^+ \bullet \mathbf{P} \bullet \mathrm{div}(\mathbf{D} \bullet \psi_{\boldsymbol{\theta}} \bullet \mathrm{grad}(\boldsymbol{\theta}))\,. \tag{9.57}$$

If we denote $\boldsymbol{\psi}_{\boldsymbol{\theta}}^+ \bullet \mathbf{F}(\boldsymbol{\psi})$ by $\mathbf{S}(\boldsymbol{\theta})$, $\boldsymbol{\psi}_{\boldsymbol{\theta}}^+ \bullet \mathbf{P}$ by $\boldsymbol{\Pi}(\boldsymbol{\theta})$, and $\mathbf{D} \bullet \boldsymbol{\psi}_{\boldsymbol{\theta}}$ by $\mathbf{D}^\star(\boldsymbol{\theta})$, we can see that now the governing equation system for the scalar field of the reacting flow is a partial differential equation system for the m-dimensional reduced cordinates vector $\boldsymbol{\theta}$:

$$\frac{\partial \boldsymbol{\theta}}{\partial t} = \mathbf{S}(\boldsymbol{\theta}) - \mathbf{v} \bullet \mathrm{grad}(\boldsymbol{\theta}) - \frac{1}{\rho} \boldsymbol{\Pi}(\boldsymbol{\theta}) \,\mathrm{div}(\mathbf{D}^\star(\boldsymbol{\theta}) \bullet \mathrm{grad}(\boldsymbol{\theta}))\,. \tag{9.58}$$

This equation system is suitable for a coupled solution together with the equations for the flow field.

9.5.2 Calculation of Low-Dimensional Manifolds

Various methods exist to calculate low-dimensional manifolds. In principle all the methods described above can be used for a generation of low-dimensional manifolds, and various methods have been devised for an efficient calculation and storage. These shall be described in the following.

9.5.2.1 Slow manifolds

Slow manifolds [74] have the advantage that they represent invariant manifolds and represent the overall slow dynamics. Nevertheless their calculation is based on a functional iteration which may have convergence problems. Therefore, Davis and Skodje [17] had the idea to start the iteration from ILDMs as starting guess, which improved the convergence considerably. Furthermore, a solution process in terms of an evolution equation for the manifolds is possible [57]. The problem with this technique is that the governing equation for the slow manifold is a partial differential equation system which is difficult to solve. Lately, the convergence problems were eliminated with the formulation of the method along the CSP ideas [30]. Similar to these approaches is the method of invariant grids [26]. Another method of approximating slow manifolds is the calculation of minimal entropy production trajectories [46].

9.5.2.2 Computational Singular Perturbation Manifolds

CSP generated manifolds can account for the variation on their dimensionality and a pre-specified accuracy [41, 42]. These manifolds are generated using an iterative algorithm, each iteration providing a higher order correction term [83, 92]. These terms have a magnitude of increasing powers of the fast/slow time scale ratio. Therefore, CSP is most successful when this time scale ratio is small [39, 43]. The disadvantage on the implementation of CSP is the computational cost of computing on-the-fly the appropriate basis vectors. Lately, a new computational technique for the computation of the basis vectors was developed, which is very efficient when large reductions are possible [30]. In addition, a new methodology was developed which employs efficient tabulation in order to avoid the basis vectors calculations [47].

9.5.2.3 Intrinsic Low-Dimensional Manifolds

Low-dimensional manifolds can also be constructed using the ILDM [51, 52] TILDM [8] or GQL [9, 11] technique. The major problem here is that ILDMs of very low dimension (≤ 2) exist only in the domain of fast chemistry close to equilibrium, making a tabulation difficult (see below). A splitting technique, however, allows to extend the ILDM into the domain of slow chemistry [10], thus keeping the overall dimension small. Furthermore, a flamelet prolongation [23] can be used too (see below).

9.5.2.4 Repro-modeling

The idea of repro-modeling is to calculate a large number of combustion scenarios (homogeneous reactor calculations, flame calculations, etc.) [80] to yield a large set of accessed compositions $\boldsymbol{\psi}$. Because of the existence of low-dimensional attractors in composition space the results will lie in the vicinity of these attractors. Therefore the accessed states are represented by a low-dimensional manifold, e.g. by approximating them by a multi-variate least squares spline. The advantage of this method is that the calculation of the accessed composition is quite straightforward. The disadvantage is that the choice of the controlling variables and the construction of the approximating function is not trivial. This method can also be combined with a relaxation process according to the ILDM-equations, yielding a simple to implement variant of the ILDM method [7].

9.5.2.5 Trajectory Generated Manifolds

Another possibility is to construct the manifolds based on homogeneous reactor calculations. An m-dimensional manifold in composition space is constructed by solving the chemical rate equations for initial conditions lying on an $m-1$ dimensional manifold of initial conditions. If $\boldsymbol{\Upsilon}$ is the set of initial conditions, then if $\boldsymbol{\psi}(t;\boldsymbol{\psi}_0)$ is the solution of $\mathrm{d}\boldsymbol{\psi}/\mathrm{d}t = \mathbf{F}(\boldsymbol{\psi})$

$$\mathbf{M} = \{\boldsymbol{\psi}(t;\boldsymbol{\psi}_0)\,|0 \leq t \wedge \{\boldsymbol{\psi}_0\} = \boldsymbol{\Upsilon}\} \tag{9.59}$$

There are several variants to choose the manifold generator $\boldsymbol{\Upsilon}$. One of them is to choose initial conditions at the boundary of the domain [71], newer variants such as the ICE-PIC method use constrained equilibrium manifolds or minimum-curvature pre-image curves (see [73]). The advantage of these methods is that there are well established methods for the calculation of the trajectories and the pre-image curves. A disadvantage is the the subsequent parameterization of the manifold is not trivial.

9.5.2.6 Flamelet Generated Manifolds, Flamelet Prolongation of ILDM

If reduced schemes for subsequent flame calculation are devised, trajectory generated manifolds have the disadvantage that for initial conditions in the domain of slow chemistry (e.g. the composition of the unburned mixtures) the rate of reaction tends to zero. In addition the coupling with molecular transport processes has to be included by a suitable projection, and in the domain of slow chemistry the diffusion processes govern the low-dimensional attractors. Therefore, following the idea of the flamelet concept, manifolds can be generated similar to trajectory generated manifolds by solving flamelet equations (here for varying boundary conditions instead of initial conditions) [87] or to extend ILDM into the domain of slow chemistry by calculating flamelets between a composition on the ILDM and the unburned mixtures [23]. Several variants exist, which are based on premixed or non-premixed flamelets. The advantage of this method is that it is quite robust to implement and accounts for the coupling of chemical kinetics with molecular transport already in the construction of the manifold. One disadvantage is the fact that an extension to higher dimensions is not straightforward, although recently such an extension has been reported [59].

9.5.2.7 Reaction-Diffusion Manifolds

We suppose that the system solution in the state space is "close" or belongs to an m_s-dimensional manifold defined by an explicit function $\boldsymbol{\psi}(\boldsymbol{\theta})$

$$\mathbf{M} = \{\boldsymbol{\psi} : \boldsymbol{\psi} = \boldsymbol{\psi}(\boldsymbol{\theta}), \boldsymbol{\psi} : R^{m_s} \rightarrow R^n\}, \tag{9.60}$$

here $\boldsymbol{\theta}$ is the m-dimensional vector of reduced coordinates. M is an invariant m-dimensional system manifold if we solve the partial differential equation system

$$\frac{\partial \boldsymbol{\psi}(\boldsymbol{\theta})}{\partial t} = \left(\mathbf{I} - \boldsymbol{\psi}_{\theta}\, \boldsymbol{\psi}_{\theta}^{+}\right) \bullet \left\{ \mathbf{F}(\boldsymbol{\psi}) - \mathbf{v} \bullet \mathrm{grad}(\boldsymbol{\psi}) - \frac{1}{\rho} \mathrm{div}\left(\mathbf{D} \bullet \mathrm{grad}(\boldsymbol{\psi})\right) \right\} = 0 \tag{9.61}$$

until a stationary solution is obtained. In this case the normal component of the vector field on the invariant manifold vanishes. Here $\boldsymbol{\psi}_{\theta}^{+}$ is the Moore-Penrose pseudo-inverse of $\boldsymbol{\psi}_{\theta}$ (see above).

Rewriting the gradients in terms of the gradients of the reduced coordinates one obtains (see [73] for a detailed discussion of these diffusion terms)

$$\frac{\partial \boldsymbol{\psi}(\boldsymbol{\theta})}{\partial t} = \left(\mathbf{I} - \boldsymbol{\psi}_{\theta}\, \boldsymbol{\psi}_{\theta}^{+}\right) \bullet \left\{ \mathbf{F}(\boldsymbol{\psi}(\boldsymbol{\theta})) - \frac{1}{\rho} \mathrm{div}\left(\mathbf{D}\boldsymbol{\psi}_{\theta} \bullet \mathrm{grad}(\boldsymbol{\theta})\right) \right\} = 0 \tag{9.62}$$

This equation system can be solved if the spatial gradients are known as functions of the reduced coordinates. There are three limiting cases:

- The gradients are zero, which means that a homogeneous system is considered only with no molecular transport. In this case the equation simplifies to the PDE for the slow manifolds [57].
- The gradients are taken from a laminar flamelet. In this case the equation reduces for a one- dimensional manifold to the flamelet equation.
- There is no chemical reaction. In this case the equation represents an evolution equation which yields the minimal surface in composition space governed by diffusion (a linear surface in the case of equal diffusivities.

One drawback of this method is that an estimate $\mathrm{grad}(\boldsymbol{\theta}) = f(\boldsymbol{\theta})$ for the gradient in terms of the reduced coordinates is needed. But there a several ways out of the problem:

- It has been shown that at least for manifolds of higher dimensions the manifold does not depend much on the gradient estimate, and rough estimates can be used.
- Known estimates for diffusion flames or premixed flames from the literature can be used (see e.g. [59]).
- Gradient estimates can be refined during the solution of the reduced set of equations [12].

9.6 Tabulation

In principle there are two different strategies for implementing reduced chemistry in reacting flow simulations. One is e.g. to use the derived quasi-steady state assumptions and partial equilibrium assumptions directly during the integration of the reacting flow equations. This can, however, be computationally demanding, because in this case partial differential equations are replaced by non-linear algebraic equations, which are in many cases not easier to solve than the differential equations. Thus, despite of a considerable reduction of the chemical kinetics, very often the computational gain is much smaller than expected. The other strategy follows in a straightforward way from the concept of manifold methods. These methods make use of the fact that the complete thermokinetic state is a known function of the reduced coordinates (see above). Therefore it is natural to store this information for subsequent use in reacting flow calculations. The necessary steps are:

- Calculate the low-dimensional manifolds using the various techniques described in Section 9.6.
- Store the results (thermokinetic state, reaction rates, information on the coupling with molecular transport) using a sophisticated tabulation scheme.
- Use an efficient table-lookup to retrieve the information during the reacting flow calculation

Although this tabulation strategy is very simple, the actual implementation is very challenging due to several reasons:

- The storage effort of most straightforward tabulation schemes increases exponentially with the dimension of the reduced coordinates ("curse of dimensionality" [2]).
- As can be seen in Fig. 9.3 only a small domain of the whole reduced coordinate space is actually accessed, and storage capacity should not be wasted by tabulating domains which are never accessed in practical applications.
- There is a need to minimize the error associated with the tabulation table-lookup
- The table-lookup has to be extremely fast and efficient.

In order to cope with these problems several strategies have been developed. In principle they can be classified in four groups.

- Tabulation in terms of a multi-dimensional grid: In this case a multi-dimensional grid is constructed and the information is stored at the different nodes. Several variants exist from regular tensor product grids over local mesh refinement to the use of so-called sparse grids (see e.g. [18, 23, 52, 87]).
- Piecewise storage of the results. These methods separate the composition space into hypercubes or polyhedrons and approximation of the composition space in the polyhedrons by a polynomial representation. One of these approaches is the PRISM ("piecewise reusable implementation of solution mapping") method [78], while another is a piecewise tabulation strategy based on orthogonal polynomials, which has been used to store ILDMs [60].
- *In situ* adaptive tabulation (ISAT) [70]: this method stores the thermokinetic state in terms of a binary search tree and is very efficient, because it tabulates only the actual accessed composition space.
- Generalized coordinates: This method uses a regular rectangular mesh constructed for generalized coordinates which is then mapped to the actual composition space. The advantage is that the mesh is adapted in a natural way to

the problem (typically the mesh is constructed during the identification of the low-dimensional manifolds), the disadvantage is that the equation system for the reacting flow has to be projected on to the generalized coordinates [1].

- Adaptive chemistry tabulation: A methodology based on Common Component Architecture is employed in order to store all CSP quantities needed for the construction of the appropriate reduced model, that delivers a pre-specified accuracy and can be incorporated into an explicit solver [47, 62]. The composition space is separated into hypercubes and a different reduced model is stored for each one. According to this method, the reduced model of two adjacent hypercubes might be of different dimensions and form.

Although much progress has been made during the last years, there is still a need to improve the existing methods.

9.7 Concluding Remarks

In this chapter we have made the attempt to outline the existing strategies for mechanism reduction, making no claim to be complete. Several topics such as the production of skeletal mechanisms (see e.g. [49, 63, 82]), the analysis of detailed reaction mechanisms using sensitivity analysis or reaction flow analysis [77, 88], or lumping techniques [34, 48, 72] have not been addressed. The numerical aspects of reduced chemistry modeling, like their use with operating splitting techniques, the explicit solvers, the numerical solution of steady state relations [22, 50, 84, 86, 91] were also not addressed.

Although much progress has been made in the development of strategies and tools for mechanism reduction, there are still many open questions (in particular with respect to accuracy and efficiency of the reduced mechanisms). However, as it was shown above, all strategies for mechanism reduction rely on common ideas, and in particular their combination will improve the quality and efficiency considerably in the future.

References

1. Bauer, J., Bykov, V., Maas, U.: Implementation of ildms based on a representation in generalized coordinates. In: P. Wesseling, E. Onate, J. Periaux (eds.) European Conference on Computational Fluid Dynamics, ECCOMAS CFD, September 5-8, 2006, Egmond aan Zee, The Netherlands (2006)
2. Bellman, R.: Dynamic programming. Princeton University Press (1957)
3. Blasenbrey, T., Maas, U.: ILDMs of higher hydrocarbons and the hierarchy of chemical kinetics. Proc. Combust. Inst. **28**, 1623–1630 (2000)
4. Bodenstein, M.: Zur kinetik des chlorknallgases. Z. Phys. Chem. **85**, 329 (1913)
5. Bodenstein, M., Lind, S.C.: Geschwindigkeit der Bildung des Bromwasserstoffs aus seinen Elementen. Z. Phys. Chem. **57**, 168 (1906)

6. Bowen, J., Acrivos, A., Oppenheim, A.: Singular perturbation refinement to quasi-steady state approximation in chemical kinetics. Chem. Eng. Sci. **18**, 177–188 (1963)
7. Buki, A., Perger, T., Turanyi, T., Maas, U.: Repro-modeling based generation of intrinsic low-dimensional manifolds. J. Math. Chem. **31**, 345–362 (2002)
8. Bykov, V., Goldfarb, I., Gol'dshtein, V., Maas, U.: On a modified version of ildm approach: asymptotic analysis based on integral manifolds. IMA J. Appl. Math. **71**, 359–382 (2006)
9. Bykov, V., Gol'dshtein, V., Maas, U.: Global quasi linearization (gql) for the automatic reduction of chemical kinetics. In: Proceedings of the European Combustion Meeting, Chania, Crete (Greece) (2007)
10. Bykov, V., Maas, U.: Extension of the ildm method to the domain of slow chemistry. Proc. Combust. Inst. **31**, 465–472 (2007)
11. Bykov, V., Maas, U.: Investigation of the hierarchical structure of kinetic models in ignition problems. Zeitschrift für Physikalische Chemie **233**, 461–479 (2009)
12. Bykov, V., Maas, U.: Problem adapted reduced models based on reaction-diffusion manifolds (redims). Proc. Combust. Inst., **32**, 561–568 (2009)
13. Chen, J.: A general procedure for constructing reduced reaction-mechanisms with given independent relations. Combust. Sci. Technol. **57**, 89–94 (1988)
14. Chiavazzo, E., Gorban, A., Karlin, I.: Comparison of invariant manifolds for model reduction in chemical kinetics. Commun. Comput. Phys. **2**, 964–992 (2007)
15. Constantin, P., Foias, C., Nicolaenko, B., Temam, R.: Integral Manifolds and Inertial Manifolds for Dissipative Partial Differential Equations. Springer-Verlag, New York (1989)
16. Contou-Carrere, M., Daoutidis, P.: Model reduction and control of multi-scale reaction-convection processes. Chem. Eng. Sci. **63**, 4012–402 (2008)
17. Davis, M., Skodje, R.: Geometric investigation of low-dimensional manifolds in systems approaching equilibrium. J. Chem. Phys. **111**, 859–874 (1999)
18. Deuflhard, P., Heroth, J., Maas, U.: Towards dynamic dimension reduction in reactive flow problems. In: J. Warnatz, F. Behrendt (eds.) Modelling of Chemical Reaction Systems, Proceedings of an International Workshop, Heidelberg, Germany, July 24-26, 1996, ISBN 3-932217-00-4, Springer (1996)
19. Ern, A., Giovangigli, V.: Multicomponent Transport Algorithms. Lecture Notes in Physics. Springer, Berlin, Heidelberg, New York (1994)
20. Fenichel, N.: Geometric singular perturbation theory for ordinary differential equations. J. Diff. Eqs. **31**, 53–78 (1971)
21. Gear, C., Kaper, T., Kevrekidis, I., Zagaris, A.: Projecting to a slow manifold: Singularly perturbed systems and legacy codes. SIAM J. App. Dyn. Sys. **4**, 711–732 (2005)
22. Gear, C.W., Kevrekidis, I.: Constraint-defined manifolds: A legacy code approach to low-dimensional computation. SIAM J. Sci. Comp. **25**, 17–28 (2005)
23. Gicquel, O., Darabiha, N., Thévenin, D.: Laminar premixed hydrogen/air counterflow flame simulations using flame prolongation of ildm with differential diffusion. Proc. Combust. Inst. **28**, 1901–1908 (2000)
24. Goettgens, J., Terhoeven, P., Korff, R.: Redmech, public domain code at techmech.itm.rwth-aachen.de (1992)
25. Golub, G., van Loan, C.: Matrix Computations. The Hopkins University Press, Baltimore, London (1989)
26. Gorban, A., Karlin, I.: Method of invariant manifold for chemical kinetics. Chem. Eng. Sci. **58**, 4751–4768 (2003)
27. Gorban, A., Karlin, I.: Invariant Manifolds for Physical and Chemical Kinetics. Lecture Notes in Phys. 660, Springer-Verlag, Berlin (2004)
28. Goussis, D., Lam, S.: A study of homogeneous methanol oxidation kinetic using CSP. Proc. Combust. Inst. **24**, 113–120 (1992)
29. Goussis, D., Najm, H.: Model reduction and physical understanding of slowly oscillating processes: the circadian cycle. SIAM Multiscale Model. Simul. **5**, 1297–1332 (2006)
30. Goussis, D., Valorani, M.: An efficient iterative algorithm for the approximation of the fast and slow dynamics of stiff systems. J. Comput. Phys. **214**, 316–346 (2006)

31. Goussis, D., Valorani, M., Creta, F., Najm, H.: Reactive and reactive-diffusive time scales in stiff reaction-diffusion systems. Prog. Comput. Fluid Dyn. **5**, 316–326 (2005)
32. Griffiths, J.F.: Reduced kinetic models and their application to practical combustion systems. Prog. Energy Combust. Sci. **21**, 25–107 (1995)
33. Hadjinicolaou, M., Goussis, D.: Asymptotic solution of stiff pdes with the CSP method: The reaction diffusion equation. SIAM J. Sci. Comp. **20**, 781–810 (1998)
34. Huang, H., Fairweather, M., Griffiths, J., Tomlin, A., Brad, R.: A systematic lumping approach for the reduction of comprehensive kinetic models. Proc. Combust. Inst. **30**, 1309–1316 (2005)
35. Jones, W.P., Rigopoulos, S.: Rate-controlled constrained equilibrium: formulation and application to non-premixed laminar flames. Combust. Flame **142**, 223–234 (2005)
36. Jones, W.P., Rigopoulos, S.: Reduced chemistry for hydrogen and methanol premixed flames via rcce. Combust. Theory Model. **11**, 755–780 (2007)
37. Kaper, H., Kaper, T.: Asymptotic analysis of two reduction methods for systems of chemical reactions. Physica D **165**, 66–93 (2002)
38. Keck, J., Gillespie, D.: Rate-controlled partial-equilibrium method for treating reacting gas mixtures. Combust. Flame **17**, 237–241 (1971)
39. Lam, S.: Using CSP to understand complex chemical kinetics. Combust. Sci. Technol. **89**, 375–404 (1993)
40. Lam, S.: Reduced chemistry-diffusion coupling. Combust. Sci. Technol. **179**, 767–786 (2007)
41. Lam, S., Goussis, D.: Understanding complex chemical kinetics with computational singular perturbation. Proc. Combust. Inst. **22**, 931–941 (1988)
42. Lam, S., Goussis, D.: Conventional Asymptotics and Computational Singular Perturbation for Simplified Kinetics Modelling. In: M.O. Smooke (ed.) Reduced Kinetic Mechanisms and Asymptotic Approximations for Methane-Air Flames, Springer Lecture Notes **384**, pp. 227–242. Springer Verlag, Berlin (1991)
43. Lam, S., Goussis, D.: The CSP method for simplifying kinetics. Int. J. Chem. Kinetics **26**, 461–486 (1994)
44. Law, C.K.: Combustion Physics. Cambridge University Press, New York (2006)
45. Law, C.: Combustion at a crossroads: Status and prospects. Proc. Combust. Inst. **31**, 1–29 (2007)
46. Lebiedz, D.: Computing minimal entropy production trajectories: An approach to model reduction in chemical kinetics. J. Chem. Phys. **120**, 6890–6897 (2004)
47. Lee, J.C., Najm, H., Lefantzi, S., Ray, J., Frenklach, M., Valorani, M., Goussis, D.: A CSP and tabulation-based adaptive chemistry model. Combust. Theory Model. **11**, 73–102 (2007)
48. Li, G., Rabitz, H.: A general analysis of approximate lumping in chemical kinetics. Chem. Eng. Sci. **45**, 977–1002 (1990)
49. Lu, T., Law, C.: A directed relation graph method for mechanism reduction. Proc. Combust. Inst. **30**, 1333–1341 (2005)
50. Lu, T., Law, C.: Toward accommodating realistic fuel chemistry in large-scale computations. Prog. Energy Combust. Sci. **35**, 192–215 (2009)
51. Maas, U.: Efficient calculation of intrinsic low-dimensional manifolds for the simplification of chemical kinetics. Computing and Visualization in Science **1**, 69–82 (1998)
52. Maas, U., Pope, S.: Implementation of simplified chemical kinetics based on intrinsic low-dimensional manifolds. Proc. Combust. Inst. **24**, 103–112 (1992)
53. Maas, U., Pope, S.: Simplifying chemical kinetics: Intrinsic low-dimensional manifolds in composition space. Combust. Flame **88**, 239–264 (1992)
54. Maas, U., Pope, S.: Laminar flame calculations using simplified chemical kinetics based on intrinsic low-dimensional manifolds. Proc. Combust. Inst. **25**, 1349–1356 (1994)
55. Maas, U., Thevenin, D.: Correlation analysis of direct numerical simulation data of turbulent non-premixed flames. Proc. Combust. Inst. **27**, 1183–1189 (1998)
56. Massias, A., Diamantis, D., Mastorakos, E., Goussis, D.: An algorithm for the construction of global reduced mechanisms with CSP data. Combust. Flame **117**, 685–708 (1999)

57. Nafe, J., Maas, U.: A general algorithm for improving ILDMs. Combust. Theory Model. **6**, 697–709 (2002)
58. Najm, H.N., Valorani, M., Goussis, D.: Analysis of methane-air edge flame structure. Combust. Theory Model. **14**, 257–294 (2010)
59. Nguyen, P., Vervisch, L., Subramanian, V., Domingo, P.: Multidimensional flamelet-generated manifolds for partially premixed combustion. Combust. Flame **157**, 43–61 (2010)
60. Niemann, H., Schmidt, D., Maas, U.: An efficient storage scheme for reduced chemical kinetics based on orthogonal polynomials. J. Eng. Math. **31**, 131–142 (1997)
61. Okino, M., Mavrovouniotis, M.: Simplification of mathematical models of chemical reaction systems. Chemical Reviews **98**, 391–408 (1998)
62. Ortega, J., Najm, H., Valorani, M., Goussis, D., Frenklach, M.: Efficient slow manifold identification for tabulation based adaptive chemistry. 5th US Combustion Meeting, San Diego, Paper C31 (2007)
63. Pepiot-Desjardins, P., Pitsch, H.: An efficient error propagation based reduction method for large chemical kinetic mechanisms. Combust. Flame **154**, 67–81 (2008)
64. Peters, N.: Numerical and Asymptotic Analysis of Systematically Reduced Reaction Schemes for Hydrocarbon Flames. In: R.G. et al. (ed.) Lecture Notes in Physics, pp. 90–109. Springer-Verlag, Berlin/New York (1985)
65. Peters, N.: Systematic Reduction of Flames Kinetics: Principles and Details. In: M. Onofri, A. Tesei (eds.) Fluid Dynamical Aspects of Combustion Theory, pp. 232–248. Longman, New York (1990)
66. Peters, N.: Multiscale combustion and turbulence. Proc. Combust. Inst. **32**, 1–25 (2009)
67. Peters, N., Rogg, B.: Reduced Kinetic Mechanisms for Applications in Combustion Systems. Springer, Berlin (1993)
68. Peters, N., Williams, F.: The asymptotic structure of stoichiometric methane-air flames. Combust. Flame **68**, 185–207 (1987)
69. Pope, S.: Computations of turbulent combustion: Progress and challenges. Proc. Combust. Inst. **23**, 591–612 (1990)
70. Pope, S.: Computationally efficient implementation of combustion chemistry using in situ adaptive tabulation. Combust. Theory Model. **1**, 41–63 (1997)
71. Pope, S., Maas, U.: Simplified chemical kinetics: Trajectory-generated low-dimensional manifolds. Internal Report, Cornell University (1993)
72. Ranzi, E., Dente, M., Goldaniga, A., Bozzano, G., Faravelli, T.: Lumping procedures in detailed kinetic modeling of gasification, pyrolysis, partial oxidation and combustion of hydrocarbon mixtures. Prog. Energy Combust. Sci. **27**, 99–139 (2001)
73. Ren, Z., Pope, S.: Transport-chemistry coupling in the reduced description of reactive flows. Combust. Theory Model. **11**, 715–739 (2007)
74. Roussel, M., Fraser, S.: On the geometry of transient relaxation. J. Chem. Phys. **94**, 7106–7113 (1991)
75. Segel, L., Slemrod, M.: The quasi-steady-state assumption: A case study in perturbation. SIAM Review **31**, 446–477 (1989)
76. Smooke, M.D. (ed.): Reduced Kinetic Mechanisms and Asymptotic Approximations for Methane-Air Flames. Lecture Notes in Physics **384**, Springer, Berlin, Heidelberg, New York (1991)
77. Tomlin, A., Turanyi, T., Pilling, M.: Mathematical tools for the construction, investigation and reduction of combustion mechanisms. In: M.J. Pilling (ed.), Oxidation Kinetics and Autoignition of Hydrocarbons. Elsevier (1997)
78. Tonse, S., Moriarty, N., Brown, N., Frenklach, M.: Prism: Piecewise reusable implementation of solution mapping. An economical strategy for chemical kinetics. Israel Journal of Chemistry **39**, 97–106 (1999)
79. Turanyi, T.: Kinal - a program package for kinetic analysis of reaction mechanisms. Computers and Chemistry **14**, 253–254 (1990)
80. Turanyi, T.: Application of repro-modelling for the reduction of chemical kinetics. Proc. Combust. Inst. **25**, 45–54 (1994)

81. Turanyi, T., Tomlin, A., Pilling, M.: On the error of the quasi-steady-state approximation. J. Phys. Chem. **97**, 163–172 (1993)
82. Valorani, M., F. Creta, F.D., Najm, H., Goussis, D.: Skeletal mechanism generation and analysis for n-heptane with CSP. Proc. Combust. Inst. **31**, 472–483 (2007)
83. Valorani, M., Goussis, D., Creta, F., Najm, H.: Higher order corrections in the approximation of low dimensional manifolds and the construction of simplified problems with the CSP method. J. Comput. Phys. **209**, 754–786 (2005)
84. Valorani, M., Goussis, D.A.: Explicit time-scale splitting algorithms for stiff problems: Autoignition of gaseous mixtures behind a steady shock. J. Comput. Phys. **169**, 44–79 (2001)
85. Valorani, M., Najm, H., Goussis, D.: CSP analysis of a transient flame-vortex interaction: Time scales and manifolds. Combust. Flame **134**, 35–53 (2003)
86. Valorani, M., Paolucci, S.: The G-Scheme: A framework for multi-scale adaptive model reduction. J. Comput. Phys. **228**, 4665–4701 (2009)
87. Van Oijen, J., De Goey, L.: Modelling of premixed laminar flames using flamelet-generated manifolds. Combust. Sci. Technol. **161**, 113–137 (2000)
88. Warnatz, J., Maas, U., Dibble, R.: Combustion. Springer Verlag, Berlin (2006)
89. Westbrook, C., Mizobuchi, Y., Poinsot, T., Smith, P., Warnatz, J.: Computational combustion. Proc. Combust. Inst. **30**, 125–157 (2005)
90. Williams, F.: Combustion Theory. Benjamin/Cummings, California (1985)
91. Yen, J., Petzold, L., Raha, S.: A time integration algorithm for flexible mechanism dynamics: The dae alpha-method. Comput. Methods Appl. Mech. Engr. **158**, 341–355 (1998)
92. Zagaris, A., Kaper, H., Kaper, T.: Analysis of the computational singular perturbation reduction method for chemical kinetics. J. Nonlinear Sci. **19**, 59–91 (2004)
93. Zagaris, A., Kaper, H., Kaper, T.: Fast and slow dynamics for the computational singular perturbation method. Multiscale Model. Simul. **2**, 613–638 (2004)
94. Zagaris, A., Kaper, H., Kaper, T.: Two perspectives on reduction of ordinary differential equations. Mathematische Nachrichten **278**, 1629–1642 (2005)

Chapter 10
The Linear-Eddy Model

Suresh Menon and Alan R. Kerstein

Abstract Regime-independent modeling is important for accurate simulation of the complex combustor designs needed to meet increasingly stringent performance requirements. One strategy for achieving robust yet affordable predictive capability is to resolve, in space and time, the relevant advection-diffusion-reaction couplings using a low-dimensional representation of turbulent advection. In the linear-eddy model (LEM), this is accomplished in one spatial dimension by introducing an instantaneous map, the 'triplet map,' that emulates the effect of an eddy turnover on property profiles along a notional line of sight. The map preserves the continuity of these profiles and obeys applicable conservation laws. Details and representative applications of the model are presented for passive and reactive scalar mixing, with emphasis on its use as a mixing-reaction closure for large-eddy simulation (LES) based on the embedding of an LEM domain in each LES control volume.

10.1 Motivation

Regime-independent modeling is a widely recognized goal of turbulent combustion modeling. This goal is driven by the need to model configurations involving various combinations of regimes such as partial premixing, extinction, re-ignition, recirculation, stratified premixed combustion, compression ignition, multi-stage ignition, and transition to detonation. Techniques involving coarse-graining, ensemble averaging, or state-space modeling face difficulties in this regard due to the lack of detailed representation of regime-specific advective-diffusive-reactive couplings.

S. Menon
School of Aerospace Engineering, Georgia Institute of Technology, USA, e-mail: suresh.menon@aerospace.gatech.edu

A. R. Kerstein
Combustion Research Facility, Sandia National Laboratories, Livermore, CA, USA, e-mail: arkerst@sandia.gov

T. Echekki, E. Mastorakos (eds.), *Turbulent Combustion Modeling*, 221
Fluid Mechanics and Its Applications 95, DOI 10.1007/978-94-007-0412-1_10,

Although progress in addressing these difficulties can be anticipated, the challenge of regime-independent modeling suggests the concurrent pursuit of an alternative approach, namely, the development of a conceptually and computationally minimal model that resolves, in space and time, some plausible representation of the coupled advancement of advection, diffusion, and reaction in turbulent reacting flow. The potential advantages of this strategy motivated the formulation, two decades ago, of the linear-eddy model (LEM) [28].

A useful starting point for introducing LEM is the numerical advancement of equations governing a 1D unsteady opposed-flow flame. This formulation can capture some but not all of the salient features of turbulent combustion. One important feature that it does not capture is the effect of rotational folding of the flame by turbulent eddy motions. Successive folding and compressive motions can cause an initially monotonic profile of, say, mixture fraction to develop multiple extrema with stoichiometric points in between, each corresponding to a flame location. At high turbulence intensity, the time scale of folding, compression, and diffusive mixing might become short enough relative to chemical time scales so that broad reaction zones, stirred internally by small eddies, are formed.

The impossibility of representing these effects in 1D by continuous-in-time motions while obeying the applicable conservation laws motivated the introduction, in LEM, of instantaneous maps. Each map can be conceptualized as representing the outcome of an individual eddy motion, although such a literal connection between maps and eddies is not required.

An LEM simulation time advances the 1D unsteady diffusion-reaction equations, including associated dilatations along the 1D domain. This advancement is punctuated by instantaneous rearrangements of property profiles by mapping operations of a specified form (see Sec. 10.2). In effect, the outcome of each map constitutes a new initial condition for further time advancement.

A numerical consequence of this procedure is that a time of map occurrence cannot be contained within a time step for advancement of the governing equations. This limits the potential advantage of implicit diffusion, so in LEM, diffusion is typically implemented explicitly, possibly with implicit chemistry depending on the stiffness of the kinetics.

Implementation of LEM as a subgrid model for large-eddy simulation (LES) has also been accomplished with the goal of applying it to more complex high Reynolds number flows. In this approach the LEM is coupled with a large-scale scalar transport method to capture both large-scale flame structures and subgrid wrinkling effects.

10.2 Triplet Map

It is convenient to define a map symbolically using the notation

$$s(y) \rightarrow s(M(y)), \tag{10.1}$$

where $s(y)$ is any property profile, y is the 1D spatial coordinate, and M is the inverse of the map, i.e., the property value at $M(y)$ is mapped to location y. The map used in LEM, termed the triplet map, has the form

$$M(y) = y_0 + \begin{cases} 3(y-y_0) & \text{if } y_0 \leq y \leq y_0 + \frac{1}{3}l, \\ 2l - 3(y-y_0) & \text{if } y_0 + \frac{1}{3}l \leq y \leq y_0 + \frac{2}{3}l, \\ 3(y-y_0) - 2l & \text{if } y_0 + \frac{2}{3}l \leq y \leq y_0 + l, \\ y - y_0 & \text{otherwise.} \end{cases} \tag{10.2}$$

This map shrinks property profiles within an interval $[y_0, y_0 + l]$ to a third of their original extent, and then fills the interval with three compressed copies of the profiles. The middle copy is reversed, which maintains the continuity of mapped properties and introduces the rotational folding effect of turbulent eddy motion (see Sec. 10.1). Property fields outside the size-l interval are unaffected.

On an unstructured adaptive 1D mesh, this spatial continuum definition of the triplet map can be implemented numerically as stated, as in an ODT adaptive-mesh implementation described in Chapter 11. Published LEM studies to date use a fixed uniform mesh, on which the triplet map is approximated as a permutation of mesh cells. It is convenient to focus on this implementation because its consistency with property conservation laws is self-evident.

Spatially discrete implementation of the triplet map is illustrated in Fig. 10.1, involving a permutation of cell indices j through $j+l-1$. Taking the map range l to be a multiple of 3 cells, the triplet map permutes the cell indices into the new order $j,\ j+3,\ j+6,\ \dots,\ j+l-3,\ j+l-2,\ j+l-5,\ j+l-8,\ \dots,\ j+4,\ j+1,\ j+2,\ j+5,\ j+8,\ \dots,\ j+l-4,\ j+l-1$. This operation reduces the separation of any pair of cells by no more than a factor of three, consistent with the scale locality of length-scale reduction by eddy motions in the turbulent cascade (see Chapter 11).

It is interesting to note that the triplet map and its uses for turbulence simulation were discussed in a recent paper by authors who developed this concept without knowledge of prior work on LEM [27]. This is perhaps indicative of the degree to which map-based advection using the triplet map is a uniquely advantageous approach for economical turbulent-mixing simulation.

10.3 Map Sizes and Frequency of Occurrence

As explained in Sec. 10.1, LEM is intended to be an alternative to models involving coarse-grained constructs such as eddy diffusivity D_T. Although eddy diffusivity is not the modality of turbulent transport implementation in LEM, it is central to LEM formulation. Eddy diffusivity parameterizes the aggregate turbulent transport induced by an ensemble of turbulent eddies. Accordingly, the ensemble of triplet maps during an LEM simulation induce aggregate transport that is quantifiable as an eddy diffusivity, here denoted D_T. Based on the triplet-map property that it induces a mean-square displacement $\frac{4}{27}l^2$ within the mapped interval, random-walk theory

$$
\begin{pmatrix}
1&0&0&0&0&0&0&0&0&0&0&0&0&0&0&0\\
0&1&0&0&0&0&0&0&0&0&0&0&0&0&0&0\\
0&0&1&0&0&0&0&0&0&0&0&0&0&0&0&0\\
0&0&0&1&0&0&0&0&0&0&0&0&0&0&0&0\\
0&0&0&0&0&0&1&0&0&0&0&0&0&0&0&0\\
0&0&0&0&0&0&0&0&0&1&0&0&0&0&0&0\\
0&0&0&0&0&0&0&0&0&0&1&0&0&0&0&0\\
0&0&0&0&0&0&0&1&0&0&0&0&0&0&0&0\\
0&0&0&0&1&0&0&0&0&0&0&0&0&0&0&0\\
0&0&0&0&0&1&0&0&0&0&0&0&0&0&0&0\\
0&0&0&0&0&0&0&0&1&0&0&0&0&0&0&0\\
0&0&0&0&0&0&0&0&0&0&0&1&0&0&0&0\\
0&0&0&0&0&0&0&0&0&0&0&0&1&0&0&0\\
0&0&0&0&0&0&0&0&0&0&0&0&0&1&0&0\\
0&0&0&0&0&0&0&0&0&0&0&0&0&0&1&0\\
0&0&0&0&0&0&0&0&0&0&0&0&0&0&0&1
\end{pmatrix}
\begin{pmatrix}16\\15\\14\\13\\12\\11\\10\\9\\8\\7\\6\\5\\4\\3\\2\\1\end{pmatrix}
=
\begin{pmatrix}16\\15\\14\\13\\10\\7\\6\\9\\12\\11\\8\\5\\4\\3\\2\\1\end{pmatrix}
$$

Fig. 10.1: Application of a triplet map with $l = 9$, formulated as a permutation matrix multiplying a vector representing a flow state. Here, the map is applied to a column vector with vertically increasing cell indices. For clarity, unity matrix elements are boldface and cells are shifted horizontally in proportion to their index values. The shifts are intended to suggest the 1D profile of a mapped variable.

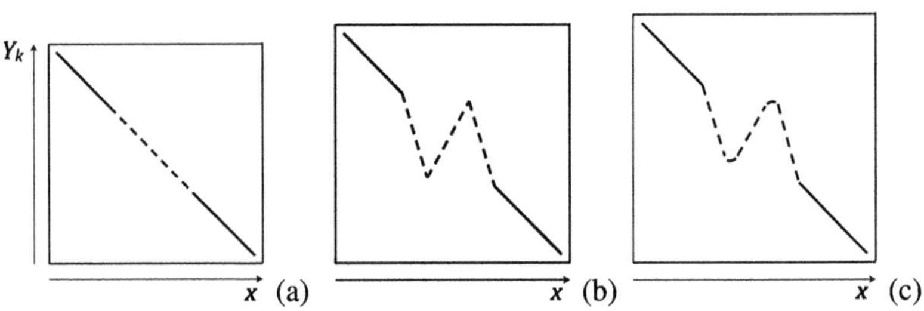

Fig. 10.2: Scalar field evolution with LEM. (a) initial field, (b) after triplet map and (c) after diffusion-reaction equation.

allows D_T to be expressed as

$$D_T = \frac{2}{27}\Lambda \int_0^{\infty} l^3 f(l)\, dl, \tag{10.3}$$

where $f(l)$ is the probability density function (PDF) of map sizes (a model input) and Λ is the frequency of map occurrences per unit domain length, so that Λ times the domain size is the rate of map occurrence on a notional spatially homogeneous domain [32]. (In previous model descriptions, the notation λ rather that Λ was used in Eq. 10.3. Here, the notation is changed for consistency with Chapter 11.) Equation 10.3 is used to solve for Λ given a D_T value and a size distribution $f(l)$ that correspond to the turbulent flow that is being simulated.

The functional form of $f(l)$ is based on the following familiar consequence of internal-range turbulence scaling. The eddy diffusivity $D_T(l)$ associated with turbulent motions of size l or less scales as $lv(l)$, where the eddy velocity $v(l)$ scales as $l^{1/3}$, so $D_T(l) \sim l^{4/3}$. The LEM analog of $D_T(l)$ is obtained by taking l to be the upper bound of the integral in Eq. 10.3 (to be distinguished from the dummy variable l in the integrand). The map-size PDF that yields this scaling is $f(l) \sim l^{-8/3}$.

The inertial range extends from the Kolmogorov microscale η to the integral scale L, which obey the scaling $L/\eta \sim \mathrm{Re}^{3/4}$, where Re is the integral-scale Reynolds number, $Re = u'L/\nu$, and u' is the turbulence intensity. Therefore, the stated scaling of f extends from η to L, and $f = 0$ outside this range. This determines the prefactor of f by the requirement that its integral over l is unity.

Here, η and L do not have precisely the same meanings in LEM as in turbulent flow, so strictly speaking these LEM quantities are related to their physical counterparts by empirical coefficients. Likewise, turbulent diffusivity values inferred from flow data typically do not precisely conform to the mathematical definition of a diffusivity, so an empirical coefficient may be needed to relate these values to the LEM parameter D_T, which has a precise definition within the model.

10.4 Application to Passive Mixing

For passive mixing involving a single scalar property s with constant molecular diffusivity κ, time advancement between map occurrences is governed by the heat equation,

$$\frac{\partial s}{\partial t} = \kappa \frac{\partial^2 s}{\partial y^2}. \tag{10.4}$$

The equation set for combustion is shown in Sec. 10.5.

For passive mixing, the family of scalar power spectra parameterized by the Schmidt number Sc is a useful comparison data set for LEM because the spectra exhibit universal behaviors with a sufficient number of features to overdetermine the model and test its performance. Spectrum comparisons have been used to set the free parameters in LEM and to demonstrate that LEM captures the salient features of turbulent mixing [5, 32]. Figure 10.2 shows a typical implementation and competition between molecular diffusion and turbulent stirring by the triplet mapping. Turbulent stirring on an initial scalar gradient mimics the folding effect of an eddy of size l with the subsequent molecular diffusion smoothing the gradient. These processes interact over a range of eddy sizes in a high Re flow resulting in statistical features of scalar mixing in excellent agreement with classical scaling laws and experimental observations.

For example, LEM as described above has been applied to mixing in grid turbulence (including three-stream mixing) [28, 33], in planar mixing layers (focusing on Sc dependence) [29], and in multi-scalar jet and homogeneous flows that exhibit differential molecular diffusion effects [5, 30, 35]. An application to mixing in homogeneous turbulence [37] revealed unexpected large-scale mixing behaviors that were subsequently verified by a pipe-flow experiment motivated by the LEM results [21–23]. Other homogeneous-turbulence studies compared scalar fluctuation decay in LEM and direct numerical simulations (DNS) [6, 43] and used LEM to study fractal scaling properties [31] and other intermittency properties [26, 32] of scalar fields mixed by turbulence.

The mixing of dry and moist air in clouds has been simulated using LEM in order to study temperature changes induced by cloud droplet evaporation and condensation and associated dynamical consequences [41, 42]. These studies, which were extended to simulate the evolution of droplet-size distributions [75], did not incorporate buoyancy effects but provided indications of larger-scale consequences of buoyancy changes during the simulations. It has been shown mathematically that triplet maps, in conjunction with a simple represention of eddy-induced motion of inertial particles, capture clustering effects thought to be crucial to the process of cloud-droplet coagulation to form raindrops [36].

10.5 Application to Reacting Flows

For reacting flows, the LEM employs more general equations for the species and temperature within the 1D context. An extension of the LEM to multi-component reacting flows can be written as [52, 53, 71, 74, 78]:

$$\rho\frac{\partial Y_k}{\partial t} = -\frac{\partial}{\partial y}(\rho Y_k V_k) + \dot{\omega}_k$$

$$\rho C_{p_{mix}}\frac{\partial T}{\partial t} = -\rho\frac{\partial T}{\partial y}\left(\sum_{k=1}^{Ns} C_{p,k} Y_k V_k\right) + \frac{\partial}{\partial y}\left(\kappa\frac{\partial T}{\partial y}\right) + \dot{\omega}_T. \qquad (10.5)$$

Here, Y_k is the $k-$th species mass fraction, $\dot{\omega}_k$ is the reaction rate, and V_k is the $k-$th species diffusion velocity.

Any type of kinetics (global or multi-step) can be included within this approach. Heat release related effects such as thermal expansion and mass transport to maintain global conservation have to be included but their specific implementation issues depend on the problem of interest [1, 47, 71, 78]. An example of the advantages of LEM reactive-flow simulation relative to other approaches is an application [13] to two simple reaction mechanisms that are widely used as test cases for closure models. LEM was shown to be in agreement with DNS of these reactions in a non-premixed configuration. LEM was then used to investigate a much wider parameter range than DNS can access affordability, yielding a data set suitable for stringent tests of proposed closures. None of the tested closures reproduced LEM product-selectivity results.

LEM has likewise been used to evaluate proposed nonpremixed-combustion closures [7, 8]. *A priori* analysis of the LEM-predicted PDF of the scalar properties has been shown to agree very well with experimental PDFs for jet diffusion flames [19]. It has also been used to explore regimes that closure models cannot yet address adequately, such as soot-radiation-turbulence coupling [79], and in applied studies such as NOx prediction in dual-stage combustion [51], jet flames [1, 47], aerosol mixing in engine exhaust [52, 78], and transient behaviors during incineration [65].

More recent studies include the study of extinction/re-ignition for a non-premixed syngas flame [67–70] for a test case simulated earlier by DNS [24, 25] using the same 21-step, 11-species skeletal syngas mechanism. Results showed the LEM can capture both the extinction and re-ignition process at approximately the same turbulent Reynolds number with the scalar dissipation rate at the stoichiometric surface predicted to be 2394 1/s, which is very close to the DNS value of 2100 1/s.

Many canonical premixed flames have also been investigated using the LEM. Freely propagating turbulent premixed flames in the flamelet regime [48, 71, 72] and in the thin-reaction-zone regime [60] have demonstrated that the changes in the flame structure can be captured without any change to the LEM. Recent stand-alone premixed LEM studies have focused on premixed combustion far from the flamelet regime. LEM has been applied to regimes of hydrogen [55] and hydrocarbon combustion relevant to homogeneous-charge compression-ignition (HCCI) combustion. An application to astrophysical flames [77] validated LEM using DNS of the well-stirred-reactor regime, then explored the more challenging 'stirred-flame' regime [34] to identify possible mechanisms for transition to detonation resulting in supernova explosions. Additionally, this study showed that the LEM turbulent burning velocity in the flamelet regime is $18D_T/L$ for high Re, a result that is useful for detecting departures from flamelet or high-Re scaling.

Recent studies employed the LEM to investigate flame-turbulence interactions for a range of conditions and fuel mixture with relatively detailed reduced kinetics [69, 70]. We use some results from this study to highlight key features of the LEM when implemented for reacting flows. For this study, the LEM scalar fields are initialized by the steady, laminar, 1D flame solution obtained for a CH_4/Air mixture at an equivalence ratio of 0.6, and a 12-step, 16-species reduced methane/air mechanism is used for the calculations. Thermally perfect gas with detailed multicomponent transport properties is employed for these simulations.

Figure 10.3 compares a typical H_2 mass fraction profile that has evolved from the initial laminar profile with stirring alone (Fig. 10.3, top left), with profiles evolved by stirring and diffusion (Fig. 10.3, top right) together, and with stirring, diffusion and reactions interacting (Fig. 10.3, bottom), which is the full LEM. It can be seen that the profile is wrinkled and exhibits high level of scalar gradients as a result of the turbulent stirring (Fig. 10.3, top left) but once the diffusion is included the gradients created by the triplet-maps are smoothed out (Fig. 10.3, top right). The peak value for the H_2 intermediate species decreases as well, which indicates that the flame is extinguishing, since reaction rates are not computed and the combined action of the turbulent stirring and diffusion pushes the flame towards extinction. However, when kinetics is also included (Fig. 10.3, bottom) the flame maintains its peak value while

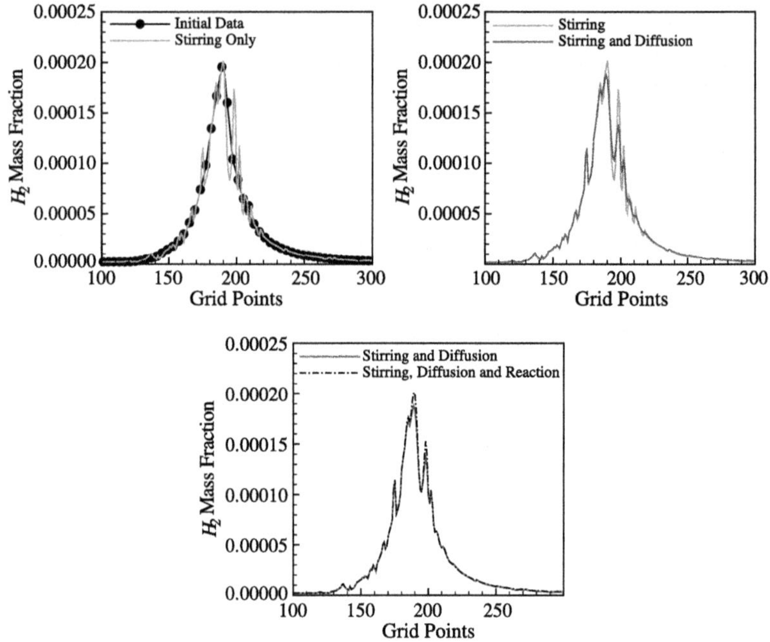

Fig. 10.3: Evolution of a scalar field (H_2) by simulating the effect of stirring (top left), stirring and diffusion (top right), and stirring, diffusion and reaction (bottom).

also showing the effect of stirring and diffusion. Thus, turbulent stirring, molecular diffusion and reaction kinetics can interact at their respective time scales in the LEM in a consistent manner. This is the unique nature of the LEM.

10.6 Application to Reacting Flows as a Subgrid Model

Extension of the LEM as a subgrid model requires some rethinking of the turbulent stirring approach described above. In the classical LES methodology large scale structures are fully resolved whereas small scale structures (subgrid-scale) are modeled by using appropriate subgrid momentum and combustion models. Thus, closures for momentum, energy and scalar transport at the resolved scales are required. This requirement is no different from the closure requirement for RANS modeling, and an earlier effort [17, 18] demonstrated a LEM based approach for RANS application. In the following, we focus primarily on the LES implementation of LEM (called LEMLES hereafter).

For LES of reacting flows, both low-Mach number [2, 3] and fully compressible approaches have been used in the past. Closure for the momentum and energy equations is achieved using a localized dynamic subgrid kinetic energy k^{sgs} model (LDKM) [16, 38–40, 49, 53, 54] in most of the studies discussed here. In addition to allowing for non-equilibrium between production and dissipation at the subgrid scales, the LDKM offers two unique advantages for LEMLES that is absent in the classical Smagorinsky type algebraic eddy viscosity mode: (a) the k^{sgs} distribution can be used to provide an estimate of the subgrid turbulence intensity $u' \approx \sqrt{2k^{sgs}/3}$, which can then be used for estimating turbulent stirring time scale and frequency in the subgrid LEM, and (b) for two-phase modeling of spray combustion, a stochastic component to the particle motion can be included to account for the effect of subgrid turbulence on particle transport [45, 53, 57]. The governing LES equations for mass, momentum and energy are obtained by using density-weighted Favre-filtering and the following discussion, although restricted to compressible flows is equally applicable in low-Mach number flows [3, 4].

The LEM-based subgrid closure for mixing and chemistry was developed [44, 50] for both premixed and non-premixed applications, and has been used for LES closure in many subsequent studies, e.g., scalar mixing [5, 46], non-premixed combustion [10, 46, 67, 69], premixed combustion [3, 4, 60, 62, 73, 76], two-phase combustion [53, 56, 57], soot formation [11, 12] and supersonic mixing and combustion [61]. This closure introduces an LEM domain in each control volume of the 3D mesh. The LEM domain size is of the order of the resolution scale of the 3D mesh, and evolves a 1D profile of the thermochemical state that serves as a representative sample of evolution within the control volume that contains it. Specializing to LES closure, the LES time-advancement cycle begins with an update of the coarse-grained LES flow state, consisting of velocity components and density. (All thermochemical information resides solely within the LEM domains.) Next, diffusive transport, chemistry, and triplet maps representing subgrid advection are implemented within each subgrid LEM domain for a time interval equal to the LES time step. Then the LEM domains communicate with each other by means of a 'splicing' procedure (see Fig. 10.6). The mass transfer across each LES mesh face during the LES time step is computed based on the LES-resolved velocity and density. The prescribed transfer is applied to the affected LEM domains by excising a piece of the domain that is upwind of the LES face and inserting it into the LEM domain that is downwind of the LES face. These processes are schematically shown in Fig. 10.4 and indicate that the LEM subgrid model can be implemented within any flow solver without any major revision to the original structure of the fluid dynamics solver.

To describe this model mathematically, we split the velocity field as: $u_i = \tilde{u}_i + (u_i')^R + (u_i')^S$. Here, $\tilde{u}_i$ is the LES-resolved velocity field, $(u_i')^R$ is the LES-resolved subgrid fluctuation (obtained from k^{sgs}) and $(u_i')^S$ is the unresolved subgrid fluctuation. Then, consider the *exact* species equation (i.e., without any explicit LES filtering) for the $k-$th scalar Y_k written in a slightly different form as:

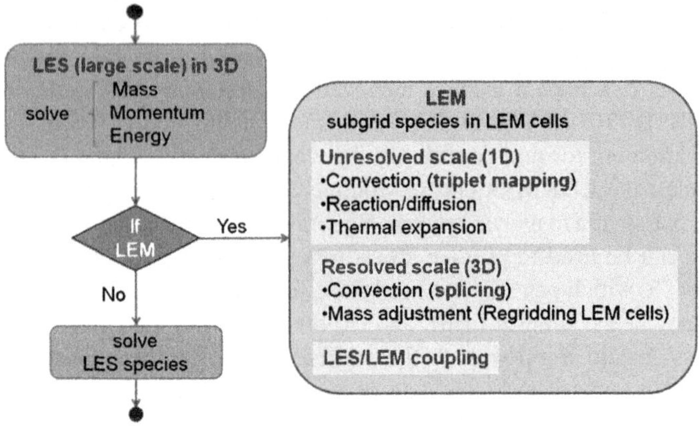

Fig. 10.4: Schematic of the LEMLES approach.

$$\rho \frac{\partial Y_k}{\partial t} = -\rho [\tilde{u}_i + (u_i')^R + (u_i')^S] \frac{\partial Y_k}{\partial x_i} - \frac{\partial}{\partial x_i}(\rho Y_k V_{i,k}) + \dot{w}_k + \dot{S}_k. \quad (10.6)$$

Here, $V_{i,k}$ are the $k-$th species diffusion velocity and $\dot{S}_k$ is a source term if there is phase change as in vaporization of liquid fuel. In LEMLES, the above equation is rewritten in a two-step process as:

$$Y_k{}^* - Y_k{}^n = \int_t^{t+\Delta t_{LES}} -\frac{1}{\rho}[\rho (u_i')^S \frac{\partial Y_k{}^n}{\partial x_i} + \frac{\partial}{\partial x_i}(\rho Y_k V_{i,k})^n - \dot{w}_k^n - \dot{S}_k^n] dt' \quad (10.7)$$

$$\frac{Y_k{}^{n+1} - Y_k{}^*}{\Delta t_{LES}} = -[\tilde{u}_i + (u_i')^R] \frac{\partial Y_k{}^n}{\partial x_i}. \quad (10.8)$$

Here, Δt_{LES} is the LES time-step. Equation 10.7 describes the subgrid LEM model, as viewed at the LES space and time scales. The integrand includes four processes that occur within each LES grid cell, and represent, respectively, (i) subgrid molecular diffusion, (ii) reaction kinetics, (iii) subgrid stirring, and (iv) phase change of the liquid fuel. These processes are modeled on a 1D domain embedded inside each LES grid where the integrand is rewritten in terms of the subgrid time and space scales. Equation 10.8 describes the large-scale 3D LES-resolved convection of the scalar field, and is implemented by a Lagrangian transfer of mass across the finite-volume cell surfaces [3, 46]. Descriptions for the subgrid processes (in Eq. 10.7) and the 3D advection process (in Eq. 10.8) are presented elsewhere [53] but are repeated here for completeness.

As shown in Eqs. 10.7 and 10.8, and in Fig. 10.4, there are two different elements to the LEMLES. We will consider the subgrid LEM (the integrand in Eq. 10.7) and the resolved-scale transport (Eq. 10.8) separately.

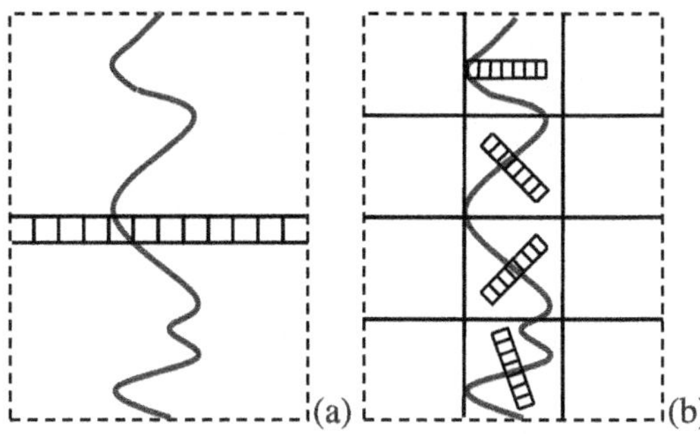

Fig. 10.5: Schematic representation of the LEM domain and the flame for (a) stand-alone LEM, (b) LEMLES.

10.6.1 The LEM Subgrid Model

The LEM is implemented within the LES cells as a subgrid model and Eqs. 10.5 are solved within each LES cell. In stand-alone LEM simulations the one-dimensional domain extends across the flame, whereas in the LEMLES approach they are embedded within each LES cell. This is schematically shown in Figs. 10.5a-b. The LEM domains are independent from each other and the 1D line is notionally aligned in the flame normal or the maximum scalar gradient direction and thus, does not represent any physical Cartesian direction.

Since all the turbulent scales below the grid are resolved in this approach, both molecular diffusion and chemical kinetics are closed in an exact manner. As a result, scalar 'subgrid' terms do not have to be closed or modeled. The subgrid pressure, p^{LEM} is assumed constant over the LEM domain, and equal to the LES value, $\overline{p}$, which is a valid assumption in the absence of strong pressure gradients. Hence, the subgrid density is computed from the equation of state at the subgrid level:

$$p^{LEM} = \rho^{LEM} T^{LEM} \sum_{k=1}^{N_{species}} Y_k^{LEM} \frac{R_u}{W_k}. \tag{10.9}$$

Here, W_k is the $k-$th species molecular weight. Conservation of mass, momentum and energy (at the LES level) and conservation of mass, energy and species (at the LEM level) are fully coupled. Chemical reaction at the LEM level determines heat release and thermal expansion at the LEM level, which at the LES level generates flow motion that, in turn, transports the species field at the LEM level. Full coupling is maintained in the LEMLES to ensure local mass conservation.

The reaction-diffusion equation on the LEM domain is solved within each LES cell with an explicit scheme. The time integration is achieved by using an operator splitting technique to account for four distinct physical processes and time scales in the LEM (molecular diffusion, chemistry, turbulent stirring and thermal expansion).

The diffusion time step is calculated as: $\Delta t_{diff} = C_{diff} \frac{\Delta s^2}{\max(D_k)}$. The maximum of D_k is used for Δt_{diff} in order to maintain the stability of the diffusion of the lightest species. The C_{diff} in the equation is a model constant, set here to 0.25 for numerical stability [63]. The chemical time step size is determined by the stiffness of the reaction mechanism. The stiffness increases as the number of radical species in a chemical kinetics mechanism increases. The chemistry is integrated for the given diffusion time step size (Δt_{diff}). A stiff ODE solver which uses adaptive time step size is employed for the integration process so that the chemical processes are resolved in their respective time scales. This approach, nevertheless, is very time consuming, and therefore, novel methods such as in-situ adaptive tabulation [11, 12] and artificial neural network [67, 69, 70] have been implemented within LEM to achieve speed-up.

The thermal expansion time scale is associated with the volumetric expansion induced by the increase in temperature through chemical heat release. In the current implementation $\Delta t_{expansion} \approx \Delta t_{diff}$, and turbulent stirring is implemented as discrete time events during the reaction-diffusion time integration. The time interval between each stirring event is: $\tau_{stir}(x) = 1/\Lambda\overline{\Delta}$, where Λ is the stirring frequency described earlier, and $\overline{\Delta}$ is the LES grid filter size. Overall, for the given LES time step, the number of stirring events is equal to $N_{stir} \approx \Delta t_{LES}/\Delta t_{stir}$.

In different implementations of this closure strategy, LEM domains can have either periodic [50] or Neumann boundary conditions [53]. In the former method the spliced pieces can be excised and inserted at arbitrary locations or a type of first-in-first-out criterion can be applied. In the latter approach the LEM domain has an input side and an output side. A common characteristic of these implementations is that the LEM domains are Lagrangian objects that have no unique spatial location or orientation within the LES control volume that contains them.

10.6.2 Large-Scale Advection of the Subgrid Field

The large scale advection is implemented in LEMLES to account for the advection of the scalar field on the resolved level between the LES cells according to Eq. 10.8. This process accounts for species transport in the 3D domain due to both the resolved LES velocity field $\widetilde{u_i}$ and the resolved subgrid kinetic energy k^{sgs}. Since the scalar structure is inherent in the subgrid LEM cells, they are transported across the LES fields by a Lagrangian transport process. This is in contrast to a conventional finite-volume or finite-difference approximation of the filtered scalar gradients across LES cells. Thus, the right hand side of Eq. 10.8 is not discretized using a conventional difference operator but rather, the convection of the subgrid scalar field (and hence the subgrid scalar gradient, $\partial Y_k/\partial s$) is explicitly carried out by the Lagrangian splicing approach.

The advection process is implemented once the subgrid evolution of the scalar field as a result of the turbulent stirring, diffusion, reaction and thermal expansion are completed at their respective time steps. The large scale advection is a result of

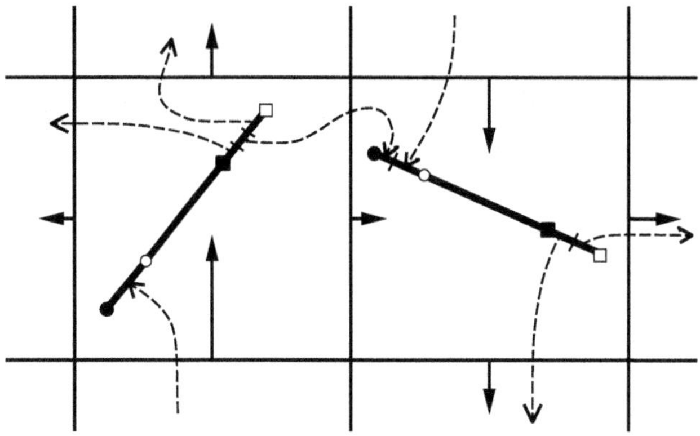

Fig. 10.6: Illustration of the splicing strategy, in which thermochemical information resides solely within LEM domains (tilted line segments) contained with control volumes of the 3D solver, but flow evolves on the coarse 3D mesh. Solid lines with arrows are flow velocities, evolved on the coarse mesh, that determine volume transfers between 1D subgrid domains. The splicing mechanism that implements these transfers is illustrated. Each LEM domain has an input end (circle) and an output end (square). Open and filled symbols demarcate the LEM domains before and after splicing. Portions transferred during splicing are separated by tick marks. Dashed curves with arrows indicate transfers between 1D domains in different control volumes. An alternative to specified input and output locations is to use periodic LEM domains. Then there are no preferred locations at which to remove and insert domain segments unless a first-in-first-out type of criterion is introduced.

both the resolved large scale ($\tilde{u}_i$) and the modeled subgrid scale velocities ($({u'}_i)^R$). In the current implementation, based on the assumption that the velocity field is isotropic on the small scales, $({u'}_i)^R$ is estimated: $({u'}_i)^R = \sqrt{\frac{2}{3}k_{sgs}}$. However, if k_{sgs} is not available this additional flux cannot be included. Regardless, this contribution is very small and in most cases can be neglected.

Equation 10.8 can be written in a finite volume discrete form as follows,

$$\frac{(\rho Y_k)^{n+1}\Delta V - (\rho Y_k)^* \Delta V}{\Delta t_{LES}} = -\sum_{j=1}^{N_f} \rho_j \left(\tilde{u}_i + (u'_i)^R\right)_j {Y_k}^* A_j \tag{10.10}$$

where ΔV is the volume of a finite volume cell, N_f is the number of cell faces, and A_j is the cell face area. The large-scale advection of the subgrid scalar *structure*, Y_k, is based on the mass flux across each cell face. Therefore, Eq. 10.10 can be rewritten by defining $m = \rho \Delta V$ and rearranging,

$$(mY_k)^{n+1} = (mY_k)^* - \Delta t_{LES} \sum_{j=1}^{N_f} (\delta m Y_k)_j \tag{10.11}$$

where $\delta m = \rho \left(\tilde{u}_i + (u_i')^R\right) A_j$, which is the mass flux that crosses a cell face, A_j. The last term in (10.11) is further decomposed into influx and outflux components. Therefore, Eq. 10.11 becomes,

$$(mY_k)^{n+1} = (mY_k)^* - \Delta t_{LES}\left[-\sum_{N_{in}}(\delta mY_k)_{in} + \sum_{N_{out}}(\delta mY_k)_{out}\right] \qquad (10.12)$$

where N_{in} and N_{out} are the number of influx and outflux faces surrounding a finite volume cell. Here, $(\delta mY_k)_{in}$ is computed by taking portions of the mass contained in LEM cells in the neighboring finite volume cells. The mass in N_{splice} LEM cells is collected based on the net in-flux and added to the LEM cells in the current finite volume cell. At the same time, $(\delta mY_k)_{out}$ is computed from the mass contained in LEM cells in the current finite volume cell and distributed to the neighboring LES cells based on the out-flux. With proper care to advect the total mass based on convective flux, proper mass conservation can be strictly enforced during this process. Since this process is also in full 3D, the advection of the *subgrid* scalar gradients by this Lagrangian process allows transport of both co-gradient and counter-gradient subgrid structure across LES cell faces. This is a unique strength of this advection process when compared to conventional gradient diffusion modeled at the LES level by a standard finite-difference or finite-volume method.

Due to compressibility, volumetric expansion and grid-stretching, N_{splice} may be different than N_{LEM} (the number of LEM cells in each LES grid volume) and this is included in the formulation. Furthermore, N_{splice} need not be an integer and the current algorithm allows for fractions of individual subgrid volumes to be transported, whereas the predetermined N_{LEM} is an integer multiple of 3, and greater than 9 to accommodate the triplet mapping procedure.

Once all the mass is transferred between the LES cells, the new LEM field can have more or less cells than that it had before the splicing started, as shown in Fig. 10.7. Also, the volume of the cells can be different from each other based on the transferred mass. Re-gridding is applied to uniformly divide all the mass between LEM cells. This process is schematically shown in Fig. 10.8. Re-gridding is strictly not needed if a variable subgrid domain is employed and is done primarily for numerical expediency. As in any simplification resulting errors have to be considered. For example, if re-gridding changes the composition in the LEM cells then it is considered artificial diffusion. This artificial diffusion can be minimized by increasing the number of LEM cells or by splicing smaller amounts of mass. While the former is essentially a grid refinement strategy, the latter is constrained by the time-step of the flow solver and cannot be arbitrarily changed. In compressible explicit flow solvers, due to the small LES time-step, the numerical diffusion effect is indeed very small. Nevertheless, this artifact of the LEM implementation should be eliminated for overall accuracy, and therefore, a more general implementation of the LEM (with variable subgrid resolution) is being pursued.

The Lagrangian advection process can track interfaces accurately. For example, Fig. 10.9 shows advection of an annular box without loss of its integrity in a uniform flow oriented along the diagonal of the simulation domain. The mass-flux based

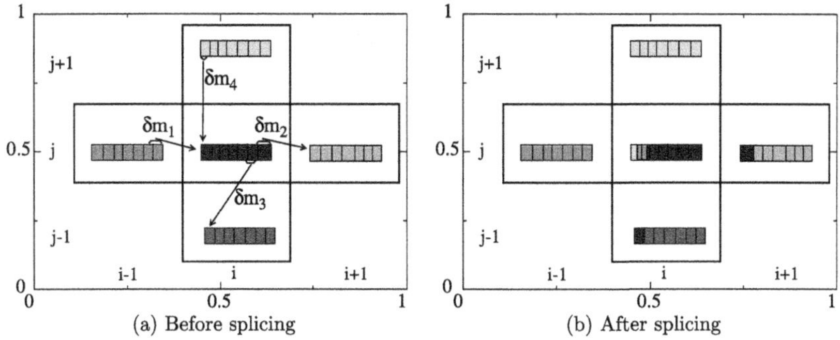

Fig. 10.7: Species field before and after the splicing of the cell (i, j).

transport process allows counter-gradient transport of scalars (since subgrid scalar structures are transported by splicing), enabling the model to avoid the pit-falls of those based on the gradient-diffusion hypothesis. Figure 10.10 shows the propagation of a circular burning flame, which includes both subgrid burning and thermal expansion effects. The circular flame is resolved on a Cartesian grid with reasonable accuracy [3].

Implementation of splicing is relatively straightforward and visually summarized in Fig. 10.7. Splicing is composed of the following steps: (1) calculate the LES flux, $\rho(\tilde{u}_i+(u'_i)^R)$ on each of the six faces of the volume, (2) determine the amount of mass to send as well as the amount of mass to receive from the neighboring LES cells, (3) arrange the LES fluxes with the largest out-flux carried out first in accordance with the premise that the 1D scalar fields are always aligned in the direction of the maximum scalar gradient, and (4) after the receipt of mass from each face commensurate with step 2, rearrange the scalar field in each cell accordingly. Re-gridding then follows, if there are heat release effects. As noted earlier, re-gridding is a numerical artifact intended for simplicity and can be eliminated with a more general formulation.

The ability of the Lagrangian advection to capture the flame structure is summarized in Fig. 10.11, which compares a 3D simulation of a turbulent premixed flame conducted using DNS and LEMDNS [62]. The latter approach implies that the LES grid is as fine as the DNS grid but the LEM was included within the subgrid. Thus, in the DNS limit the splicing process occurs only due to the resolved velocity field and the LEM subgrid processes are reduced to only diffusion and kinetics (i.e., stirring is turned off). The excellent agreement between a DNS and the LEMDNS shows the physically consistent and correct implementation of the scalar evolution by the two-step procedure (Eqs. 10.7 and 10.8).

A recently developed alternative LEM-based subgrid closure called LEM3D [64] has a well-defined spatial structure. In fact, its structure is the same as in ODTLES [66], a method for 3D flow advancement involving an array of coupled 1D domains, using the one-dimensional-turbulence (ODT) model (see Chapter 11).

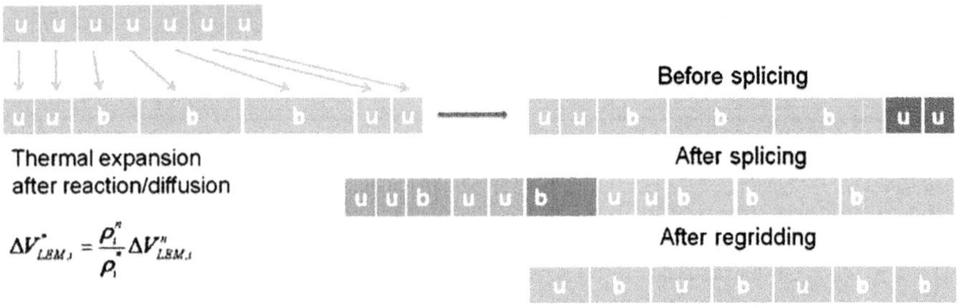

Fig. 10.8: Schematic representation of the subgrid scalar field after thermal expansion and re-gridding. Here u and b indicate an unburned and a burned LEM cell, respectively.

Fig. 10.9: Propagation of an annular box in a velocity field aligned along the diagonal.

An advantage of LEM3D relative to the splicing strategy is that it avoids the imposition of Neumann boundary conditions on LEM within each 3D control volume, which is the currently preferred splicing formulation. A related advantage is that a triplet map need not be contained within one 3D control volume, nor need it be limited in size relative to the control volume. An advantage of the splicing strategy is that it is readily implemented within an arbitrary structured or unstructured mesh. LEM3D is most easily implemented on a Cartesian mesh, with some possibility of generalization to generalized curvilinear coordinates. Owing to the novelty and limited evaluation of LEM3D to date, further discussion of LEM-based subgrid closure focuses on the splicing strategy.

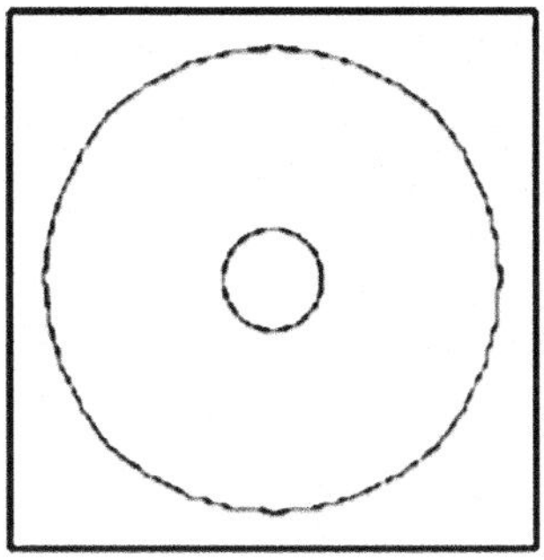

Fig. 10.10: Propagation of a circular flame front. Reprinted from [3] with permission from Taylor and Francis.

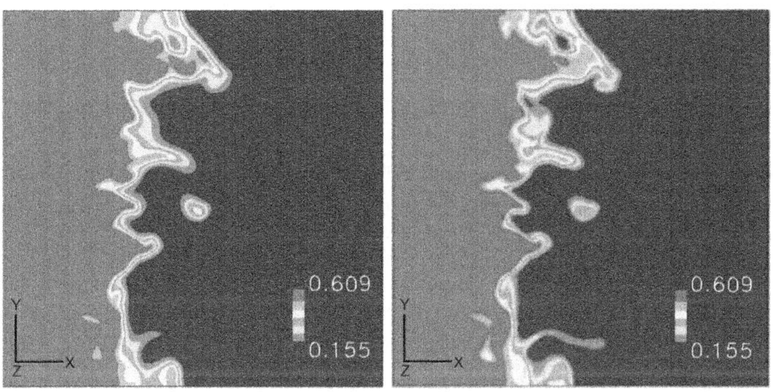

Fig. 10.11: DNS and LEMDNS of premixed flame-turbulence interaction. Reprinted from [62] with permission from the Combustion Institute.

10.7 LEMLES Applications to Reacting Flows

As cited above, the LEMLES has been used quite extensively for a range of problems from simple canonical flame-turbulence interactions to complex flows in gas turbine combustors. In the following we touch upon some key results primarily to highlight the predictive ability of the LEMLES approach. An underlying theme in all these comparisons is that the basic LEMLES approach is identical for all cases. The only changes occurring are the changes in the LES geometry, test conditions and appropriate boundary conditions. In the following, we summarize some key results primarily to highlight the capability of the approach. Cited references have more detailed analysis and interpretations.

We begin with application to premixed combustion behind a triangular bluff body [59] that has been extensively studied in the past by various LES methods and closures. Here, we compare a simple subgrid EBU closure with the LEMLES to high-

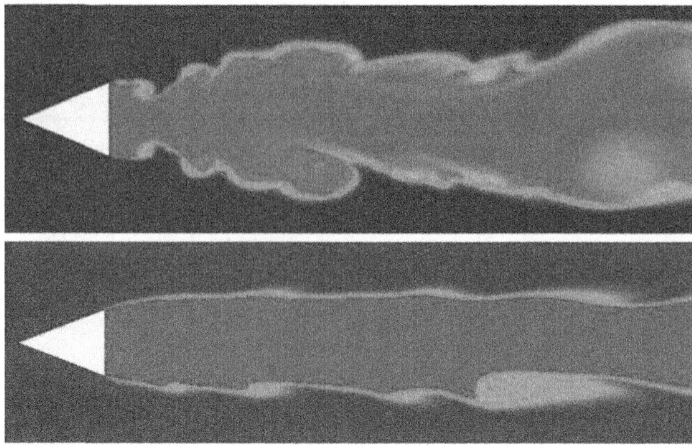

Fig. 10.12: Flame structure predicted behind a triangular bluff body. (a) LEMLES and (b) EBULES. Reprinted from [59].

light some features. EBULES is a very simple closure for the reaction kinetics and therefore is used extensively in the literature. It strengths and limitations are well known and it is not the intent here to focus on these issues. Furthermore, other flame closure methods [14, 15] have also proven quite accurate for this test case and therefore, the current comparison is primarily to provide a reference. As shown in Fig. 10.12 and Fig. 10.13 show some key differences between the two predictive methods. The LEMLES flame structure is more wrinkled and its spreading is increased due to its increase burning rate. This is reflected in the mean temperature radial profile and the better agreement with the Reynolds stress measurements. Other premixed studies [3, 4, 9, 60, 62] have demonstrated that LEMLES has the ability to not only capture the flame propagation speed and structure, it also enables better prediction of the flame-turbulence interactions.

A more recent study focused on the ability of the LEMLES approach to capture complex flame structures in both premixed and non-premixed combustion in a special combustor. The Stagnation Point Reverse Flow (SPRF) is a low NOx combustor [76] that employs exhaust gas recirculation to achieve stable combustion in lean conditions while minimizing NOx and CO emissions. In this combustor, flame stabilization is achieved via a high-temperature downstream stagnation region. The schematic of the device shown in Fig. 10.14 consists of a generic can combustor with the inlet and the exhaust on the same end of the combustor. This design enables the EGR-preheated air to provide a high temperature and stable combustion environment. Modeling the flow and flame structure in this device is complicated by the interaction between various features of flow confinement, stagnation, jet entrainment and product preheating and dilution through interaction of the incoming mixture with reverse co-flow. Experimentally, this combustor was shown to operate in both premixed and non-premixed combustion mode with low emissions but

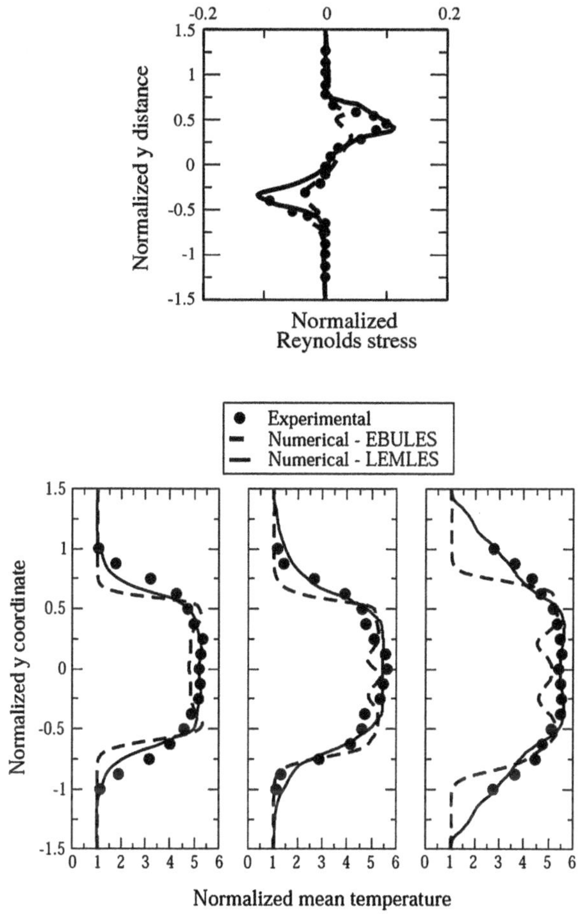

Fig. 10.13: The time-averaged radial profiles behind the bluff body. (a) Reynolds stress, and (b) mean temperature at three axial locations behind the bluff body. Reprinted from [59].

with very different flame structures, and therefore, it is a challenge for numerical prediction.

The recent study [76] simulated both premixed and non-prmeixed combustion using LEMLES and compared with other subgrid models such as the subgrid eddy break-up, artificially thickened flame and steady flamelets. Here, only representative results of the EBULES and LEMLES are discussed, although comparison with other models were reported earlier. Figure 10.15 compares the predictions with the averaged chemiluminescence field (note that although not clear here, the data does show that this flame is attached [20]). The EBULES flame structure is not clearly defined whereas the LEMLES predicts roughly the same flame length and reveals a flame

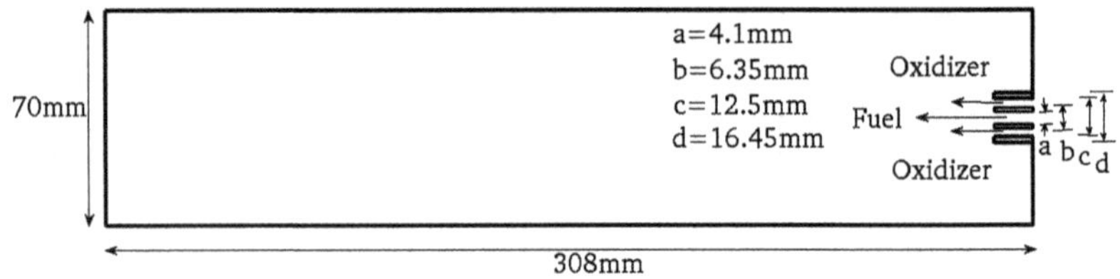

Fig. 10.14: Schematic and typical dimensions of the SPRF combustor.

anchored to the injector with heat release along the shear layer and in the downstream region of the combustor in agreement with the data. Comparison with data is reported elsewhere [76] and shows excellent agreement at nearly all locations.

When the combustor is operated in the non-premixed mode (with the fuel injected through the inner tube) the flame is lifted (Fig. 10.16). The LEMLES prediction is also a lifted flame whereas the EBULES did not capture this effect. The LEMLES flame structure and shape is somewhat different from the measurements but this is attributed to the reduced 1-step kinetics employed [76]. Regardless, the overall agreement with data is encouraging considering that the same model is employed for both premixed and non-premixed combustion without any *ad hoc* adjustments.

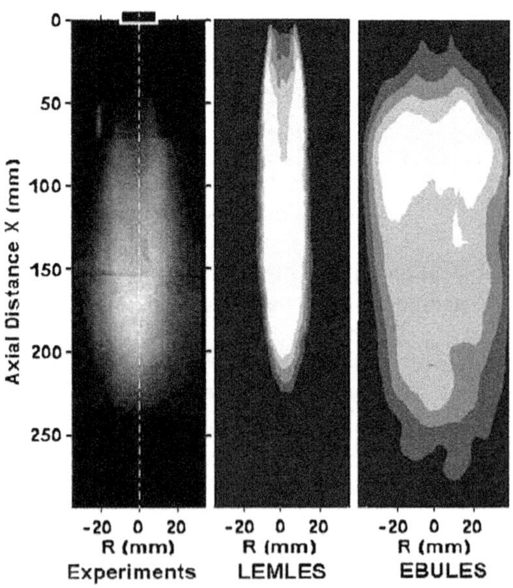

Fig. 10.15: Heat release in the premixed SPRF configuration. (a) Experiments, (b) LEMLES and (c) EBULES. Reprinted from [76] with permission from the Combustion Institute.

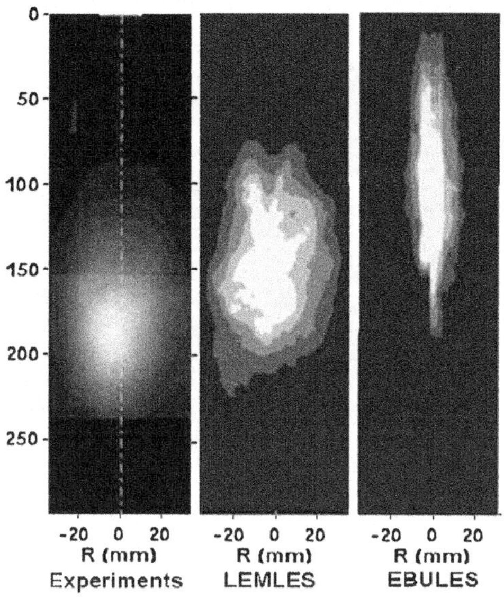

Fig. 10.16: Heat release in the non-premixed SPRF configuration. (a) Experiments, (b) LEMLES and (c) EBULES. Reprinted from [76] with permission from the Combustion Institute.

Finally, an application in spray combustion is discussed briefly although more details are given elsewhere [53, 56, 57]. The LDI combustor consists of a 60-degree, six helical swirl-vaned inlet that leads to a venturi, followed by a short divergent diffuser section that ends at the dump plane of a square combustion chamber. The swirler has outer diameter of 22.5 mm with the inner diameter of 8.8 mm. The calculated swirl number is 1.0 and Fig. 10.17(a) shows the schematic of the swirl generating blades in this combustor. The entire combustor (including the swirler assembly) is simulated with and without break-up modeling of the kerosene-air combustion system [56, 57]. Figure 10.17(b) shows the spray in the vicinity of the injector showing the breakup process. Details are in these cited references but here we focus on the nature of flame holding and the typical characteristics of the flame to highlight the ability of the LEMLES approach.

Centerline mean streamwise velocity is shown in Fig. 10.18(a). A prominent central re-circulation zone (CRZ) along the axis is observed for the non-reacting [56] (not shown) and reacting cases. This CRZ is created by the swirling inflow due to a radial pressure gradient caused by the centrifugal effect, which in turn gives rise to axial (and adverse) pressure gradient. For high swirl numbers, a strong coupling between axial and tangential velocity occurs and the adverse pressure gradient is strong enough to overcome the axial motion of the fluid. This establishes a recirculation zone, a form of vortex breakdown in the central region. This is the primary aerodynamic flame holding and stabilizing mechanism in gas turbine combustors.

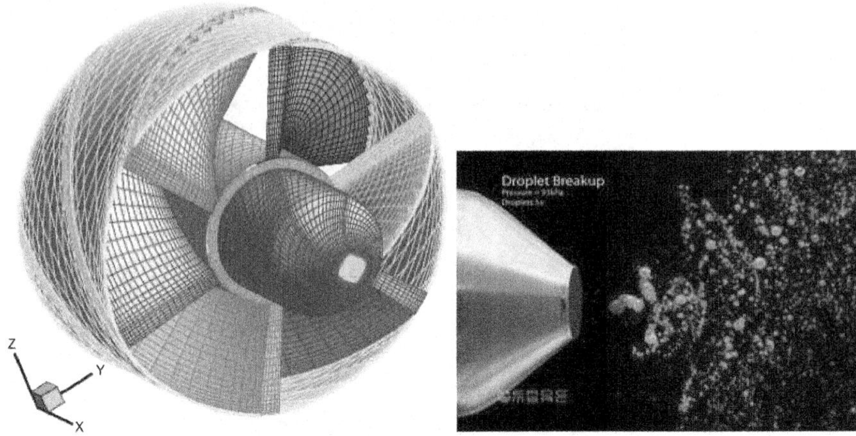

Fig. 10.17: The LDI swirler assembly and an instantaneous view of the droplets in the near -vicinity of the injector showing breakup. Reprinted from [57] with permission from the Combustion Institute.

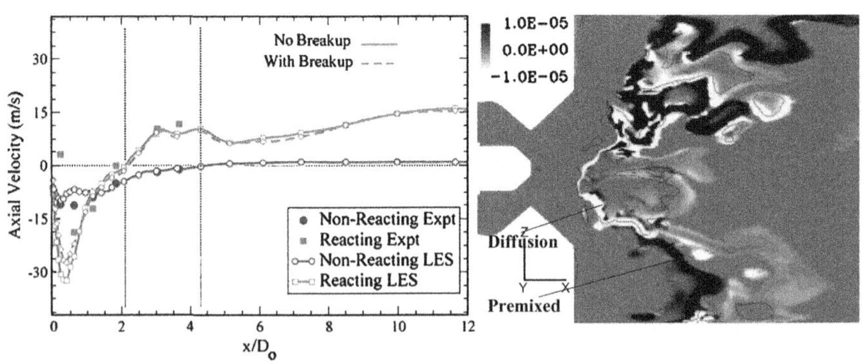

Fig. 10.18: Centerline velocity decay and flame structure in the LDI combustion. (a) Centerline axial velocity, (b) Flame index contours, with the black lines indicating the stoichiometric mixture fraction contour. Reprinted from [56] with permission from the Combustion Institute.

Instantaneous flame structure is analyzed using the flame index, FI=$\nabla Y_F.\nabla Y_{O_2}$. To determine the flame regime, an indexed reaction rate is defined based on the flame index as: $\dot{\omega}_F^* = |\dot{\omega}_F|\frac{FI}{|FI|}$, and is shown in Fig. 10.18(b). Here, $\dot{\omega}_F$ is the fuel destruction rate. The stoichiometric equivalence ratio is shown as a thin line in the same figure. The flame is premixed when the FI (and consequently $\dot{\omega}_F^*$) is positive and diffusion when the FI is negative. In the central region, presence of fuel vapor in proximity of recirculating hot gases devoid of oxidizer generates a diffusion flame, as seen by light colored "V" shaped flame surface. This is confirmed by the coincidence between the flame and the stoichiometric line. Further outwards in the radial direction, significant dark colored contours are seen, indicating a premixed flame. Along the outer edges there is sufficient time for fuel-air mixing to complete before ignition, and thus, both non-premixed and premixed flames occur adjacently near the top half of the combustor with diffusion burning of fuel vapor evaporated from the particles that have gone through the primary flame without completely losing their identity. Since the LEMLES approach does not make any *a priori* assumptions regarding the nature of the flame it is able to capture the multi-faceted flame structure in these combustors.

10.8 Summary and Future Prospects

As summarized here the LEM has been used both as a stand-alone model and as a subgrid model for LES. Application of LEM as a RANS subgrid model has also been reported in some earlier and recent studies. The ability of this modeling strategy to study a wide range of reacting flows without any model adjustments (when combined with the LDKM closure for momentum there are no ad hoc adjustable parameters in the method) is one of the key strengths of this modeling approach. The robustness for a wide range of applications comes with an increase in cost. Efficient parallel implementation (the LEM subgrid model is highly parallel in its nature) can reduce the cost significantly but it is still expensive when compared to a steady flamelet or EBULES approach. On the other hand, the LEM-based closure can be used to go from one simulation regime to another without changing the model and this is a critical strength essential to study complex flame dynamics as in combustion instability, LBO and extinction-re-ignition processes. More recent successes in parametrizing the LEM in a neural network [67, 68, 70] offers a new potential for incorporating this model's ability in a cost effective LES strategy.

Acknowledgement

The research at Georgia Tech is supported in part by Office of Naval Research, NASA Glenn Research Center, Air Force Office of Scientific Research, General Electric Aircraft Engine Company and Pratt & Whitney. This work in SNL is partially supported by the U.S. Department of Energy, Office of Basic Energy Sciences, Division of Chemical Sciences, Geosciences, and Energy Biosciences. Sandia National Laboratories is a multi-program laboratory operated by Sandia Corporation, a Lockheed Martin Company, for the United States Department of Energy under contract DE-AC04-94-AL85000.

References

1. Calhoon, W.H., Menon, S., Goldin, G.: Comparison of reduced and full chemical mechanisms for nonpremixed turbulent h2-air het flames. Combust. Sci. Technol. **104**, 115–141 (1995)
2. Chakravarthy, V., Menon, S.: Modeling of turbulent premixed flames in the flamelet regime. In: Proceedings of first International Symposium on Turbulent and Shear Flow Phenomena, pp. 189–194. Begel House (1999)
3. Chakravarthy, V., Menon, S.: Large-eddy simulations of turbulent premixed flames in the flamelet regime. Combust. Sci. Technol. **162**, 175–222 (2000)
4. Chakravarthy, V., Menon, S.: Subgrid modeling of premixed flames in the flamelet regime. Flow Turbul. Combust. **65**, 23–45 (2000)
5. Chakravarthy, V., Menon, S.: Linear-eddy simulations of Reynolds and Schmidt number dependencies in turbulent scalar mixing. Phys. Fluids **13**, 488–499 (2001)
6. Cremer, M.A., McMurtry, P.A., Kerstein, A.R.: The effect of turbulence length-scale distribution on scalar mixing in homogeneous turbulent-flow. Phys. Fluids **6**, 2143–2153 (1994)
7. DesJardin, P.E., Frankel, S.H.: Assessment of turbulent combustion submodels using the Linear Eddy Model. Combust. Flame **104**, 343–357 (1996)
8. DesJardin, P.E., Frankel, S.H.: Linear-eddy modeling of nonequilibrium turbulent reacting flows with nonpremixed reactants. Combust. Flame **109**, 471–481 (1997)
9. Eggenspieler, G., Menon, S.: Combustion and emission modeling near lean blow-out in gas turbine engines. Prog. Comput. Fluid Dyn. **5**, 281–297 (2005)
10. El-Asrag, H., Menon: Large eddy simulation of a bluff-body stabilized swirling non-premixed flames. Proc. Combust. Inst. **31**, 1747–1754 (2007)
11. El-Asrag, H., Menon, S.: Simulation of soot formation in turbulent premixed flame. Combust. Flame **150**, 108–126 (2007)
12. El-Asrag, H., Menon, S.: Large eddy simulation of soot formation in a turbulent non-premixed jet flame. Combust. Flame **156**, 385–395 (2009)
13. Frankel, S.H., McMurtry, P.A., Givi, P.: Linear eddy modeling of reactant conversion and selectivity in turbulent flows. AIChE J. **41**, 259–270 (1995)
14. Fureby, C., Löfström, C.: Large-eddy simulations of bluff body stabilized flames. Proc. Combust. Inst. **25**, 1257–1264 (1994)
15. Fureby, C., Möller, S.I.: Large-eddy simulation of reacting flows applied to bluff body stabilized flames. AIAA J. **33**, 2339–2347 (1995)
16. Genin, F., Menon, S.: Studies of shock/turbulent shear layer interaction using large-eddy simulation. Computer Fluids **39**, 800–819 (2010)
17. Goldin, G.M., Calhoon, W.H., Menon, S.: A Linear eddy mixing model for steady non-premixed turbulent combustion. AIAA-1995-0379 (1995)
18. Goldin, G.M., Menon, S.: Scalar pdf construction model for turbulent non-premixed combustion. Combust. Sci. Technol. **125**, 47–72 (1997)
19. Goldin, G.M., Menon, S.: A comparison of scalar pdf turbulent combustion models. Combust. Flame **113**, 442–453 (1998)
20. Gopalakrishnan, P., Bobba, M.K., Seitzman, J.M.: Controlling mechnamisms for low nox emissions in a non-premixed stagnation point reverse flow combustor. Proc. Combust. Inst. **31**, 3401–3408 (2007)
21. Guilkey, J.E., Kerstein, A., McMurtry, P.A., Klewicki, J.C.: Effects of initial conditions on scalar statistics in pipe mixing. AIChE J. **43**, 1947–1954 (1997)
22. Guilkey, J.E., Kerstein, A.R., McMurtry, P.A., Klewicki, J.C.: Mixing mechanisms in turbulent pipe flow. Phys. Fluids **9**, 717–723 (1997)
23. Guilkey, J.E., Kerstein, A.R., McMurtry, P.A., Klewicki, J.C.: Long-tailed probability distributions in turbulent-pipe flow mixing. Phys. Rev. E **56**, 1753–1758 (1997)
24. Hawkes, E.R., Sankaran, R., Chen, J.H., Kaiser, S.A., Frank, J.H.: An analysis of lower-dimensional approximations to the scalar dissipation rate using direct numerical simulations of plane jet flames. Proc. Combust. Inst. **32**, 1455–1463 (2009)

25. Hawkes, E.R., Sankaran, R., Sutherland, J.C., Chen, J.H.: Scalar mixing in direct numerical simulations of temporally evolving plane jet flames with skeletal CO/H2 kinetics. Proc. Combust. Inst. **31**, 1633–1640 (2007)
26. Jaberi, F.A., Miller, R.S., Madnia, C.K., Givi, P.: Non-gaussian scalar statistics in homogeneous turbulence. J. Fluid Mech. **313**, 241–282 (1996)
27. Kalda, J., Morozenko, A.: Turbulent mixing: the roots of intermittency. New J. Phys. **10**, 093003 (2008)
28. Kerstein, A.R.: Linear-eddy model of turbulent scalar transport and mixing. Combust. Sci. Technol. **60**, 391–421 (1988)
29. Kerstein, A.R.: Linear-eddy model of turbulent transport II: Application to shear layer mixing. Combust. Flame **75**, 397–413 (1989)
30. Kerstein, A.R.: Linear-eddy model of turbulent transport. Part 3. Mixing and differential molecular diffusion in round jets. J. Fluid Mech. **216**, 411–435 (1990)
31. Kerstein, A.R.: Linear-eddy modeling of turbulent transport. Part V: Geometry of scalar interfaces. Phys. Fluids A **3**, 1110–1114 (1990)
32. Kerstein, A.R.: Linear-eddy modeling of turbulent transport. Part 6. Microstructure of diffusive scalar mixing fields. J. Fluid Mech. **231**, 361–394 (1991)
33. Kerstein, A.R.: Linear-eddy modeling of turbulent transport. Part 7. Finite-rate chemistry and multi-stream mixing. J. Fluid Mech. **240**, 289–313 (1992)
34. Kerstein, A.R.: Prandtl-number dependence of turbulent flame propagation. Phys. Rev. E **64**, 066306 (2001)
35. Kerstein, A.R., Cremer, M.A., McMurtry, P.A.: Scaling properties of differential molecular diffusion effects in turbulence. Phys. Fluids **7**, 1999–2007 (1995)
36. Kerstein, A.R., Krueger, S.K.: Linear eddy simulations of mixing in a homogeneous turbulent-flow. Phys. Rev. E **73**, 025302 (2006)
37. Kerstein, A.R., McMurtry, P.A.: Low wave-number statistics of randomly advected passive scalars. Phys. Rev. E **50**, 2057 (1994)
38. Kim, W.W., Menon, S.: A new incompressible solver for large-eddy simulations. Int. J. Num. Fluid Mech. **31**, 983–1017 (1999)
39. Kim, W.W., Menon, S.: Numerical simulations of turbulent premixed flames in the thin-reaction-zones regime. Combust. Sci. Technol. **160**, 119–150 (2000)
40. Kim, W.W., Menon, S., Mongia, H.C.: Large-eddy simulation of a gas turbine combustor flow. Combust. Sci. Technol. **143**, 25–62 (1999)
41. Krueger, S.K.: Linear eddy modeling of entrainment and mixing in stratus clouds. J. Atmos. Sci. **50**, 3078–3090 (1993)
42. Krueger, S.K., Su, C.W., McMurtry, P.A.: Modeling entrainment and finescale mixing in cumulus clouds. J. Atmos. Sci. **54**, 2697–2712 (1997)
43. McMurtry, P.A., Gansauge, T.C., Kerstein, A.R., Krueger, S.K.: Linear eddy simulations of mixing in a homogeneous turbulent-flow. Phys. Fluids A **5**, 1023–1034 (1993)
44. McMurtry, P.A., Menon, S., Kerstein, A.R.: A new subgrid model for turbulent combustion: Application to hydrogen-air combustion. Proc. Combust. Inst. **24**, 271–278 (1993)
45. Menon, S.: Computational and modeling constraints for large-eddy simulations of turbulent combustion. Int. J. Engine Res. **1**, 209–227 (2000)
46. Menon, S., Calhoon, W.H.: Subgrid mixing and molecular transport modeling for large-eddy simulations of turbulent reacting flows. Proc. Combust. Inst. **26**, 59–66 (1996)
47. Menon, S., Calhoon, W.H., Goldin, J.R., Kerstein, A.R.: Effects of molecular transport on turbulent-chemistry interactions in hydrogen-argon-air jet diffusion flame. Proc. Combust. Inst. **25**, 1125–1131 (1994)
48. Menon, S., Kerstein, A.R.: Stochastic simulation of the structure and propagation rate of turbulent premixed flames. Proc. Combust. Inst. **24**, 443–450 (1992)
49. Menon, S., Kim, W.W.: High Reynolds number flow simulations using the localized dynamic subgrid-scale model. AIAA-96-0425 (1996)

50. Menon, S., McMurtry, P., Kerstein, A.R.: A linear eddy mixing model for large eddy simulation of turbulent combustion. In: B. Galperin, S. Orszag (eds.) LES of Complex Engineering and Geophysical Flows, pp. 287–314. Cambridge University Press, Cambridge, UK (1993)
51. Menon, S., McMurtry, P.A., Kerstein, A.R., Chen, J.Y.: A new mixing to predict NOx production in turbulent hydrogen-air jet flame. J. Prop. Power **10**, 161–168 (1994)
52. Menon, S., Pannala, S.: Subgrid combustion simulations of reacting two-phase shear layers. AIAA-98-3318 (1998)
53. Menon, S., Patel, N.: Subgrid modeling for LES of spray combustion in large-scale combustors. AIAA J. **44**, 709–723 (2006)
54. Menon, S., Yeung, P.K., Kim, W.W.: Effect of subgrid models on the computed interscale energy transfer in isotropic turbulence. Computers Fluids **25**, 165–180 (1996)
55. Oevermann, M., Schmidt, H., Kerstein, A.R.: Investigation of autoignition under thermal stratification using linear eddy modeling. Combust. Flame **155**, 370–379 (2008)
56. Patel, N., Kirtas, M., Sankaran, V., Menon, S.: Simulation of spray combustion in a lean direct injection combustor. Proc. Combust. Inst. **31**, 2327–2334 (2007)
57. Patel, N., Menon, S.: Simulation of spray-turbulence-flame interactions in a lean direct injection combustor. Combust. Flame **153**, 228–257 (2008)
58. Pope, S.B.: Pdf methods for turbulent reactive flows. Prog. Energy Combust. Sci. **11**, 119–192 (1985)
59. Porumbel, I., Menon, S.: Large-eddy simulation of bluff body stabilized premixed flames. AIAA-2006-0152 (2006)
60. Sankaran, V., Menon, S.: Structure of premixed flame in the thin-reaction-zones regime. Proc. Combust. Inst. **28**, 203–210 (2000)
61. Sankaran, V., Menon, S.: LES of scalar mixing in supersonic shear layers. Proc. Combust. Inst. **30**, 2835–2842 (2005)
62. Sankaran, V., Menon, S.: Subgrid combustion modeling of 3-d premixed flames in the thin-reaction-zone regime. Proc. Combust. Inst. **30**, 575–582 (2005)
63. Sankaran, V., Porumbel, I., Menon, S.: Large-eddy simulation of a single-cup gas turbine combustor. AIAA-2003-5083 (2003)
64. Weydahl, T.: A framework for mixing-reaction closure with the linear-eddy model. Ph.D. Thesis, Norwegian University of Science and Technology, Trodheim, Norway (2010)
65. Schenck, H.W., Wendt, J.O.L., Kerstein, A.R.: Mixing characterization of transient puffs in a rotary kiln incinerator. Combust. Sci. Technol. **116**, 427–453 (1996)
66. Schmidt, R.C., Kerstein, A.R., McDermott, R.: ODTLES: A multi-scale model for 3D turbulent flow based on one-dimensional turbulence modeling. Comput. Meth. Appl. Mech. Engg. **199**, 865–880 (2009)
67. Sen, B.A., Hawkes, E., Menon, S.: Large eddy simulation of extinction and reignition with artificial neural networks based chemical kinetics. Combust. Flame **157**, 566–578 (2010)
68. Sen, B.A., Menon, S.: Artificial neural networks based chemistry-mixing subgrid model for LES. AIAA-2009-0241 (2009)
69. Sen, B.A., Menon, S.: Turbulent premixed flame modeling using artificial neural network based chemical kinetics. Proc. Combust. Inst. **32**, 1605–1611 (2009)
70. Sen, B.A., Menon, S.: Linear eddy mixing and artificial neural networks for LES subgrid chemistry closure. Combust. Flame **157**, 62–74 (2010)
71. Smith, T., Menon, S.: One-dimensional simulations of freely propagating turbulent premixed flames. Combust. Sci. Technol. **128**, 99–130 (1996)
72. Smith, T.M., Menon, S.: Model simulations of freely propagating turbulent premixed flames. Proc. Combust. Inst. **26**, 299–306 (1996)
73. Smith, T.M., Menon, S.: Large-eddy simulations of turbulent reacting stagnation point flows. AIAA-97-0372 (1997)
74. Smith, T.M., Menon, S.: Subgrid combustion modeling for premixed turbulent reacting flows. AIAA-98-0242 (1998)

75. Su, C.W., Krueger, S.K., McMurtry, P.A., Austin, P.H.: Linear eddy modeling of droplet spectral evolution during entrainment and mixing in cumulus clouds. Atmos. Res. **47**, 41–58 (1998)
76. Undapalli, S., Menon, S.: LES of premixed and non-premixed combustion in a stagnation point reverse flow combustor. Proc. Combust. Inst. **32**, 1537–1544 (2009)
77. Woosley, S.E., Kerstein, A.R., Sankaran, V., Aspden, A.J., Roepke, F.K.: Type Ia supernovae: Calculations of turbulent flames using the linear eddy model. Astrophys. J. **704**, 255–273 (2009)
78. Wu, J., Menon, S.: Aerosol dynamics in the near-field engine exhaust plumes. J. Appl. Meteo. **40**, 795–809 (2001)
79. Zimberg, M.J., Frankel, S.H., Gore, J.P., Sivathanu, Y.R.: A study of coupled turbulent mixing, soot chemistry, and radiation effects using the linear eddy model. Combust. Flame **113**, 454–469 (1998)

Chapter 11
The One-Dimensional-Turbulence Model

Tarek Echekki, Alan R. Kerstein, and James C. Sutherland

Abstract The one-dimensional turbulence (ODT) model represents an efficient and novel multiscale approach to couple the processes of reaction, diffusion and turbulent transport. The principal ingredients of the model include a coupled deterministic solution for reaction and molecular transport and a stochastic prescription for turbulent transport. The model may be implemented as stand-alone for simple turbulent flows and admits various forms for the description of spatially developing and temporally developing flows. It also may be implemented within the context of a coupled multiscale solution using the ODTLES approach. This chapter outlines the model formulation, and applications of ODT using stand-alone solutions and ODTLES.

11.1 Motivation

In his survey of turbulent combustion modeling, Norbert Peters [25] emphasizes the difficulties that can arise due to the interactions between turbulence and chemistry over a wide range of length and time scales. Assumptions about inertial-range scaling of the turbulent cascade are not necessarily applicable, and there are few if any formal or conceptual constructs to which the modeler can turn when these scalings do not apply. He notes important empirical evidence that gross features of turbulent combustion often conform to inertial-range phenomenology, particularly with regard to its most important consequence for combustion: the length and time scale

Tarek Echekki
North Carolina State University, Raleigh, NC, USA, e-mail: techekk@ncsu.edu

Alan R. Kerstein
Sandia National Laboratories, Livermore, CA, USA, e-mail: arkerst@sandia.gov

James C. Sutherland
The University of Utah, Salt Lake City, UT, USA, e-mail: James.Sutherland@utah.edu

T. Echekki, E. Mastorakos (eds.), *Turbulent Combustion Modeling*,
Fluid Mechanics and Its Applications 95, DOI 10.1007/978-94-007-0412-1_11,

separation between the dominant large-scale turbulent motions and the small scales of molecular transport and chemical kinetics in flames. Many existing models rely on this scale separation, while the linear-eddy model (LEM: see Chapter 10) and one-dimensional turbulence (ODT) do not. To the extent that scale separation is not obeyed in turbulent combustion processes, LEM and ODT can be especially suitable for modeling these processes.

Peters discusses the role of scale invariance as well as scale separation in the inertial range. The latter is more general: scale separation is possible in the absence of scale invariance, but there is no scale invariance without scale separation (see below). Scale invariance of spectral energy transfer in turbulence, in conjunction with (and in fact, relying on) the negligible influence of viscous dissipation in the inertial range (an idealization, but valid in this context), imply that the characteristic eddy time scale $\tau(l)$ at length scale l obeys the dependence $\tau(l) \sim l^{2/3}$. This implies power-law dependences of eddy velocity, energy, and diffusivity, and ultimately, scale separation between inertial-range and dissipation-range processes.

LEM resolves all advective, diffusive, and chemical length and time scales of turbulent combustion, and hence does not rely on scale separation. It represents thermal expansion by means of dilatation of the 1D domain, but it does not model other aspects of feedback from combustion to turbulent motions, such as viscosity variations and turbulence generation by expansion. In particular, the probability distribution function (pdf) from which the maps representing turbulent eddies are sampled has a fixed functional form whose construction is guided by the inertial-range diffusivity scaling (Chapter 10). To the extent that turbulence-chemistry interactions result in deviations from this scaling, the fidelity of LEM can be impaired. Other influences, such as rapid transients caused by initial conditions or time-varying boundary conditions, can also cause significant deviations from inertial-range scalings.

As noted, there is little theoretical guidance on how to model the deviations from inertial-range scalings that might be caused by these influences or the effects of these deviations on combustion dynamics. Within the 1D stochastic approach based on the triplet-map representation of turbulent eddies (see Chapter 10), incorporation of these influences is therefore approached from an entirely different perspective.

This perspective is introduced by first considering direct numerical simulation (DNS) of turbulent combustion. In DNS, no theory or guiding principle is needed to capture combustion-induced deviations from inertial-range scaling because the underlying equations of motion are solved. Thus, the consequences of turbulence-chemistry interactions are outcomes of the simulated flow evolution that do not require prior analysis or modeling. To the extent that the local, time-resolved interactions between turbulent eddy motions and combustion processes can be represented robustly with the 1D stochastic framework, that framework can likewise capture their consequences as outcomes of simulated evolution.

A formulation with this capability becomes, in effect, a broadly predictive model of turbulent flow evolution rather than a model focused, as LEM is focused, on simulation of mixing and reaction in a parametrically specified turbulent environment. Thus, the quest for a robust turbulent combustion model leads ultimately to reconsideration of turbulence modeling more generally. This is the context in which ODT was originally formulated. The stages of its development for application to free and confined shear flows, variable-density flows, buoyant stratified flows, and multiphase flows, as well as combustion and other reacting flows, have been documented in references cited below.

To introduce ODT, a formulation applicable to constant-property flow is outlined. Algorithmic as well as physical modeling considerations are discussed. Extensions needed for combustion applications and representative examples are described. Use of ODT for subgrid modeling in 3D simulations of constant-property flow and of combustion is discussed, and future prospects are assessed.

11.2 Constant-Property ODT

11.2.1 Model Formulation

ODT is introduced with reference to LEM by formulating LEM in a notation that carries over directly to ODT. An eddy rate distribution $\lambda(y_0,l;t)$ is defined by setting $\lambda(y_0,l;t)\,dy\,dl$ equal to the expected number, per unit time, of eddies for which the lower boundary of the eddy is in $[y_0,y_0+dy]$ and the eddy size is in the range $[l,l+dl]$. In terms of Λ and $f(l)$ defined in Chapter 10, $\lambda(y_0,l;t)=\Lambda f(l)$. Thus, the units of λ are 1/(length$^2\times$time) and its integral over eddy sizes l is Λ.

In LEM, λ has no dependence on y_0 or t unless one chooses to hard-wire such dependence into Λ or $f(l)$, which has been done in some instances, e.g., [13]. ODT is formulated to incorporate such dependence in a way that reflects the relationship between the likelihood of an eddy, quantified by the eddy time scale τ, and the local flow state. This requires the introduction of a local, instantaneous representation of the flow state in ODT, causing ODT to be a fundamentally different type of model than LEM, whose turbulent state is characterized parametrically.

The flow representation in ODT consists of 1D profiles of one or more velocity components whose evolution, in the simplest constant-property formulation, is qualitatively the same as in LEM applied to a constant-density passive scalar. For example, with one velocity component $u(y,t)$, time advancement is governed by

$$\frac{\partial u}{\partial t}=\nu\frac{\partial^2 u}{\partial y^2}, \tag{11.1}$$

where ν is the kinematic viscosity. This advancement is punctuated by 'eddy events,' each of which consists of a triplet map (defined and explained in Chapter 10) applied to the u profile, possibly (depending on the formulation) followed by another operation that is described shortly. Absent the latter, the evolution of u is equivalent to the evolution of a constant-property passive scalar in LEM, except for the crucial distinction that $\lambda(y_0,l;t)$ is now a specified function of the current flow state $u(y,t)$. Several formulations of this dependence have been introduced during the course of ODT development. The formulation discussed here,

$$\lambda = \frac{C\nu}{l^4}\sqrt{\left(\frac{u_K l}{\nu}\right)^2 - Z}, \tag{11.2}$$

is a specialization of the formulation described in [17], which has a more detailed explanation of the motivation and features of ODT than can be provided here.

In Eq. 11.2, C and Z are free parameters whose roles are explained shortly, and for any property profile $s(y,t)$,

$$s_K \equiv \frac{1}{l^2}\int s(M(y))K(y)\,dy \tag{11.3}$$

defines s_K. Here, $M(y)$ is defined by the formal mathematical representation of the triplet map,

$$s(y) \rightarrow s(M(y)), \tag{11.4}$$

which indicates that the value of property s at $M(y)$ is mapped to location y. Thus, M is the operational inverse of the triplet map, which is formally convenient because M is a single-valued map but the triplet map is triple-valued. The 'kernel' $K(y) \equiv y - M(y)$ is the map-induced displacement of the point that is mapped to y by the triplet map. $K(y)$ thus depends on map parameters y_0 and l, but this dependence is suppressed in the condensed notation used here.

To explain the role of the kernel, the more general form of the eddy event in ODT is introduced. Namely, Eq. 11.4 is generalized to

$$s(y) \rightarrow s(M(y)) + c_s K(y), \tag{11.5}$$

which indicates that the eddy event applies the triplet map to property s and then adds the kernel times a coefficient c_s to $s(y)$. The kernel addition is applied only to velocity components and is intended to add or substract kinetic energy without changing the total momentum, which is assured for constant-density flow because $K(y)$ integrates to zero. This provides a mechanism for energy redistribution among velocity components when the formulation contains more than one component, enabling the model to simulate the tendency of turbulent eddies to drive the flow toward isotropy. Total energy must be conserved, imposing a constraint on the values of the coefficients c_s. The additional constraints needed to uniquely determine all the coefficients are obtained by requiring the kernel addition to produce an energy distribution within the eddy interval $[y_0, y_0 + l]$ that is as close to isotropic as possible. There are other reasonable criteria for determining the coefficients that might be advantageous in some cases [19]. In applications such as buoyant stratified flows or turbulent advection of immiscible liquids, eddy events might induce changes of the gravitational or surface-tension potential energy, requiring equal-and-opposite changes of kinetic energy. Through the kernel operation, conservation of total energy in ODT couples flow evolution to dynamically active scalars such as density in buoyant flows, as demonstrated in ODT studies of these flows [5, 15–17, 20, 35, 36, 38–40].

By construction, Eq. 11.5 conserves momentum for constant-density flow. For variable-density flow, a more general treatment is needed. Momentum conservation is no longer automatic, so extra degrees of freedom are introduced to enforce momentum conservation, further generalizing the eddy event as follows:

$$s(y) \rightarrow s(M(y)) + b_s J(y) + c_s K(y), \tag{11.6}$$

where $J(y) \equiv |K(y)|$ and the additional coefficients b_s are determined by requiring the eddy event to conserve the y-integrated momentum of all velocity components s. For details, see [1].

Most combustion applications of ODT pre-date this variable-density formulation, and several pre-date the introduction of the kernel formalism in Eqs. 11.2 and 11.5, so they do not include the kinetic-energy and variable-density phenomenology that can now be incorporated into ODT. The less complete treatment is adequate for many combustion applications, much as LEM, which is considerably simpler, is broadly useful for turbulent combustion simulation. The variable-density formulation is not discussed further here, but its future use, where appropriate, in ODT simulations of turbulent combustion is encouraged.

Before proceeding further, the guiding principle that motivates the model constructs introduced thus far is explained. Empirical evidence and formal analysis support the viewpoint that the turbulent cascade tends to be local in scale space, meaning that individual eddy motions such as vortex stretching typically shrink flow features in turbulence by order-one geometrical increments, such that the wide range of flow scales in turbulence is the cumulative outcome of many incremental scale reductions rather than a smaller number more drastic reductions.

Enforcement of this scale-locality principle in ODT is the basis of much of the model formalism. The triplet map decreases flow scales by no more than a factor of three. No other measure-preserving map induces less scale reduction. (Measure preservation assures that applicable conservation laws are obeyed.)

Energy changes during the eddy event are likewise consistent with scale locality. The kernel used for this purpose must be zero at the eddy endpoints (to prevent discontinuities) and must integrate to zero, so it must have at least two extrema. The function $K(y)$ consists of three linear segments over size-$l/3$ spatial intervals. Thus it introduces structure consistent with the scale reduction by the triplet map. The map is applied before the kernel because the kernel followed by the map would introduce structure at scale $l/9$.

Eddy selection as well as eddy implementation is guided by scale locality, in this case meaning that size-l motions are driven by size-l influences. Eddy phenomenology (i.e., mixing-length concepts applied to an individual eddy) suggests that λ should be of order $1/(l^2\tau)$, where τ is the eddy turnover time, or equivalently, $V(l)/l^3$, where $V(l)$ is the velocity difference between the eddy endpoints. The latter estimate, with numerical coefficients absorbed in the free parameter C of Eq. 11.2, was used in the original ODT formulation [15] and many subsequent applications. When kernels were introduced, the estimate $V(l) \sim u_K$ was adopted, which connects eddy selection to energy-based eddy implementation using an expression that measures velocity variations over an order-l distance, and hence is consistent with scale locality. In Eq. 11.2, the expression for λ involves a square root that contains a term proportional to u_K^2, hence a kinetic-energy term. In formulations involving multiple velocity components or other energy contributions such as gravitational potential energy, these contributions are additive under the square root.

A specific connection between u_K and eddy energetics is identified through consideration of the possible range of kernel-induced energy changes. Kernel addition can reduce the u kinetic energy within the eddy to zero only if the spatial profile of u within the eddy is proportional to the kernel function, so that kernel addition can make u identically zero within the eddy. Otherwise, there is a maximum amount of energy that can be extracted from the u profile by kernel addition that is less than the total u kinetic energy within the eddy. This maximum, termed the 'available energy' of the u component, is $(27/8)\rho l u_K^2$, where ρ is the density [19]. This connects u_K to flow energetics in various ways. For example, implementation of the isotropy criterion involves assignment of the coefficients c_s so as to equalize component available energies. In buoyant stratified flows, the available energy is the maximum kinetic energy that can be extracted in order to compensate for an equal-and-opposite change of gravitational potential energy. If there is less than the needed amount of available energy, then the eddy is energetically prohibited, so its λ value is set equal to zero.

The indication of a prohibited eddy is that the quantity in the square root in Eq. 11.2 is negative. The parameter Z is introduced so that an eddy can be prohibited even if there is net available energy. As indicated by the normalization of u_K in that equation, Z in effect sets a threshold Reynolds number for eddy turnover. Nonzero Z is not always required for good model performance, but in some instances it improves the results sufficiently to justify the introduction of an additional adjustable parameter. In some instances, Z is assigned a small positive value solely for computational efficiency. It prevents the implementation of unphysically small eddies that, if implemented, would have no noticeable effect on results of interest.

C is the main adjustable parameter of ODT. It scales the eddy event rate, and hence the simulated turbulence intensity, for a given flow configuration. In transient flows, it controls overall time development, e.g., the spreading rate of free shear flows.

Just as there can be a need to assign a positive Z value to suppress small eddies, there can be a need to suppress unphysically large eddies that would otherwise occur. This need arises primarily in simulations of free shear flows with laminar co-flows or free streams. Eddies much larger than the width of the turbulent region can have enough available energy to enable their occurrence. (For a planar mixing layer, the difference between the free-stream velocities can provide enough available energy irrespective of any turbulence.) Such eddies violate the scale-locality requirement that the scale of the flow features that provide the available energy for eddy occurrence should be of the order of the eddy size.

Several large-eddy-suppression procedures have proven useful. One that is found to work particularly well for jets and jet diffusion flames [7, 27] requires that τ (which is l/u_K in the formulation described here) must be less than the elapsed flow-advancement time times an adjustable coefficient, otherwise the eddy is prohibited. The introduction of an additional free parameter is found to be well justified by the resulting model performance. An alternative that was used to simulate mixing layers [1] and buoyant fire plumes [31] is the scale-reduction method, in which the eddy is divided into three equal parts and each of these must be found to have enough available energy for eddy turnover (based on a small Z value that negates any numerical noise contribution; results are insensitive to the chosen value), otherwise the eddy is prohibited. If not prohibited based on this test, the eddy is processed in the usual manner.

11.2.2 Numerical Implementation

Based on Eq. 11.3, u_K and thus λ depend on y_0 and l, and this dependence is time varying due to the time advancement of $u(y,t)$. Therefore at each instant there is a new eddy rate distribution from which individual eddy events are to be sampled. Computation of, and sampling from, this two-parameter distribution on an ongoing basis is computationally unaffordable. Therefore, the thinning algorithm [22] for efficient sampling from nonstationary Poisson processes (which is a generalization of the von Neumann rejection method) is employed. A fixed eddy rate distribution $\hat{\lambda}$ is constructed so as to oversample all eddies, i.e., it exceeds the true λ value for all y_0, l, and t. When an eddy is sampled from the fixed distribution, the true λ value for that eddy based on the flow state at that instant is computed and the eddy is accepted with probability $\lambda/\hat{\lambda}$, otherwise rejected. This approach strongly influences many aspects of algorithm formulation and coding.

ODT has been implemented numerically using both uniform and adaptive meshes. On a uniform mesh, the triplet map is implemented as a permutation of mesh cells. On an adaptive mesh, the mathematical definition of the triplet map on the spatial continuum is applied. Properties are assumed constant within each cell, so the continuum triplet map is applied to piecewise constant continuum property profiles. This involves mapping cell faces, which creates new faces because the map is multivalued, and assigning cell property values accordingly.

The uniform-mesh implementation is described in many of the publications that have been cited, and a uniform-mesh code and documentation are available for download [9]. The adaptive-mesh implementation is explained and applied in [31].

11.2.3 Generalizations and Couplings

The adaptive mesh facilitates several generalizations of ODT that are difficult to implement on a uniform mesh. It allows Lagrangian rather than Eulerian implementation of advection (in the conventional sense), which is useful for incorporating thermal expansion and for implementing spatial (streamwise) rather than temporal advancement of ODT. Spatial advancement conserves property fluxes rather than properties. Flux conservation requires an implementation of continuity that involves dilatations along the 1D domain, which are convenient to implement in a Lagrangian manner. The numerics of Eulerian spatial advancement are especially challenging for variable-property flows. This motivated the introduction of an adaptive mesh to simulate the vertical spatial development of a fire plume [31]. Variable-property spatial advancement has been implemented on a uniform mesh by performing a Lagrangian sub-step and then interpolating the displaced mesh back to the fixed uniform mesh [1].

The adaptive mesh also facilitates ODT implementation in cylindrical geometry, in which triplet maps must conserve $r\,dr$ rather than dy. In this case, the triplet map is not readily approximated by permuting the cells of a uniform mesh. An adaptive-mesh implementation of a cylindrical spatial formulation has been used to simulate round jet diffusion flames [21]. An earlier cylindrical LEM formulation on a uniform mesh [14] conserves ensemble averages but is not locally conservative, which is less desirable but adequate for some purposes.

A spatially advancing ODT realization can be interpreted as a 2D flow snapshot. On this basis, a spatially advancing fire-plume simulation [31] has been used to compute 2D radiation fields, which are then used to specify the background radiation field for the next simulated realization. This alternation between ODT and the radiation computation was iterated to statistical convergence to obtain a coupled flow-radiation solution.

Another physical process that has been coupled to the ODT flow simulation is inertial-particle response to turbulent motions (one-way coupling). This formulation has been used to simulate wall deposition in channel flow [34].

For some applications, full spatial resolution is unaffordable even in 1D. Therefore various approaches to subgrid closure in ODT have been developed and applied [20, 24].

Chapter 10 describes ways in which LEM domains have been coupled to under-resolved 3D flow simulations to provide mixing and chemistry closure. ODT has the capability to provide subgrid momentum closure as well, as demonstrated in applications to channel flow [33] and homogeneous decaying turbulence [32]. Various formulations of ODT-based 3D flow simulation have been proposed [17, 18, 23]. Formulations that have been used for combustion simulation are described in Sect. 11.3.3.

11.2.4 Features of the ODT Representation of Turbulent Flow

The ODT representation of a time-developing Kelvin–Helmholtz instability, illustrated in Fig. 11.1, indicates some of the flow features captured by the model. This illustration is based on the ODT formulation of [19].

The rendering shows that the width of the active mixing zone grows primarily by the relatively infrequent occurrence of a large event extending beyond the current range of the mixing zone, with some additional contribution by the more numerous small events. This process is consistent with the dominant role of large engulfing motions and the secondary role of small-scale nibbling in turbulent entraining flows under neutral-buoyancy conditions. (The effect of density stratification on the ODT representation of turbulent entrainment has been investigated [1, 15].)

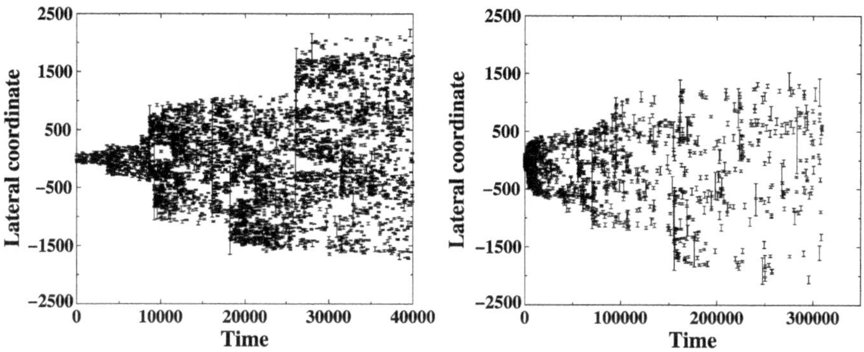

Fig. 11.1: Graphical representation of the sequence of eddy events during a simulated ODT realization of a time-developing Kelvin–Helmholtz instability (left panel) and a time-developing planar wake (right panel) [19]. The Kelvin–Helmholtz and wake simulations are initialized using step-function and top-hat initial velocity profiles, respectively. The space and time units in this illustration are arbitrary. In the plots, each eddy is represented by an error bar whose vertical span corresponds to the eddy range $[y_0, y_0 + l]$, and whose horizontal location corresponds to the time of eddy occurrence. Reprinted from [19] with permission. Copyright © 2001, Cambridge University Press.

Bunching of events, especially after the occurrence of a large event, reflects the interactions between the eddy events and the evolving velocity profile that induce the model analog of the turbulent cascade. Each eddy event compresses and folds the velocity profile within the range of the eddy. This increases the local shear and thus the available energy that determines the frequency of subsequent eddies within that range. A feedback process is thus induced that promotes the occurrence of successively smaller eddies. Eventually, velocity fluctuation length scales are reduced sufficiently so that damping of the fluctuations by concurrent viscous transport dominates the production of fluctuations by eddies. Viscous damping thus terminates the local burst of eddy activity.

A planar-wake simulation is also shown in Fig. 11.1. In the Kelvin–Helmholtz simulation, vigorous turbulence, indicated by the number and size range of eddies as the flow evolves, is sustained by the shear imposed on the flow by the free-stream conditions (far-field velocity difference). The wake, however, evolves in a uniform background. As the initial velocity perturbation is dispersed by eddies and dissipated by concurrent viscous evolution, the turbulence intensity decreases, affecting the eddy frequency and size range and slowing the growth of the turbulent zone. These qualitative impressions are supported by the quantitative consistency of ODT simulation statistics with the known similarity scalings for these flows [19].

11.3 Applications of ODT in Combustion

Like its predecessor, LEM (see Chapter 10), the ODT model may be implemented either as a stand-alone model or within the context of a 3D solution, such as LES. The stand-alone ODT model may also serve a similar role to direct numerical simulations (DNS) for the construction of libraries for turbulence-chemistry interactions [28–30]. ODT stand-alone models are limited in scope to flows with one dominant flow direction, where a boundary-layer like solution may be adopted. Similar to LEM, the implementation of ODT within the context of more complex flows may be achieved through the coupling of ODT with a coarse-grained simulation approach, such as LES [3].

There are many variants of the ODT model in the literature. The modeling approach is typically explained by combining the discretization, solution algorithm, and governing equations. In Sect. 11.3.1, we present a unified method by which all of the approaches in the literature may be derived, along with a brief discussion of the equations ultimately used by in the various approaches.

Section 11.3.2 then presents a sampling of results from stand-alone ODT simulations of turbulent combustion.

11.3.1 Governing Equations

This section presents a brief discussion of the various forms of the governing equations presently solved in ODT. A more detailed exposition can be found in [37].

A generic balance equation for an intensive property ψ in a control volume (CV) V enclosed by surface S can be written as

$$\int_{\mathsf{V}(t)} \frac{\partial \rho \psi}{\partial t} \mathrm{dV} + \int_{\mathsf{S}(t)} \rho \psi (\mathbf{v}_r + \mathbf{v}_s) \cdot \mathbf{a} \mathrm{dS} = - \int_{\mathsf{S}(t)} \boldsymbol{\Phi}_\psi \cdot \mathbf{a} \mathrm{dS} + \int_{\mathsf{V}(t)} \sigma_\psi \mathrm{dV}, \quad (11.7)$$

where $\mathbf{v}_s$ is the velocity of the surface S, $\mathbf{v}$ is the mass-averaged velocity, $\mathbf{v}_r = \mathbf{v} - \mathbf{v}_s$ is the velocity of the fluid relative to the surface, ρ is the density, $\boldsymbol{\Phi}_\psi$ is the mass diffusive flux of ψ, and σ_ψ is the volumetric rate of production of ψ. Table 11.1 defines the terms ψ, $\boldsymbol{\Phi}_\psi$, and σ_ψ for various quantities. These equations are closed with an appropriate equation of state relating the local pressure to the composition, density and temperature.

Table 11.1: Definition of terms in (11.7) for some common governing equations. Here $\boldsymbol{\tau}$ is the stress tensor, $\mathbf{g}$ is the gravitational acceleration vector, Y_k is the mass fraction of species k, $\mathbf{j}_k$ is the species diffusive flux vector, and $\mathbf{q}$ is the heat flux vector. Other equations may be added as necessary. This is just a partial representation of commonly solved equations.

Equation	ψ	Non-convective Flux, $\boldsymbol{\Phi}_\psi$	Source Term, σ_ψ
Continuity	1	0	0
Momentum	$\mathbf{v}$	$p\mathbf{I}-\boldsymbol{\tau}$	$\rho\mathbf{g}$
Species	Y_k	$\mathbf{j}_k$	σ_k
Total Internal Energy	e_0	$p\mathbf{v}+\boldsymbol{\tau}\cdot\mathbf{v}-\lambda\nabla T+\sum_{k=1}^{n} h_k\mathbf{j}_k$	$\rho\mathbf{g}\cdot\mathbf{v}$
Internal Energy	e	$\mathbf{q}$	$\boldsymbol{\tau}:\nabla\mathbf{u}-p\nabla\cdot\mathbf{u}$
Enthalpy	h	$\mathbf{q}$	$\frac{\partial p}{\partial t}+\mathbf{u}\cdot\nabla p+\boldsymbol{\tau}:\nabla\mathbf{u}$

In the following, we present the various forms of the governing equations in use for ODT. Much of the treatment of the governing equations for ODT in the literature combine the governing equations with the numerical algorithm. The following does not address numerical solution techniques for the equations; rather, we focus on a unified approach for arriving at the various forms of the governing equations implied by present ODT approaches in the literature.

11.3.1.1 Temporally Evolving Lagrangian Formulation

The first ODT formulations employed a temporally evolving formulation in a Lagrangian frame of reference. In this case, we have $\mathbf{v}=\mathbf{v}_s$ so that $\mathbf{v}_r=0$. Writing (11.7) in one dimension and using the continuity equation ($\psi=1$) to convert it to the weak form yields

$$\frac{\mathrm{d}\psi}{\mathrm{d}t}=\frac{1}{\rho}\left[-\frac{\partial\Phi_{\psi,y}}{\partial y}+\sigma_\psi\right]. \tag{11.8}$$

In (11.8), $\frac{\mathrm{d}\psi}{\mathrm{d}t}$ represents the local change of ψ as it moves at velocity v_s, $-\frac{1}{\rho}\frac{\partial\Phi_{\psi,y}}{\partial y}$ is the change in ψ due to diffusion, and $\frac{\sigma_\psi}{\rho}$ is the change in ψ due to consumption/production.

Implementations of this approach use moving meshes and finite-volume schemes. The CV surface positions can be determined by solving (11.8) for $\psi=v$ (the lateral fluid velocity) and an ODE for position,

$$\frac{\mathrm{d}y}{\mathrm{d}t}=v=v_s. \tag{11.9}$$

Rather than solving (11.8) for $\psi=v$, however, most ODT formulations employing the temporally evolving Lagrangian formulation instead use a discrete form of the continuity equation written in Eulerian coordinates, $\frac{\partial\rho}{\partial t}=-\frac{\partial v_s}{\partial y}$ together with the assumption of constant pressure and an equation of state to solve for v_s for use in (11.9).

11.3.1.2 Temporally Evolving Eulerian Formulation

Recently, an Eulerian temporally evolving formulation for the ODT equations was proposed [26, 27]. In the Eulerian frame of reference, we fix the CV surface positions so that $\mathbf{v}_s = 0$ and $\mathbf{v}_r = \mathbf{v}$ so that (11.7) becomes, in differential form,

$$\frac{\partial\rho\psi}{\partial t}=-\nabla\cdot\rho\psi\mathbf{v}-\nabla\cdot\boldsymbol{\Phi}_\psi+\sigma_\psi. \tag{11.10}$$

The one-dimensional differential form of (11.10) is

$$\frac{\partial\rho\psi}{\partial t}=-\frac{\partial\rho\psi v}{\partial y}-\frac{\partial\Phi_{\psi,y}}{\partial y}+\sigma_\psi. \tag{11.11}$$

The velocity v in (11.11) represents the local fluid velocity in the y-direction, and $\frac{\partial\rho\psi}{\partial t}$ represents the local change in $\rho\psi$ at a given point in space and time. Current approaches using the Eulerian form have solved the compressible form of these equations [26, 27]. The equations solved are given by (11.10) and Table 11.1, where $\psi=1$ is solved for ρ, $\psi=v$ and $\psi=u$ are solved for the lateral and streamwise momentum components, $\psi=e_0$ is solved for the total internal energy, $\psi=Y_k$ is solved for the species mass fractions, and an equation of state is used to relate T, p, ρ, and Y_k.

11.3.1.3 Space-Time Mapping

The equations discussed above (both the Lagrangian and Eulerian forms) provide solutions with (t,y) as independent variables. Frequently, however, we require solutions that evolve spatially (e.g., when comparing with data from a spatially evolving jet). This requires a space-time mapping, achieved by solving an ODE for streamwise position,

$$\frac{\mathrm{d}x}{\mathrm{d}t}=\bar{u}, \tag{11.12}$$

where $\bar{u}$ is a suitably chosen average velocity for advection of the ODT domain in the streamwise direction. This creates an approximate 'average' location of the line in space. Of course, the line would actually tend to bend due to variations in u. This is explicitly ignored by adopting an average velocity, $\bar{u}$, that advects the line downstream. One possible choice for $\bar{u}$ is

$$\bar{u}(t)=u_\infty+\frac{\int_{-\infty}^{\infty}\rho\,(u-u_\infty)^2\,dy}{\int_{-\infty}^{\infty}\rho\,(u-u_\infty)\,dy}. \tag{11.13}$$

An alternative approach to using the approximate space-time mapping (11.12) is to reformulate the governing equations with (x,y) as independent variables rather than (t,y). This is considered in the following sections for both the Lagrangian and Eulerian frames of reference.

11.3.1.4 Spatially Evolving Lagrangian Formulation

The spatially evolving Lagrangian formulation is obtained from (11.7) by choosing $\mathbf{v}_s \cdot \mathbf{x} = 0$ and $\mathbf{v}_s \cdot \mathbf{y} = v$. Then assuming steady state, 2D, and writing the differential equation in weak form (using the continuity equation), (11.7) becomes

$$\frac{\mathrm{d}\psi}{\mathrm{d}x} = -\frac{1}{\rho u}\left[\frac{\partial \Phi_{\psi,y}}{\partial y} - \sigma_\psi\right]. \tag{11.14}$$

In deriving (11.14), we have neglected the streamwise diffusive term, $\frac{\partial \Phi_{\psi,x}}{\partial x}$. This term is neglected primarily for practical algorithmic reasons. However, in application to spatially evolving jets, diffusion in the lateral direction will likely dominate any diffusion in the downstream direction. Nevertheless, this is an assumption in the spatial ODT formulations.

This approach has been adopted by Ricks et al. [31] to perform spatially evolving simulations of buoyant pool fires including soot transport and radiation (see Sec. 11.3.2).

11.3.1.5 Spatially-Evolving Eulerian Formulation

The spatially evolving form of the equations for ODT in Eulerian form is obtained from (11.10) by assuming steady state and variation only in x and y. Then using the continuity equation to write it in weak form and neglecting $\frac{\partial \Phi_{\psi,x}}{\partial x}$ as we did in (11.14), we find

$$\frac{\partial \psi}{\partial x} = -\frac{1}{\rho u}\left[\rho v \frac{\partial \psi}{\partial y} + \frac{\partial \Phi_{\psi,y}}{\partial y} - \sigma_\psi\right]. \tag{11.15}$$

11.3.2 Stand-Alone ODT Simulations

As a stand-alone model, ODT has been implemented for the study of jet diffusion flames [7, 11, 12, 21, 27–30], buoyant fire plumes [31], flame spread [35], and autoignition in jet flows [6]. In these studies different model formulations have been implemented, which illustrate the versatility of the ODT modeling framework. We start with the temporally-evolving Lagrangian formulation, which has been adopted for high-Reynolds number jets, but may be implemented as well for compressible turbulent shear layers and wall-bounded flows. We then illustrate the temporally evolving Eulerian formulation for temporally evolving shear layers [27] and the temporally evolving Lagrangian formulation for buoyant fire plumes [31]. In all formulations, a deterministic solution, involving diffusion, reaction and advective transport operators in the Eulerian formulation, in conjunction with a stochastic implementation of turbulent advection, is implemented.

The temporally evolving Lagrangian formulation is based on the solution of equation (11.8) for the streamwise (x) component of momentum along with equations for energy and species. Figure 11.2 shows temperature contours for a wall fire from [35] from a single (left) and 300 (right) realizations. This was solved using the temporal formulation with (11.12) to provide an approximate space-time mapping. Close examination reveals that small scale triplet mapping events are first observed at approximate heights of 25 cm and at a distance of approximately 2 cm away from the wall corresponding to the high values of temperature (and velocity, not shown). As the flow further accelerates, progressively larger eddy stirring events are shown to occur causing larger scale macro-mixing and the engulfment of the surrounding air. This transition of energy from small to large scales of motion is also consistent with recent observations from experiments and LES predictions of large scale plumes [4].

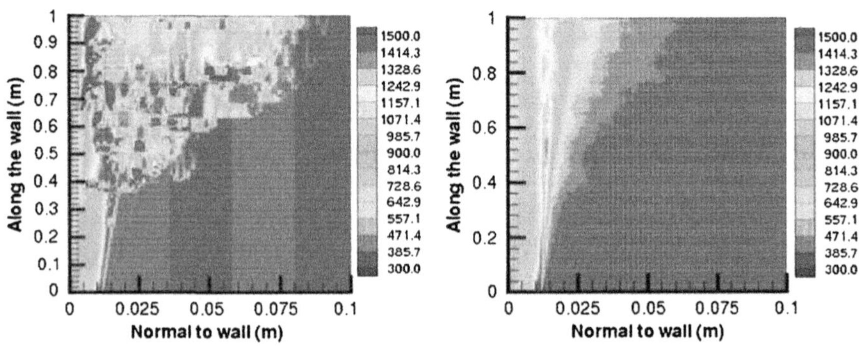

Fig. 11.2: 2D renderings of temperature corresponding to a single realization (left) and averaged (right) over 300 realizations of a wall fire. From [35].

Results from a piloted jet flame simulation with extinction and reignition are shown in Figure 11.3 for the same formulation. The results illustrate how the ODT model is able to predict extinction and reignition in piloted turbulent non-premixed flames. Two zones may be identified in the jet flame. The first corresponds to a region extending approximately fifteen diameters downstream from the inlet that illustrates a transition from piloted burning to extinction. This transition is followed by a gradual reignition as shown by the increased OH mass fraction. The 2D rendering of stirring events also shows that stirring events are initiated at the interfaces between the fuel jet and the pilot flow and the interfaces between the pilot flow and the co-flow air, and that the size of eddies progressively increases as a function of downstream distance, emulating the progressive growth of the shear layers. Evidence of the existence of an 'energy cascade' in the ODT solutions is demonstrated by the presence of smaller eddies that trail larger eddies, with regions of spatial intermittency, as these smaller eddies dissipate (see also Fig. 11.1).

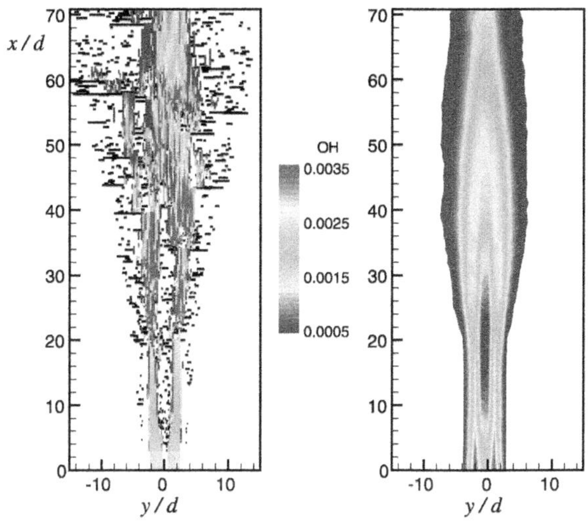

Fig. 11.3: 2D renderings of OH mass fraction corresponding to a single realization (left) and averaged (right) over 100 realizations of Sandia piloted methane-air flame F. The 1D domain corresponds to the horizontal axis. Its temporal evolution is converted to a downstream distance based on (11.12) and (11.13). Reprinted from [30] with permission from Taylor and Francis.

The formulation proposed by Punati et al. [26, 27] solves the Eulerian form of the governing equations described in Sec. 11.3.1. Specifically, (11.10) is solved with $\psi = \left(\rho\ \rho v\ \rho u\ \rho e_0\ \rho Y_i\right)$, where u is the streamwise velocity and v is the velocity component in the direction of the the ODT line orientation. See Table 11.1 for definitions of the diffusive fluxes in these equations. These equations are solved together with the ideal gas equation of state, detailed CO/H_2 oxidation kinetics, and mixture-averaged transport to make direct comparison with DNS data of a planar, temporally evolving CO/H_2-air nonpremixed jet [10]. The DNS dataset includes extinction and reignition, with the onset of extinction at a characteristic jet time of $\tau \approx 20$ and reignition occurring at around $\tau \approx 30$. This calculation allowed direct comparison between the ODT and DNS data. Initial conditions were extracted directly from the DNS data, and all treatment of diffusion, thermodynamics, and chemical kinetics was equivalent.

Figure 11.4 shows the average and RMS velocity and mixture fraction profiles for the ODT and DNS simulations. The spreading rate is well captured by ODT. The RMS profiles are captured reasonably, but the ODT under-represents the magnitude of the RMS fluctuations. Similar trends hold for all species (including minor species), with the exception that extinction is over-predicted by the ODT simulations.

Figure 11.5 shows the evolution of the probability density functions conditioned on lean and rich mixture fractions for the temperature and scalar dissipation rate. The temperature PDF illustrates that the ODT predicts an earlier onset of extinction than the DNS. Specifically, at $\tau = 6$ there is already evidence of extinction in the ODT data. However, the extinction-reignition process is captured relatively well despite these differences.

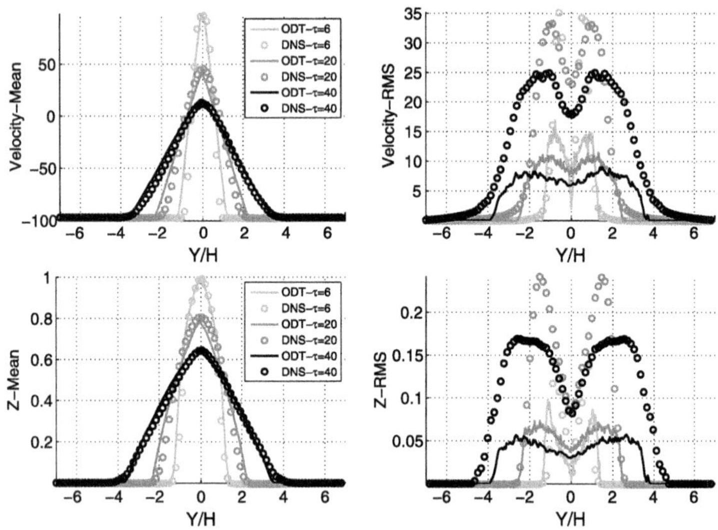

Fig. 11.4: Evolution of the streamwise velocity (top) and mixture fraction (bottom) showing the mean (left) and RMS (right) for the ODT (lines) and DNS (circles) data. Results are shown for characteristic jet times from $\tau = 6$ to $\tau = 40$. From [27].

The data shown in Figs. 11.4 and 11.5 were obtained from 400 ODT realizations, and each realization required approximately two CPU hours. In contrast, the DNS calculations (which were three-dimensional) required several million CPU hours. Although ODT cannot capture uniquely multidimensional effects that DNS can, it does represent many of the physical processes present in true three-dimensional turbulent flow at a fraction of the cost of DNS, and thus serves as a very useful tool in combustion modeling.

Ricks et al. [31] simulated a buoyant fire plume using ODT by solving (11.14) (the spatially evolving form of the governing equations in Lagrangian form) as outlined in Sect. 11.3.1, including transport equations for solid-phase soot particles as

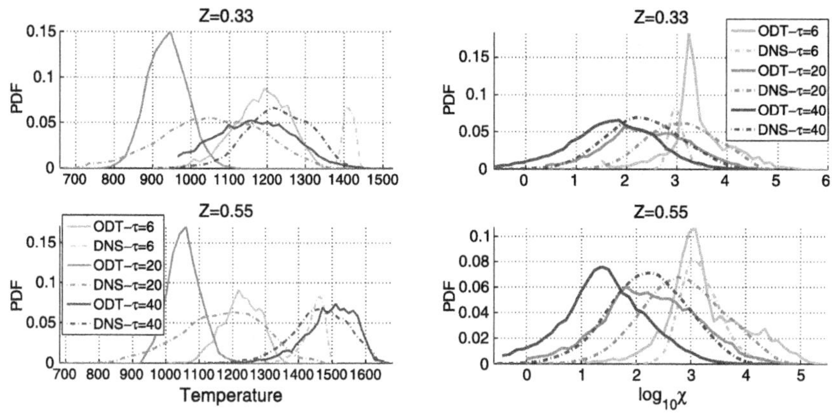

Fig. 11.5: Conditional pdf of temperature (left) and scalar dissipation rate (right). Results are shown for lean and rich mixture fractions and three different times during the jet evolution. DNS data is shown dot-dash lines and ODT data is shown with solid lines. From [27].

well as gas-phase transport equations. However, Ricks et al. [31] adopt a simplified approach for the representation of the gas phase species using the flamelet assumption and transporting the gas phase mixture fraction. In this formulation, the two independent variables correspond to the lateral (along the ODT domain) coordinate, y, and the streamwise spatial coordinate, x. The 1D nature of the solution enables the implementation of a host of models for soot evolution (including soot oxidation by OH), and transport (including thermal diffusion), radiation and gas phase chemistry on large-scale computational domains.

Figure 11.6 shows 2D renderings of the temperature corresponding to two separate realizations of the ODT simulation of a fire plume by Ricks et al. [31]. The ODT domain is aligned with the horizontal direction (x); a marching algorithm is implemented to evolve the ODT solution in the vertical (y) direction on a computational domain of 2 m $\times$ 3 m. The necking just above the base of the flame is due to the spatial form of the continuity enforcement, which induces inward lateral flow in order to compensate for the buoyancy-induced increase in the streamwise mass flux. Additional statistics on soot evolution and radiation effects may be found in [31].

11.3.3 Hybrid ODTLES

Similarly to LEM, ODT may be coupled to a 3D coarse-grained simulation approach, such as LES, for chemistry and mixing closure. Moreover, there are different strategies for LES and ODT coupling based on Eulerian and Lagrangian for-

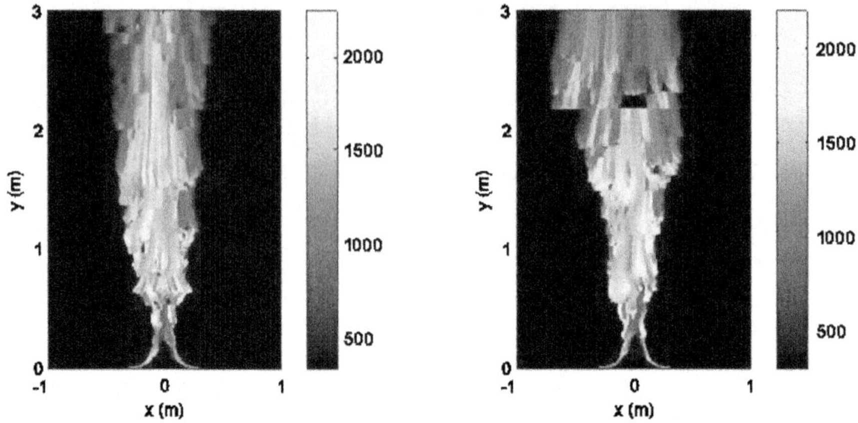

Fig. 11.6: 2D renderings of temperature (in K) corresponding to two separate realizations of a buoyant fire plume simulation. Reprinted from [31] with permission from Taylor and Francis.

mulations. In both formulations, ODT domains or elements are embedded in 3D solutions to resolve subfilter-scale momentum and scalar statistics. In the Eulerian formulation, ODT elements are fixed in space. Advective transport contributions in this formulation are represented by both large-scale transport resolved by LES and subfilter-scale transport represented by ODT stirring events. The simplest Lagrangian formulation may be implemented by attaching ODT solutions along the normal to the flame brush. In this formulation, the ODT elements are advected along with this brush. Similar strategies have been successfully adopted with LEM as discussed in Chapter 10.

In contrast to LEM, the coupling of LES with ODT may present a number of additional advantages:

- ODT has the capability to provide closure for momentum. However, one may choose to provide closure for scalars only and allow for a standard model for momentum closure (as discussed below).
- In ODT, the coupling of momentum and scalars is implemented on the fine time and length scales of ODT solutions; this coupling is very crucial near physical boundaries (e.g. walls) where both scalar and momentum boundary conditions may be implemented.
- Historically, large-scale transport with LES-LEM has been implemented using 'splicing' events, which extract segments from a LEM solution in one LES grid and transfers them to another LEM solution in a neighboring LES cell. The LES-ODT formulation of Cao and Echekki [3] proposes an alternative representation for large-scale transport based on ODT domains extending beyond a single LES cell.

The ODTLES model formulation is illustrated using the Eulerian formulation by Cao and Echekki [3]. The ODTLES formulation is based on two simulations that are implemented in the same computational domain. The first is a 3D LES for mass and momentum transport. The second is based on fine-grained simulations implemented on an ensemble of 1D ODT elements, which are embedded in the LES domain. Here, we describe a formulation in which ODT elements are distributed in a 3D Cartesian lattice as shown in Fig. 11.7 where the LES is solved on a structured Cartesian grid as well. However, a more complex layout may be adopted. Moreover, the formulation is used primarily for reactive scalars closure. For momentum closure, a 'standard' LES closure model for subgrid stresses may be adopted.

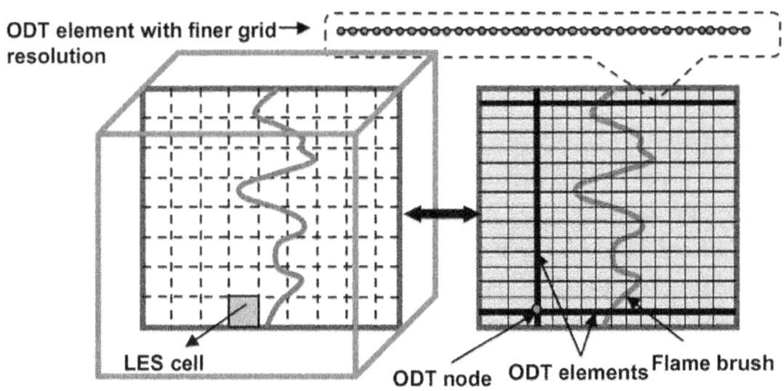

Fig. 11.7: Layout of ODT elements on a Cartesian grid in LES. Adapted from [3].

The ODT governing equations are solved on each individual ODT element. The temporal and spatial resolution requirements in ODT are similar to those needed for direct numerical simulations. The coordinate system on which the governing equations are based is a Cartesian coordinate system with one component along the ODT domain, x_1, and two additional orthogonal components, x_2 and x_3. The spatial coordinate, x_1, replaces the ODT domain coordinate, y, in previous discussions. When laid out on a Cartesian lattice, the direction x_1 represents the axis that is aligned with the ODT element; while the other coordinates x_2 and x_3 represent the remaining axes. The velocity field is split into a filtered (resolved in LES) component and a residual component:

$$u_i = \tilde{u}_i + u_i^* \tag{11.16}$$

where $\tilde{u}_i$ is the filtered velocity in the ith direction. The contribution of transport due to this velocity component is denoted as large-scale transport. The second term on the right-hand side, u_i^*, is the residual term of the velocity field in the ith direction. This latter term is modeled using the stochastic turbulent stirring events in ODT. The contribution of transport due to this velocity component is denoted as subfilter-scale transport. The variable-density governing equations on each ODT element of momentum, temperature, and species mass fractions are:

- Momentum

$$\frac{\partial u_i}{\partial t} = + \left[\frac{1}{\rho} \frac{\partial \tau_{i1}}{\partial x_1} \right] + \left\{ -\frac{1}{\rho} \frac{\partial p}{\partial x_i} - \tilde{u}_j \frac{\partial u_i}{\partial x_j} + \frac{1}{\rho} \left(\frac{\partial \tau_{i2}}{\partial x_2} + \frac{\partial \tau_{i3}}{\partial x_3} \right) \right\} \quad (11.17)$$

- Temperature

$$\frac{\partial T}{\partial t} = \left[-\frac{1}{\rho c_p} \left(\frac{\partial q_1}{\partial x_1} + \sum_{k=1}^{N} h_k \omega_k \right) \right] - \left\{ \tilde{u}_j \frac{\partial T}{\partial x_j} + \frac{1}{\rho c_p} \left(\frac{\partial q_2}{\partial x_2} + \frac{\partial q_3}{\partial x_3} \right) \right\} \quad (11.18)$$

- Species ($k = 1, \ldots, N$)

$$\frac{\partial Y_k}{\partial t} = \left[\frac{1}{\rho} \left(-\frac{\partial j_{k1}}{\partial x_1} + \omega_k \right) \right] - \left\{ \tilde{u}_j \frac{\partial Y_k}{\partial x_j} + \frac{1}{\rho} \left(\frac{\partial j_{k2}}{\partial x_2} + \frac{\partial j_{k3}}{\partial x_3} \right) \right\} \quad (11.19)$$

In equations (11.17)-(11.19), the index j represents the sum over all three directions of the advective terms. The diffusive fluxes, j_{k1}, j_{k2} and j_{k3} correspond to mass diffusion fluxes of the k species in the x_1, x_2 and x_3 directions, respectively. They may be expressed as ρV_{k1}, ρV_{k2} and ρV_{k3}, where V_{ki} is the mass diffusion velocity of species k in the ith direction. q_1, q_2, and q_3 correspond to the components of the heat flux vector in the x_1, x_2 and x_3 directions, respectively. These components represent the contributions of heat conduction, heat transport by mass diffusion, the Dufour effect and radiative heat transport. The ODT governing equations feature contributions which are resolved on the ODT domain (terms inside brackets '[ ]'). These are the same source and transport terms present in the stand-alone ODT equations. The resolved contributions include (1) molecular transport with gradients along the 1D elements, (2) chemical and heat source terms, and (3) the subfilter-scale momentum and scalar transport; this latter term is represented by the ODT stochastic stirring events discussed in Sect. 11.2.1 and implemented on a range of length scales. Other contributions require gradients along the normal components to the ODT domain (terms inside brackets '{}'). The unresolved contributions include: (1) large-scale transport (advective transport based on the filtered velocity components), (2) molecular diffusion with gradients along x_2 and x_3, and (3) the pressure gradient terms in the momentum equations.

The coupling of LES and ODT solutions is implemented both temporally and spatially. The ODT integration treats reaction-diffusion, subfilter-scale transport (stirring events) and filtered-advection as parallel events that are integrated with their own time steps, and which are fractions of the LES time step. During the temporal integration of the two solutions, statistics are transmitted from one solution scheme to another. For ODT, the LES velocity field $\tilde{u}$ is evaluated from the LES solution of the momentum equations and interpolated onto the ODT elements' finer grids. For LES, a number of variables may be filtered from ODT solutions, including closure for the mass density, $\overline{\rho}$. In what follows strategies adopted for the integration of the various terms in the ODT equations are briefly discussed:

11.3.3.1 Molecular Processes

Molecular processes in the ODT governing equations include (1) reaction, (2) diffusion along the ODT elements' directions, and (3) diffusion along the directions normal to the ODT elements. The integration of the first two contributions is similar to their implementation in the stand-alone ODT formulation. The third contribution must be modeled. The treatment of the unresolved terms may be implemented either deterministically or stochastically. In the Cao and Echekki [3] work, the representation of non-resolved diffusive contributions is achieved deterministically by scaling the resolved diffusive transport terms by a factor to represent the filtered contribution of mass transport from the unresolved transport. For example, if there is no preferred gradient, such as in the presence of a flame brush, a factor of 3 is adopted.

11.3.3.2 Representation of Subfilter-Scale Stresses and Scalar Fluxes

The stochastic contributions represent 3D subfilter-scale advective transport of momentum and scalars (i.e. subfilter-scale stresses and scalar fluxes) resulting from the residual velocity components. For momentum additional contributions may be attributed to pressure scrambling [19]. The stochastic terms are implemented through discrete triplet map events, which are implemented concurrently with other processes within ODT. The rules for stirring events are identical to those applied in stand-alone ODT. The range of length scales for the selected eddies is prescribed prior to the simulation based on a choice of L_{min} and L_{max}, which represent the smallest and largest eddies allowed. The value of L_{min} plays a similar role to the Kolmogorov length scale, and corresponds to length scales where viscous dissipation is predominant and stirring events are less likely to occur. The value of L_{max} determines the cut-off length scale beyond which turbulent advective transport is represented using the filtered advective terms. These parameters are additional model parameters to C and Z prescribed earlier for the ODT-implementation. The cumulative contribution from stirring events over time represents the subfilter-scale stresses and fluxes.

11.3.3.3 Large-Scale Transport

The large-scale transport of momentum and scalars in ODT is represented by the operators $\tilde{u}_j \frac{\partial u_i}{\partial x_j}$, $\tilde{u}_j \frac{\partial T}{\partial x_j}$ and $\tilde{u}_j \frac{\partial Y_k}{\partial x_j}$ in the ODT governing equations. The implementation of large-scale transport represents a fundamental challenge for the following reasons: 1) Advective transport is a 3D process; thus, at least two directions are not resolved on the ODT time scale or on the ODT 1D elements, 2) Non-linear contributions from advection processes pose important constraints on scalar boundedness. In the Cao and Echekki [3] formulation, advective fluxes are constructed at 'nodes' that represent the intersection in space of three or more ODT elements. At these nodes, the solutions of the velocity and scalar equations are updated; then, ODT solutions between these nodes are updated through single-component advection along the corresponding ODT element. In a Cartesian lattice of ODT elements, these nodes represent the intersection of three orthogonal 1D elements. Below, we describe the implementation based on this Cartesian lattice configuration. The large-scale transport process is implemented as a separate process concurrently with reaction-diffusion and stirring in two steps: 1) node advection, and 2) inter-node relaxation.

Node Advection is implemented as follows. At the prescribed time step for large-scale transport, the solution at each node is evaluated based on gradients represented by the 3 ODT elements intersecting at the node. This process is implemented in two steps. First, the solution is updated at each node for the momentum and scalars using the following governing equations:

$$\frac{\partial \phi^l}{\partial t} = -\tilde{u}_1^1 \frac{\partial \phi^1}{\partial x_1} - \tilde{u}_2^2 \frac{\partial \phi^2}{\partial x_2} - \tilde{u}_3^3 \frac{\partial \phi^3}{\partial x_3} \tag{11.20}$$

In this expression, the dependent variable ϕ corresponds to any one of the variables in the solution vector. The subscripts, 1, 2 or 3, correspond to the components of the velocity vector; while, the superscripts, 1, 2 or 3, correspond to the direction of the ODT domain. Although each node is updated with the same right-hand side that represents contributions from the three directions of ODT elements, its value at the end of the update is different because of the values of ϕ^l are different at the start of the update from the 3 contributing ODT elements. A second step involves the averaging of these 3 solutions at the nodes as follows:

$$\phi^l = \frac{\phi^1 + \phi^2 + \phi^3}{3} \tag{11.21}$$

Inter-Node Relaxation involves a relaxation of the solution between the nodes based on the updated internal boundary conditions at the nodes. This relaxation is accomplished through an integration of the solution at grid cells between the nodes using a single-component advective flux according to the following relation:

$$\frac{\partial \phi^l}{\partial t} = -\kappa \tilde{u}_l^l \frac{\partial \phi^l}{\partial x_l} \tag{11.22}$$

In this expression, κ represents a relaxation coefficient, which governs the rate at which the inter-node solution is updated to reflect changes at the nodes. Because ODTLES is a statistical approach, a range of values for κ may be adopted to yield reasonable statistics for the scalars and momentum solutions.

Similarly to the approach by Schmidt et al. [32], a correction to the ODT velocity solution is implemented such that the filtered ODT velocity field matches the solution from LES.

The above Eulerian formulation was implemented by Cao and Echekki [3] for the modeling of non-homogeneous ignition in a random mixture-fraction field with preheated oxidizer and of the same configurations show that the model represents adequately turbulent transport through the contributions of subfilter-scale transport and large-scale transport. Important effects of preferential diffusion as well as turbulence intensity are identified in reactive-scalar conditional statistics.

Figure 11.8 shows isocontours of filtered reaction progress variable (normalized temperature) at a prescribed value of 0.5 (for flame tracking) at different times based on the ODTLES simulations. The figure shows the formation of discrete ignition kernels at favorable mixture conditions, their growth and their merger at later times. The reaction progress variable is obtained by filtering the ODT solutions; the size of the initial nascent kernels is smaller than the LES grid.

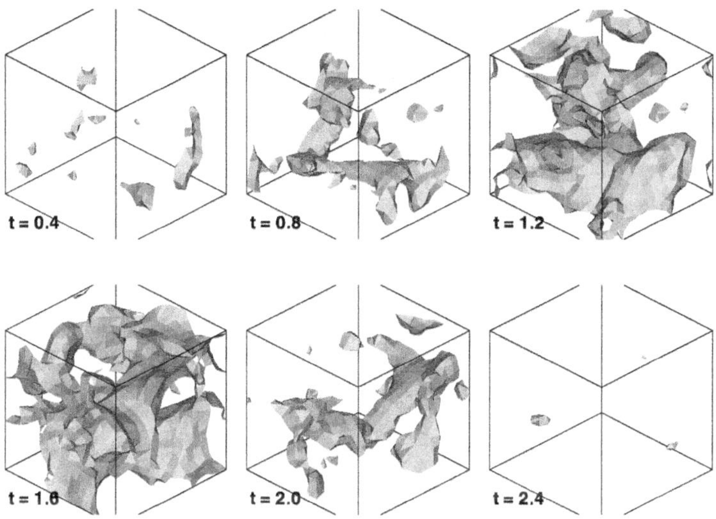

Fig. 11.8: Evolution of flame kernel based on filtered reaction progress variables during non-homogeneous mixture ignition in isotropic turbulence. The figure shows the formation of ignition kernels at conditions favorable to the onset of ignition. Additional kernels are formed at less favorable conditions for autoignition after more delay. The kernels eventually grow to interact at later stages and merge to form larger kernels, until the entire mixture is burned. Reprinted from [3] with permission from Taylor and Francis.

Figure 11.9 shows the evolution of the heat-release rate conditional statistics at two Lewis numbers, 0.5 and 2, representing the ratio of the thermal diffusivity to the species mass diffusivities. The lower Lewis number heat-release rate profiles exhibit higher peaks initially and then lower peaks eventually as the combustion progresses from fuel-lean conditions to richer conditions. The difference between the two cases reflects the strong dependence of the heat release rate on temperature, which is affected by the Lewis number. Lower Lewis numbers indicate slower diffusion of heat relative to species. Therefore, the initial formation of the corresponding kernels favors kernels that shielded from heat loss. However, the same mechanism may prevent the ignition of the unburned layers next to the ignition kernels and their propagation. Cases a and b shown in comparison with DNS statistics correspond to two different and coarse LES resolutions (case a is twice as resolved as case b). The two cases are in very good agreement with the DNS statistics and show that the ODTLES formulation predicts reasonably well the contributions from large-scale and subfilter-scale transport.

The ODTLES formulation has been extended recently to the study of turbulent premixed flames by Echekki and Park [8]. A Lagrangian formulation has been implemented more recently by Balasubramanian [2] for the study of a buoyant fire plume. In this formulation, the ODT elements are attached to a filtered mixture fraction surface with a fixed value corresponding to the stoichiometric value.

11.4 Concluding Remarks

Here and in Chapter 10, a strategy for turbulent combustion modeling has been outlined that involves a conceptually and computationally minimal representation of the local unsteady evolution of the coupled processes of advection, diffusion, and reaction. ODT, described in this chapter, incorporates a representation of the dependences of the occurrence of eddy motions on the mechanisms that drive these motions. In addition to capturing important effects of the unsteady couplings, this feature results in a formulation that is, in many respects, a self-contained predictive model of turbulent flow. This is perhaps a natural consequence of the effort to capture the couplings relevant to combustion; for a model to do this well, it must capture much of the phenomenology of turbulence.

The main limitation of ODT in this regard is its restriction to one spatial dimension. It is thus complementary to LES, which captures large-scale 3D motions but does not resolve flame structure and evolution. Coupling of ODT to LES has been described. The successes of the LEMLES formulations for the modeling of practical combustion flows (see Chapter 10) also support the potential of ODTLES as a viable modeling approach for similar problems. More importantly, both LEMLES and ODTLES may be viewed as frameworks with which multiphysics and multiphase problems may be addressed. In addition to the momentum and standard scalar equations for combustion problems, additional transport equations may be implemented within these ODTLES frameworks, including particle transport and multiscale descriptions of radiative transport in participating media.

One focus of current efforts is the coupling of arrays of ODT domains so as to obtain a self-contained 3D flow simulation (with the smallest scales resolved only in 1D), thus eliminating the need for a distinct coarse-grained 3D flow solver [17]. This modeling strategy is termed autonomous microstructure evolution (AME). Another focus involves Lagrangian implementation of the ODTLES framework based on ODT elements attached to the flame brush.

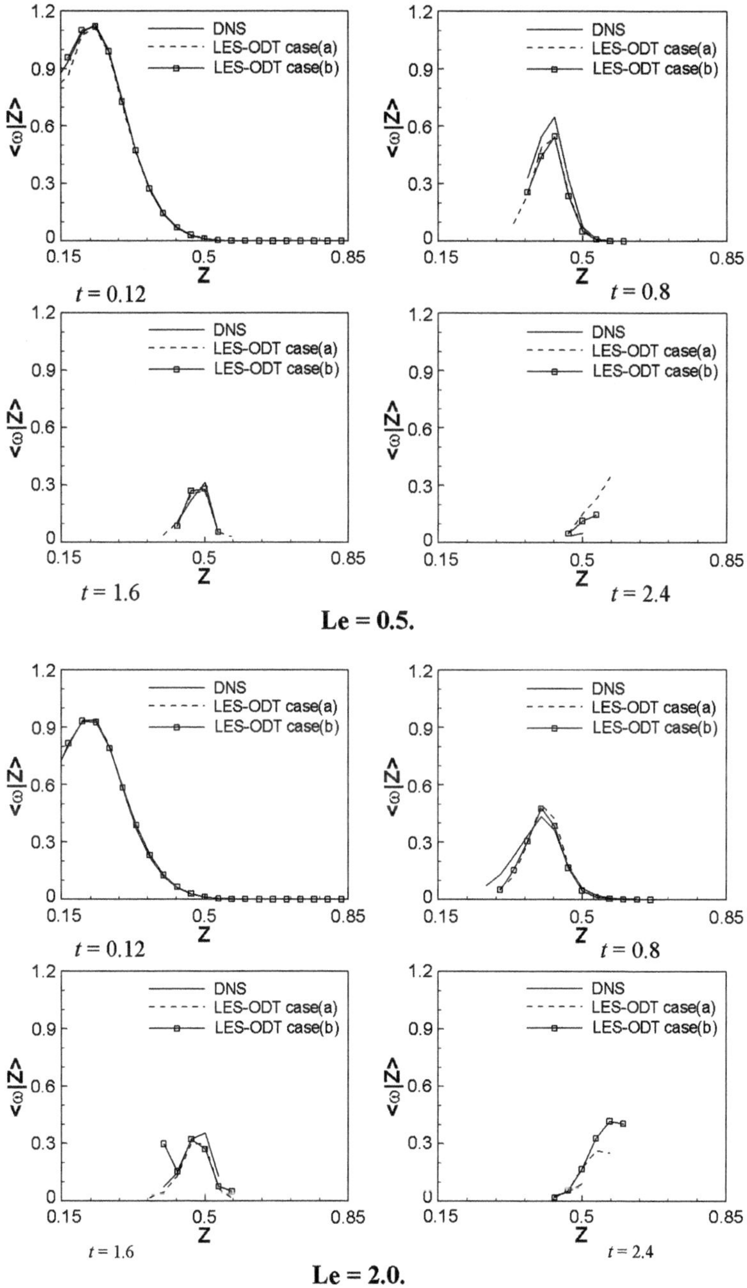

Fig. 11.9: Evolution of the conditional means of the heat release rate conditions during the ignition of a non-homogeneous mixture of fuel and preheat oxidizer at Le = 0.5 and 2.0. Reprinted from [3] with permission from Taylor and Francis.

Acknowledgement

Dr. T. Echekki acknowledges support from the Air Force Office of Scientific Research through grants F49620-03-1-0023 monitored by Dr. Julian Tishkoff and FA9550-09-1-0492 monitored by Dr. Fariba Fahroo and the National Science Foundation through grant DMS-0915150 monitored by Dr. Junping Wang. Dr. A. Kerstein's research was partially supported by the U.S. Department of Energy, Office of Basic Energy Sciences, Division of Chemical Sciences, Geosciences, and Energy Biosciences. Sandia National Laboratories is a multi-program laboratory operated by Sandia Corporation, a Lockheed Martin Company, for the United States Department of Energy under contract DE-AC04-94-AL85000. Dr. J. Sutherland acknowledges support from the Department of Energy under award number FC26-08NT0005015.

References

1. Ashurst, W.T., Kerstein, A.R.: One-dimensional turbulence: Variable-density formulation and application to mixing layers. Phys. Fluids **17**, 025107 (2005)
2. Balasubramanian, S.: A novel approach for the direct simulation of subgrid-scale physics in fire simulations. Master's Thesis, Department of Mechanical and Aerospace Engineering, North Carolina State University (2010)
3. Cao, S., Echekki, T.: A low-dimensional stochatic closure model for combustion large-eddy simulation. J. Turbul. **9**, 1–35 (2008).
4. DesJardin, P.E., O'Hern, T.J., Tieszen, S.R.: Large eddy simulations and experimental measurements of the near field of a large helium-air plume. Phys. Fluids **16**, 1866–1883 (2004)
5. Dreeben, T.D., Kerstein, A.R.: Simulation of vertical slot convection using one-dimensional turbulence. Int. J. Heat Mass Transf. **43**, 3823–3834 (2000)
6. Echekki, T., Gupta, K.: Hydrogen autoignition in a turbulent jet with preheated co-flow air. Int. J. Hydrogen Energy **34**, 8352–8377 (2009)
7. Echekki, T., Kerstein, A.R., Chen, J.-Y., Dreeben, T.D.: One-dimensional turbulence simulation of turbulent jet diffusion flames. Combust. Flame **125**, 1083–1105 (2001)
8. Echekki, T., Park, J.: The LES-ODT model for turbulent premixed flames, AIAA-2010-0207, The 48th AIAA Aerospace Sciences Meeting, Orlando, FL, January 4–7 (2010)
9. `http://groups.google.com/group/odt-research`
10. Hawkes, E.R., Sankaran, R., Sutherland, J.C., Chen, J.H.: Scalar mixing in direct numerical simulations of temporally-evolving plane jet flames with detailed CO/H_2 kinetics. Proc. Combust. Inst. **31**, 1633–1640 (2007)
11. Hewson, J.C., Kerstein, A.R.: Stochastic simulation of transport and chemical kinetics in turbulent $CO/H_2/N_2$ flames. Combust. Theory Model. **5**, 559–697 (2001)
12. Hewson, J.C., Kerstein, A.R.: Local extinction and reignition in nonpremixed turbulent $CO/H_2/N_2$ jet flames. Combust. Sci. Technol. **174**, 35–66 (2002)
13. Kerstein, A.R.: Linear eddy modeling of turbulent transport. 2. Application to shear layer mixing. Combust. Flame **75**, 397–413 (1989)
14. Kerstein, A.R.: Linear-eddy modeling of turbulent transport. 3. Mixing and differential molecular-diffusion in round jets. J. Fluid Mech. **216**, 411–435 (1990)
15. Kerstein, A.R.: One-dimensional turbulence: Model formulation and application to homogeneous turbulence, shear flows, and buoyant stratified flows. J. Fluid Mech. **392**, 277–334 (1999)

16. Kerstein, A.R.: One-dimensional turbulence. Part 2. Staircases in double-diffusive convection. Dyn. Atmos. Oceans **30**, 25–46 (1999)
17. Kerstein, A.R.: Lect. Notes Phys. **756**, 291–333 (2009)
18. Kerstein, A.R.: One-dimensional turbulence: A new approach to high-fidelity subgrid closure of turbulent flow simulations. Computer Phys. Commun. **148**, 1–16 (2002)
19. Kerstein, A.R., Ashurst, Wm.T., Wunsch, S., Nilsen, V.: One-dimensional turbulence: Vector formulation and application to free shear flows. J. Fluid Mech. **447**, 85–109 (2001)
20. Kerstein, A.R., Wunsch, S.: Simulation of a stably stratified atmospheric boundary layer using one-dimensional turbulence. Bound. Layer Meteorol. **118**, 325–356 (2006)
21. Krishnamoorthy, N.: Reaction models and reaction state parameterization for turbulent non-premixed combustion. Ph.D. Thesis, University of Utah, Salt Lake City (2008)
22. Law, A.M., Kelton, W.D.: *Simulation Modeling and Analysis*, 3rd Ed. (McGraw-Hill, New York 2000)
23. McDermott, R.J.: Toward one-dimensional turbulence subgrid closure for large-eddy simulation. Ph.D. Thesis, University of Utah, Salt Lake City (2005)
24. McDermott, R.J., Kerstein, A.R., Schmidt, R.C., Smith, P.J.: The ensemble mean limit of the one-dimensional turbulence model and application to finite-volume large-eddy simulation. J. Turbul. **6**, 1–33 (2005)
25. Peters, N., *Turbulent Combustion* (Cambridge Univ. Press, Cambridge 2000)
26. Punati, N., Sutherland, J.C.: Application of an Eulerian one dimensional turbulence model to simulation of turbulent jets, U.S. Joint Sections of the Combustion Institute, Ann Arbor, MI, May (2009)
27. Punati, N., Sutherland, J.C., Kerstein, A.R., Hawkes, E.R., Chen, J.H.: An evaluation of the one-dimensional turbulence model: Comparison with direct numerical simulations of CO/H_2 jets with extinction and reignition. Proc. Combust. Instit., to appear (2011)
28. Ranganath, B., Echekki, T.: One-dimensional turbulence-based closure for turbulent non-premixed flames. Prog. Comput. Fluid Dyn. **6**, 409–418 (2006)
29. Ranganath, B., Echekki, T.: One-dimensional turbulence-based closure with extinction and reignition. Combust. Flame **154**, 23–46 (2008)
30. Ranganath, B., Echekki, T.: ODT closure with extinction and reignition in piloted methane-air jet diffusion flame. Combust. Sci. Technol. **181**, 570–596 (2009)
31. Ricks, A.J., Hewson, J.C., Kerstein, A.R., Gore, J.P., Tieszen, S.R., Ashurst, Wm.T.: A spatially-developing one-dimensional turbulence (ODT) study of soot and enthalpy evolution in meter-scale buoyant turbulent flames. Combust. Sci. Technol. **182**, 60–101 (2010)
32. Schmidt, R.C., Kerstein, A.R., McDermott, R.: ODTLES: A multi-scale model for 3D turbulent flow based on one-dimensional turbulence modeling. Comput. Methods. Appl. Mech. Engg. **199**, 865–880 (2010)
33. Schmidt, R.C., Kerstein, A.R., Wunsch, S., Nilsen, V.: Near-wall LES closure based on one-dimensional turbulence modeling. J. Comput. Phys. **186**, 317–355 (2003)
34. Schmidt, J.R., Wendt, J.O.L., Kerstein, A.R.: Non-equilibrium wall deposition of inertial particles in turbulent flow. J. Stat. Phys. **137**, 233–257 (2009)
35. Shihn, H., DesJardin, P.E.: Simulation of vertical wall fires with one-dimensional turbulence modeling. Spring Technical Meeting, The Combustion Institute/Canadian Section, Ontario, Canada, May 9–12 (2004)
36. Shihn, H., Desjardin, P.E.: Near-wall modeling of an isothermal vertical wall using one-dimensional turbulence. Int. J. Heat Mass Transf. **50**, 1314–1327 (2007).
37. Sutherland, J.C., Punati, N., Kerstein, A.R.: A unified approach to the various formulations of the one-dimensional turbulence model. Technical Report ICSE100101, The University of Utah Institute for Clean and Secure Energy, Salt Lake City, UT, 2010. Available from (accessed June 2010): `http://repository.icse.utah.edu/dspace/handle/123456789/9861`
38. Wunsch, S.: Stochastic simulations of buoyancy-reversal experiments. Phys. Fluids **15**, 1442–1456 (2003)

39. Wunsch, S., Kerstein, A.R.: A model for layer formation in stably-stratified turbulence. Phys. Fluids **13**, 702–712 (2001)
40. Wunsch, S., Kerstein, A.R.: A stochastic model for high Rayleigh-number convection. J. Fluid Mech. **528**, 173–205 (2005)

Chapter 12
Unsteady Flame Embedding

Hossam A. El-Asrag and Ahmed F. Ghoniem

Abstract Direct simulation of all the length and time scales relevant to practical combustion processes is computationally prohibitive. When combustion processes are driven by reaction and transport phenomena occurring at the *unresolved* scales of a numerical simulation, one must introduce a dynamic subgrid model that accounts for the multiscale nature of the problem using information available on a resolvable grid. Here, we discuss a model that captures unsteady flow-flame interactions—including extinction, re-ignition, and history effects—via embedded simulations at the subgrid level. The model efficiently accounts for subgrid flame structure and incorporates detailed chemistry and transport, allowing more accurate prediction of the stretch effect and the heat release. In this chapter we first review the work done in the past thirty years to develop the flame embedding concept. Next we present a formulation for the same concept that is compatible with Large Eddy Simulation in the flamelet regimes. The unsteady flame embedding approach (UFE) treats the flame as an ensemble of locally one-dimensional flames, similar to the flamelet approach. However, a set of elemental one-dimensional flames is used to describe the turbulent flame structure directly at the subgrid level. The calculations employ a one-dimensional unsteady flame model that incorporates unsteady strain rate, curvature, and mixture boundary conditions imposed by the resolved scales. The model is used for closure of the subgrid terms in the context of large eddy simulation. Direct numerical simulation (DNS) data from a flame-vortex interaction problem is used for comparison.

Hossam A. El-Asrag
Massachusetts Institute of Technology, Cambridge MA 02139, e-mail: helasrag@mit.edu

Ahmed F. Ghoniem
Massachusetts Institute of Technology, Cambridge MA 02139, e-mail: ghoniem@mit.edu

T. Echekki, E. Mastorakos (eds.), *Turbulent Combustion Modeling*,
Fluid Mechanics and Its Applications 95, DOI 10.1007/978-94-007-0412-1_12,

12.1 Introduction

Even with the current trend of exponential growth in computing capabilities [64], numerical simulation of multi-dimensional turbulent reacting flows is computationally prohibitive making it necessary to rely on under-resolved simulations and the implementations of local models that capture the unresolved processes as well as their interactions with the resolved ones. The increase in the hardware computational power allows us to improve our models by including more physics and more detailed kinetics. Detailed numerical simulation of turbulent reacting flows is challenging because of the unsteadiness of the flow and the multiscale nature of the different interacting physical processes. The wide spectrum of length and time scales in turbulent reactive flows limits the utility of direct numerical simulations (DNS) to being a research tool. The practical alternative is under-resolved calculations in which the large scales are resolved on a coarse grid and the impact of the small scales is modeled.

Large eddy simulation (LES) [51] is one such approach. The extension of these approaches to combustion poses a tremendous modeling challenge, since the scales where the reactants mix and burn are often below those resolved directly. This is especially true for premixed flames, where the flame thickness is often of the order of the smallest turbulent length scales [43]. In particular, it is often important to account for the interactions between unsteady processes and the premixed flame, e.g., in modeling the acoustic instabilities [1].

Furthermore, unsteady flow features in the vicinity of the flame surface affect its propagation speed significantly. For example, the flame surface wrinkling is found to be a strong function of the hysteresis and memory effect generated by the boundary conditions and the flame/turblence interactions [8]. Moreover, unsteady effects can introduce scalars counter-gradient or gradient transport, depending on the flame structure and its interaction with the vortices. For example, low Lewis number flames are found to exhibit counter-gradient transport [3]. The Lewis number represents the relative contribution of thermal and molecular diffusivity. The interaction between these two transport phenomena is more significant under time-dependent conditions and controls the flame local temperature in the presence of the curvature and strain rate. The local curvature and strain rate are functions of turbulence and hence are unsteady. In flames with Lewis number larger than unity and local positive curvature, reactants diffuse towards the reaction zone at slower rate than heat is conducted out, which decreases the burning rate and might lead to non-equilibrium phenomenon as flame extinction. On the other hand, flames with Lewis number smaller than unity are subjected to higher thermo-diffusive instabilities [3, 32].

Detailed kinetics simulations [9] confirm that species with low diffusion coefficient are more susceptible to unsteady effects. Moreover, experimental [58] as well as numerical [27] data confirm that as turbulent intensity increases the strain rate effect is more predominant and the curvature distribution has a broader standard deviation than for lower strained flames. These flame/turbulence interactions occur mainly on the small scales which are unsteady.

Multi-scale combustion modeling has been applied successfully in the context of LES. The one-dimensional turbulence (ODT) model [10, 29] and linear eddy model (LEM) [11, 15, 16, 28, 38, 57] have been employed in premixed and non-premixed configurations (see Chapters 10 and 11 for more details). Other combustion closure models include the transported filtered density function (FDF) methods [54] for modeling turbulent non-premixed flames, the artificially thickened flame (ATF) model [62] for premixed flames, the flame-surface density models (FSD) [2, 24], and the presumed PDF flamelet approaches [42, 51].

In the flamelet approach, and because of the fast chemistry assumption, scalar transport in the tangential direction is neglected with respect to the normal direction. Using this assumption, the scalar transport equations are transformed in the mixture fraction space and the flamelet equations are derived by asymptotic analysis [51]. A set of 1-D solutions, based on solving the 1-D steady flamelet equations, are tabulated then mapped into the fully 3-D turbulent environment by convoluting the solutions with a PDF distribution. The flamelet approach is widely used for premixed flames combined with the level set approach [31, 50, 52]. Other flamelet type combustion subgrid models are the Flamelet Prolongation of ILDM (FPI) [19], the Flamelet Generated Manifolds (FGM) [63], and the Flamelet Progress Variable (FPV) [26, 47].

Few attempts, however, have been made to extend the flamelet approach to include unsteady effects. The steady state assumption usually employed in conventional flamelet type approaches is not suitable if slow physical and chemical processes are involved such as NO_x formation or radiation, or if unsteady processes are encountered as flame extinction and re-ignition. The two known models that account for the flamelets' unsteady effects are the Lagrangian Flamelet Model (LFM) [48] and the Eulerian Flamelet Model (EFM) [49]. In LFM a single flamelet is introduced at the nozzle exit and is allowed to transport downstream with the flow. The flamelet in the mixture fraction space is solved simultaneously with the flow field calculations. And is integrated using a Lagrangian time scale that is computed from the filtered flow velocity conditioned on the stoichiometric mixture fraction. This approach was further modified as the Eulerian Flamelet Model (EFM) [49], where the flamelet equations are solved in the Eulerian form rather than in the mixture fraction space.

The unsteady flame embedding approach (UFE), described in this work models the unsteady interaction between the resolved and the unresolved scales using unsteady strained flames computed along a set of 1-D lines within the coarse grids of the LES. The unsteady strain rate and the boundary conditions acting on the local flame are updated temporally with the solution advancement. This allows for consideration of history effects, as well as accurate modeling of the unsteady flame stretch at the small scales. The model utilizes the flame embedding concept introduced earlier [20, 23].

In Sec. 12.2 we review the work done in the past thirty years to develop the flame embedding concept. In Sec. 12.3 the elemental flame model formulation is described. Following that the numerical approach used to solve the flow along the one-dimensional flame is shown in Sec. 12.4. In Sec.12.5 we introduce a new UFE subgrid model components for LES applications. Validation of the subgrid model is then described in Sec. 12.6. The chapter ends with the conclusions and future work in Sec. 12.7.

12.2 Historical Perspective on the Flame Embedding Concept

According to the laminar flamelet concept [65], if the smallest turbulent length scale is larger than the flame thickness, the flame can be regarded as a thin sheet that is convoluted by turbulent eddies. Early work by Chorin [6] and Ghoniem et al. [20] utilized this concept and implemented a flame interface tracking scheme to model premixed combustion. In these papers the elementary turbulent eddies and their cumulative effect was modeled by Chorin's random vortex method [5] and the position of the interface between the burned and unburned gases was tracked by a flame propagation and advection algorithm. A volume-of-fluid approach was used to define the volume ratio of the burned and unburned gases (f) inside each cell, and the flame interface was constructed based on these ratios. The two fluids were advected according to the velocity field at the cell boundaries. The impact of the self-propagation of a laminar flame normal to itself was approximated using Huygens principle to compute the normal propagation speed due to consumption of the unburned gases. This approximation was extended [21] to include the effect of exothermicity and volumetric expansion. The model was applied to a backward facing step stabilized flame in a combustion tunnel and compared well with the experimental data.

The early work used a constant laminar flame speed. However, in highly turbulent flows the effect of curvature induces different hydrodynamic stability effects based on the flame topology and its interaction with the surrounding vortices. The above model was then extended [22] by computing the burning velocity as function of the consumption speed and the flame front curvature. The numerical results are compared with the experimental data on the flame propagation in a quiescent mixture inside a circular chamber, initiated by a central ignition source. The flow was assumed inviscid, two-dimensional with adiabatic walls. The results show that the heat conduction, coupled with front curvature, leads to the stabilization of short wavelength perturbations. However, long wavelength perturbations cannot be damped by this mechanism, and the coherence of the flame front must be attributed to other mechanisms. Two mechanisms were identified: the stability of the front due to the interaction between the distortion caused by the perturbations growth with the convexity of the flame front and the expansion of the wavelength as the flame expands, and the consumption of the concave parts of the flame.

Up to this point, the above modeling approach had not considered the unsteady effect the turbulence exerts on the reaction zone. Under realistic conditions, the amplitude and frequency of an oscillating strain can impact unsteady chemical processes, such as ignition and quenching. To study these unsteady effects, Ghoniem et al. [23] used a one-dimensional flame model and one-step Arrhenius-kinetics to quantify the effects of steady and periodic strains on a diffusion flame, where the one-dimensional unsteady flame equations [41] were solved. A boundary-layer analysis for the flame structure is performed, i.e, due to the small flame thickness, normal gradients are much stronger than tangential gradients, so that the latter are ignored in the diffusion terms. The resulting set of equations represent the one-dimensional flow for an axisymmetric stagnation point flame configuration. Applied to a flat one-dimensional non-premixed flame configuration, the results show that the ignition delay time required to reach steady burning increases exponentially with strain. With a more sustained ignition source, the delay time is reduced, and the quenching strain increases. The burning rate was also found to depend on the strain rate fluctuations amplitude and frequency. These fluctuations were found to cause local or complete quenching based on the amplitude and frequency. Another interesting observation is that compressive strain inhibits reactants diffusion into the reaction zone, and leads to partial extinction of the flame, which soon reignites as positive strain resumes. The study shows that modeling the unsteady processes is important especially in turbulent flows, in which rapidly varying modes of instability may develop simultaneously.

The combustion model was generalized later by Kino et al. [30] to simulate a two-dimensional reacting shear layer at the conditions of fast chemistry. The combustion model consolidated the two-dimensional unsteady strained flame equations described earlier, with the vorticity equation and acted as one-step forward towards generalizing the flame model to multi-dimensional flow conditions. The resulting model is incorporated into adaptive, Lagrangian vortex element techniques. The evolution of external flow field is computed by tracking the motion of the vortex elements, while flame front topology is described by a collection of thin flamelets, which are used to compute the unsteady response of the local flame structure to the imposed strain. This elemental flame model embedded in a Lagrangian vortex element technique was further extended to account for detailed chemical kinetics and multi-species transport with different Lewis numbers by Petrov and Ghoniem [46]. A reduced four-step mechanism for methane was employed to study the evolution of a reacting shear layer. An example of these calculations is shown here. The combustion was initiated at time zero by embedding elemental flames whose structures correspond to a steady strain of 100 1/s. The results at $t = 2.25$ and 3 ms, i.e., after two and three eddy turnover times, are shown for the temperature, heat release rate in Fig. 12.1, and the production rates for CO and CO_2 in Fig. 12.2.

The unsteady strained flame model was used to examine the effect of temporal gradients on combustion in a premixed methane/air mixture [36]. The efficiency of the numerical solution was improved by using an inexact Newton backtracking algorithm to solve the nonlinear algebraic equations. This algorithm will be discussed later in details. The results show that equivalence ratio variations with timescales lower than 10 ms have significant effects on the burning process, including reaction zone broadening, burning rate enhancement, and extension of the flammability limit toward leaner mixture. The study stresses the importance of modeling the unsteady effects to simulate turbulent flames.

The flame embedding approach was validated by comparing its results with data extracted from a set of two-dimensional simulations of premixed flame-vortex inter-

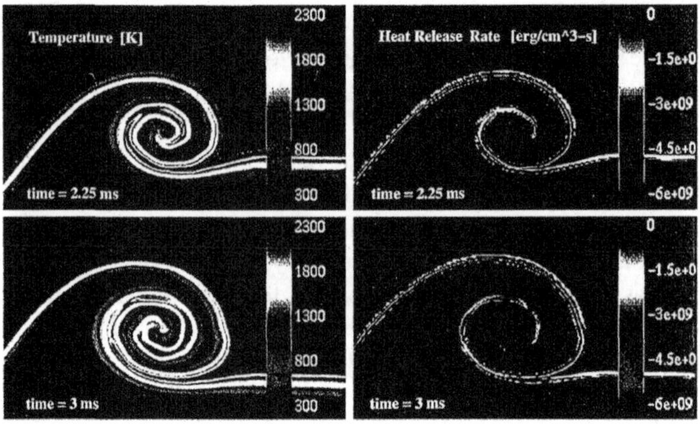

Fig. 12.1: The evolution of the temperature and heat release rate inside a large vortex in the spatially developing reacting mixing layer: top stream: air moving at 100 m/sec, bottom stream methane moving at 50 m/sec. From [46].

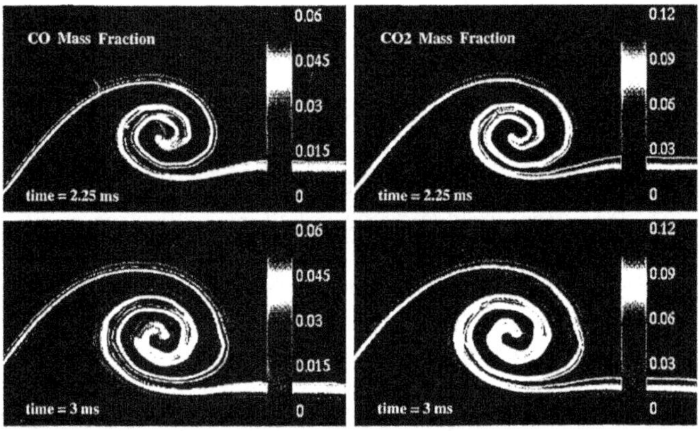

Fig. 12.2: The evolution of the CO and CO_2 mass fractions inside a large vortex in the spatially developing reacting mixing layer: top stream: air moving at 100 m/sec, bottom stream methane moving at 50 m/sec. From [46].

actions for a matrix of vortex strengths and length scales [37]. Elements along the flame front, or elemental flames from the direct solution were selected and compared with the solution obtained from the flame embedding results. The results show good agreement when the actual strain rate at the reaction zone of the one-dimensional flame was made to match that of the two dimensional flame.

Figure 12.3 (from Marzouk et al. [37]) compares the instantaneous structure of the elemental flame with that of the two-dimensional flame vortex interaction solution along the centerline at $t = 2.0$ ms. In the elemental flame calculation, the time-dependent strain rate $a(t)$ was chosen to ensure matching of mean reacting strain rates. This comparison highlights the challenge inherent in using the elemental flame approach to predict the two-dimensional flame-vortex interaction. Though the mean strain rates in the two flames are equal, the strain rate distributions are qualitatively different, with the strain rate (ε) across the flame falling through a plateau in the two-dimensional flame but rising monotonically in the one-dimensional flame. While this disagreement may cause the remaining profiles to differ slightly, the elemental flame model is able to characterize the heat release rate in the two-dimensional reacting flow with good accuracy. A further comparison between the results of the elemental flame structure and the two-dimensional solution along the centerline is shown in Figure 12.4; here, selected major species and radicals are shown at $t = 1.0$ ms (Case 4 in [37]), while matching the mean reacting strain rates in both solutions. The strain rate, temperature, and heat release profiles are qualitatively similar to those in Figure 12.3. The species profiles match well, with slight differences visible mostly in the preheat zone of the flame.

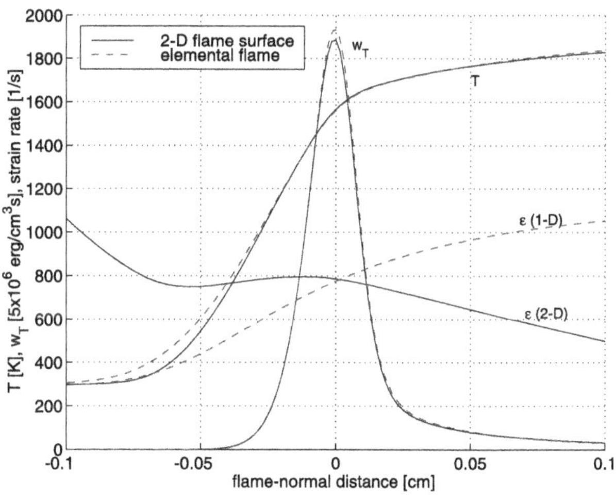

Fig. 12.3: Structure comparison between the one- and two-dimensional flame elements, $t = 2.0$ ms, showing the temperature T, the heat release rate $\dot{w}_T$, and the strain rate. Both elements are at the same mean *reacting* strain. From [37].

12.3 Elemental Flame Model Formulation

The flame strained in a stagnation point flow, shown in Fig. 12.5, has long been proposed as a reference flame model for turbulent combustion in the flamelet regime [7, 12, 13, 23, 45, 56, 59]. We use this model here but allow the strain rate to vary as a function of time, to capture the physics of unsteady flame response. This model

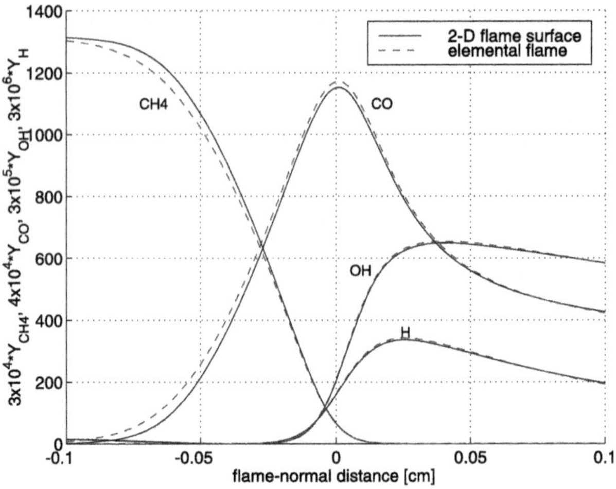

Fig. 12.4: Structure comparison between the one- and two-dimensional flame elements, $t = 1.0$ ms, showing various major and minor species. Both elements are at the same mean *reacting* strain. From [37].

is labeled the *elemental flame* [44, 45] to distinguish it from the conventional quasi-steady flamelet approach used in closure modeling.

One-dimensional governing equations for the elemental flame are obtained as follows. A boundary layer approximation is applied across the flame, and a solution is considered along the stagnation streamline, $x = 0$; y is the coordinate normal to the flame surface. The outer flow—a stagnation point potential flow with velocity field $u_{-\infty} = a(t)x$, $v_{-\infty} = -(j+1)a(t)y$—yields the pressure gradient as a function of the imposed strain rate:

$$\frac{\partial p}{\partial x} = -\rho_u u_{-\infty} a - \rho_u \frac{u_{-\infty}}{a} \frac{da}{dt} \tag{12.1}$$

Note that $j = 0$ for a planar flow and $j = 1$ for an axisymmetric flow, in which case r may be substituted for x in the expressions above. The notation ρ_u in (12.1) emphasizes that the density of the *unburned* mixture is used to define the pressure gradient. If the densities of both incoming streams are equal, $\rho_u = \rho_{-\infty} = \rho_\infty$, this distinction is moot. In the premixed flame, however, expansion resulting from heat release within the flame requires a products-side stream entering with lower density—and to maintain a constant pressure gradient, a higher effective strain rate. Defining the strain rate parameter $a(t)$ on the unburned side, in accordance with the expression for pressure gradient above, thus insures consistency across all flame configurations.

Introducing the similarity variable $U \equiv u/u_{-\infty}$, the notation $V \equiv \rho v$, and substituting the pressure gradient expression into the equation for momentum conservation inside the boundary layer, we obtain the following equations for species, energy, momentum, and mass conservation, respectively:

$$\rho \frac{\partial Y_k}{\partial t} + V \frac{\partial Y_k}{\partial y} + \frac{\partial}{\partial y}(\rho Y_k \mathscr{V}_k) - \dot{w}_k W_k = 0 \tag{12.2}$$

$$\rho \frac{\partial T}{\partial t} + V \frac{\partial T}{\partial y} - \frac{1}{c_p} \frac{\partial}{\partial y}\left(\lambda \frac{\partial T}{\partial y}\right) + \frac{1}{c_p}\left(\sum_k^K \rho c_{p,k} Y_k \mathscr{V}_k\right) \frac{\partial T}{\partial y} + \frac{1}{c_p} \sum_k^K \dot{w}_k H_k = 0 \tag{12.3}$$

$$\rho \frac{\partial U}{\partial t} + \rho U \frac{1}{a} \frac{da}{dt} + \rho U^2 a + V \frac{\partial U}{\partial y} - \frac{\partial}{\partial y}\left(\mu \frac{\partial U}{\partial y}\right) - \rho_u \left(\frac{1}{a} \frac{da}{dt} + a\right) = 0 \tag{12.4}$$

$$\frac{\partial \rho}{\partial t} + \frac{\partial V}{\partial y} + (j+1)\rho U a = 0 \tag{12.5}$$

where the diffusion velocity is:

$$\mathscr{V}_k = -\frac{1}{X_k} D_{km} \frac{\partial X_k}{\partial y} \tag{12.6}$$

and the mixture-averaged diffusion coefficient D_{km} is defined in terms of binary diffusion coefficients $\mathscr{D}_{jk}$ as [65]:

$$D_{km} = \frac{1 - Y_k}{\sum_{j \neq k}^K X_j / \mathscr{D}_{jk}} \tag{12.7}$$

Here Y_k is the mass fraction of species k, while W_k and $\dot{w}_k$ are the molar weight and molar production rate, respectively. In the remaining equations, c_p is the specific heat of the mixture, λ is the thermal conductivity, H_k is the molar enthalpy of the k-th species, ρ_u is the density of the reactants mixture, and μ is the dynamic viscosity of the mixture. Note that thermal diffusion velocities are neglected. The low Mach number assumption has been employed, and hence density is calculated as a function of the temperature, species mass fractions, and the spatially uniform thermodynamic pressure via the ideal gas equation of state.

While the velocity u is zero along the stagnation streamline, the momentum equation (12.4) in U is retained to govern variation of the strain rate through the flame, where, by definition,

$$aU = \left.\frac{\partial u}{\partial x}\right|_{x=0} \tag{12.8}$$

yields the strain rate profile.

Boundary conditions for the species and energy equations consist of defining the composition and temperature of the two incoming streams of the stagnation point flow:

$$y = \pm\infty : \; Y_k = Y_{k,\pm\infty}(t), T = T_{\pm\infty}(t). \tag{12.9}$$

The continuity equation requires only one boundary condition; typically, this boundary condition would specify zero velocity at the stagnation point, $V(y=0) = 0$. Numerical considerations discussed in the next section, however, suggest that we impose a boundary condition at $y = -\infty$, and leave the stagnation point definition to fix the origin of the y-axis.

The momentum conservation equation requires two boundary conditions. At an unburned stream, $u = u_{-\infty}$, so the boundary condition is by definition $U = 1$. Setting the spatial gradients in (12.4) to zero yields an ODE for U at the burned-stream boundary, denoted by U_b:

$$\frac{dU_b}{dt} = -U_b^2 a - U_b\left(\frac{1}{a}\frac{da}{dt}\right) + \frac{\rho_u}{\rho_b}\left(\frac{1}{a}\frac{da}{dt} + a\right) \tag{12.10}$$

Here ρ_b is the density of the burned mixture. For a premixed flame, the strain rate in the incoming products-side mixture thus responds dynamically to the imposed strain rate. This far-field boundary condition also places an important requirement on the size of the computational domain; the flame must be far enough from the $+\infty$ and $-\infty$ boundaries for spatial gradients in U to vanish. In the case of steady strain rate, the burned-stream boundary condition on U reduces to

$$U_b = \sqrt{\frac{\rho_u}{\rho_b}}. \tag{12.11}$$

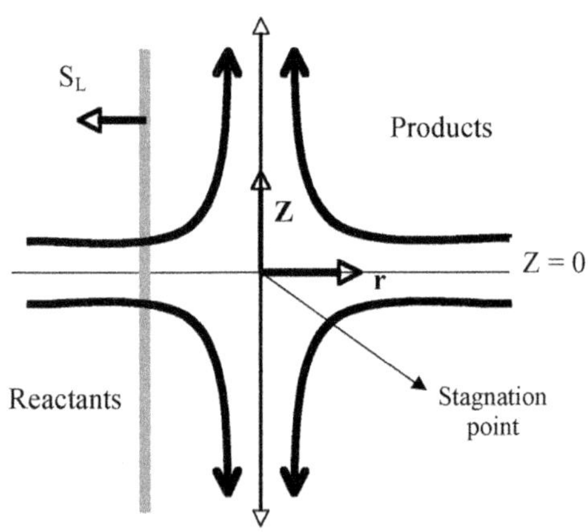

Fig. 12.5: Premixed flame in a stagnation point flow; the elemental flame model.

12.4 Numerical Solution for the Elemental Flame Model

Numerical solutions of the governing equations are obtained via a fully-implicit finite difference method, as necessitated by the stiffness of detailed kinetics. A first-order backward Euler formulation is used. All the governing equations are solved simultaneously, and thus the continuity equation (12.5) is in a sense an algebraic constraint on the implicit system.

A first-order upwind discretization is applied to all convective terms, while diffusion terms are discretized to second-order accuracy using centered differences. An 'upwind' discretization is used in the continuity equation as well, although it is not properly a transport equation and $\partial V/\partial y$ is not a convective term *per se*. Nonetheless, taking the positive sign outside $\partial V/\partial y$ to suggest a positive upwind velocity, the following discretization of the continuity equation adds dissipation of the appropriate sign:

$$\frac{\partial \rho}{\partial t}+\frac{\partial V}{\partial y}+\rho U a=0 \rightsquigarrow \frac{\rho_j^{n+1}-\rho_j^n}{\Delta t}+\frac{V_j^{n+1}-V_{j-1}^{n+1}}{y_j-y_{j-1}}+\rho_j^{n+1}U_j^{n+1}a^{n+1}=0 \quad (12.12)$$

A boundary value on the mass flux V must be chosen at $y=-\infty$. The boundary value can be set to any reasonable number based on the size of the computational domain and the strain rate. The solution to the problem matches the mass flux profile to the flame location, as reflected in the profiles of T, Y_k, ρ, and U.

In computations with unsteady strain rate, this boundary condition on V must be updated intermittently. The strain rate parameter a can easily vary one or two orders of magnitude in a given computation. The following method is applied to update the mass flux profile in this case. At the start of time step $n+1$, an initial guess for V^{n+1} is obtained by integrating the continuity equation with a^{n+1}, U^n, and ρ^n:

$$\frac{\rho_j^n-\rho_j^{n-1}}{\Delta t}+\frac{V_j^{n+1,\,guess}-V_{j-1}^{n+1,\,guess}}{y_j-y_{j-1}}+\rho_j^n U_j^n a^{n+1}=0 \qquad (12.13)$$

In one step, this expression updates the boundary value on V at $y=-\infty$ and generates a new guess for V^{n+1}. All the spatial discretizations are performed on a non-uniform adaptive grid, permitting a dynamic clustering of grid points in regions where spatial gradients are strong, and thus ensuring adequate resolution through the reaction-diffusion zone over all the integration time.

The time step for integration is constant, typically chosen on the order of 1 μs. At each time step, discretization reduces the governing PDEs to a set of nonlinear algebraic equations. The nonlinear system can be written as:

$$F(x)=0, \quad F:\mathbb{R}^n \rightarrow \mathbb{R}^n \qquad (12.14)$$

The output of the function F is a vector containing residuals of the discretized governing equations, while x is the solution vector, containing profiles of each fundamental variable: Y_k, T, U, and V.

A set of numerical methods to solve (12.14) efficiently and robustly [35] using an inexact Newton iteration [40] is used to accelerate convergence. The inexact Newton condition essentially restates the exact Newton condition from the perspective of an iterative linear solver. An iterative method is used to find an approximate solution to the exact Newton condition, $F'(x_i)s_i = -F(x_i)$, and η_i specifies the tolerance to which this solution (s_i) is found.

In our implementation, the inexact Newton method is coupled with a safeguarded backtracking globalization to improve its domain of convergence [14]. If the step s_i of the inexact Newton condition does not sufficiently reduce $\|F\|$, the step is reduced by a scalar factor θ, essentially "backtracking" along the search direction. Backtracking continues until the condition on $\|F\|$ is met, for in a sufficiently small neighborhood of the trial solution x_i, the linear model must indicate the correct downward path; the Newton equation is consistent.

12.5 UFE LES Sub-grid Combustion Model

In this section, the elements of the UFE sub-grid combustion model along with the closure for the filtered LES equations are introduced.

The flame embedding approach is broadly compatible with the flamelet assumption [42]. If chemistry is fast relative to mixing, reactions occur within a thin zone and the flame acquires a well defined burning velocity. In this limit, turbulent eddies do not penetrate the thin reaction zone, and a laminar flame structure can be assumed. Thus, a turbulent flame can be regarded as a stretched laminar flame that is embedded in an ensemble of vortices of different size and intensity. The net impact of these vortices is to convolute the flame surface on multiple scales while exerting a time-dependent strain on it. The components of the UFE model are shown in the schematic diagram in Fig. 12.6. A high order accurate level set algorithm is used to track the motion of the flame front. The solution of the level set equation provides information regarding the flame surface location and orientation inside each cell. The level set that defines the flame surface is specified using a surrogate progress variable $C = C^*$. For this purpose, a progress variable equation is filtered and solved along with the filtered Navier Stokes equations. The closure for the source terms in the progress variable equation, the energy equation, and the level set equation is implemented by solving directly on the subgrid level a set of unsteady strained stagnation point flames. The level set subroutine and the 1-D strained flame code provide the information needed for closure of the source terms for the LES solver. Consequently, in our approach the level set is used in two distinct ways; numerically to track the flame surface and as a tool to construct a subgrid model for combustion closure. The coupling is done by integrating within each cell the solution of the strained flame along the flame surface. A systematic diagram explaining the above UFE closure algorithm is shown in Fig. 12.7. At each time step the above procedure is applied for each cell in the LES mesh. More details on the LES approach and the subgrid model can be found in El-Asrag et al. [18].

A direct subgrid closure algorithm is proposed for the turbulent flame speed, the filtered energy, and progress variable source terms. At each time step, an average sub-filter flame front is constructed. First, we introduce an inner grid inside each cell. Then, by using the filtered level set-field values $\check{G}$ at the cell corners,

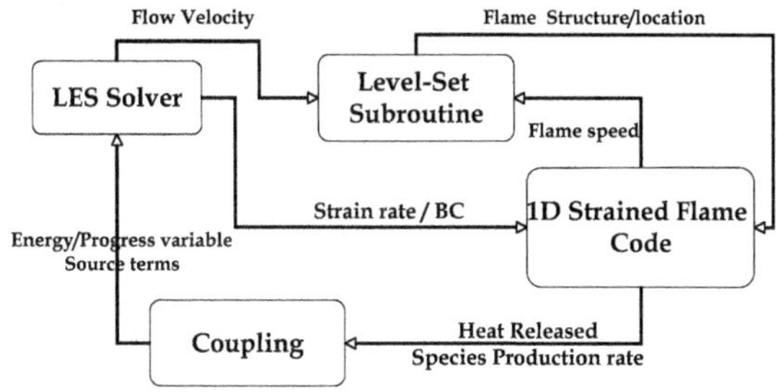

Fig. 12.6: The components of the unsteady flame embedding algorithm.

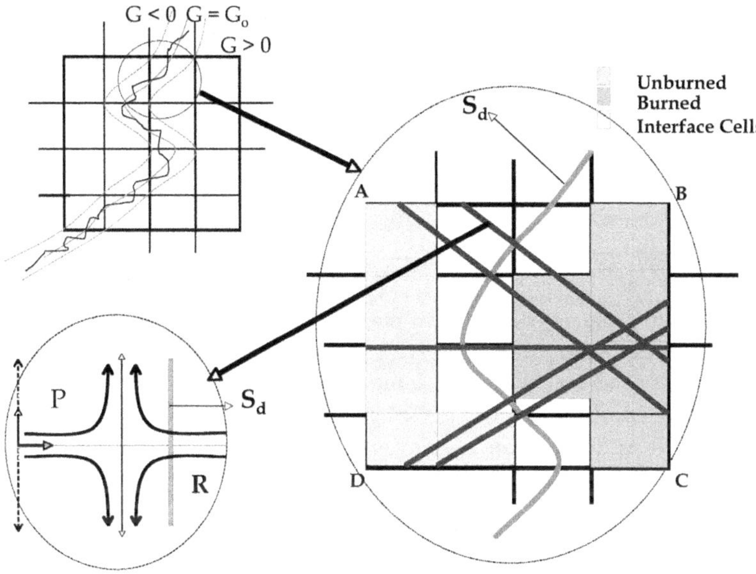

Fig. 12.7: A systematic diagram explaining the UFE closure algorithm. For each cell in the LES mesh in the upper left corner, the level set values on the cell corners are used to provide information about the flame structure as well as the normal directions. One-dimensional lines are constructed at the location $C = C^{\star}$ and a 1-D strained flame is solved along these lines. The information along this 1-D line is then used for closure of the LES equations.

we undergo bilinear interpolation on the inner mesh using a second order formula. The flame sub-filter structure is constructed by joining the values corresponding to $G = G_o$. The next step is to construct a set of 1-D lines at the intersections of the flame front within the inner mesh points in the direction normal to the flame front defined as $\mathbf{n} = -\nabla\check{G}/|\nabla\check{G}|$. Along each line a stagnation point flow problem is solved for the elemental flame model described in Sec. 12.3.

For a general scalar Φ, the sub-grid contribution $\overline{\Phi}^{sgs}$ to the filtered value $\overline{\Phi}$ over the flame surface will be computed in two steps. In each cell, after constructing the 1-D lines normal to the flame 2-D surface, the stagnation point flow problem is solved using the local filtered strain rate and boundary conditions. The scalars profiles are then integrated numerically over each line:

$$\Phi_i = \sum_{j=1}^{n-1} \frac{1}{2} \left[\Phi(j+1) + \Phi(j)\right] \cdot \left[x(j+1) + x(j)\right], \tag{12.15}$$

where n is the number of grid points along the 1-D line, $\mathbf{x}$ is the local coordinate, and i is an index that refers to the line number inside the cell volume. The number of 1-D lines inside each cell can be considered as equivalent to the number of flamelets generated *a priori* for tabulation in the laminar flamelet approach. The second step is to integrate over the flame surface to get the flame surface integrated contribution to the filtered field for each cell volume conditioned on $C = C^*$. The conditioned values are computed using line integral in 2-D and surface integral in 3-D space. Consequently, for m 1-D lines along the flame surface, the line integral can be evaluated as follows using Riemann sum directly over the flame surface.

$$\overline{\Phi}|C^* = \sum_{i=1}^{m} (\Phi_i \cdot \Delta S_i) \tag{12.16}$$

where ΔS_i is the distance between two points on the flame surface at the sub-grid level and Φ_i is computed from Eq.12.15. To account for the unresolved subgrid fluctuations, the conditioned integral solution over the flame surface $\overline{\Phi}|C^*$ is convoluted with a PDF that describes the subgrid fluctuations for the filtered progress variable $\widetilde{C}$. A Beta distribution is adapted here following the work in the literature [39, 47, 51, 60]. Finally the subgrid contribution to the filtered flow field is computed as:

$$\overline{\Phi}^{sgs} = \int_0^1 \overline{\Phi}|C^* \widetilde{P}(C^*; \mathbf{x}, \mathbf{t})\, dC^*, \tag{12.17}$$

where $\widetilde{\mathrm{P}}(C^*; \mathbf{x}, \mathbf{t})$ is the mass weighted filtered probability density function of the progress variable C. The mean of the Beta distribution is computed from filtered progress variable $\widetilde{c}$ equation:

$$\frac{\partial \bar{\rho}\widetilde{c}}{\partial t} + \frac{\partial \bar{\rho}\widetilde{u}\widetilde{c}}{\partial x_j} + \frac{\partial}{\partial x_j}\bar{\rho}\left(\widetilde{uc} - \widetilde{u}\widetilde{c}\right) = \overline{\dot{\omega}} + \frac{\partial}{\partial x_j}\bar{\rho} D_C \frac{\partial \widetilde{c}}{\partial x_j}, \tag{12.18}$$

where D_C is the progress variable diffusivity and the PDF variance is computed algebraically [47, 51]:

$$\widetilde{C''^2} = C_V \overline{\Delta^2} \frac{\partial \widetilde{C}}{\partial x_i}, \tag{12.19}$$

where the coefficient C_V is computed dynamically [47].

To this end, the only remaining task is to compute Φ for the closure of filtered progress variable, energy, and level set equations. Along each 1-D embedded line the flame location that corresponds to $\widetilde{C} = C^*$ is chosen as the location of maximum heat release. For closure of the level set equation source term, Φ is the filtered flame speed $\widehat{S}_\Delta$ conditioned at $\widetilde{C} = C^*$. The filtered flame speed $\widehat{S}_\Delta$ is expressed as [4, 9]:

$$\widehat{S}_{\Delta} = \widehat{S}_r^{sgs} + \widehat{S}_n^{sgs} + \widehat{S}_t^{sgs}, \tag{12.20}$$

where the subgrid reaction rate, normal diffusion, and curvature contributions to the displacement speed are $\widehat{S}_r^{sgs}$, $\widehat{S}_n^{sgs}$, and $\widehat{S}_t^{sgs}$, respectively. These values are defined on the flame surface conditioned on $\widetilde{C} = C^*$ as follows:

$$\begin{aligned}
\widehat{S}_r^{sgs} &= \left.\frac{\widehat{\dot{\omega}}}{\widehat{\rho}|\nabla\widetilde{C}|}\right|_{\widetilde{C}=C^*} \\
\widehat{S}_n^{sgs} &= \left.\frac{\mathbf{n}\cdot\nabla\left(\widetilde{\rho}\overline{D}\mathbf{n}\cdot\nabla\widetilde{C}\right)}{\widehat{\rho}|\nabla\widetilde{C}|}\right|_{\widetilde{C}=C^*} \\
\widehat{S}_t^{sgs} &= 2\overline{D}\widehat{\kappa}\Big|_{\widetilde{C}=C^*},
\end{aligned} \tag{12.21}$$

where the filtered curvature is defined as $\widehat{\kappa} = \frac{1}{2}\nabla\cdot\mathbf{n}$. In Eq. 12.21 $\widehat{\dot{\omega}}$ and $\widehat{\rho}$ are the progress variable production rate and density along the 1-D line, and $\overline{D}$ is the mean diffusivity. The conditioned flame speed is then integrated using a similar expression to Eq.12.16. For closure of the energy and the reaction progress variable source terms, Φ_i will be equal to the production rate equivalent to the set of species that represent the reactive progress variable. Finally the heat release is computed as $\dot{\omega}_h = \sum_{k=1}^{N_S} \dot{\omega}_k h_k$, where h_k is the species enthalpy.

12.6 Numerical Results

In this section the model validation is introduced. Flame/vortex interaction is a well known canonical problem frequently used for the validation of turbulent premixed flame models. The flame stretch is essentially governed by the flow characteristics in the normal direction to the flame front. Therefore this setup, even in 2-D, is an efficient model to describe flame stretch and curvature [53]. Here, we first describe briefly the numerical approach employed for solving the Navier Stokes equations and the problem setup. Then finally the comparisons between the UFE model and the DNS data.

For accurate prediction of the sub-grid scalars, high-fidelity codes based on higher order stable numerical schemes are needed to avoid contamination of the resolved scales by the numerical error. An implicit fully compressible fifth order accurate in space and time code is developed [17, 18] for this purpose. A hybrid explicit Runge-Kutta/Implicit algorithm is employed [34, 55]. This scheme has been shown to reduce the stiffness produced by highly stretched grids [61], an important property for wall-bounded flows where grid stretching is vital to capture the viscous and heat transfer effects near the wall. Consequently, this allows for higher stability and CFL condition. In our validation studies, up to 100 times the explicit CFL is achieved for simple test cases. The algorithm employs collocated finite-difference discretization on a structured grid. Time integration is performed by combining an

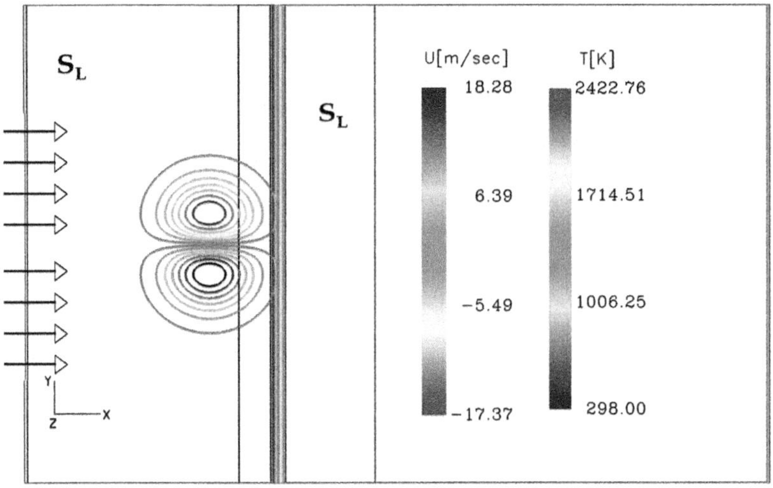

Fig. 12.8: The flame vortex interaction initial setup. From [18].

explicit Runge-Kutta scheme with the solution of an implicit scheme for the governing equations. The convective nonlinear Euler terms are discretized by a fifth order WENO algorithm [25] that relaxes to third order upwind on the boundaries. The viscous terms are discretized using central second/fourth order accurate finite difference formulation. The Navier-Stokes Characteristic Boundary Conditions (NSCBC) approach [53] is employed for the inflow/outflow boundary conditions.

The problem setup is shown in Fig. 12.8. An inflow/outflow configuration is used in the axial direction and symmetric boundary conditions for the upper and lower boundaries. The inflow speed matches the flame laminar burning speed S_L. The vortex strength U_{max} and size R_C are adjusted to operate the flame in the thin reaction zone regime. The vortex is initialized using the following stream function ψ [53]:

$$\psi = U_{max}\,\exp\left(-\frac{x^2+y^2}{2R_C^2}\right). \tag{12.22}$$

The flame is initialized using a planar premixed flame solution for a stoichiometric equivalence ratio. A single step chemistry is used $\mathbf{R} \rightarrow \mathbf{P}$, where the production rate is computed from the following expression:

$$\dot{\omega} = \mathfrak{B}\rho Y_R \exp\left(\frac{-\beta(1-\theta)}{1-\alpha(1-\theta)}\right), \tag{12.23}$$

here $\mathfrak{B}$ is the pre-exponential factor, α is the temperature factor, and β is the reduced activation energy. The pre-exponential factor is adjusted to achieve the required flame speed. The set of non-dimensional parameters that characterize the problem is shown in Table 12.1.

Table 12.1: Flame/vortex interaction operating conditions and parameters [a].

Le	Sc	Pr	Re_{S_L}	Re_ν	$\frac{U_{max}}{S_L}$	Ka	Da	S_L	$\frac{R_c}{\delta_F}$
1.0	0.73	0.73	48	259	38	74	0.263	0.48	10

[a] Where Le is the Lewis number, Sc is the Schmidt number, $\mathrm{Pr} = \mathrm{Sc}/\mathrm{Le}$ is the Prandtl number, $\mathrm{Ka} = \left((U_{max}/S_L)^3/(R_C/\delta_F)\right)^{0.5}$ is the Karlovitz number, and $\mathrm{Da} = (S_L\, R_C)/(U_{max}\delta_F)$ is the Damköhler number. Here δ_F is the laminar flame thickness computed based on the adiabatic flame temperature T_{ad}, the unburned gas temperature T_o and the maximum temperature gradient $|\frac{dT}{dx}|_{max}$ as $\delta_F = \frac{T_{ad}-T_o}{|\frac{dT}{dx}|_{max}}$. Re_{S_L} is the Reynolds number based on the inflow or the laminar flame speed, $\mathrm{Re}_\nu = (U_{max}R_C)/(2\nu)$ is the turbulent Reynolds number based on the vortex size R_C and core velocity U_{max}.

Two test cases are compared. Test case(1) is fully resolved DNS, where the grid spacing Δx is adjusted to resolve the flame thickness by 8 grid points and the ratio of $\Delta x/\eta = 1.2$. These ratios are recommended for DNS simulations in the thin reaction zone (TRZ) regime [33]. The second test case uses UFE as a sub-grid closure for an under-resolved mesh, where the grid spacing to the Kolmogorov length scale ratio is around 10.

Some instantaneous results are shown next. Figure 12.9 shows the reaction rate contours combined with the vorticity spectrum at two time instants for DNS and UFE. The general turbulence/flame interaction is well captured. The level-set $(G = G_o)$ is shown by the middle thick line. This line represents the flame-surface at $C = C^*$. As the figure shows the flame reaction zone is well captured and the flame displacement speed is well predicted by the UFE model. This is attributed to the model ability to account for the history effects by computing the displacement speed dynamically on the flame surface.

Figure 12.10 shows four consecutive time instants for the flame/vortex interaction and the heat release contours. Vorticity is generated by the dilatation effect as the flame propagates normal to itself and the misalignment of the density and the pressure contours as the flame propagates towards the fresh gases (i.e. baroclinic vorticity). As the vortex hits the flame surface, vorticity is generated with an opposite direction downstream the flame surface.

Figure 12.11 shows the velocity vector plot combined with the progress variable iso-contours. As shown, the turbulent vortices of length scales smaller than the flame thickness penetrate in the preheat zone and cause distortion of the progress variables iso-contours. The contours are not parallel since the displacement speed is different for different iso-levels. This observation confirms that the flame operates in the thin reaction zone regime.

The averaged volumetric consumption speed of the flame at each time instant is plotted in Fig. 12.12. The burning speed is computed from $S_T = \left(\frac{1}{\rho_u A_p}\right)\int_V \dot{\omega}\, dV$ [3],

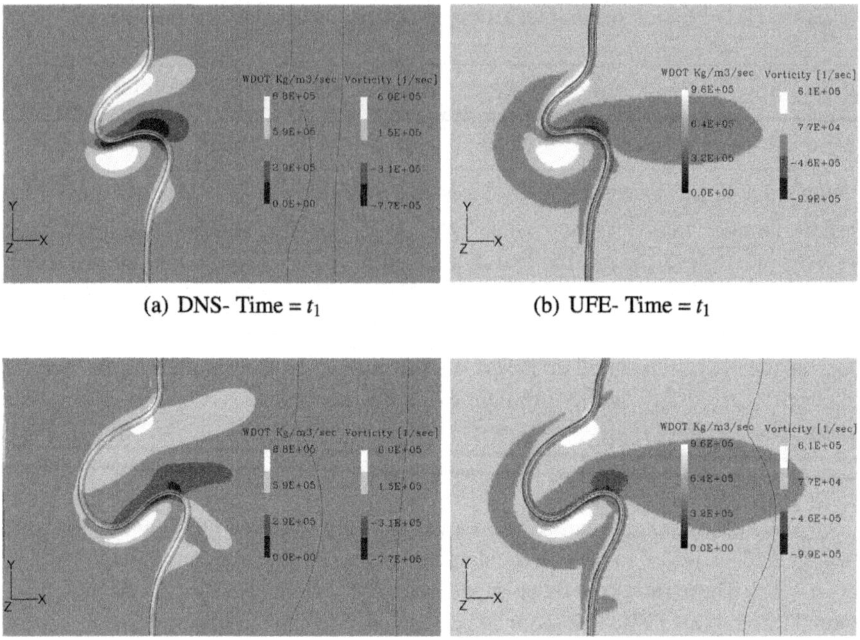

(a) DNS- Time = t_1 (b) UFE- Time = t_1

(c) DNS- Time = t_2 (d) UFE- Time = t_2

Fig. 12.9: Comparison of the reaction rate contours combined with the vorticity spectrum (background) at two time instants between DNS and UFE. The contour of $G = G_o$ line is shown in the UFE case. From [18].

where A_p is the projected area/length of the flame surface in the mean flow direction. This quantity is averaged over space and plotted at each time step. The figure shows good comparison with the DNS data. The flame speed increases monotonically with time due to the stretch effect and the increase in the flame surface area. However, this rate declines with time due to the viscous dissipation effects as no flame quenching is observed here for a stoichiometric unity Lewis number flame.

The PDF of the progress variable shows the well-known bi-modal Brass-Moss-Libby (BML) shape in Fig. 12.13. The data compares well with the DNS results. For premixed flames the most two common states are the fully burned and the unburned [53]. However, only finite-rate models can capture the intermediate burned states. The PDF of curvature is also shown in Fig.12.14. In agreement with the literature [4] the curvature has a global zero mean value, with the burned side more skewed towards negative curvature and the unburned side towards positive curvature.

The above curvature behavior shown in Fig. 12.14 is expected to affect the sign of the displacement speed, where more probability of a negative displacement speed is observed towards the unburned side since the upper flame part is propagating to-

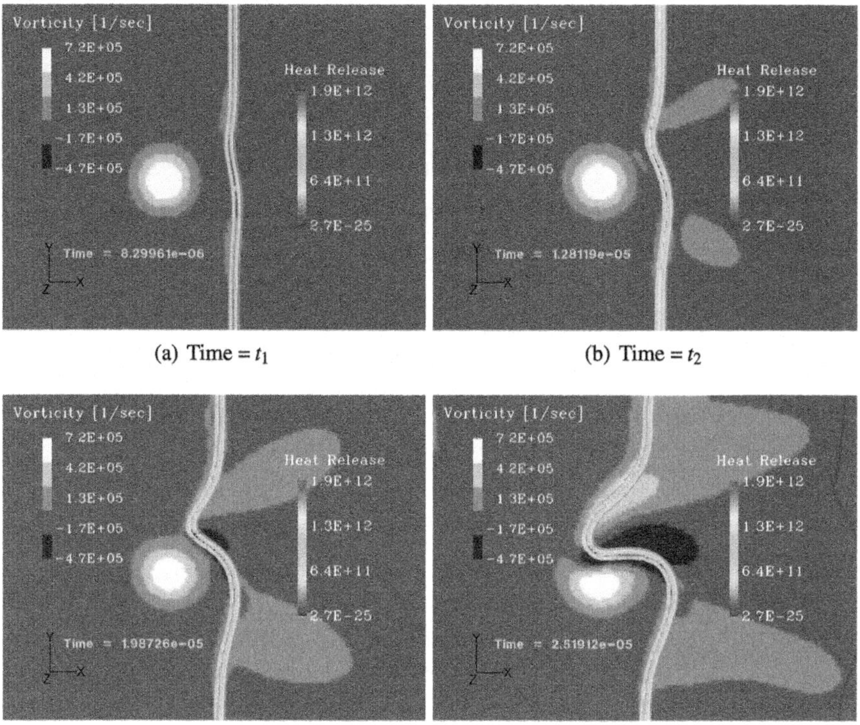

(a) Time = t_1

(b) Time = t_2

(c) Time = t_3

(d) Time = t_4

Fig. 12.10: Vorticity spectrum (background) and heat released contours at four consecutive time scales. From [18].

wards the inflow. This is shown in Fig. 12.15(b). In agreement with the curvature behavior, the displacement speed also shows zero mean value, and as the progress variable increases (moving towards the unburned gases), the displacement speed has more negative values. Figure 12.15(a) shows the ensemble average displacement speed normalized by the laminar flame speed in the mean progress variable space. The figure shows that the displacement speed decreases as we go towards the unburned side due to the curvature effect as indicated earlier. When the displacement speed is density weighted, however, it shows nearly constant behavior, as the effect of expansion due to heat release compensates the increase/decrease of the displacement speed by the curvature effect [4].

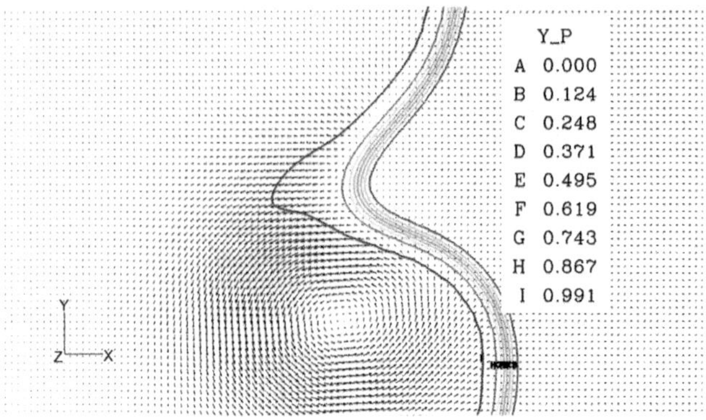

Fig. 12.11: The velocity vector plot combined with progress variable contours showing the penetration of vortices in the preheat zone. From [18].

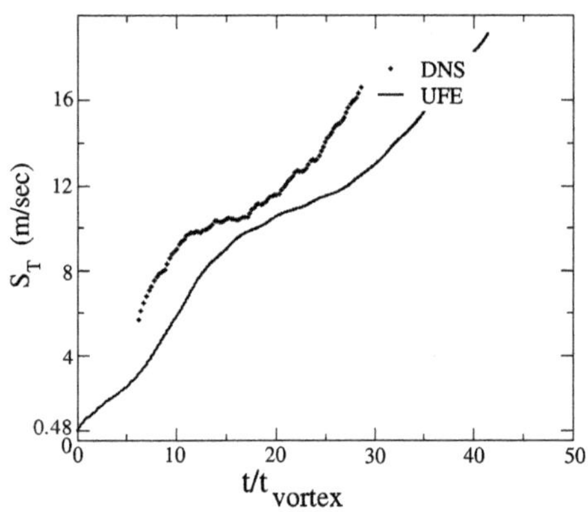

Fig. 12.12: Comparison of the volumetric mean flame speed (m/s) between DNS (•) and UFE (−). From [18].

12.7 Conclusions

An unsteady sub-grid combustion closure model is developed based on the flame embedding concept. The UFE model solves directly on a set of 1-D problems embedded in the normal direction to the flame surface using the resolved information on the LES mesh. The model problem for the 1-D line is the stagnation point flow configuration, where a formulation of variable strain rate is solved along the stagnation line. The model is validated with a vortex/flame interaction problem. Promising results were shown by comparing with the DNS data. The results show that the

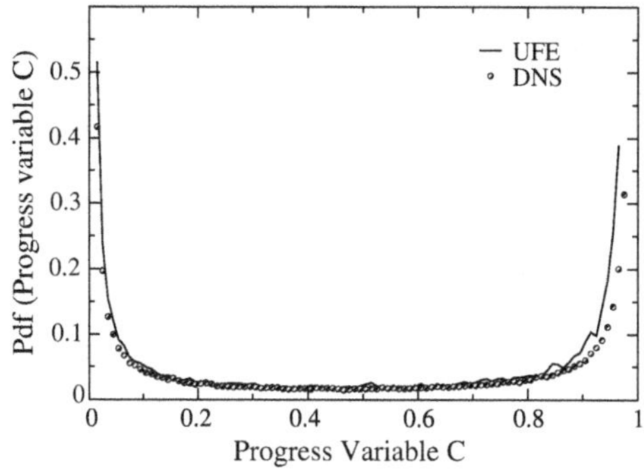

Fig. 12.13: Comparison of the progress variable PDF(C) between DNS (•) and UFE (–). From [18].

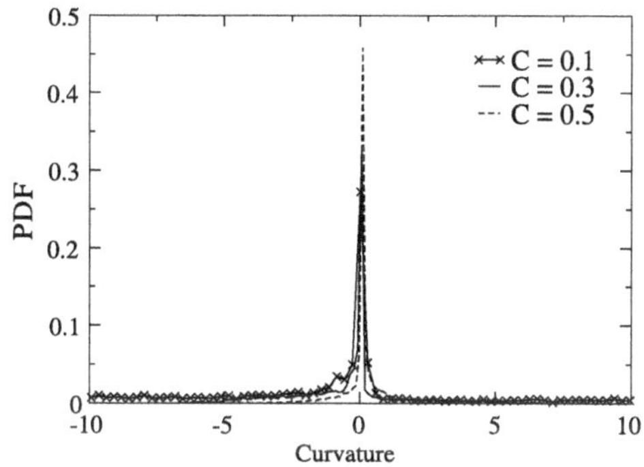

Fig. 12.14: The curvature PDF(κ) for different progress variable iso-contours. From [18].

unsteady turbulent flame propagation and the flame turbulence interaction is well captured. Future work will investigate the role of the scalars subgrid fluctuations and the model extension to simulate problems that exhibit extinction and reignition events.

Acknowledgment

This work is supported by the King Abdullah University of Science and Technology (KAUST), Award No. KUS-11-010-01. The authors also gratefully acknowledge the inspiring discussions and valuable comments from Suresh Menon, Heinz Pitsch, Tarek Echekki, Evatt R. Hawkes, R. Stewart Cant, Christopher Rutland, Youssef Marzouk, and Jean-Christophe Nave.

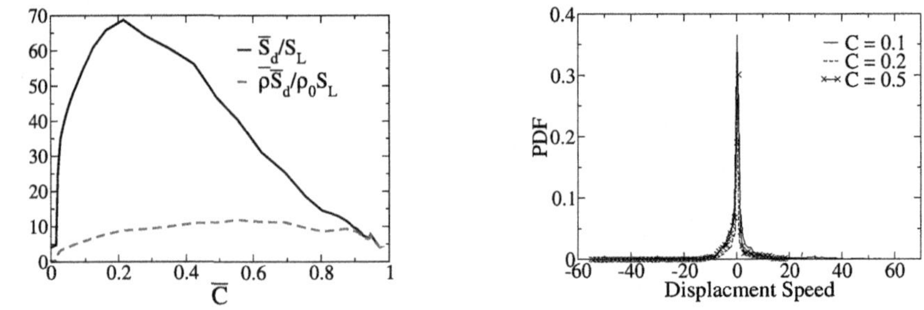

(a) $\overline{S}_d\left(\overline{C}\right)$. (b) $PDF(S_d)$

Fig. 12.15: The ensemble average displacement speed in the progress variable space (a) and the PDF of the displacement speed (b). From [18].

References

1. Annaswamy, A.M., Ghoniem, A.F.: Active control of combustion instability: theory and practice. IEEE Control Syst. Mag. **22**, 37–54 (2002)
2. Boger, M., Veynante, D., Boughanem, H., Trouve, A.: Direct numerical simulation analysis of flame surface density concept for large eddy simulation of turbulent premixed flames. Proc. Combust. Inst. **27**, 917–25 (1998)
3. Chakraborty, N., Cant, R.S.: Effects of Lewis number on turbulent scalar transport and its modeling in turbulent premixed flames. Combust. Flame **156**, 1427–1444 (2009)
4. Chakraborty, N., Cant, R.S.: Unsteady effects of strain rate and curvature on turbulent premixed flames in an inflow-outflow configuration. Combust. Flame **137**, 129–147 (2004)
5. Chorin, A.J.: Vortex sheet aproximation of boundary layers. J. Comput. Phys. **27**, 428–442 (1978)
6. Chorin, A.J.: Flame advection and propagation algorithm. J. Comput. Phys. **35**, 1–11 (1980)
7. Darabiha, N.: Transient behaviour of laminar counterflow hydrogen-air diffusion flames with complex chemistry. Combust. Sci. Technol. **86**, 163–181 (1992)
8. Driscoll, J.: Turbulent premixed combustion: Flamelet structure and its effect on turbulent burning velocities. Prog. Energy Combust. Sci. **34**, 91–134 (2008)
9. Echekki, T., Chen, J.H.: Unsteady strain rate and curvature effects in turbulent premixed methane-air flames. Combust. Flame **106**, 184–190 (1996)
10. Echekki, T., Kerstein, A.R., Dreeben, T.D., Chen, J.Y.: 'One-dimensional turbulence' simulation of turbulent jet diffusion flames: model formulation and illustrative applications. Combust. Flame **125**, 1083–1105 (2001)
11. Eggenspieler, G., Menon, S.: Large-eddy simulation of pollutant emission in a doe-hat combustor. J. Propuls. Power **20**, 1076–1085 (2004)
12. Egolfopoulos, F.N.: Dynamics and structure of unsteady, strained, laminar premixed flames. Proc. Combust. Inst. **25**, 1365–1373 (1994)
13. Egolfopoulos, F.N., Campbell, C.S.: Unsteady counterflowing strained diffusion flames: diffusion-limited frequency response. J. Fluid Mech. **318**, 1–29 (1996)
14. Eisenstat, S.C., Walker, H.F.: Globally convergent inexact Newton methods. SIAM J. Optim. **4**, 393–422 (1994)
15. El-Asrag, H.A., Lu, T., Law, C., Menon, S.: Simulation of soot formation in turbulent premixed flames. Combust. Flame **150**, 108–126 (2007)
16. El-Asrag, H.A., Menon, S.: Large eddy simulation of soot formation in a turbulent non-premixed jet flame. Combust. Flame **156**, 385–395 (2009)

17. El-Asrag, H.A., Nave, J.C., Ghoniem, A.F.: An algorithm for accurate prediction of turbulent burning velocity for under-resolved premixed flames. In: Eastern States Section of the Combustion Institute. College Park, Maryland, Oct. 18–21 (2009)
18. El-Asrag, H.A., Nave, J.C., Ghoniem, A.F.: Unsteady flame embedding (UFE) subgrid model for turbulent premixed combustion simulations. In: 48th Aerospace Sciences Meeting and Exhibit, AIAA-2010-201. AIAA, Orlando, FL (2010)
19. Fiorina, B., Gicquel, O., Vervisch, L., Carpentier, S., Darabiha, N.: Approximating the chemical structure of partially premixed and diffusion counterflow flames using fpi flamelet tabulation. Combust. Flame **140**, 147–160 (2005)
20. Ghoniem, A.F., Chorin, A.J., Oppenheim, A.K.: Numerical modeling of turbulent combustion in premixed gases. Proc. Combust. Inst. **18**, 1375–1383 (1981)
21. Ghoniem, A.F., Chorin, A.J., Oppenheim, A.K.: Numerical modeling of turbulent flow in a combustion tunnel. Phil. Trans. R. Soc. Lond. **A304**, 303–325 (1982)
22. Ghoniem, A.F., Knio, O.M.: Numerical simulation of flame propagation in constant volume chambers. Proc. Combust. Inst. **21**, 1313–1320 (1986)
23. Ghoniem, A.F., Soteriou, M.C., Knio, O.M.: Effect of steady and periodic strain on unsteady flamelet combustion. Proc. Combust. Inst. **24**, 223–230 (1992)
24. Hawkes, E.R., Cant, R.S.: A flame surface density approach to large-eddy simulation of premixed turbulent combustion. Proc. Combust. Inst. **28**, 51–58 (2000)
25. Henrick, A.K., Aslam, T.D., Powers, J.M.: Highly accurate numerical simulations of pulsating one-dimensional detonations. In: 43rd AIAA Aerospace Sciences Meeting and Exhibit, Reno, NV, AIAA-2005-1311 (2005)
26. Ihme, M., Cha, C.M., Pitsch, H.: Prediction of local extinction and re-ignition effects in non-premixed turbulent combustion using a flamelet/progress variable approach. Proc. Combust. Inst. **30**, 793–800 (2005)
27. Jenkins, K., Cant, R.S.: Curvature effects on flame kernels in a turbulent environment. Proc. Combust. Inst. **29**, 2023–2029 (2002)
28. Kerstein, A.R.: Linear-eddy modeling of turbulent transport. II: Application to shear layer mixing. Combust. Flame **75**, 397–413 (1989)
29. Kerstein, A.R.: One-dimensional turbulence: model formulation and application to homogeneous turbulence, shear flows, and buoyant stratified flows. J. Fluid Mech. **392**, 277–334 (1999)
30. Knio, O.M., Shi, X., Ghoniem, A.F.: Lagrangian simulation of a thin non-premixed flame in the field of an asymmetric layer. Combust. Flame **106**, 41–61 (1996)
31. Knudsen, E., Pitsch, H.: A dynamic model for the turbulent burning velocity for large eddy simulation of premixed combustion. Combust. Flame **154**, 740–760 (2008)
32. Law, C.K.: Combustion Physics, 1 edn. Cambridge University Press, New York, NY 10014-2473, USA (2006)
33. Lodato, G., Domingo, P., Vervisch, L., Veynante, D.: Scalar variances: LES against measurements and mesh optimization criterion: scalar gradient: a three-dimensional estimation from planar measurements using DNS. Proceedings of the Summer Program, pp. 387–398 (2008)
34. MacCormack, R.W.: Private communication
35. Marzouk, Y.M.: The effect of flow and mixture inhomogeneity on the dynamics of strained flames. S.M. thesis, Massachusetts Institute of Technology, Department of Mechanical Engineering (1999)
36. Marzouk, Y.M., Ghoniem, A.F., Najm, H.N.: Dynamic response of strained premixed flames to equivalence ratio gradients. Proc. Combust. Inst. **28**, 1859–1866 (2000)
37. Marzouk, Y.M., Ghoniem, A.F., Najm, H.N.: Towards a flame embedding model for turbulent combustion simulation. AIAA J. **41**, 641–652 (2003)
38. Menon, S., Kerstein, A.: Stochastic simulation of the structure and propagation rate of turbulent premixed flames. Proc. Combust. Inst. **24**, 443–450 (1992)
39. Nguyen, P.D., Vervisch, L., Subramanian, V., Domingo, P.: Multidimensional flamelet-generated manifolds for partially premixed combustion. Combust. Flame **157**, 43–61 (2010)

40. Pernice, M., Walker, H.F.: NITSOL: A Newton iterative solver for nonlinear systems. SIAM J. Sci. Comput. **19**, 302–318 (1998)
41. Peters, N.: Laminar flamelet concepts in turbulent combustion. Proc. Combust. Inst. **21**, 1231–1250 (1986)
42. Peters, N.: Turbulent combustion, 4 Edn. Cambridge University Press (2006)
43. Peters, N.: Multiscale combustion and turbulence. Proc. Combust. Inst. **32**, 1–25 (2009)
44. Petrov, C.: Numerical simulation of reacting flows with complex chemistry using flame embedding. Ph.D. thesis, Massachusetts Institute of Technology, Department of Mechanical Engineering (1997)
45. Petrov, C., Ghoniem, A.F.: The transient response of strained laminar-premixed flames. Combust. Flame **102**, 401–417 (1995)
46. Petrov, C., Ghoniem, A.F.: Numerical simulation of reacting flows with multi-step kinetics. In: The 35th Aerospace Sciences Meeting and Exhibit, Reno, Nevada, January, AIAA-97-0291 (1997)
47. Pierce, C.D., Moin, P.: Progress-variable approach for large-eddy simulation of non-premixed turbulent combustion. J. Fluid Mech. **504**, 73–97 (2004)
48. Pitsch, H.: Unsteady flamelet modeling of differential diffusion in turbulent jet diffusion flames. Combust. Flame **123**, 358–374 (2000)
49. Pitsch, H.: Improved pollutant predictions in large-eddy simulations of turbulent non-premixed combustion by considering scalar dissipation rate fluctuations. Proc. Combust. Inst. **29**, 1971–1978 (2002)
50. Pitsch, H.: A consistent level set formulation for large-eddy simulation of premixed turbulent combustion. Combust. Flame **143**, 587–598 (2005)
51. Pitsch, H.: Large-eddy simulation of turbulent combustion. Annu. Rev. Fluid Mech. **38**, 453–82 (2006)
52. Pitsch, H., de Lageneste, L.D.: Large-eddy simulation of premixed turbulent combustion using a level-set approach. Proc. Combust. Inst. **29**, 2001–2008 (2002)
53. Poinsot, T., Veynante, D., Candel, S.: Quenching processes and premixed turbulent combustion diagrams. J. Fluid Mech. **228**, 561–606 (1991)
54. Pope, S.: Computations of turbulent combustion. Proc. Combust. Inst. **23**, 591–612 (1990)
55. Rossow, C.C.: Efficient computation of compressible and incompressible flows. J. Comput. Phys. **220**, 879–899 (2007)
56. Rutland, C., Ferziger, J.: Unsteady strained premixed laminar flames. Combust. Sci. Technol. **73**, 305–326 (1990)
57. Sen, B., Menon, S.: Linear eddy mixing based tabulation and artificial neural networks for large eddy simulations of turbulent flames. Combust. Flame **157**, 62–74 (2010)
58. Shepherd, I., Cheng, R., Plessing, T., Kortschik, C., Peters, N.: Premixed flame front structure in intense turbulence. Proc. Combust. Inst. **29**, 1833–1840 (2002)
59. Stahl, G., Warnatz, J.: Numerical investigation of time-dependent properties and extinction of strained methane- and propane-air flamelets. Combust. Flame **85**, 285–299 (1991)
60. Subramanian, V., Domingo, P., Vervisch, L.: Large eddy simulation of forced ignition of an annular bluff-body burner. Combust. Flame **157**, 579–601 (2010)
61. Swanson, R., Turkel, E., Rossow, C.C.: Convergence acceleration of Runge-Kutta schemes for solving the Navier-Stokes equations. J. Comput. Phys. **224**, 365–388 (2007)
62. Thibaut, D., Candel, S.: Numerical study of unsteady turbulent premixed combustion: Application to flashback simulation. Combust. Flame **113**, 53–65 (1998)
63. Vreman, A.W., Albrecht, B.A., van Oijen, J.A., de Goey, L.P.H., Bastiaans, R.J.M.: Premixed and nonpremixed generated manifolds in large-eddy simulation of Sandia flame D and F. Combust. Flame **153**, 394–416 (2008)
64. Westbrook, C.K., Mizobuchi, Y., Poinsot, T.J., Smith, P.J., Warantz, J.: Computational combustion. Proc. Combust. Inst. **30**, 125–157 (2005)
65. Williams, F.A.: Combustion Theory: The fundamental theory of chemically reacting flow systems. Addison-Wesley, USA (1965)

Chapter 13
Adaptive Methods for Simulation of Turbulent Combustion

John Bell and Marcus Day

Abstract Adaptive mesh refinement (AMR) is an effective approach for simulating fluid flow systems that exhibit a large range of numerical resolution requirements. For example, an AMR simulation could dynamically focus maximum numerical resolution near a propagating flame structure, while simultaneously placing coarser computational zones near relatively large flow structures in the exhaust region downstream of the flame. However, since turbulent reacting flow applications already tend to be significantly complex, an AMR implementation might quickly become prohibitively intricate. In this chapter, we discuss basic AMR algorithm design principles that can be applied in a straightforward way to build up extremely efficient multi-stage solution strategies. As an example, we discuss an adaptive projection scheme for low Mach number flows, which was used to analyze flame-turbulence interactions in a full-scale simulation of a turbulent premixed burner experiment using detailed chemistry and transport models.

13.1 Introduction

We have entered a new era in turbulent combustion calculations; we can now simulate laboratory-scale turbulent reacting flows with sufficient fidelity that the computed time-dependent multi-dimensional solution should be expected to agree with experimental measurements. Such calculations are possible only through the effective use of modern high-performance computing architectures, and even then only by exploiting many of the multi-scale aspects of the reacting flow system.

John B. Bell
Center for Computational Sciences and Engineering, Lawrence Berkeley National Laboratory, Berkeley, CA, USA, e-mail: `jbbell@lbl.gov`

Marcus S. Day
Center for Computational Sciences and Engineering, Lawrence Berkeley National Laboratory, Berkeley, CA, USA, e-mail: `msday@lbl.gov`

T. Echekki, E. Mastorakos (eds.), *Turbulent Combustion Modeling*,
Fluid Mechanics and Its Applications 95, DOI 10.1007/978-94-007-0412-1_13,

Adaptive mesh refinement provides a mechanism for exploiting spatial and temporal variability in resolution requirements for given problem. With an adaptive algorithm, resolution can be changed dynamically to match local features of the flow. For example, finer resolution can be employed in a neighborhood of a flame to resolve the details of the flame structure. Fine resolution can also be used around complex fluid dynamical features such as shear layers or regions of intense turbulence while using coarse resolution where the flow has little structure such as in the products region of a premixed flame.

This chapter discusses the construction of structured, hierarchical adaptive mesh refinement schemes for modeling reacting flows. We will consider both a fully compressible formulation that is applicable to generic reacting flow simulations and a low Mach number formulation that exploits the temporal separation of scales between acoustic wave propagation and fluid motion typical of a broad range of turbulent combustion scenarios. The discussion will focus on issues related to AMR for the low Mach number model; developing an adaptive compressible flow solver requires only a subset of the ideas. We will restrict our consideration to single-phase gaseous combustion and, for the sake of exposition, assume an ideal gas equation of state. The generalization to more complex equations of state is discussed in Bell et al. [9].

The objective here is to provide a pedagogical overview of the basic design concepts used to develop block-structured AMR algorithms. The mathematical formulation is presented first, followed by key AMR concepts. Low Mach number AMR is then presented, followed by a discussion of implementation issues and software design. Finally, applications are discussed.

13.2 Mathematical Formulation

Gas phase combustion problems can be modeled by the multicomponent reacting compressible Navier-Stokes equations. Ignoring external body forces and assuming an ideal gas law, the governing equations expressing conservation of species mass, momentum and total energy are:

$$\begin{aligned}
\frac{\partial \rho Y_m}{\partial t} + \nabla \cdot \rho Y_m U &= \nabla \cdot \mathscr{F}_m + \omega_m, \\
\frac{\partial \rho U}{\partial t} + \nabla \cdot \rho U U + \nabla p &= \nabla \cdot \tau, \\
\frac{\partial \rho E}{\partial t} + \nabla \cdot (\rho E + p) U &= \nabla \cdot (\mathscr{Q} + \tau \cdot U)
\end{aligned}$$

where ρ is the density, U is the velocity, $E = \sum e_m(T) Y_m + 1/2 U \cdot U$ is the total energy, Y_m is the mass fraction of species m, T is the temperature, and ω_m is the net mass production rate for species m due to chemical reactions. Also, $\mathscr{Q}$ is the heat flux, τ is the stress tensor, and $\mathscr{F}_m$ is the diffusion flux of the mth species. Note that the internal energy, e_m, incorporates the potential chemical energy that is released by exothermic reactions in the flame. For these equations, we have $\sum Y_m = 1$, $\sum \mathscr{F}_m = 0$ and $\sum \omega_m = 0$, so that the sum of the species transport equations gives the conservation of total mass

$$\frac{\partial \rho}{\partial t} + \nabla \cdot \rho U = 0$$

These evolution equations are supplemented by an equation of state, $p = p(\rho, Y_m, T)$. For the mixture of ideal gases considered here the equation of state is given by

$$p = \rho RT = \rho \mathscr{R} T \sum_m \frac{Y_m}{W_m}$$

where W_m is the molecular weight of species m. We note that the numerical solution approaches discussed here have been applied to a much more general equation of state (e.g., see [9]).

The compressible flow equations admit two types of waves, material waves that propagate at the fluid velocity U, and acoustic waves that propagate relative to the moving fluid at the speed of sound, c. One possible approach to simulating reacting flow is to discretize the compressible system directly and numerically resolve all of the time scales. However, for a wide range of reacting flow phenomena, including most practical combustion systems, the fluid velocity is considerably smaller than the sound speed, and this disparity in scales can be exploited to compute much more efficiently. For typical laboratory flame experiments $U \approx 3 - 30$ m/s, while the sound speed in the hot product gases is about 1000 m/s. Flows in this regime are referred to as low Mach number since the Mach number, $M = |U|/c \ll 1$.

The low Mach number combustion formulation was first introduced by Rehm and Baum [34] and was later derived rigorously from low Mach number asymptotic analysis by Majda and Sethian [24]. The basic steps of the analysis are to first nondimensionalize the system with respect to the time and length scales of the flow and then to expand the resulting terms in powers of the Mach number M. Examining the behavior as $M \to 0$, one can show that in an unconfined domain, the pressure can be decomposed as

$$p(x,t) = p_0 + \pi(x,t)$$

where p_0 is the ambient thermodynamic pressure and π is a perturbational pressure field that satisfies $\pi / p_0 \sim \mathscr{O}\left(M^2\right)$. (In a more general setting, p_0 may be a function of time.) With this decomposition, p_0 defines the thermodynamic state; thermodynamic quantities are independent of π. The flow model in this regime becomes

$$\begin{aligned} \frac{\partial \rho Y_m}{\partial t} + \nabla \cdot \rho U Y_m &= \nabla \cdot \mathscr{F}_m + \omega_m, \\ \frac{\partial \rho U}{\partial t} + \nabla \cdot \rho U U + \nabla \pi &= \nabla \cdot \tau, \\ \frac{\partial \rho h}{\partial t} + \nabla \cdot \rho U h &= \nabla \cdot \mathscr{Q}, \end{aligned} \tag{13.1}$$

where the equation of state constrains the evolution. Note that the energy variable used in this formulation is the mass-averaged enthalpy, defined as $h(T,Y_m) = \sum_m Y_m h_m(T) Y_m$, where

$$h_m(T) = \int_{298K}^{T} c_{p,m}(T)dT + h_m(298K). \tag{13.2}$$

Here, $h_m(298K)$ is the standard heat of formation of species m at 298 K, and $c_{p,m}$ is its heat capacity at constant pressure. Empirical fits for both $h_m(T)$ and $c_{p,m}(T)$ are readily available. By combining the enthalpy equation and the species conservation equation, we can derive an alternative form of the energy equation in terms of T given by

$$\rho c_p\left(\frac{\partial T}{\partial t} + U\cdot\nabla T\right) = \nabla\cdot\mathscr{Q} - \sum_m h_m\left(\nabla\cdot\mathscr{F}_m + \omega_m\right)$$

where $c_p = \sum_m Y_m c_{pm}$ is the specific heat of the mixture at constant pressure. Although we advance the energy equation in terms of enthalpy, the temperature equation is useful in defining the constraint.

This low Mach number model retains compressibility effects due to chemical heat release and other thermal processes, but eliminates acoustic wave propagation entirely. The perturbation pressure, π, plays the role of a Lagrange multiplier to constrain the evolution so that the thermodynamic pressure is equilibrated everywhere instantaneously. Note that the form of these equations is no longer an initial value problem; the constrained system forms a differential algebraic equation (DAE) system that is considerably more difficult to evolve numerically.

For the low Mach number combustion model, we do not work with the constraint given by the equation of state directly. Instead, we differentiate the equation of state in the Lagrangian frame and use the evolution equations for ρ, Y_m and the auxiliary equation for T to define a constraint on the velocity:

$$\begin{aligned}
\nabla\cdot U &= -\frac{1}{\rho}\frac{D\rho}{Dt} = \frac{1}{T}\frac{DT}{Dt} + \frac{\mathscr{R}}{R}\sum_m \frac{1}{W_m}\frac{DY_m}{Dt} \\
&= \frac{1}{\rho c_p T}\left(\nabla\cdot\mathscr{Q} - \sum_m h_m \nabla\cdot Y_m\cdot\nabla h_m\right) + \frac{1}{\rho}\sum_m \frac{W}{W_m}\nabla(\rho D_m \nabla Y_m) \\
&\quad + \frac{1}{\rho}\sum_m\left(\frac{W}{W_m} - \frac{h_m(T)}{c_p T}\right)\dot{\omega}_m \\
&\equiv S
\end{aligned}$$

In section 13.4 we will outline an approach to solving the constrained low Mach number system based on a projection formulation. However, we first present a number of basic concepts inherent to our approach to adaptive-grid discretizations.

13.3 AMR Basic Concepts

There are several distinct approaches to developing adaptive mesh refinement algorithms. The approach taken here uses a block-structured hierarchical form of refinement, hereafter referred to simply as AMR. AMR was first developed by Berger and Oliger [12] for hyperbolic partial differential equations. A conservative version of this methodology for gas dynamics was developed by Berger and Colella [11] and extended to three dimensions by Bell et al. [5]. This approach was extended to variable-density incompressible flow by Almgren et al. [1]. Pember et al. [31] generalized the approach to low Mach number combustion with simplified chemistry and transport. Day and Bell [19] extended the method to treat detailed chemistry and transport.

13.3.1 Creating and Managing the Grid Hierarchy

In AMR, the local mesh refinement strategy is designed to exploit the advantages of uniform-grid PDE discretizations (uniform well-characterized errors, high accuracy per floating-point operation, memory-efficient stencil evaluations, etc.). The fundamental data object is a logically rectangular subdomain of uniformly spaced grid cells. A given level of refinement, ℓ, in the AMR hierarchy is represented as a union of such subdomains with a common Δx_ℓ. The refinement ratio, $r_\ell = \Delta x_\ell / \Delta x_{\ell+1}$ (typically 2 or 4) defines the reduction in grid spacing with increasing levels in the AMR hierarchy. The computational domain for an AMR calculation is rectangular

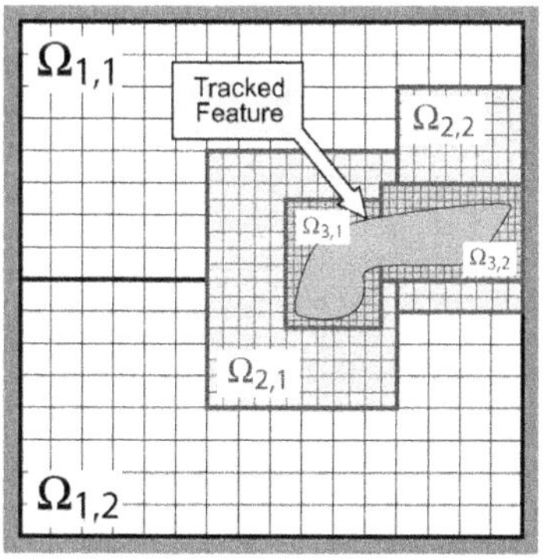

Fig. 13.1: Typical AMR grid structure in 2D surrounding a representative feature. The computational domain (gray box) is tiled by the union of the coarsest (black) grids, $\ell_1 = \cup\, \Omega_{1,i}$. The finest level, $\ell_3 = \cup\, \Omega_{3,i}$ encloses the feature of interest, and is properly nested within ℓ_2.

and is completely tiled by the coarsest AMR level, ℓ_1. Finer levels tile successively smaller regions of the domain and satisfy the following *proper nesting* requirements (see Figure 13.1):

- Each subdomain at level ℓ_j, $\Omega_{j,i}$, starts and ends at the corners of a cell in the next coarser level, ℓ_{j-1}
- There must be at least $n_{buf} > 0$ level ℓ_j cells between the bounding edge of level ℓ_{j+1} and any uncovered ℓ_{j-1} cell.

where a cell in level ℓ_j is said to be "uncovered" if there are no cells in level ℓ_{j+1} at that location. Note that fine levels may extend to the domain boundary and that there is no requirement that level ℓ_{j+1} grids be fully contained within a single level ℓ_j grid (e.g., see $\Omega_{3,2}$ in Figure 13.1).

A key feature of the AMR approach is that interfaces between adjacent levels in the hierarchy are aligned with the coordinate directions *and* coincide with cell boundaries on both levels. As we will see, this greatly simplifies the construction of composite (multi-level) solution methods. Also, the subdomains, $\Omega_{\ell,j}$ at each refinement level typically contain a large number of cells, typically 16-64 cells in each direction. Compared with cell-by-cell local refinement strategies, AMR affords significant advantages in terms of access to a wide range of well-characterized discretizations for uniform grids over most of the computational domain. Nonuniform regions are limited to the substantially smaller codimension-one regions at the interfaces between levels. Finally, the AMR grid structure presents a natural parallelization strategy for distributed-data computing systems, which we discuss briefly later in this chapter.

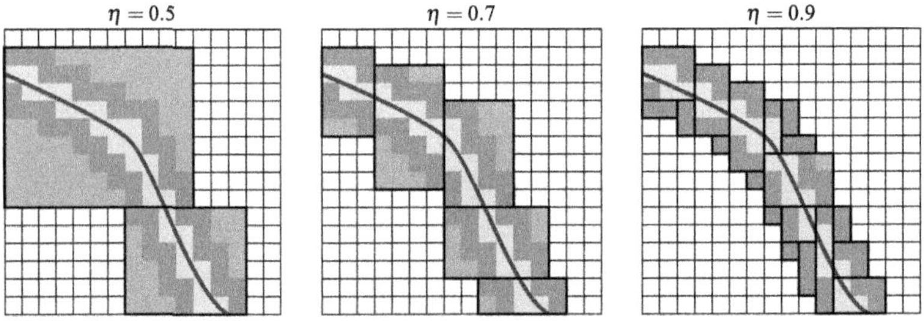

Fig. 13.2: Grid generation to refine a structure (blue line). Cells tagged for refinement include those containing the structure (blue cells) and those in a buffer zone surrounding them (pink cells). New rectangular grid boxes are constructed to enclose the tagged cells with a gridding efficiency, η=(# cells tagged)/ (# cells newly boxed).

Although the AMR grid hierarchy does not "move" with respect to the Eulerian frame, the structure is dynamic in time. At discrete intervals between time steps, subsets of the hierarchy are destroyed and regenerated in response to the evolving solution (see Figure 13.2). During such a *re-grid* operation, an error estimation procedure identifies individual cells at level ℓ_j requiring additional refinement (blue cells in Figure 13.2, for example). These cells are clustered together and surrounded with a buffer layer (pink cells in Figure 13.2), and then grouped into rectangular boxes and refined to generate new grids at level ℓ_{j+1}. Data on the new patches is filled by a copy-on-intersect operation from data on the previous level ℓ_{j+1} grids. In regions where the procedure generates entirely new cells at that level, data is filled by interpolation from next coarser level. This re-grid procedure must simultaneously recompute the grid structure at all finer levels, $\ell_{j+2}\dots\ell_{max}$ in order to guarantee that the new hierarchy does not violate proper nesting requirements. In a typical application regeneration of the AMR grid hierarchy accounts for less than 1% of the total run time. The initial grid hierarchy is created using a similar strategy, except that the state data is filled by user-supplied functions for initial data at each level. The re-grid frequency required for a specific application is determined by the dynamics of refined features on the Eulerian grid, and is controlled by the error tagging procedure and through a number of adjustable parameters, including buffer zone thickness, nesting buffer width, n_{buf}, and grid efficiencies (η in Figure 13.2).

13.3.2 AMR Discretization

There are several approaches to solving partial differential equations on a hierarchical AMR grid structure. One approach would be to simply write discrete approximations that operate directly on the irregular grids. Here, we consider a different paradigm in which the levels are advanced independently with a uniform-grid scheme, and then synchronized to account for the irregular interface between levels.

In the remainder of this section, we will discuss how to develop finite-volume discretization schemes on AMR grids. The key ideas will be discussed for two-level systems in one spatial dimension with and without subcycling in time. We will briefly discuss the issues related to generalizing to multiple space dimensions and extensions to more than two levels.

13.3.3 Hyperbolic Conservation Laws

We first consider a system of hyperbolic conservation laws

$$U_t + F_x = 0$$

discretized with a time-explicit finite volume scheme:

$$\frac{U_i^{n+1} - U_i^n}{\Delta t} = \frac{F_{i-1/2}^{n+1/2} - F_{i+1/2}^{n+1/2}}{\Delta x} \tag{13.3}$$

where the numerical fluxes are explicitly computed from the solution at time t^n, i.e., $F^{n+1/2} = F(U^n, \Delta t)$. The explicit treatment of fluxes implies that Δt is restricted by the CFL limit. The ideas presented here can easily be generalized to any other discretized PDE systems that are based on flux differencing. Conservative flux-based methods for the compressible Navier-Stokes equations, for example, fit within this framework.

We want to consider how to modify the basic uniform-grid scheme to update U^n on a locally refined grid, as depicted in Figure 13.3. The goal is to define a "composite" (multi-level) solution using a numerical flux algorithm that is unaware of the multi-level nature of the data. Initially, we will assume that both the coarse and fine grid use the same time step. In this case, we need to specify how to compute the numerical flux at each interface in Figure 13.3. We define an averaging (or

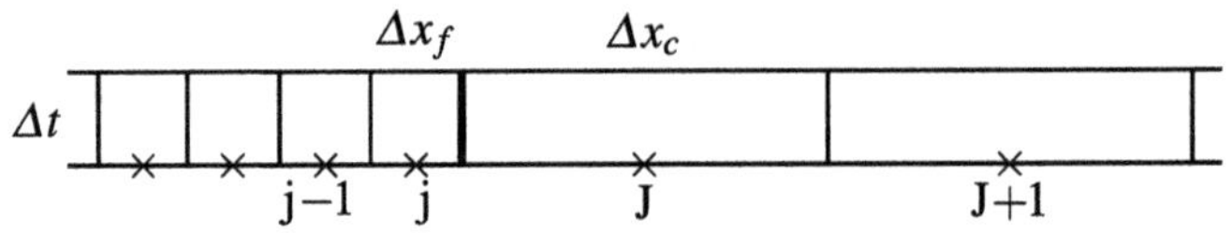

Fig. 13.3: A two-level 1D system with a coarse/fine interface, $c-f$, between fine cell j, and coarse cell J.

restriction) operator $\mathscr{A}$ that maps data at the fine grid resolution to the coarse grid resolution. We also define an interpolation operator $\mathscr{I}$ that interpolates data from the coarse resolution to fine resolution. With these operators, we can define the flux on the coarse grids away from the coarse / fine grid interface by using $\mathscr{A}(U^{f,n})$ to define data needed to compute the coarse fluxes in regions covered by fine grid. For example, if the numerical flux uses two values on each side of the interface, then to compute the flux at edge $J + 1/2$ we define an effective coarse value at a fictitious coarse cell $J - 1$ and use that value to compute the numerical flux. Similarly, we can use $\mathscr{I}(U^{c,n})$ to construct data at the fine resolution from the coarse data near the boundary of the fine grid as needed to compute fluxes at the fine resolution. To complete the specification of a composite grid solution, we only need to specifiy the flux at the coarse/fine boundary. We define that flux to be the flux computed using the fine grid data and interpolated coarse grid data. This completes the specification of a well-defined flux at each interface that enable us to advance the composite solution.

We can reinterpret this algorithm in such a way that the coarse and fine grids can be advanced independently and subsequently synchronized. This interpretation provides an algorithm to advance the composite solution in terms of an AMR grid hierarchy as defined earlier. Recall that the coarsest AMR level tiles the entire computational domain, including region that is tiled by the finer grids. Although this region is "covered" and thus not formally part of the composite solution, it provides a convenient location to store $\mathscr{A}(U^{f,n})$. Operationally, we store $\mathscr{A}(U^{f,n})$ over the entire covered portion of the coarse grid so that the coarse grid integration need not be aware of where the fine grid is located. We can then apply the uniform grid algorithm to advance the solution on the entire coarse grid in Figure 13.4.

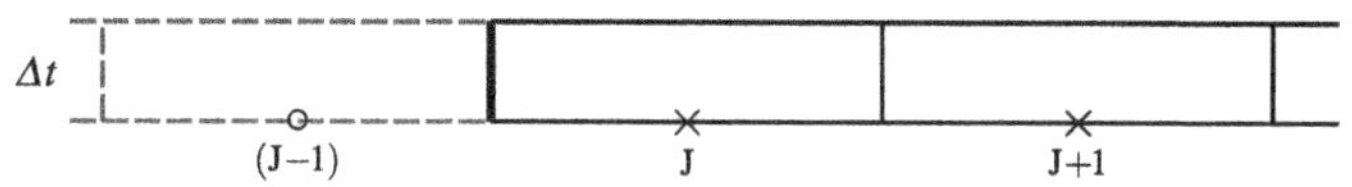

Fig. 13.4: The coarse integration. Covered cells (indexed in parentheses) are first filled using $\mathscr{A}(U^{f,n})$, and then become part of the uniform-grid data at the coarse level.

In order to advance the fine level, we use $\mathscr{I}$ to generate ficticious fine data in a buffer zone along the boundary of the fine grids. These cells are often referred to as "ghost" cells (see Figure 13.5). The ghost cells provide sufficient data to compute fluxes needed advance the solution on the fine grid, including fluxes at the coarse / fine boundary needed to advance fine cells adjacent to the boundary.

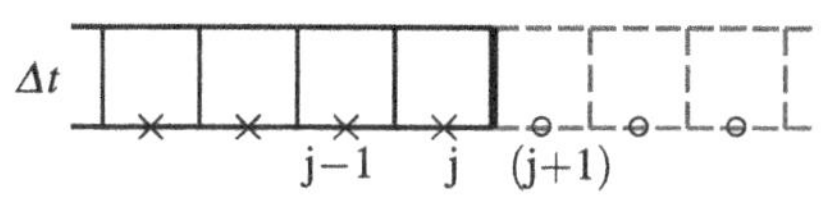

Fig. 13.5: The fine grid integration. Ghost cells (indexed in parentheses) are filled using $\mathscr{I}(U^{c,n})$.

At this point we have values for U^{n+1} that are defined on both the coarse and fine levels. The only difference between this provisional solution and our desired composite solution is that the two levels have not been computed with the same flux at the $c-f$ interface. In particular, the coarse grid cell U^n_J was advanced with the coarse flux $F^{n+1/2}_{J-1/2}$ instead of the fine grid flux $F^f_{j+1/2}$ as specified in the definition of the composite solution given above. We can correct this discrepancy by modifying U^{n+1}_J by

$$\Delta x_c U^{n+1}_J := \Delta x_c U^{n+1}_J - \Delta t^f F^c_{J-1/2} + \Delta t^f F^f_{j+1/2}.$$

This inter-level synchronization step is referred to as *refluxing*.

13.3.3.1 Sub-cycling

We can extend the notion of refluxing to enable us to subcycle in time, provided that Δt^c is an integer multiple of Δt^f. First we advance the coarse grid with Δt^c using the uniform grid algorithm as shown in Figure 13.6. Next, we advance the fine grid r steps with $\Delta t^f = \Delta t^c / r$. Ghost data (in parentheses) is required on the boundary of the fine grid for each subcycled step. Here the data must be interpolated in time and space in order to provide consistent boundary conditions for the fine levels at intermediate times, as shown in Figure 13.7. The only modification to the algorithm that is necessary for sub-cyling is that the refluxing correction to the coarse cell at the boundary of the fine grid is now summed over the r fine grid time steps

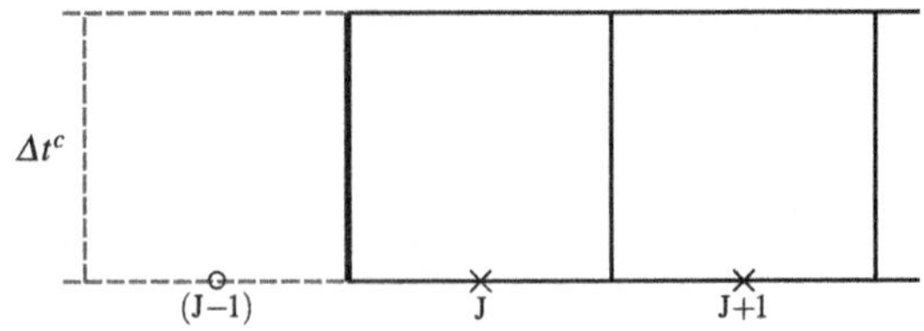

Fig. 13.6: Time advance of the coarse level near a coarse/fine interface in a subcycled integration. Ghost data (in parentheses) is obtained by the restriction operator, $\mathscr{A}(U^{f,n})$.

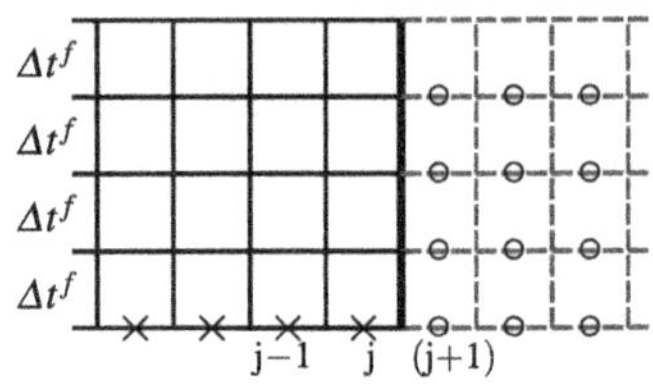

Fig. 13.7: Time advance of the fine level near a coarse/fine interface in a subcycled integration. Ghost data (in parentheses) is obtained by the interpolation operator, $\mathscr{I}(U^{c,n}, U^{c,n+1})$, and interpolation in space *and* time.

$$\Delta x_c U_J^{n+1} := \Delta x_c U_J^{n+1} - \Delta t^c F^c_{J-\frac{1}{2}} + \sum \Delta t^f F^f_{j+\frac{1}{2}} \tag{13.4}$$

From this discussion, we can abstract out a process for developing PDE discretizations on a locally refined grid using a uniform-grid flux-based discretization method, and arrive at a set of discretization design principles:

1. Define what is meant by the *solution* on the composite grid hierarchy.
2. Identify the errors that result from solving the equations on each level of the hierarchy "independently".
3. Solve correction equation(s) to "fix" the solution.
4. For subcycling, average the correction in time.

Within this setting, the coarse grid supplies Dirichlet data as boundary conditions for the fine grids. The errors then take the form of a flux mismatch at the coarse/fine interface. These ideas can be extended to multiple dimensions simply by defining suitable operators for $\mathscr{A}$ and $\mathscr{I}$. Because of the locality property of the refluxing operation, the algorithms discussed above can be applied recursively when there are more than two levels.

13.3.4 Elliptic

Next we illustrate how to apply the design principles described above to construct an algorithm to solve an elliptic equation on an AMR grid. We first consider how to define an appropriate composite solution to

$$-\phi_{xx} = \rho$$

on the AMR grid shown in Figure 13.3. A second-order discretization at all fine grid points $i \neq j$ can be written based on centered difference approximations as

$$-\frac{1}{\Delta x_f}\left(\frac{(\phi_{i+1}-\phi_i)}{\Delta x_f} - \frac{(\phi_i - \phi_{i-1})}{\Delta x_f}\right) = \rho_i,$$

and at all coarse grid points $I \neq J$ as

$$-\frac{1}{\Delta x_c}\left(\frac{(\phi_{I+1}-\phi_I)}{\Delta x_c} - \frac{(\phi_I - \phi_{I-1})}{\Delta x_c}\right) = \rho_I.$$

These discretizations are analogous to the flux form used in Equation 13.3 with ϕ_x playing the role of the numerical flux. At the $c-f$ boundary, the computation ϕ_x^{c-f} is slightly more complicated. The strategy is to build a polynomial interpolant, $I^e(\phi)$ using both coarse and fine data neighboring the $c-f$ interface to define an effective value for ϕ_{I^e} at $j+1$. Then approximate

$$\phi_x^{c-f} = \frac{\phi_{I^e} - \phi j}{\Delta x_f} \tag{13.5}$$

We can then discretize the equation using

$$-\frac{1}{\Delta x_f}\left(\phi_x^{c-f} - \frac{(\phi_j - \phi_{j-1})}{\Delta x_f}\right) = \rho_j$$

at $i = j$ and

$$-\frac{1}{\Delta x_c}\left(\frac{(\phi_{J+1}-\phi_J)}{\Delta x_c} - \phi_x^{c-f}\right) = \rho_J$$

at $I = J$. Provided the interpolant $I^e(\phi)$ is sufficiently accurate, this set of expressions over the composite grid defines a suitable discrete approximation to the elliptic equation. In particular, in the case of the second-order stencil used away from the $c-f$ boundary, the interpolant I^e needs to be sufficiently accurate that the local truncation error at the coarse fine boundary is first-order accurate to ensure second-order accuracy of the composite solution.

As before, we now consider what happens when we discretize the system on the coarse and fine grids in two separate steps, and we use our composite discretization to construct the appropriate synchronization. In particular, we first solve the coarse-grid problem

$$-\frac{1}{\Delta x_c}\left(\frac{(\overline{\phi}_{I+1}-\overline{\phi}_I)}{\Delta x_c}-\frac{(\overline{\phi}_I-\overline{\phi}_{I-1})}{\Delta x_c}\right)=\rho_I$$

at *all* coarse grid points I, including those covered by the fine cells. We then solve

$$-\frac{1}{\Delta x_f}\left(\frac{(\overline{\phi}_{i+1}-\overline{\phi}_i)}{\Delta x_f}-\frac{(\overline{\phi}_i-\overline{\phi}_{i-1})}{\Delta x_f}\right)=\rho_i$$

at all fine grid points, $i \neq j$. At $i = j$, we use the "correct" stencil defined using the Eq. 13.5.

The composite solution, $\overline{\phi}^c$ and $\overline{\phi}^f$, obtained from solving the levels separately represents the "provisional" composite solution, which satisfies the composite equations everywhere except at J. As before, the error is manifest in the difference between ϕ_x^{c-f} and $-\frac{(\overline{\phi}_J-\overline{\phi}_{J-1})}{\Delta x_c}$. If we let $e = \phi - \overline{\phi}$, then

$$-\Delta^{c-f} e = 0$$

except at $I = J$, where

$$-\Delta^c e = \frac{1}{\Delta x_c}\left(\phi_x^{c-f}-\frac{(\overline{\phi}_J-\overline{\phi}_{J-1})}{\Delta x_c}\right)$$

where Δ^{c-f} is the composite discrete Laplacian defined above.

We solve this linear system for the error, e, and correct the provisional solution by setting

$$\phi^c = \overline{\phi}^c + e^c$$

$$\phi^f = \overline{\phi}^f + e^f$$

Note that while the flux mismatch is localized to the $c-f$ interface, the correction, e is not. However, because our model problem is linear, this multi-step approach exactly recovers the solution to the original composite system. Also note that the correction equation is an elliptic system on the composite grid, and is in fact no simpler to solve than the original composite problem. The benefit to this formulation however, is that it allows us to use subcycling in time for algorithms that solve an elliptic equation as part of the more complex discretization. An example of this will be discussed in the next section when we construct a semi-implicit discretization of a parabolic PDE. Also, in many circumstances we appeal to the properties of *elliptic regularity* to reduce the complexity/cost of the composite approximation. In particular, the right hand side of the synchronization equation vanishes on the region covered by the fine grid so that on the fine grid e is a discrete harmonic. Thus, e is very smooth in the fine grid region and often will be well-represented on the coarse grid. In practice, this means that the synchronization need only be computed at the coarse level and interpolated to the fine resolution.

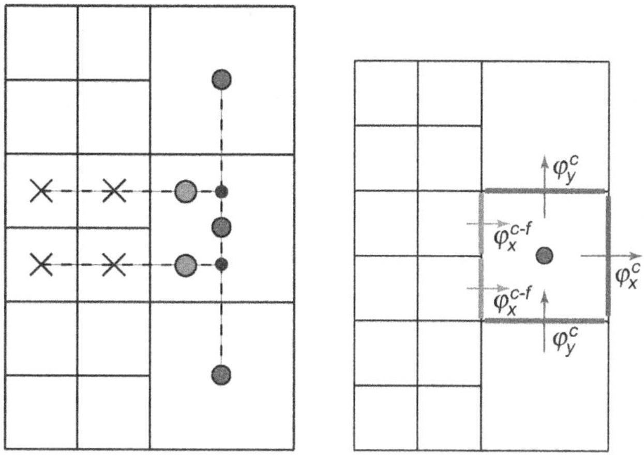

Fig. 13.8: The left image illustrates the interpolation stencils used to interpolate data to the location needed to evaluate ϕ_x on the fine grid, denoted by the green dot. The image on the right shows how a composite elliptic operator is defined in terms of a flux integral at the coarse / fine boundary.

These options also extend to more than two levels of refinement. As above one can form and solve a composite correction equation over the entire hierarchy or one can solve only on the coarse grid and interpolate to the finer grids. An additional option is to solve the composite operator on levels ℓ and $\ell+1$ and interpolate to finer levels.

In defining the composite solution for the elliptic case, we have used a slightly different paradigm than in the explicit hyperbolic case. In particular, the composite solution does not have any dependence on the coarse data covered by the fine grid. Although developing a discretization in this form in possible, the implicit coupling would require that the dependencies associated with averaging operator $\mathscr{A}$ be reflected in the overall discretization, making for extremely large stencils, particularly in multiple dimensions. In multiple dimensions, we use the interplation scheme depicted in Figure 13.8. We first interpolate values from coarse grid cells indicated in red to define values at the blue points using quadratic interpolation. We then interpolate from the blue points and the fine cells marked $\times$ to define values at the green points needed to evaluate the stencil, again quadratically. We note that in this form, the boundary flux ϕ_x^{c-f} is a function of coarse value at the red points and the fine values at the $\times$ points on the fine grid. The multidimensional composite discretization at the coarse cell adjacent to the fine grid is then defined by the divergence of the composite flux at the boundary and standard fluxes at the other interfaces as shown in Figure 13.8. With these definitions, the local truncation error is uniformly first order accurate and the solution is uniformly second order.

13.3.5 Parabolic Systems

In this section we discuss how to combine AMR discretizations for hyperbolic and elliptic systems in order to develop an AMR discretization of the parabolic model equation

$$u_t + f_x = \varepsilon u_{xx}$$

on the AMR grid depicted in Figure 13.3. We consider a semi-implicit algorithm in which we combine a time-explicit treatment advective terms and Crank-Nicolson treatment of diffusion. In particular,

$$\frac{u_i^{n+1} - u_i^n}{\Delta t} + \frac{f_{i+1/2}^{n+1/2} - f_{i-1/2}^{n+1/2}}{\Delta x_{loc}} = \frac{\varepsilon}{2}\left((\Delta^{c-f} u^{n+1})_i + (\Delta^{c-f} u^n)_i\right)$$

where Δx_{loc} is the local value of Δx. Here the hyperbolic flux $f^{n+1/2} = f^{n+1/2}(u^n)$ is computed using the composite hyperbolic flux defined above and Δ^{c-f} denotes the composite discrete Laplacian. This defines a suitable definition of a composite solution.

As before, we consider what happens if we advance the coarse and fine grids separately. We first advance the coarse solution, and then use the result to generate Dirichlet conditions for the fine grid advance. Let $\overline{u}^{n+1}$ denote the provisional composite solution formed in this way. Let $e^{n+1} = u^{n+1} - \overline{u}^{n+1}$ represent the difference between the provisional and exact composite solutions. Substituting $\overline{u}^{n+1}$ into the composite discretization, we see that e^{n+1} satisfies

$$\left(I - \frac{\varepsilon \Delta t}{2} \Delta^{c-f}\right) e^{n+1} = \frac{\Delta t}{\Delta x_c} (\delta f + \delta D)$$

where

$$\Delta t\, \delta f = \Delta t \left(-\overline{f}_{J-1/2} + f_{j+1/2}\right)$$

$$\Delta t\, \delta D = \frac{\varepsilon \Delta t}{2}\left(\left(\overline{u}_{x,J-1/2}^{c,n} + \overline{u}_{x,J-1/2}^{c,n+1}\right) - \left(u_x^{c-f,n} + u_x^{c-f,n+1}\right)\right)$$

As before, the source term is localized to the coarse cell at $c-f$ boundary and takes the form of a mismatch in coarse and fine fluxes. Updating $u^{n+1} = \overline{u}^{n+1} + e^{n+1}$ recovers the exact composite solution. As in the elliptic case, the implicit coupling in the parabolic equation leads to an implicit coupling in the synchronization equation.

The parabolic algorithm can also be extended to incorporate subcycling in time. To define the provisonal solution we first advance coarse grid over the time interval Δt^c. We then advance fine grid over the interval Δt^f a total of r times, using Dirichlet boundary data interpolated from the coarse grid. As in the explicit hyperbolic case discussed above, this interpolation must be in space and time in order to provide consistent boundary data for the fine grid at intermediate time levels during the subcycling.

The synchronization equation for the subcycled case now corrects the solution over the entire coarse interval; the source terms for the refluxing are summed over the r fine grid time steps. The parabolic synchronization equation is given by

$$\left(I - \frac{\varepsilon \Delta t^c}{2} \Delta^c\right) e^{n+1} = \frac{\Delta t^c}{\Delta x_c} (\delta f + \delta D)$$

where Δ^c is the coarse grid Laplacian.

$$\begin{aligned}
\Delta t^c \, \delta f &= -\Delta t^c \, \overline{f}_{J-1/2} + \sum \Delta t^f f_{j+1/2} \\
\Delta t^c \, \delta D &= \frac{\varepsilon \Delta t^c}{2} \left(\overline{u}^{c,n}_{x,J-1/2} + \overline{u}^{c,n+1}_{x,J-1/2} \right) \\
&\quad - \sum \frac{\varepsilon \Delta t^f}{2} \left(u_x^{c-f,n} + u_x^{c-f,n+1} \right)
\end{aligned}$$

In this case, for simplicity, we are only solving the correction equation on the coarse grid; corrections to the fine grid are interpolated from the coarse data. This reflects the notion that how we define the composite solution is not so straightforward and the refluxing equation does not exactly recover an "exact" composite solution. However, it does provide a second-order accurate, conservative and stable solution when the underlying basic operators are themselves second-order accurate and stable. Extension of these ideas to multiple dimensions and more than two levels of refinement are a straightforward extension of idea discussed earlier.

Summarizing the construction of the AMR discretization, there are a couple of key design goals. For the first-order hyperbolic and second-order elliptic and parabolic equations, the coarse grids provide Dirichlet boundary conditions to the fine grids. The synchronization steps ensure proper mathcing of the fluxes, and take the mathematical form of the original equation; e.g., a simple hyperbolic scheme leads to a simple local reflux corrections; an implicit parabolic equation leads to an implicit parabolic correction.

13.4 AMR for Low Mach Number Combustion

An AMR implementation for an explicit compressible Navier-Stokes solver is a straightforward application of some of the ideas discussed above. Developing an AMR algorithm for the low Mach number equations is considerably more complex. Here we will only sketch the basic ideas and refer the reader to Almgren et al. [1] and Day and Bell [19] for the details. We first discuss a single grid integration algorithm and then discuss some of the key issues associated with an AMR implementation.

Our basic discretization strategy is based on a projection formulation in which we evolve the system without strictly enforcing the constraint and then project the resulting solution back onto the constraint. Structurally, the low Mach number equations evolve momentum, density, species and enthalpy subject to a divergence constraint on the velocity field, which constrains the evolution of the thermodynamics variables so that the equation of state is satisfied. Thus, the constrained evolution in the low Mach number model is similar in structure to the incompressible Navier Stokes equations. For incompressible flows, projection-based fractional step methods, which parallel standard DAE methodologies [4, 13], have proven to provide an efficient discretization strategy [3, 8, 17]. Our goal then is to define a generalized projection methodology for low Mach number reacting flows. This generalization requires that we address two key differences between the incompressible flow equations and the low Mach number system. First, the low Mach number system includes finite amplitude density variations; and, second, the constraint on the velocity field is inhomogeneous.

Two different projection-based sequential algorithms have been proposed. One of these approaches, developed by McMurtry et al. [26] and Rutland and Ferziger [36], advances the thermodynamic variables and then uses the conservation of mass equation to constrain the evolution. Imposing the constraint in this form requires the solution of a Poisson equation. Although this approach does not fit within a mathematical projection framework, is has been successfully used by a number of authors to model reacting flows. See, for example, [27–29, 33, 37].

We use a different approach based on a generalized projection framework first introduced in Bell and Marcus [10]. The basic idea is that, subject to boundary conditions, any vector field, V can be decomposed as

$$V = U_d + \frac{1}{\rho}\nabla\phi$$

where U_d is divergence free and U_d and $\frac{1}{\rho}\nabla\phi$ are orthogonal with respect to a suitable inner product. This decomposition is an exact analog to the standard Hodge decomposition in a ρ-weighted inner product; e.g.,

$$\int (U_d \cdot \frac{1}{\rho}\nabla\phi)\rho \; dm = 0$$

Using this inner product, we can define a ρ-based projection, $\mathbf{P}_\rho$ such that $\mathbf{P}_\rho V = U_d$ with $||\mathbf{P}_\rho|| = 1$ and $\mathbf{P}_\rho^2 = \mathbf{P}_\rho$.

This projection operator allows us to address the finite amplitude density variations in the low Mach number system. The Majda and Sethian analysis [24] shows the flow compressibility, S, can be represented in terms of the gradient of a potential,

$$\nabla \cdot \nabla\xi = S$$

Using this form, we can generalize the ρ-weighted vector field decomposition to write any velocity field as

$$V = U_d + \nabla\xi + \frac{1}{\rho}\nabla\phi$$

We can then define

$$U = \mathbf{P}_\rho(V - \nabla\xi) + \nabla\xi$$

so that $\nabla \cdot U = S$ and $\mathbf{P}_\rho(\frac{1}{\rho}\nabla\phi) = 0$. This construction, which specifically addresses finite amplitude density variations, provides the basis for a robust projection algorithm to evolve the low Mach number equations.

The basic idea of the variable-ρ projection algorithm is to advance the thermodynamic variables using discretized conservation equations, and generate a provisional velocity field with lagged approximation to the constraint. We then use the vector field decomposition to extract the component of the velocity update that satisfies the modified constraint on the velocity divergence based on the time-advanced thermodynamic variables. More specifically, we advance the species mass densities and enthalpy equations using a flux-based conservative discretization,

$$\frac{\rho^{n+1}\chi^{n+1} - \rho^n\chi^n}{\Delta t} + \nabla \cdot (\rho U^{ADV}\chi)^{n+1/2} = D_\chi + R_\chi \text{ for } \chi = h, Y_m \quad .$$

Here we use a Crank-Nicolson discretization of the diffusion terms, and a specialized second-order Godunov algorithm to compute the advective derivatives. As part of the Godunov algorithm we compute an advective velocity field U^{ADV} on cell interfaces that has been projected so that it satisfies the constraint as well. The chemical rate equations are decoupled in an operator split form. In particular, we first advance the chemistry over the interval, $\Delta t/2$. We then advance the advection and diffusion components over Δt, and then finish with a second advance of the chemistry by $\Delta t/2$. This operator-split strategy allows us to decouple the pointwise chemical kinetics so that we can use stiff ODE integration methodologies to advance the kinetics equations, while preserving the second-order accuracy of the integration.

In a similar way, we compute a provisional velocity based on a lagged approximation to the dynamic pressure, π

$$\frac{U^* - U^n}{\Delta t} = -[U^{ADV} \cdot \nabla U]^{n+1/2} - \frac{1}{\rho^{n+1/2}}\nabla\pi^{n-1/2} + \frac{1}{\rho^{n+1/2}}\nabla \cdot \frac{\tau^n + \tau^*}{2} \quad .$$

The updated thermodynamic variables are then used to compute the constraint at the new time level S^{n+1}. To extract the component satisfying the divergence constraint we solve the variable-coefficient linear system

$$\nabla \cdot \left(\frac{1}{\rho}\nabla\phi\right) = \nabla \cdot \mathbf{V}^* - S^{n+1}$$

for ϕ, where $\mathbf{V}^* = \mathbf{U}^* + (\Delta t/\rho^{n+1/2})\nabla\pi^{n-1/2}$, and set

$$\pi^{n+1/2} = \phi \quad \text{and} \quad \mathbf{U}^{n+1} = \mathbf{V}^* - \frac{\Delta t}{\rho^{n+1/2}}\nabla\phi \tag{13.6}$$

This procedure implements the generalized vector field decomposition discussed above but exploits linearity to perform only a single elliptic solve to enforce the constraint. This final projection is based on node-centered ϕ data and a conformal bilinear finite element construction on a locally refined grid. As with the finite-volume elliptic solvers discussed above, the nodal solver can be decomposed into a solution operator on each of the refinement levels independently, followed by a correction procedure. See Reference [1] for details. The projection operator defined in this way is not a discrete projection; such approaches are referred to as approximate projection algorithms. For approximate projections, the projection step can be constructed using a number of analytically equivalent forms for V^*. The choice identified in Equation 13.6 was demonstrated to be best from a perspective of robustness and accuracy (see Reference [2] for details).

The algorithm presented above discretely conserves species mass (up to reactions) and enthalpy. However, because we use a linearized form of the constraint, the solution can drift off the constraint surface, p_0 =*constant*. To correct for this drift, we include a relaxation term to the compressibility expression used to compute U^{ADV}. This relaxation term is constructed to force the solution back toward the constraint without violating discrete conservation.

The adaptive time-step algorithm advances grids at different levels using time steps appropriate to that level based on CFL considerations. The procedure can most easily be thought of as a recursive algorithm, in which to advance level ℓ, $1 \leq \ell \leq \ell_{max}$ the following steps are taken:

- Advance level ℓ in time as if it is the only level. Supply boundary conditions for U, ρ, Y_m, h and π from level $\ell - 1$ if level $\ell > 1$, and from the physical domain boundaries.
  - Compute U^{ADV} including projection to enforce the constraint
  - Advance ρ, Y_m, h
  - Compute provisional U^*
  - Evaluate constraint, S^{n+1}
  - Apply generalized projection to compute U^{n+1} and $\pi^{n+1/2}$
- If $\ell < \ell_{max}$
  - Advance level $(\ell + 1)$ r times with time step $\Delta t^{\ell+1} = \frac{1}{r}\Delta t^{\ell}$ as indicated above.
  - Synchronize the data between levels ℓ and $\ell + 1$, and interpolate corrections to higher levels if $\ell + 1 < \ell_{max}$.
    - Solve elliptic synchronization equation for U^{ADV}
    - Compute changes for advection terms from explicit reflux and changes to U^{ADV}
    - Solve parabolic synchronization with right hand side representing reflux corrections from advection and diffusion
    - Apply synchronization projection to correct final velocity field. The right hand side includes terms representing the discrepancy from projecting levels independently and corrections to the velocity from the advection and diffusion synchronization.

13.5 Implementation Issues and Software Design

The combination of adaptive mesh refinement and the projection-based low Mach number formulation can considerably reduce the computational cost of reacting flow simulations; however, the computational demand can still be significant, particularly when modeling turbulent flame phenomena with detailed chemistry and transport. Consequently, we must be able to implement the adaptive low Mach number algorithm described above so that we can effectively utilize high-performance parallel computers. Before discussing the implementation in detail, we first comment on the impact of some of the choices we made in developing the basic algorithm on the design of the software. Our basic discretization strategy decomposes the problem into different mathematical components to treat advection, diffusion, chemical reactions, and projections. We use an explicit treatment of advection so that the implicit solves needed for diffusion and the projection represent discrete approximations to self-adjoint elliptic partial differential equations. Consequently, we can solve the requisite linear systems using geometric multigrid. Also, we have decomposed the dynamics so that the chemistry is advanced independent of the other processes. As a result the chemistry can be treated locally on a point-by-point basis.

Our choice of AMR strategy has a significant impact on software design. By adopting a block-structured form of AMR, the solution at each level in the hierarchy is naturally represented in terms of data defined on a collection of logically rectangular grid patches each containing a large number of points. Thus, the data is represented by a modest collection of relatively large regular data objects as compared to a point-by-point refinement strategy. This type of approach allows us to amortize the irregular aspects of an adaptive algorithm over large regular operations on the grid patches. This organization of data into large aggregate grid patches also provides a model for parallelization of the AMR methodology.

Our adaptive methodology is embodied in a hybrid FORTRAN software system. In this framework, memory management and flow control are expressed in the C++ portions of the program and the numerically intensive portions of the computation are handled in FORTRAN. The software is written using a layered approach, with a foundation library, BoxLib, that is responsible for the basic data container abstractions at the lowest levels, and a framework library, AMRLib, that marshals the components of the AMR discretization. Support libraries built on BoxLib are used as necessary to implement utility components, such as interpolation of cell and interface data between levels, and linear solvers used in the projections and diffusion solves.

In BoxLib, the fundamental parallel data abstraction is a MultiFab. A MultiFab is the union of a set of distributed blocks of FORTRAN-compatible data arrays. Each block is defined on a Box, which is defined by a pair of integer coordinate tuples that identify the lower and upper bounds of the region in a global index space. MultiFabs at each level of refinement are distributed independently. The software supports two data distribution schemes, as well as a dynamic switching scheme that decides which approach to use based on the number of grids at a level and the number of processors. The first scheme is based on a heuristic knapsack algorithm as described in Crutchfield [18] and in Rendleman [35]. The second is based on the use of a Morton-ordering space-filling curve. MultiFab operations are performed with an *owner computes* rule with each processor operating independently on its local data. For operations that require data owned by other processors, the MultiFab operations are preceded by a data exchange between processors.

Each processor contains *meta-data* that is needed to fully specify the geometry and processor assignments of the MultiFabs. At a minimum, this requires the storage of an array of boxes for each AMR level, the refinement ratio between levels, and the physical location of the origin of the base grid coordinate indices. In the parallel implementation, meta-data also includes the processor distribution of data, which is used to dynamically evaluate the necessary communication patterns to share data amongst the processors.

13.5.1 Performance of Adaptive Projection

Compared to a time-explicit numerical integration of the compressible reacting flow equations, the adaptive projection methodology for low Mach number reacting flows is enormously complex. The integration algorithm is a multi-stage procedure involving a number of elliptic/parabolic solves each time step. Dynamic regridding operations track developing features in the flow, and require frequent redistribution of the data and workload, and the workload itself is non-uniform due to the inhomogeneous nature of the reaction chemistry. Not only is it interesting to assess the parallel scalability of this algorithm to thousands of processors, it is also important to gauge the effectiveness of the low Mach number algorithm with respect to time-explicit simulation approaches directly, which historically have shown nearly ideal scaling behavior.

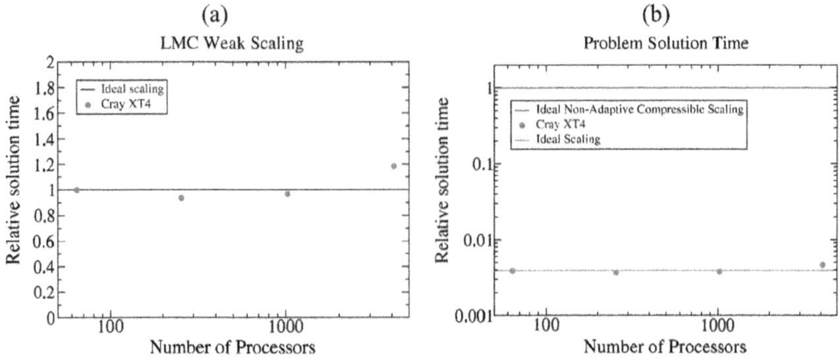

Fig. 13.9: Parallel performance of the low Mach number adaptive algorithm. (a) Weak scaling behavior of the adaptive low Mach number algorithm. (b) Scaling of the time-to-solution, relative to a compressible solver.

Here we consider a weak scaling study in which the relative amount of work per processor remains constant as we increase the number of processors. The study here is based on the propagation of a doubly-periodic wrinkled premixed methane-air flame in three dimensions using the GRI-Mech 3.0 [23] chemical mechanism for methane combustion. By replicating the problem in the periodic directions, we are able to scale the problem size without modifying the problem characteristics so that we can reliably test the behavior of the full AMR algorithm. Normalized computational times versus number of processors is presented in Figure 13.9(a). This figure shows a modest increase in execution time of approximately 20% as we increase the number of processors from 64 to 4096 on the XT4-type architecture. A more interesting metric than simple scalability is the relative performance of the low Mach number methodology to that of a time-explicit scheme. Figure 13.9(b) shows the time-to-solution for the same range of processor counts using our low Mach number algorithm compare to the idealized performance of a non-adaptive compressible reacting flow solver [6]. Note that for this study, we did not run the compressible code at all of the resolutions; we simply extrapolated the performance from 64 processors of assuming ideal scaling. The data shows that the low Mach number methodology is more than a factor of 200 faster than the compressible method. It is important to point out additionally that this estimate is quite conservative, since for many interesting 3D problems the finest level typically covers a far smaller fraction of the domain (the example presented below, for example, required the highest resolution on less than 4% of the domain).

13.6 Application – Lean Premixed Hydrogen Flames

The AMR algorithm design principles and BoxLib software library have been used to build a wide variety of PDE integration schemes, with applications in compressible and incompressible turbulence (jets, shear layers), astrophysics, porous media, fluctuating hydrodynamics, and reacting flows. The adaptive projection scheme has been particularly successful in studies of low Mach number reacting flows, such as astrophysical explosions and terrestrial combustion. In this section, we summarize a recent combustion study that serves to highlight the applicability of this AMR scheme to realistic systems with complex transport and chemistry models at the laboratory scale.

13.6.1 Background

Within the combustion community there is considerable interest in developing fuel-flexible burners that can be used to stabilize lean premixed flames in a stationary turbine designed for power generation. Low-swirl burner technology, originally introduced by Cheng and co-workers [14] as tool for studying the fundamental properties of lean, premixed turbulent flames, has the potential for meeting this need. Burners based on modifications of the original design have been used by a number

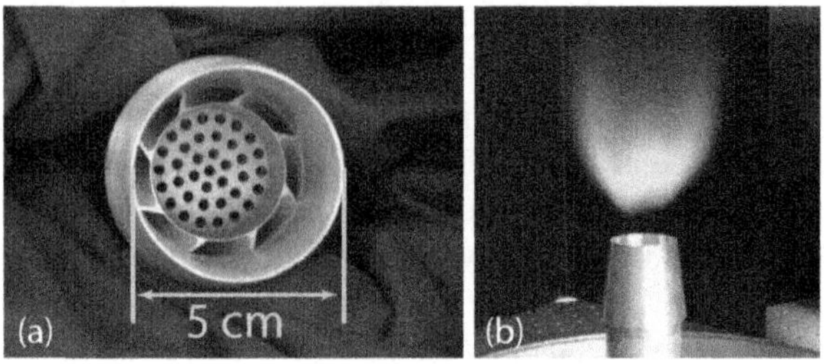

Fig. 13.10: (a) Low-swirl nozzle, showing vanes and turbulence generation plate. (b) typical turbulent low-swirl methane flame stabilized in divergent flow above the low-swirl nozzle. Images courtesy of Robert Cheng, LBNL.

of research groups [15, 25, 30, 32], and have the potential for use in the design of next-generation, lean premixed combustion systems, including those burning lean hydrogen at both at atmospheric and elevated pressures [16].

The low-swirl burner concept is extremely simple: premixed fuel exits a pipe after passing through a turbulence generation plate and an annular set of curved vanes as shown in Figure 13.10. The vanes impart a swirl component to the flow over a narrow layer near the pipe wall, and a detached premixed flame anchors in the diverging flow above the pipe exit. Turbulence in fuel stream wrinkles the flame, which enhances the overall rate of combustion in the device; the flames stabilize where the mean burning speed matches the axial flow velocity. Application of these types of burners, particularly for alternative fuels, depends on improving our understanding of basic flame structure, stabilization mechanisms, emissions and responses to changes in fuel. Numerical modeling has the potential to address some of these issues, but simulation of these types of burners has proven to be difficult because of the large range of spatial and temporal scales in the system; the bulk of the analysis to date has been experimental.

As noted in Bell et al. [7], the detailed structure of lean premixed flames becomes particularly important and difficult to simulate when burning hydrogen. Lean hydrogen-air flames burn in cellular structures—localized regions of intense burning, separated by regions of local extinction. In this regime, the flame surface is broken into discontinuous segments. This type of structure introduces severe difficulties in applying standard turbulence/chemistry interaction models, which are based on the presence of a highly wrinkled but continuous flame surface that propagates locally as an idealized laminar flame structure. In the absence of a suitably general model for the turbulent combustion of lean hydrogen-air mixtures, numerical simulations must incorporate sufficient detail in the chemical kinetics, differential species transport and turbulent fluid dynamics to capture all the important couplings in these flows.

13.6.2 Models and Setup

For this problem, we treat the fluid as a mixture of perfect gases. We use a mixture-averaged model for differential species diffusion, ignoring Soret, Dufour and radiative transport processes (see [21] for a complete discussion of this approximation). In this case, the diffusive fluxes in Equations 13.1 can be written

$$\mathscr{F}_m = \rho D_m \nabla Y_m, \quad \tau = 2\eta_{mix}\mathscr{S} - \frac{2}{3}\eta_{mix}\nabla\cdot U, \quad \mathscr{Q} = \lambda_{mix}\nabla T - h_m\mathscr{F}_m$$

where $\mathscr{S}$ is the symmetric part of strain tensor, D_m is the mixture-averaged mass diffusion coefficient for species m, and η_{mix} and λ_{mix} are the mixture-averaged viscosity and thermal conductivity, respectively. A lean hydrogen-air inlet fuel mixture (ϕ=0.37) was modeled with the hydrogen sub-mechanism of GRI-Mech 2.11. There are 9 chemical speices and 27 fundamental Arrhenius chemical reactions. The transport coefficients and thermodynamic relationships are obtained from EGLib [22].

An idealized flat unstretched steady 1D (the so-called "laminar flame") configuration provides scale factors that characterize this hydrogen-air flame: the thermal thicknesses is approximately 800 μm; the full half-width maximum of the fuel consumption layer is approximately 0.5 mm. The computational domain for this study measures 25 cm^3 (we aassume that this will place the computational boundaries sufficiently far from the flame as to not significantly affect its dynamics). The base mesh for the simulation is a uniform grid of 256^3 cells and we use 3 additional levels of factor-of-two grid refinement to track regions of high vorticity (turbulence) and reactivity (combustion). The flame is contained entirely within the finest level with an effective resolution of 2048^3. In previous work, we demonstrated that this level of resolution is adequate to capture the detailed structure of the flame including the peak fuel consumption, the thermal field and major species. Note that this level occupies less than 4% of the entire computational domain. Although evolution of the flow outside of the fine grid region certainly impacts on the flame dynamics, there are no features in the flow that require high resolution.

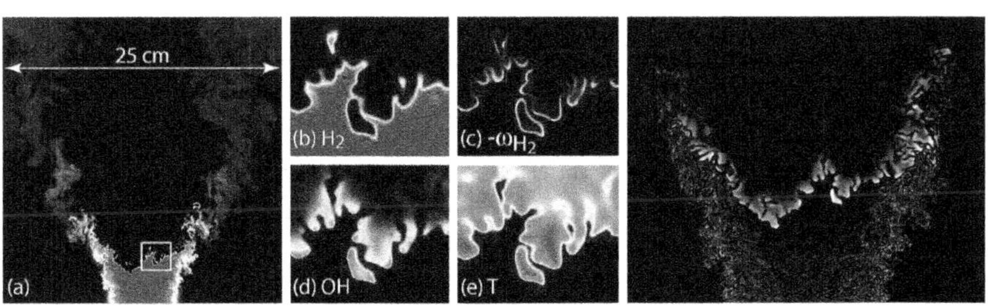

Fig. 13.11: Vertical slice of solution centered on nozzle axis (Frames (b)-(e) over 2.5 cm white box in (a)). (a-b) Mole fraction of H_2, (c) H_2 Consumption rate, (d) Mole fraction of OH, (e) temperature, and (f) Composite image (width = 8cm) depicts OH concentration (in orange) and the vorticity magnitude (in grey).

Flow from the low-swirl nozzle enters our computational domain centered on the bottom boundary. Profiles for the velocity components at the nozzle incorporate an experimental characterization provided by Petersson et al. [32]. On the inlet face outside the nozzle, a 35 cm/s upward coflow of cold air is specified. Turbulent fluctuations in the nozzle flow were prepared in an auxiliary simulation to have the experimentally measured intensity and integral length scale, and were added to the mean flows as a time-dependent boundary function. The remaining boundaries are outflow. The domain was intitially filled with air at standard conditions, except for a small volume of hot air above the nozzle. As the simulation progresss, the flame ignites, and propagates downstream in the flow until reaching a quasi-stationary position in the radially divergent flow field. The additional levels of grid refinement were added and the simulation continued until reaching a new quasi-steady configuration.

13.6.3 Simulation Results

In Figure 13.11, we show a cross-section through the middle of the simulation that provides a picture of the overall structure of the flame. In the upper part of the slice plane in Figure 13.11(a), one sees a faint cloud of unburnt fuel at low concentrations (blue). This represents fuel that has been sufficiently diluted with air that it is below the flammability limit. Figure 13.11(b-e) show various other fields near the flame surface. Turbulent fluctuations from two primary sources interact with the flame (see Figure 13.11(f)): plate turbulence from the inlet advects downstream and wrinkles the flame in the central core; a shear layer from the swirl entrains coflowing air from the sides and reduces fuel concentrations below flammability.

One of the principal diagnostics used in the experiments is OH-PLIF (planar laser-induced fluorescence) based on imaging the fluorescence of OH radicals excited by a tuned laser sheet. In Figure 13.12, we show a typical vertical slice of the OH concentration from the simulation alongside typical PLIF images from the low-swirl burner experiment at similar conditions. Figure 13.12(a) shows the profile of OH over the (25 cm)^2 slice, while Figures 13.12(b) and (d) show progressive enlargements of the data corresponding to the field of view (FOV) of the typical OH-PLIF data shown in Figures 13.12(c) and (e). The figure shows that the simulation captures with remarkable fidelity the primary features of these flames, including their distribution of sizes, shapes and global structure. The simulation also captures the observed variability of the OH signal (brightness on the experimental images) along the flame surface. Previous work on turbulent hydrogen flames shows that the high diffusivity of H_2 can lead to local enrichment of the fuel mixture along the front. This local enrichment can lead to intensification of local burning, which leads to increased OH that can be seen experimentally as a brigher signal in the OH-PLIF (and red regions in the simulation slice data).

From analysis of the simulation data, we can obtain a detailed characterization of the local flame structure that is not feasible from the experimental data. We begin

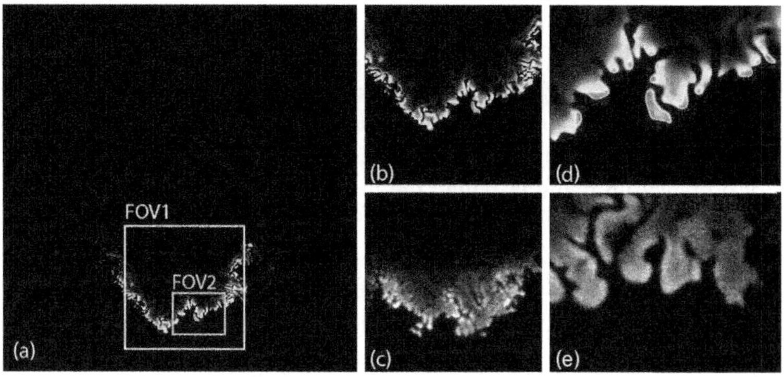

Fig. 13.12: (a) Typical slice of OH concentration on vertical midplane from the simulation. Red indicates high relative values, and are correlated with the flame front. The white box ("FOV1") represents the field of view of OH-PLIF measurements from the representative low swirl experiment focused on the large-scale flame structure. The smaller yellow box ("FOV2") represents the experimental field-of-view OH-PLIF measurements focused more tightly on the local structure of the flame surface (width of FOV2 is approximately 26 mm). (b) Zoom of profile in (a) over FOV1, (c) typical experimental OH-PLIF image over FOV1, (d) Zoom of profile in (a) over FOV2, and (e) typical experimental OH-PLIF image over FOV2.

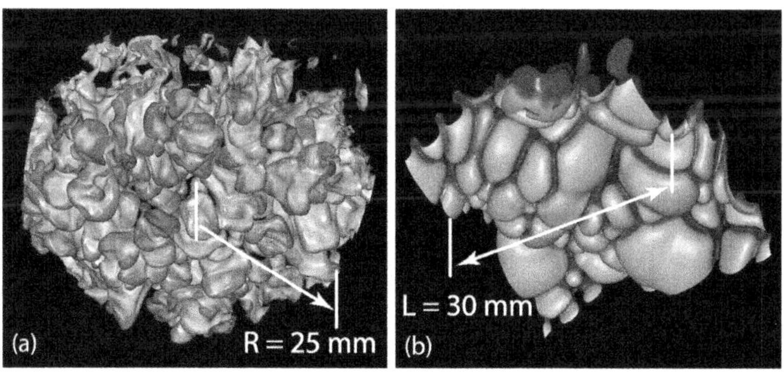

Fig. 13.13: Flame isotherm, T= 1144 K, colored by the local rate of H_2 consumption. (a) Low-swirl flame isotherm, conditioned on $r <$ 25mm, z <8mm. (b) Representative isotherm of an idealized ϕ=0.37 H_2-air flame propagating freely in uniform flow.

with an examination of the fuel consumption rate. As with the OH-PLIF data, the fuel consumption shows considerable variability along the flame front. To quantify this behavior, we extract the T=1144 K surface, which corresponds to that of the peak fuel consumption in the flat laminar flame at ϕ=0.37. In Figure 13.13, this isotherm is colored by local fuel consumption in the core of the burner. For comparison, we also show the analogous image for a freely propagating hydrogen flame (taken from the study in [20]). Both images are based on the same color map, and the length scales are as indicated in the figure. Comparing these images, one can see that in the turbulent flame, there are finer structures than in the non-turbulent case. The turbulent flame shows more ridge-like, cylindrical structures compared to the freely propagating flame in which the features are more spherical. Finally, the burning rate over most of the flame surface is considerably higher in the turbulent configuration, particularly where the flame is tightly folded.

To quantify the relationship between flame curvature and local burning speed we compute a local consumption-based flame speed along the isotherm. For this construction, we triangulate the isotherm, trimming away sections whether the fuel consumption is less than half the laminar flame value, and construct a local coordinate

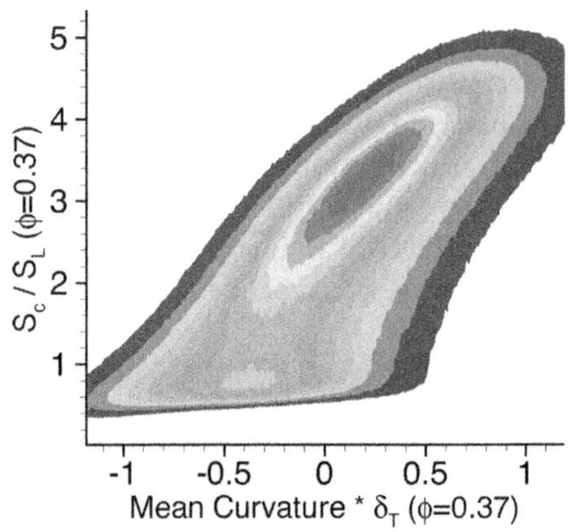

Fig. 13.14: Joint probability distribution function of s_c, with mean curvature.

system in a neighbor hood of the flame. We can then define a local consumption-based flame speed on the triangulated isotherm surface by integrating fuel consumption through a region normal to the flame. (See [20] for a detailed description of this construction.) In Figure 13.14 we show a joint PDF of mean curvature and s_c, normalized the laminar burning speed, s_L, of a flat steady hydrogen-air flame at ϕ=0.37. This figure shows a strong correlation between local burning speed and curvature. Regions of intense burning correspond to regions of positive curvature (positive curvature corresponds to regions where the center of curvature is on the products side of the flame interface). Also note that the dominant behavior corresponds to local burning speeds that range from 2 to 4 times the laminar flame speed. Even at zero curvature, the most probable local burning speed is three times the laminar flame speed. This type of behavior illustrates the challenges in developing suitable turbulence / chemistry interaction models for lean hydrogen flames and underscores the need to perform simulations of laboratory-scale turbulent flames.

13.7 Summary

In this chapter, we have discussed the development of parallel, adaptive solvers for reacting flows. The particular class of adaptive methods we consider are hierarchical block-structured methods that support subcycling in time. We have discussed the key ideas needed to design adaptive methods within this context and illustrated how those approaches are instantiated for various simple classical partial differential equations. We then showed how those basic discretization concepts can be interwoven to construct an adaptive low Mach number simulation capability. The basic algorithmic constructs associated with block-structured AMR motivate the design of a software framework to support parallel implementation of these algorithms. The overall synthesis of these ideas leads to an enhanced simulation capability that has made it possible to simulate realistic turbulent flames without using explicit models for turbulence or turbulence/chemistry interaction.

Acknowledgments

This work was support by the DOE Office of Advanced Scientific Computing Research under the U.S. Department of Energy under contract No. DE-AC02-05CH11231. The computations presented were performed on Franklin at NERSC as part of an INCITE award.

References

1. Almgren, A.S., Bell, J.B., Colella, P., Howell, L.H., Welcome, M.L.: A conservative adaptive projection method for the variable density incompressible Navier-Stokes equations. J. Comput. Phys. **142**, 1–46 (1998)
2. Almgren, A.S., Bell, J.B., Crutchfield, W.Y.: Approximate projection methods: Part I. inviscid analysis. SIAM J. Sci. Comput. **22**, 1139–1159 (2000)
3. Almgren, A.S., Bell, J.B., Szymczak, W.G.: A numerical method for the incompressible Navier-Stokes equations based on an approximate projection. SIAM J. Sci. Comput. **17**, 358–369 (1996)
4. Ascher, U., Petzold, L.R.: Projected implicit Runge Kutta methods for differential algebraic systems. SIAM J. Num. Anal. **28**, 1097–1120 (1991)
5. Bell, J., Berger, M., Saltzman, J., Welcome, M.: A three-dimensional adaptive mesh refinement for hyperbolic conservation laws. SIAM Journal on Scientific and Statistical Computing **15**, 127–138 (1994)
6. Bell, J., Day, M., Kuhl, A.L.: Numerical simulations of shock-induced mixing and combustion. In: 19th ICDERS. Hakone, Japan (July 27 – August 1, 2003)
7. Bell, J.B., Cheng, R.K., Day, M.S., Shepherd, I.G.: Numerical simulation of Lewis number effects on lean premixed turbulent flames. Proc. Combust. Inst. **31**, 1309–1317 (2007)
8. Bell, J.B., Colella, P., Glaz, H.M.: A second-order projection method for the incompressible Navier-Stokes equations. J. Comput. Phys. **85**, 257–283 (1989)
9. Bell, J.B., Day, M.S., Rendleman, C.A., Woosley, S.E., Zingale, M.A.: Adaptive low Mach number simulations of nuclear flame microphysics. J. Comput. Phys. **195**, 677–694 (2004)

10. Bell, J.B., Marcus, D.L.: A second-order projection method for variable density flows. J. Comput. Phys. **101**, 334–348 (1992)
11. Berger, M.J., Colella, P.: Local adaptive mesh refinement for shock hydrodynamics. J. Comput. Phys. **82**, 64–84 (1989)
12. Berger, M.J., Oliger, J.: Adaptive mesh refinement for hyperbolic partial differential equations. J. Comput. Phys. **53**, 484–512 (1984)
13. Brenan, K.E., Campbell, S.L., Petzold, L.R.: Numerical Solution of Initial-Value problems in Differential-Algrebraic Equations. SIAM, Philadelphia, PA (1996)
14. Chan, C.K., Lau, K.S., Chin, W.K., Cheng, R.K.: Freely propagating open premixed turbulent flames stabilized by swirl. Proc. Combust. Inst. **24**, 511–518 (1992)
15. Cheng, R.K.: Velocity and scalar characteristics of premixed turbulent flames stabilized by weak swirl. Combust. Flame **101**, 1–14 (1991)
16. Cheng, R.K., Littlejohn, D., Strakey, P.A., Sidwell, T.: Laboratory investigations of a low-swirl injector with h_2 and ch_4 at gas turbine conditions. Proc. Combust. Inst. **32**, 21–46 (2009)
17. Chorin, A.J.: Numerical solution of the Navier-Stokes equations. Math. Comp. **22**, 745–762 (1968)
18. Crutchfield, W.Y.: Load balancing irregular algorithms. Tech. Rep. UCRL-JC-107679, Lawrence Livermore National Laboratory (1991)
19. Day, M.S., Bell, J.B.: Numerical simulation of laminar reacting flows with complex chemistry. Combust. Theory Model. **4**, 535–556 (2000)
20. Day, M.S., Bell, J.B., Bremer, P.T., Pascucci, V., Beckner, V.E.: Turbulence effects on cellular burning structures in lean premixed hydrogen flames. Combust. Flame **156**, 1035–1045 (2009)
21. Ern, A., Giovangigli, V.: Multicomponent Transport Algorithms, *Lecture Notes in Physics* **24**, Springer-Verlag, Berlin (1994)
22. Ern, A., Giovangigli, V.: EGLIB: A General-Purpose Fortran Library for Multicomponent Transport Property Evaluations. J. Comput. Phys. **120**, 105–116 (2005)
23. Frenklach, M., Wang, H., Goldenberg, M., Smith, G.P., Golden, D.M., Bowman, C.T., Hanson, R.K., Gardiner, W.C., Lissianski, V.: GRI-Mech—an optimized detailed chemical reaction mechanism for methane combustion. Tech. Rep. GRI-95/0058, Gas Research Institute (1995). `http://www.me.berkeley.edu/gri_mech/`
24. Majda, A., Sethian, J.A.: The derivation and numerical solution of the equations for zero Mach number combustion. Combust. Sci. Technol. **42**, 185–205 (1985)
25. Mansour, M., Chen, Y.C.: Experimental Thermal Fluid Sci. **32**, 1390–1395 (2008)
26. McMurtry, P., Jou, W.H., Riley, J., Metcalfe, R.: Direct numerical simulations or a reacting mixing layer with chemical heat release. AIAA J. **24**, 962–970 (1986)
27. Najm, H.N., Knio, O.M., Paul, P.H., Wyckoff, P.S.: A study of flame observables in premixed methane-air flames. Combust. Sci. Technol. **140**, 369–403 (1998)
28. Najm, H.N., Wyckoff, P.S.: Premixed flame response to unsteady strain rate and curvature. Combust. Flame **110**, 92–112 (1997)
29. Najm, H.N., Wyckoff, P.S., Knio, O.M.: A semi-implicit numerical scheme for reacting flow. I. Stiff chemistry. J. Comput. Phys. **143**, 381–402 (1998)
30. Nogenmyr, K., Peterson, P., Bai, X.S., Nauert, A., Olofsson, J., Brackman, C., Seyfried, H., Zetterberg, J., Li, Z.S., Richter, M., Dreizler, A., Linne, M., Alden, M.: Large eddy simulation and experiments of stratified lean premixed methane/air turbulent flames. Proc. Combust. Inst. **31**, 1467–1475 (2007)
31. Pember, R.B., Howell, L.H., Bell, J.B., Colella, P., Crutchfield, W.Y., Fiveland, W.A., Jessee, J.P.: An adaptive projection method for unsteady, low-Mach number combustion. Combust. Sci. Technol. **140**, 123–168 (1998)
32. Peterson, P., Olofsson, J., Brackman, C., Seyfried, H., Zetterberg, J., Richter, M., Alden, M., Linne, M., Cheng, R., Nauert, A., Geyer, D., Dreizler, A.: Simultaneous PIV/OH PLIF, Rayleigh thermometry/OH PLIF and stereo PIV measurements in a low-swirl flame. Appl. Opt **46**, 3928–3936 (2007)

33. Qian, J., Tryggvason, G., Law, C.K.: Front tracking method for the motion of premixed flames. J. Comput. Phys. **144**, 52–69 (1988)
34. Rehm, R.G., Baum, H.R.: The equations of motion for thermally driven buoyant flows. N. B. S. J. Res. **83**, 297–308 (1978)
35. Rendleman, C.A., Beckner, V.E., Lijewski, M., Crutchfield, W.Y., Bell, J.B.: Parallelization of structured, hierarchical adaptive mesh refinement algorithms. Comput. Vis. Sci. **3**, 147–157 (2000)
36. Rutland, C., Ferziger, J.: Simulations of flame-vortex interactions. Combust. Flame **84**, 343–360 (1991)
37. Zhang, S., Rutland, C.J.: Premixed flame effects on turbulence and pressure-related terms. Combust. Flame **102**, 447–461 (1995)

Chapter 14
Wavelet Methods in Computational Combustion

Robert Prosser and R. Stewart Cant

Abstract Discretisation schemes based on the use of wavelet methods offer many potential advantages for the numerical simulation of combustion. In many cases of interest, flame structures are thin relative to the largest length scales of the problem and most length scales of the flow field, and so lend themselves to simulation using adaptive-mesh methods. Wavelet methods are naturally adaptive, in that the coefficients of the wavelet transform are non-zero only in regions where there is significant variation present in the solution. Hence, simple thresholding can be employed to make valuable savings in storage and in execution time. In this chapter, the basic principles of wavelet methods are established. Orthogonal and biorthogonal wavelet formulations are described and their advantages and disadvantages are discussed. An illustration of a wavelet-based discretisation scheme is provided using the Navier-Stokes momentum equation as an example. The same wavelet approach is applied to the simulation of a one-dimensional laminar premixed flame for which an asymptotic solution exists. Comparisons are made between the computational and analytical results and the accuracy of the wavelet approach is assessed. Extensions to higher dimensions are discussed. Finally, the current state of development of wavelet methods is outlined and conclusions are drawn.

14.1 Introduction

High-fidelity numerical simulation of turbulent combustion is a very demanding task. Techniques such as Direct Numerical Simulation (DNS) and Large Eddy Simulation (LES) require the flow field to be represented in three spatial dimensions

---

Robert Prosser
School of MACE, University of Manchester, Manchester M60 1QD, UK, e-mail: robert.prosser@manchester.ac.uk

R. Stewart Cant
Cambridge University, Cambridge, CB2 1PZ, UK, e-mail: rsc10@eng.cam.ac.uk

T. Echekki, E. Mastorakos (eds.), *Turbulent Combustion Modeling*,
Fluid Mechanics and Its Applications 95, DOI 10.1007/978-94-007-0412-1_14,

and cannot make use of statistical symmetries of the problem in the same manner as the traditional Reynolds-Averaged Navier Stokes (RANS) approach. Moreover, the flow field is evolving in time, and hence it is essential to use time-accurate solution methods in order to capture the flow, the flame and the full complexity of their interactions. All of this is computationally expensive, especially when the requirement is for full spatial and temporal resolution of all relevant phenomena, as is the case in DNS.

In turbulent flow, the computational expense follows directly from the range of length and time scales that must be represented. In combustion problems, it is most often the case that there are further length and time scales even shorter than those of the turbulent flow field. The very smallest scales are associated with the diffusion-reaction layers deep within the flame structure, and these tend to be highly localised in space, at least on an instantaneous basis. Here the computational expense arises mainly from the tendency of the flame structure to move around within the domain due to the effects of advection by the flow field and propagation due to heat conduction and molecular diffusion. This means that sufficient computational mesh support must be provided to ensure proper spatial resolution everywhere in the domain. The time-advancement algorithm must be applied at every spatial mesh point, and hence the computational cost is set.

There have been many attempts to exploit the localised nature of the flame structure using adaptive mesh refinement (AMR) techniques [9, 28]. Here, the computational mesh is refined locally in order to provide high resolution only in regions of the domain (such as within the flame) where there are steep spatial gradients. Conversely, the mesh can be made less dense in regions where there is little activity in the solution. Since the flame structure typically occupies only a small part of the total volume of the domain, considerable computational cost savings can be made without a net loss of resolution. Considerable progress has been made and the utility of the approach has been demonstrated. Nevertheless it is difficult to apply such techniques while retaining high-order accuracy, and there are major issues concerning their efficient implementation on massively-parallel computers.

Ideally, an adaptive spectral method is required which would allow for very high spatial accuracy coupled with a capability for dynamic local mesh refinement. Methods based on Fourier transforms have been used for many years in simulations of turbulence, offering excellent accuracy but lacking the flexibility and spatial localisation necessary for use in an adaptive manner. More recently, wavelet transforms have emerged as a possible framework in which to build a class of numerical methods offering high accuracy combined with solution adaption. Wavelets, unlike the more familiar Fourier basis functions, are localised in both the spatial and spectral domains, and have many other interesting properties which make them an attractive prospect for use in future combustion simulations.

This article introduces wavelet analysis in the context of combustion DNS, outlines the relevant mathematical background and describes the application of wavelet techniques within a suitable numerical solution method. Results are presented which indicate the power of the approach, and conclusions are drawn.

14.2 Wavelet Transforms

14.2.1 Orthogonal Wavelets

A wavelet transform can be interpreted in much the same way as a Fourier transform. Where the two differ is in the choice of the basis function employed. In the Fourier setting, the basis is chosen to be the complex exponential. In the wavelet transform, there are actually two basis functions: the *scaling function* and the *wavelet*. The scaling function can be interpreted as a low resolution band pass filter, while the wavelet represents the complementary high resolution band pass filter [21].

The wavelet decomposition is based on the repeated application of a two-scale relationship. In particular, we have [8]

$$\bigoplus_{i=-\infty}^{\infty} \mathbf{W}_i = L^2(\mathbb{R})$$
$$\mathbf{V}_J = \mathbf{V}_{J-1} \oplus \mathbf{W}_{J-1}$$
$$\mathbf{V}_i \perp \mathbf{W}_i \qquad (14.1)$$

where $\bigoplus$ represents the *direct sum*. Equation 14.1(a) indicates that $L^2(\mathbb{R})$ can decomposed into a family of *wavelet spaces* $\mathbf{W}_i$. Each wavelet space forms half of a partnership with a *scaling function space* $\mathbf{V}_i$. i characterizes the resolution of the spaces with $i \to \infty$ as the resolution is refined. The basis for the scaling function space is denoted $\phi_{i,k}(x) \equiv \phi\left(2^i x - k\right)$, and the basis for the wavelet space is $\psi_{i,k} \equiv \psi\left(2^i x - k\right)$. Factors of $\sqrt{2}$ can appear in these definitions, depending on the choice of normalization. On the real line, both the wavelet and scaling function exhibit scale and translation invariance.

The projection of $f(x) \in L^2(\mathbb{R})$ is accomplished by taking the inner product with respect to the basis, i.e.

$$f(x) = \sum_{i=-\infty}^{\infty} Q_i(f)(x),$$

where Q_i is the projector onto $\mathbf{W}_i$ which, in the case of *orthogonal* basis functions can be written as

$$Q_i(f)(x) = \sum_{k\in\mathbb{Z}} < f(u), \psi_{i,k}(u) > \psi_{i,k}(x). \qquad (14.2)$$

The two scale relation defined by Eq. 14.1 allows the wavelet projection to be written in terms of scaling function projections;

$$Q_i(f)(x) = \left(P_{i+1}(f) - P_i(f)\right)(x)$$

where

$$P_i(f)(x) = \sum_{k\in\mathbb{Z}} < f(u), \phi_{i,k}(u) > \phi_{i,k}(x). \qquad (14.3)$$

Equation 14.1(b) provides the basis for the practical implementation of a wavelet decomposition for finite domains. Assume for simplicity that we have a finite dimensional representation of some periodic function $f(x) \in [0,1)$, which is sampled on 2^J grid points $x_{J,k} = k2^{-J}$, $0 \le k < 2^J$. If $f(x)$ is approximated by $P_J(f)(x)$ then an application of the two scale relation allows us to write

$$P_J(f) = (P_{J-1} + Q_{J-1})(f)(x). \tag{14.4}$$

The orthogonality of $\mathbf{V}_i$ and $\mathbf{W}_i$ implies that the two projectors P_{J-1} and Q_{J-1} lead to complementary representations of $P_J(f)$ on reduced dimensional spaces (in this case, each representation is defined on grids of resolution 2^{J-1}). In the first pass of the transform then, a vector of length 2^J is replaced with 2 vectors of length 2^{J-1}, but the key observation here is that the wavelet projection $Q_{J-1}(f)$ is *sparse*, by which we mean that many of the 2^{J-1} coefficients arising from the wavelet projection are close to zero. The defining property of the wavelet transform is that the resulting coefficients are only non-trivial when the analyzing wavelets are close to regions of rapid change. Such a feature makes wavelets a natural tool with which to explore flame structures, wherein two relatively static regions (reactants and products, or fuel and oxidiser) are separated by a region of rapid change (the flame). The sparsity of the wavelet representation is exploited via *thresholding*, in which those wavelet coefficients with a magnitude less than a user specified threshold ε are discarded with a minimal loss of accuracy (in a sense that can be made precise, i.e. see [12]). Hence, the two scale representation with thresholding allows us to replace a vector of dimension 2^J, with one comprising $O(2^{J-1})$ components—$P_{J-1}(f)$ contains 2^{J-1} non-zero components, and Q_{J-1} typically contains a much reduced number of 'large' wavelet coefficients. The actual number of retained coefficients depends on the smoothness of the analyzed function $f(x)$.

Equation 14.4 can be repeatedly applied to $P_{J-1}(f)$ to obtain

$$\begin{aligned} P_J(f)(x) &= (P_{J-2} + Q_{J-2} + Q_{J-1})(f)(x) \\ &\vdots \\ &= P_0(f)(x) + \sum_{i=0}^{J-1} Q_i(f)(x) \end{aligned} \tag{14.5}$$

Equation 14.5 is the finite dimensional equivalent of Eq. 14.1(a). In practical calculations involving wavelets on the interval, the lower limit appearing in the sum of Eq. 14.5 is larger than zero. This reflects the need to keep the support of the wavelets smaller than the discretized interval. One final point to note is that the mapping $f(x) \rightarrow P_J(f)(x)$ requires an initial projection quadrature. This quadrature is usually approximate—many authors choose instead to use the sample values of the initial discretisation as the set of scaling function coefficients.

The implementation of the wavelet transform can take a number of forms. For some basis functions, the transform can be accomplished by a modified FFT [23]. In the most common approach (which makes use of Daubechies compact wavelet

[8]), the transform takes the form of a repeated finite difference-like operation (once for each space on each resolution). The weights associated with these operations are the *quadrature mirror filter coefficients* defining the wavelet. Examples of the implementation of wavelet transforms can be found in e.g. [24].

14.2.2 Biorthogonal Wavelet Transforms

Much of the utility of the wavelet transform emerges from the unique properties of the chosen wavelet, but there are also a number of problems with the orthogonal representations described in the previous section. In the orthogonal setting, many different choices of wavelet exist, some of which are more suitable for CFD applications than others. From a purely practical point of view many choices of wavelets do not have compact support (i.e. [23])—this implies that the inner products in Eqs. 14.2 and 14.3 effectively contain infinitely many quadrature filter coefficients and become difficult to evaluate. Such wavelets can be approximated as having compact support, but the resulting approximations lose their exact (to machine precision) orthogonality. Families of orthonormal wavelets with compact support do exist—the most famous examples being those of Daubechies [8]—but the orthogonal restraint leads to wavelets which are asymmetric.

For the simulation of fluid mechanics problems, asymmetric bases are undesirable as they introduce chirality into the numerical approximations for the governing equations [19, 20]. In addition the initial projection quadrature, which maps $f(x) \to P_J(f(x))$, is non-trivial and usually irreversible (to machine precision). In the setting of a collocation numerical scheme (or indeed any approach to the approximation of non-linear PDEs) the continual mapping to and from transformed representations leads to strong chirality and—eventually—instability in the numerical solution [25].

One approach to bypass the problems associated with chirality is to modify the wavelets by relaxing the constraint on orthogonality, to produce *biorthogonal* wavelet systems. Most of the relations described in the previous section hold for biorthogonal systems, but *two* sets of basis functions are required—the so-called *primal* and *dual* bases [6]. The scaling function and wavelet projectors are then written as

$$P_J(f) = \sum_{k \in Z} < f(u), \widetilde{\phi}_{J,k}(u) > \phi_{J,k}(x)$$

$$Q_J(f) = \sum_{k \in Z} < f(u), \widetilde{\psi}_{J,k}(u) > \psi_{J,k}(x),$$

where the tildes refer to the dual quantities. Biorthogonal wavelets contain sufficient flexibility to provide compact bases with symmetry—such considerations are particularly important in the construction of edge wavelets for bounded intervals [7]. The problem of the initial projection quadrature remains.

14.2.3 Second Generation Wavelets

A variant of the biorthogonal wavelet can be obtained via so-called *second generation wavelets*, pioneered by Sweldens and Donoho [12, 14, 27, 29, 30]. These wavelets are derived without recourse to Fourier transforms, and provide a simpler framework in which to define wavelets for more general settings than the real line. In addition, the classical biorthogonal wavelets of Cohen et al. [7] can be derived using the second generation approach. Finally, second generation wavelets circumvent the problems associated with an inexact initial projection quadrature.

There are two classes of wavelets proposed in the second generation framework: *interpolating wavelets* and *average interpolating wavelets* (i.e. see [27]) In this work, we choose the family of interpolating wavelets derived from the fundamental solutions discussed by Deslauriers and Dubuc [11]. In the second generation interpolating wavelet approach, the dual scaling function is defined by

$$\tilde{\phi}_{J,k} = \delta\left(x - k2^{-J}\right),$$

where $\delta(\cdot)$ is the Dirac delta function. The scaling functions and wavelets have compact support and are symmetric in the interior of a domain, in particular

$$\begin{aligned} \text{support}\left(\phi_{j,k}\right) &= \left[2^{-j}(m-(N-1)), 2^{-j}(m+(N-1))\right], \\ \text{support}\left(\psi_{j,k}\right) &= [2^{-(j+1)}((2k+1)-(N-1)), 2^{-(j+1)}((2k+1)+(N-1))]. \end{aligned}$$

In this sense, we say that the coefficients $s_{j,k}$ and $d_{j,k}$ are respectively associated with the grid point $k2^{-1}$ and $(2k+1)2^{-(j+1)}$. Similar arguments can be made for the near boundary constructions, but the definitions become more algebraically complicated.

The dual scaling function (and indeed, the dual wavelet) are not members of $L^2(\mathbb{R})$, and the resulting transforms are limited in their range of application to smooth functions [12]. An alternative way of saying this is that the resultant wavelets do not strictly qualify as wavelets, since they do not satisfy the admissibility condition [8]

$$\int \psi(x)\,dx \neq 0.$$

The result of this inequality is that the wavelets are prone to aliasing, and do not conserve the integral of the original function:

$$\int P_i(f)\,dx \neq \int P_m(f)\,dx \quad i \neq m. \tag{14.6}$$

For classes of smooth function, this problem does not appear to be especially significant [12]; such conditions are typically encountered in the DNS of reacting flow wherein the smoothness of the resolved profiles—coupled with the limited number of allowable subspace decompositions—appears to minimize the effect of aliasing.

The projection of $f(x)$ onto the scaling function space leads to scaling function coefficients that are samples of the original function;

$$\begin{aligned} P_J(f)(x) &= \sum_{k\in\mathbb{Z}} < f(u), \delta\left(u - k2^{-J}\right) > \phi_{J,k}(x) \\ &= \sum_{k\in\mathbb{Z}} f\left(k2^{-J}\right)\phi_{J,k}(x) = \sum_{k\in\mathbb{Z}} f_{J,k}\phi_{J,k}(x). \end{aligned}$$

It follows that $f_{J-1,k} = f_{J,2k}$ and hence successive scaling function approximations are subsamples of the original discretisation. The wavelet transform is given by

$$\begin{aligned} Q_i(f)(x) &= \sum_{k\in Z} d_{i,\alpha}\psi_{i,\alpha}(x) \\ &= \left(P_{i+1}(f) - P_i(f)\right)(x) \\ &= \sum_{k\in\mathbb{Z}} f_{i+1,k}\phi_{i+1,k}(x) - \sum_{m\in\mathbb{Z}} f_{i,m}\phi_{i,m}(x) \\ d_{i,\alpha} &= Q_i(f)\big|_{x=(2\alpha+1)2^{-(i+1)}} \\ &= f_{i+1,2a+1} - \sum_{m\in\mathbb{Z}} \phi\left(\alpha + \frac{1}{2} - m\right) f_{i,m}. \end{aligned} \tag{14.7}$$

Clearly, the sparsity of the representation is governed by the properties of $\phi(\cdot)$. Furthermore, if the scaling function has compact support, it follows that $\phi(\alpha + 1/2 - m)$ comprises only finitely many non-zero entries, and Eq. 14.7 can be implemented as a finite difference-like operation.

14.3 Wavelets as a Method for DNS

We describe a collocation approach to the integration of the governing equations. The collocation strategy—where the wavelets are used merely to calculate derivatives [5]—provides a natural approach, since solution algorithms couched in terms of the transformed variables face formidable difficulties in evaluating the nonlinear terms associated with both convection and chemical reaction. For a 1-D flow domain Ω, we initially discretize the governing equations on to a regular mesh $G \subset \Omega$. As the solution evolves, we want to solve the equations only on those parts of the domain $G^{\varepsilon} \subset G$ where strong variations in the flow behaviour occur. Such behaviour might be found in, say, turbulent shear layers, or rapid temperature changes associated with chemical reaction. Wavelets are very good at picking up these regions of change, so a natural choice for the unstructured grid might be

$$G^{\varepsilon} = \{x_{i,\alpha} : |d_{i,\alpha}| > \varepsilon\}$$

where $d_{i,\alpha}$ is the wavelet coefficient associated with the point $x_{i,\alpha}$. If we try to calculate the derivatives on this mesh using traditional means, then we are left with the classical hanging node problem—those points at the outskirts of the unstructured mesh do not have sufficient neighbours to calculate their associated derivatives accurately. In the multiresolution approach however, we first calculate the wavelet expansion of the solution on G^{ε}. By construction, the coefficients on $G\backslash G^{\varepsilon}$ are small in magnitude (defined here as $|d_{i,\alpha}| < \varepsilon$) and may be omitted when calculating the approximate derivative. The resulting derivative can then be inverted on to the unstructured grid, and the solution time advanced as normal. The difficulty in this approach lies now in calculating the wavelet transform on the unstructured mesh. Methods for achieving this are available (i.e. [26]), but are too involved to describe here.

For ease of discussion, we will describe a wavelet collocation discretization of the momentum equation in what follows. The principles remain quite general however, and can be applied to the entire coupled reacting Navier-Stokes system. In one-dimensional problems and neglecting density variations, the momentum equation can be written as

$$\frac{\partial u}{\partial t} + u\frac{\partial u}{\partial x} + \rho^{-1}\frac{\partial p}{\partial x} = \rho^{-1}\frac{\partial \tau_{xx}}{\partial x}.$$

Our goal is to express this equation in terms of projected variables $P_J(\rho)$ and $P_J(u)$. Projecting the entire equation directly onto $\mathbf{V}_J$ leads to

$$\frac{\partial P_J(u)}{\partial t} + P_J\left(u\frac{\partial u}{\partial x}\right) + \rho^{-1}P_J\left(\frac{\partial p}{\partial x}\right) = \rho^{-1}P_J\left(\frac{\partial \tau_{xx}}{\partial x}\right). \tag{14.8}$$

The unsteady term does not provide any difficulties, since the time derivative commutes with the projector. Problems arise with the convective terms because: (a) we require an approximation for the derivative itself and (b) there is a non-linear product to incorporate.

As a model for the effect of non-linearities, consider a one dimensional velocity field $u(x)$ projected onto $\mathbf{V}_J$:

$$P_J(u) = \sum_{m\in Z} u_{J,m}\phi_{J,m}(x).$$

Consider further the non-linear term $P_J(u)P_J(u)$, which can be obtained as part of the convective term in Eq. 14.8. The explicit representation of this term is

$$P_J(u)P_J(u) = \sum_{m\in Z}\sum_{n\in Z} u_{J,m}u_{J,n}\phi_{J,m}(x)\phi_{J,n}(x) \notin \mathbf{V}_J. \tag{14.9}$$

In order to be consistent with the time derivative term, we must re-project Eq. 14.9 to obtain

$$\begin{aligned} P_J(P_J(u)P_J(u)) &= \sum_{\alpha\in Z}\sum_{m\in Z}\sum_{n\in Z} u_{J,m}u_{J,n} < \phi_{J,m}(\widetilde{x})\phi_{J,n}(\widetilde{x}), \widetilde{\phi}_{J,\alpha}(\widetilde{x}) > \phi_{J,\alpha}(x), \\ &= \sum_{\alpha\in Z} u_{J,\alpha}u_{J,\alpha}\phi_{J,\alpha}(x). \end{aligned} \tag{14.10}$$

More important is the effect of this non-linearity on spaces other than $\mathbf{V}_J$. For example, projecting Eq. 14.9 on to $\mathbf{W}_J$, we find

$$Q_J\left(P_J\left(u\right)P_J\left(u\right)\right) = \sum_{\alpha\in Z}\sum_{m\in Z}\sum_{n\in Z} u_{J,m}u_{J,n} < \phi_{J,m}\left(\widetilde{x}\right)\phi_{J,n}\left(\widetilde{x}\right), \widetilde{\psi}_{J,\alpha}\left(\widetilde{x}\right) > \psi_{J,\alpha}\left(x\right). \tag{14.11}$$

The normal consequence of the biorthogonality between $\mathbf{V}_J$ and $\mathbf{W}_J$ (i.e. the relation $Q_J\left(P_J\left(\cdot\right)\right) = 0$) does not hold here, since we have two scaling functions. Hence

$$< \phi_{J,m}\left(\widetilde{x}\right)\phi_{J,n}\left(\widetilde{x}\right), \widetilde{\psi}_{J,\alpha}\left(\widetilde{x}\right) > \neq 0 \tag{14.12}$$

in general, even without explicitly evaluating Eq. 14.11, we see that the nonlinearity has extended the representation of $u\left(x\right)$ from $\mathbf{V}_J$ to $\mathbf{V}_{J+1}$ by populating $\mathbf{W}_J$. Similar arguments can be made for other wavelet spaces. This behaviour is the wavelet analogue of the spectral spreading caused by the convection driven convolution arising in Fourier based pseudo-spectral methods [5], and has been studied in some detail by Beylkin [3].

If $u\left(x\right)$ is sufficiently smooth, then the high resolution wavelet coefficients will be small in magnitude. Furthermore, their interactions (governed by the coupling matrix in Eq. 14.12) will also produce only small effects. It is natural then for a DNS exploiting wavelet thresholding to assume $\mathbf{V}_J$ is of a sufficiently high resolution that the additional components arising from the non-linearity are small with reference to the size of the non-trivial coefficients. We can then replace each of the terms in Eq. 14.8 by their projected counterparts i.e.

$$\begin{aligned} P_J\left(u\frac{\partial u}{\partial x}\right) &\simeq P_J\left(u\right)P_J\left(\frac{\partial u}{\partial x}\right) \\ &\simeq P_J\left(u\right)P_J\left(\frac{\partial}{\partial x}P_J^{-1}P_J\left(u\right)\right) \end{aligned} \tag{14.13}$$

where the identity operator has been decomposed as $I = P_J P_J^{-1}$. As before, this expression will contain terms in both $\mathbf{V}_J$ and $\mathbf{W}_J$. To express the resulting term consistently, then, we write equation 14.13(b) as

$$P_J\left(P_J\left(u\right)\left(P_J\left(\frac{\partial}{\partial x}\right)P_J^{-1}P_J\left(u\right)\right)\right).$$

Using the interpolating properties of the basis, this becomes

$$P_J\left(P_J\left(u\right)\left(P_J\left(\frac{\partial}{\partial x}\right)P_J^{-1}P_J\left(u\right)\right)\right) = P_J\left(u\right)\left(P_J\left(\frac{\partial}{\partial x}\right)P_J^{-1}P_J\left(u\right)\right),$$

and we acknowledge that some information transmitted by this operation to $\mathbf{W}_J$ is lost. Similar expressions can be derived for the simpler, linear viscous and pressure terms.

14.3.1 The Wavelet Representation of the Derivative

The term

$$P_J\left(\frac{\partial}{\partial x}\right)P_J^{-1} \tag{14.14}$$

provides the representation of the differential operator on $\mathbf{V}_J$; Fig. 14.1 shows the $O\left(2^J\right)$ non-zero elements for this operator.

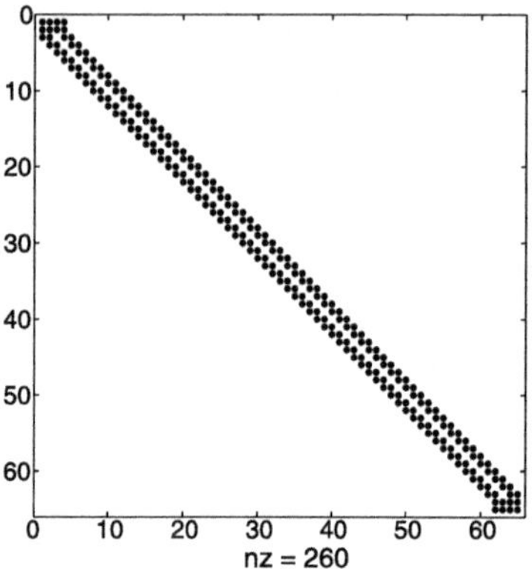

Fig. 14.1: Structure of $\mathbb{D}$ approximating $\frac{d}{dx}$ on $\mathbf{V}_J$. Reprinted from [26] wither permission from the Institution of Mechanical Engineers.

The derivative itself can be calculated as

$$\begin{aligned}\left(P_J\frac{\partial}{\partial x}P_J^{-1}\right)P_J\left(f\right)\big|_{x_{J,\alpha}} &= < \widetilde{\phi}_{J,\alpha}, \frac{\partial}{\partial x}\phi_{J,\beta} > P_J\left(f\right)\big|_{x_{J,\beta}}, \\ &= \mathbb{D}_{\alpha,\beta}\, P_J\left(f\right)\big|_{x_{J,\beta}}.\end{aligned}$$

In practice, the band diagonal structure of $\mathbb{D}_{\alpha,\beta}$ lends itself to a finite difference like implementation.

In order to take advantage of the sparsity in the wavelet representation of the flow field, the representations of both the operator and the flow field need to be expressed in terms of a multi-scale decomposition. Repeatedly applying Eq. 14.1 to Eq. 14.14 leads to the so-called standard decomposition of $\partial/\partial x$ [2, 4]:

$$\mathbb{D}^w = \left(P_j + \sum_{i=j}^{J} Q_i\right)\left(\frac{\partial}{\partial x}\right)\left(P_j + \sum_{i=j}^{J} Q_i\right)^{-1}.$$

This operator comprises a family of band diagonal sub-matrices which are ordered as in Fig. 14.2.

In terms of implementation, we have

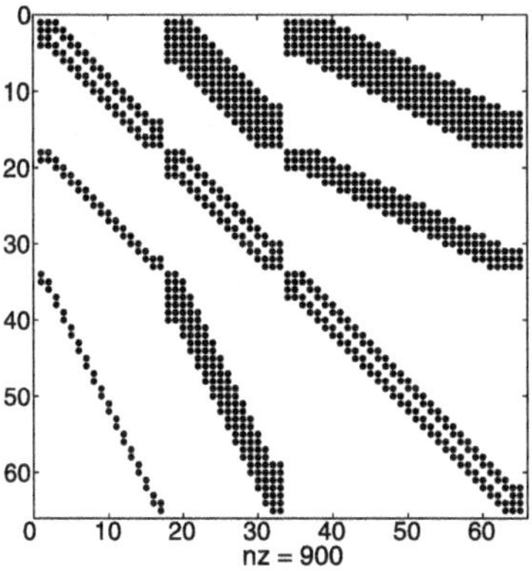

Fig. 14.2: Structure of $\mathbb{D}^w$ approximating $(P_{J-2} + \sum_{i=J-2}^{J-1} Q_i)\frac{d}{dx}(P_{J-2} + \sum_{i=J-2}^{J-1} Q_i)^{-1}$. Reprinted from [26] with permission from the Institution of Mechanical Engineers.

$$\left(P_j + \sum_{i=j}^{J} Q_i\right) \frac{\partial f}{\partial x}\bigg|_{x_{J,\alpha}} \simeq \mathbb{D}^w_{\alpha,\beta} \left\{P_j + \sum_{i=j}^{J} Q_i\right\} (f)|_{x_{J,\beta}}.$$

$\mathbb{D}^w_{\alpha,\beta}$ is referred to as the standard decomposition of $\frac{\partial}{\partial x}$ [4] and typically comprises $O\left(2^J \log_2\left(2^J\right) = J \log 2^J\right)$ non-zero elements.

14.3.2 Higher Dimensional Discretizations

A natural generalisation to the one dimensional wavelets explored in this paper is achieved by deriving multidimensional analogues from first principles (i.e. [22]). For most practical purposes, however, multidimensional generalisations are derived via tensor products of one dimensional transforms, i.e. for a two dimensional discretisation, we define $\mathbf{V_J^{(2)}}$ using

$$\begin{aligned} \mathbf{V_J^{(2)}} &= \mathbf{V_{J-1}^{(2)}} \oplus \mathbf{W_{J-1}^{(2)}} \\ \mathbf{V_J^{(2)}} &- \mathbf{V_J^{(x)}} \otimes \mathbf{V_J^{(y)}}. \end{aligned} \tag{14.15}$$

By applying a two scale decomposition in each spatial direction to Eq. 14.15(b), we obtain

$$\mathbf{V_J^{(2)}} = \left(\mathbf{V_{J-1}^{(x)}} \oplus \mathbf{W_{J-1}^{(x)}}\right) \otimes \left(\mathbf{V_{J-1}^{(y)}} \oplus \mathbf{W_{J-1}^{(y)}}\right) \tag{14.16}$$

which when expanded, leads to the multidimensional wavelet space

$$\mathbf{W}^{(2)}_{\mathbf{J-1}} = \left(\mathbf{W}^{(x)}_{\mathbf{J-1}} \otimes \mathbf{V}^{(y)}_{\mathbf{J-1}}\right) \oplus \left(\mathbf{V}^{(x)}_{\mathbf{J-1}} \otimes \mathbf{V}^{(y)}_{\mathbf{J-1}}\right) \oplus \left(\mathbf{W}^{(x)}_{\mathbf{J-1}} \otimes \mathbf{W}^{(y)}_{\mathbf{J-1}}\right). \qquad (14.17)$$

Clearly, the wavelet space actually contains three contributions; the 'pure' wavelet space $\left(\mathbf{W}^{(x)}_{\mathbf{J-1}} \otimes \mathbf{W}^{(y)}_{\mathbf{J-1}}\right)$, and two 'cross' spaces: $\left(\mathbf{W}^{(x)}_{\mathbf{J-1}} \otimes \mathbf{V}^{(y)}_{\mathbf{J-1}}\right)$ and $\left(\mathbf{V}^{(x)}_{\mathbf{J-1}} \otimes \mathbf{W}^{(y)}_{\mathbf{J-1}}\right)$. Equation 14.16 may be further decomposed by splitting $\mathbf{V}^{(x)}_{\mathbf{J-1}}$ and $\mathbf{V}^{(y)}_{\mathbf{J-1}}$ into their respective scaling function and wavelet spaces of resolution $J-2$. The appearance of the two dimensional decomposition shares many similarities to the standard decomposition of $\frac{d}{dx}$. An example is provided in Figs. 14.3 and 14.4, which show an instantaneous realisation of the kinetic energy from a two dimensional 'turbulent' flow, and its associated multiresolution decomposition. In this case, $J = 8$, and there are 4 subspaces in each of the $x-$ and $y-$ directions. The lower left corner of Fig. 14.4 depicts an approximation to the original signal on $\mathbf{V}^{(2)}_{\mathbf{4}}$. Each of the subsequent subblocks growing out to the top right of the figure are associated with $\mathbf{W}^{(x)}_{\mathbf{i}} \otimes \mathbf{W}^{(y)}_{\mathbf{i}}$. The blocks above this diagonal correspond to spaces of the form $\mathbf{V}^{(x)}_{\mathbf{i}} \otimes \mathbf{W}^{(y)}_{\mathbf{m}}$ while those below correspond to $\mathbf{W}^{(x)}_{\mathbf{i}} \otimes \mathbf{V}^{(y)}_{\mathbf{m}}$. The particular point to note is the sparsity in the wavelet representation: away from the coarsest representations, the magnitudes of the wavelet coefficients are close to zero.

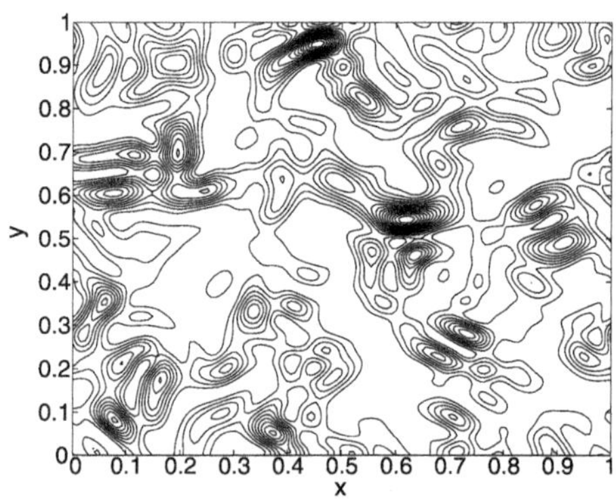

Fig. 14.3: Kinetic energy contours for 2D turbulence.

The practical application of wavelets to the Direct Numerical Simulation of reacting flows faces a number of difficulties. The non-linear terms associated with the chemical reaction rate (and, to a lesser extent, the convective term) effectively preclude the evolution of the simulation in the transform domain. Consequently, the transport equations must be time advanced in physical space. The wavelets are then used as part of a collocation strategy: (a) to inform the grid tracking algorithm of where points are required, and; (b) to calculate the derivatives on this reduced grid. The algorithmic difficulty is then one of correctly calculating the wavelet transform on a generally sparse grid.

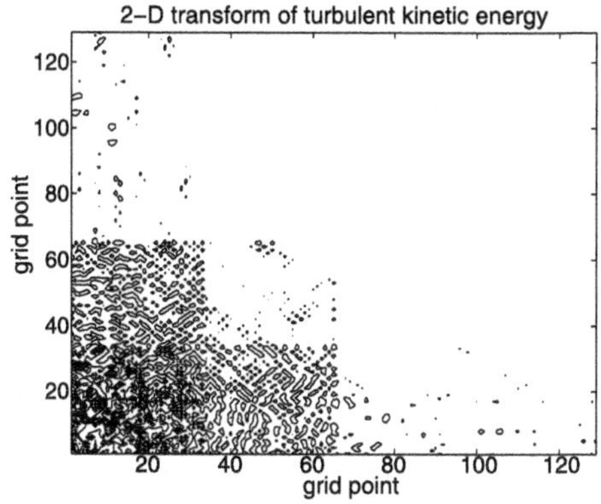

Fig. 14.4: 2D Wavelet transform of turbulent kinetic energy.

Most wavelet transforms are 'top-down' algorithms; the wavelet transform is obtained by the repeated application of quadrature mirror filters in a finite difference like algorithm. Each application of the filters to a high resolution signal produce two complementary low resolution signals (the scaling function coefficients and the wavelet coefficients). On a sparse grid, the so-called hanging node problem appears—grid points retained at the edge of a region of rapid change do not contain sufficient neighbour nodes to calculate their wavelet coefficients. The hanging node problem can be circumvented by a number of methods in 1-D (i.e. [31]), but the extension to 2-D is difficult. In the latter case, there is no unique way in which to order the wavelet coefficients, and multidimensional transforms lose their commutivity in the presence of a hard non linear threshold (i.e. a 2-D transform comprising $x-$ followed by $y-$ transforms will yield different coefficients to a 2-D transform comprising $y-$ followed by $x-$ transforms). Practical experience has shown that—particularly in low Mach number flows—the error induced by the hanging node problem manifests itself as a small perturbation in the dependent variables. These perturbations are picked up by the grid tracking algorithm as physical entities, and subsequently allowed to evolve. On the *new* hanging nodes thus formed, new perturbations are introduced and the grid grows again to accommodate these new features. The process repeats until the full grid is retained. Circumventing this problem forms part of ongoing research efforts.

14.4 An Application of Wavelets to Reacting Flows

14.4.1 Governing Equations

The equations governing a chemically reacting flow in one dimension are given by

$$\begin{aligned}
\frac{\partial \rho}{\partial t} + \frac{\partial}{\partial x}(\rho u) &= 0 \\
\frac{\partial \rho u}{\partial t} + \frac{\partial}{\partial x}(\rho u u + p) &= \frac{\partial}{\partial x}(\tau_{xx}) \\
\frac{\partial \rho E}{\partial t} + \frac{\partial}{\partial x}(u(\rho E + p)) &= -\frac{\partial q}{\partial x} + \frac{\partial}{\partial x}(u\tau_{xx}) \\
\rho \frac{\partial \rho Y_l}{\partial t} + \frac{\partial}{\partial x}(\rho u Y_l) &= \omega_l + \frac{\partial}{\partial x}\left(\rho D \frac{\partial Y_l}{\partial x}\right) \qquad l = 1,2,\ldots,N_s - 1.
\end{aligned} \tag{14.18}$$

where $N_s > 1$ is the number of species, τ_{xx} is the viscous stress and the heat flux vector q is defined as

$$q = -\lambda \frac{\partial T}{\partial x} - \sum_{l=1}^{N_s} h_l \rho D \frac{\partial Y_l}{\partial x}. \tag{14.19}$$

For simplicity, the chemical reaction is assumed to comprise the single step

$$\text{Reactants} \rightarrow \text{Products}$$

in which case, $N_s = 2$ and the thermochemical state of the gas is characterized by a *progress variable*. The progress variable is interpreted as a normalized product mass fraction, and takes a value of 0 in the reactants and 1 in the products. $c = Y_1$ here and $Y_2 = 1 - Y_1$. The reaction rate controlling the production of c is given by [32]

$$\omega_1 = \rho B^* (1 - c) \exp\left(\frac{-\beta\left(1 - \widehat{T}\right)}{1 - \alpha\left(1 - \widehat{T}\right)}\right),$$

with $\omega_2 = -\omega_1$. B^* is the pre-exponential factor (taken to be $285.1 \times 10^{-3} s^{-1}$ here), $\beta (= 6)$ is the Zeldovich number and $\widehat{T}$ is the reduced temperature;

$$\widehat{T} = \frac{T - T_0}{T_{ad} - T_0}.$$

T_0 and T_{ad} are the unburned reactant and adiabatic product temperature, respectively. α is related to the heat release of the fuel, and is set here to 0.8. This corresponds to an adiabatic flame temperature of 1500 K for an inlet temperature of 300 K. The reaction rate has been adjusted to give a laminar flame speed of $\sim$ 30 cm/s, a value typical of many hydrocarbon-air flames. The inlet velocity is set equal to the laminar flame speed, so that a stationary flame profile is obtained. This specification gives a simulation Mach number of $O\left(10^{-3}\right)$ based on the laminar flame speed. The thermal conductivity is a modified form of the equation proposed by Echekki et al. [13]:

$$\lambda = \lambda_0 c_p \left(\frac{T}{T_0}\right),$$

where $\lambda_0 = 2.58 \times 10^{-5}$ kg/(ms). The temperature dependence of the conductivity has been chosen such that the resulting temperature and species profiles can be derived analytically using asymptotic methods [32]. The viscosity and mass diffusion are calculated via the joint assumptions of constant Prandtl number ($= 0.75$) and constant unit Lewis number. The other thermodynamic quantities are assumed to be constant and are set equal to the values given for air. The stagnation internal energy is obtained using

$$E = \sum_{l=1}^{N_s} e_l Y_l + \frac{u^2}{2},$$

where e_l is the species internal energy, comprising the internal energy of formation e_l^0, and a sensible component:

$$e_l = e_l^0 + \int_{T_0}^{T} c_v\left(T'\right) dT'.$$

The specific heats and the molecular weights of the components are assumed to be constant, with $c_p = 1005$ J/(kg K), $\gamma = 1.4$ and $W = 28.96$ kg/kmol. The pressure is calculated from the thermal equation of state

$$p = \rho \mathscr{R}^0 T \sum_{l=1}^{N_s} \frac{Y_l}{W_l}, \tag{14.20}$$

where $\mathscr{R}^0$ (=8314.5 J/(kmol K)) is the universal gas constant.

The governing equations retain full compressibility, and hence acoustic waves will need to be accurately captured by the wavelets. This has been done deliberately, since the accurate resolution of the pressure profile provides a stringent challenge for the discretization.

14.5 Results

Figures 14.5(a)-(d) show a benchmark solution for a 1-D laminar flame. The solution has been calculated using $N = 4$ interpolating wavelets and initially with wavelet thresholding switched off (i.e. using a full grid). The discretization comprises initially $2^8 + 1$ (i.e. $\mathbf{V}_J = \mathbf{V}_8$) grid points; the extra grid point comes from end effects introduced via discretizing an interval.

Figure 14.6 shows a comparison between the benchmark solution and an analytic profile obtained by high activation energy asymptotics [32]; the agreement between the two is excellent, and thereby establishes the credibility of wavelets as a means of DNS.

Figure 14.7 shows the reaction rate profile and the retained grid points, the latter being obtained by thresholding the wavelet series based on the magnitudes of the respective dependent (conservative) variables. The automatic clustering of the grid points in the region of the flame are the principal driver for the development of wavelet methods.

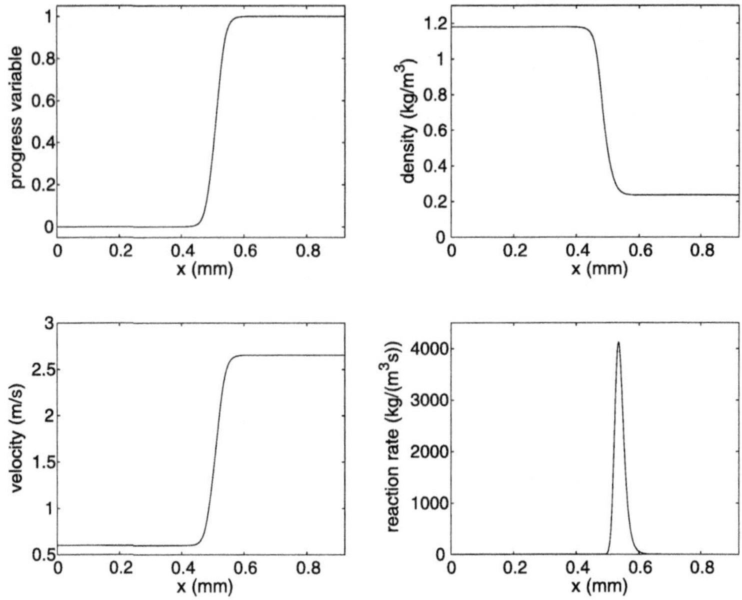

Fig. 14.5: Benchmark laminar flame profile.

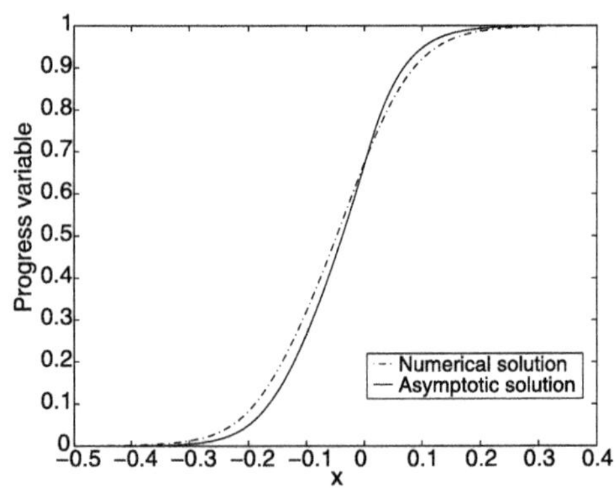

Fig. 14.6: Progress variable profile obtained analytically and numerically with $c = 0$. Reprinted from [26] wither permission from the Institution of Mechanical Engineers.

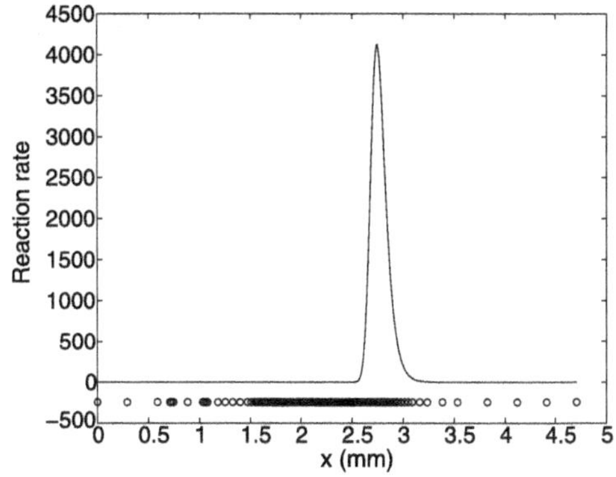

Fig. 14.7: Reaction rate. ○ denotes location of retained grid points.

The effects of increasing threshold ε on the temperature profile have been found to be essentially invisible. Of more interest is the effect of thresholding on the resolution of the pressure profile. It is well known that low Mach number systems exhibit stiffness (i.e. see Klein [18]); this stiffness manifests itself here as a pressure profile that is extremely sensitive to numerical noise. Figures 14.8, 14.9 and 14.10 show the pressure profile obtained with increasing thresholds. The dynamic pressure change is itself very small, and so consequently is quite sensitive to any perturbation.

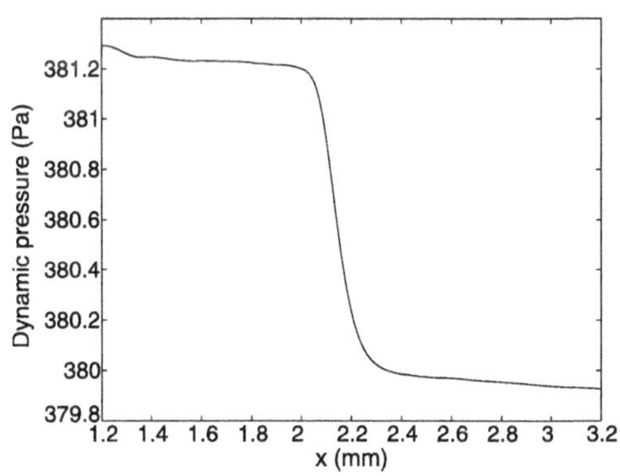

Fig. 14.8: Dynamic pressure profile for thresholded solution, with $\varepsilon = 10^{-6}$. Reprinted from [26] wither permission from the Institution of Mechanical Engineers.

An estimate for the effect of wavelet thresholding on pressure can be established for low Mach number systems by examining the thermodynamic relationship between the conserved variables and the pressure. Using standard thermodynamic relations, the change in pressure for a mixture of N_s ideal gases can be related to changes in the conserved variables by

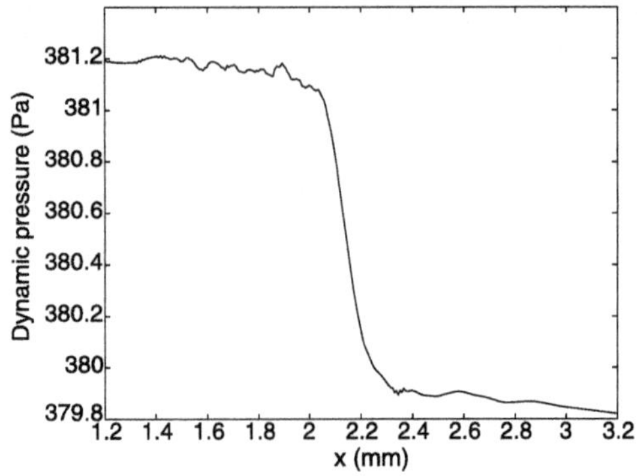

Fig. 14.9: Dynamic pressure profile for thresholded solution, with $\varepsilon = 10^{-5}$. Reprinted from [26] wither permission from the Institution of Mechanical Engineers.

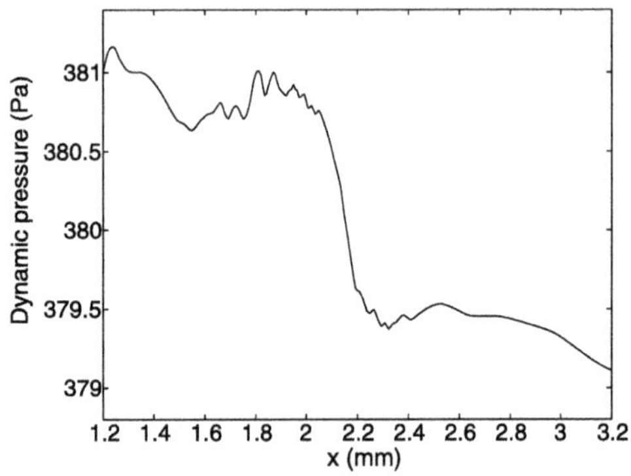

Fig. 14.10: Dynamic pressure profile for thresholded solution, with $\varepsilon = 10^{-4}$. Reprinted from [26] wither permission from the Institution of Mechanical Engineers.

$$dp = (\gamma - 1)\left[d(\rho E) - ud(\rho u) + \frac{1}{2}u^2 d\rho\right] - \sum_{\alpha=1}^{N_s}\left((\gamma-1)e_\alpha - R_\alpha T\right)d(\rho Y_\alpha). \tag{14.21}$$

A better understanding of the relative effects can be obtained by non-dimensionalising Eq. 14.21 to obtain

$$\begin{aligned} dp &= (\gamma - 1)\left[d(\rho E) - (\gamma-1)M^2\left(ud(\rho u) - \frac{1}{2}u^2 d\rho\right)\right] \\ &\quad - \sum_{\alpha=1}^{N_s}\left(\widetilde{e}_\alpha - \frac{R_\alpha T}{(\gamma-1)}\right)d(\rho Y_\alpha), \end{aligned} \tag{14.22}$$

where the pressure has been non-dimensionalized with respect to $\gamma\rho^0 RT_0$; the remaining quantities have been non-dimensionalized with respect to u_0, $(c_p)_0$ and T_0. Consider a discretization which decomposes $\mathbf{V}_J$ onto $\mathbf{V}_{J-1}$ and $\mathbf{W}_{J-1}$ for clarity. At some point during the integration, the solution is transformed onto $\mathbf{V}_{J-1}$ and $\mathbf{W}_{J-1}$, and then a threshold is applied. The resultant reduced series is then used to reconstruct the dependent variables. This procedure is a model for an adaptive solution algorithm where in this case, the missing coefficients are replaced with a zero in the transform domain.

By construction in this simple example, errors of $O(\varepsilon)$ are incurred in the energy, the momentum, the density and the species mass fraction profiles. Using equation 14.22, we write

$$\begin{aligned}\Delta p = (\gamma-1)\left[\Delta(\rho E) - (\gamma-1)M^2\left(u\Delta(\rho u) - \frac{1}{2}u^2\Delta\rho\right)\right] \\ - \sum_{\alpha=1}^{N_s}\left(\widetilde{e}_\alpha - \frac{R_\alpha T}{(\gamma-1)}\right)\Delta(\rho Y_\alpha)\end{aligned} \tag{14.23}$$

where the Δ symbol is used to denote the departure from the benchmark solution as a result of thresholding:

$$\begin{aligned}\Delta p &= p - p^\varepsilon \\ \Delta(\rho E) &= (\rho E) - (\rho E)^\varepsilon \\ \Delta(\rho) &= \rho - \rho^\varepsilon \\ \Delta(\rho Y_\alpha) &= (\rho Y_\alpha) - (\rho Y_\alpha)^\varepsilon .\end{aligned}$$

It is clear that both the momentum and density driven effects will be very small $\left(O\left(M^2\right) \sim 10^{-6}\varepsilon\right)$ in comparison to those effects induced by errors in the energy and species mass fraction, which are both of $O(\varepsilon)$. Consequently, in any adaptive numerical strategy it is important to use at least ρE and (ρY_α) as the key variables upon which the unstructured grid is built. If the previous analysis is extended to a multiresolution decomposition, essentially the same argument holds, but with the magnitudes of the errors $\Delta(\cdot)$ increased by a factor of approximately $O(J)$ where J refers to the index of the original discretisation.

Interestingly, a similar problem is observed if asynchronous time evolution methods are employed (i.e. the schemes proposed by Bacry et al. [1]). It appears that the coupling engendered by the pressure (through the energy equation) links phenomena across many different scales, all of which interact with equal importance.

14.6 Conclusions

Wavelets have enjoyed considerable popularity in recent years. Their utility stems from the scale and position dependent decompositions they provide, which in turn furnishes flow descriptions with a level of fidelity beyond that offered by Fourier transforms (which are spatially delocalized). Considerable effort has gone into the exploitation of wavelets for reacting flow simulation. They provide a natural framework to produce adaptive computations, although some technical challenges remain regarding inertial and chemistry driven non-linearity (this problem however is prevalent to some extent with any numerical scheme). We have discussed in some detail one possible line of attack for reacting flows involving interpolating wavelets. Others are available (i.e. the vaguelettes of Schneider et al. [15–17]). We have chosen to describe the interpolating wavelet approach because of the closeness of the discretisation to existing engineering calculation methods (i.e. it needs no initial projection quadrature, nor operator modified vaguelettes, etc.)

In addition, the discussions presented here regarding non-linearities and differential operators are not restricted to the arena of DNS. We have seen how new wavelet spaces may be populated during a nonlinear interaction. Furthermore, there is a clear path from this emergent population to all other scales through the medium of the differential operator. We may interpret mappings of the form $\mathbf{W}_i \to \mathbf{W}_k$ $k > i$ within the operator as being akin to an inertial cascade processes, while the converse represents backscatter. Hence, wavelets provide a scale and position dependent method of identifying and predicting the emergent phenomena in a turbulent flow and of estimating how that phenomena affects the resolved field. A modelling strategy may exploit this by estimating the magnitudes of the wavelet coefficients emerging from the non-linearity, and using the (known) values of the operator to estimate the effects on the resolved scales. Such studies have been undertaken by a number of workers (i.e. [10]), but further work is yet required.

References

1. Bacry, E., Mallat, S.G., Papanicolaou, G.: A wavelet based space-time adaptive numerical method for partial differential equations. Math. Modelling Num. Anal. **26**, 793–834 (1992)
2. Beylkin, G.: On the representation of operators in bases of compactly supported wavelets. SIAM J. Num. Anal. **6**, 1716–1740 (1992)
3. Beylkin, G.: On the fast algorithm for multiplication of functions in the wavelet bases. International Conference on Wavelets and Applications (1992)
4. Beylkin, G., Coifman, R., Rokhlin, V.: Fast wavelet transforms and numerical algorithms I. Commun. Pure Appl. Math. **44**, 141–183 (1991)
5. Canuto, C., Hussaini, M.Y., Quarteroni, A., Zang, T.: Spectral Methods in Fluid Dynamics, Springer-Verlag (1988)
6. Cohen, A., Daubechies, I., Feauveau, J.C.: Biorthogonal bases of compactly supported wavelets. Commun. Pure Appl. Math. **45**, 485–460 (1992)
7. Cohen, A., Daubechies, I., Vial, P.: Wavelets on the interval and fast wavelet transforms. Appl. Comp. Harmonic Analysis **1**, 54–81 (1993)
8. Daubechies, I.: Ten lectures on wavelets, SIAM, Philadelphia (1992)
9. Day, M., Bell, J., Bremer, P.-T., Pascucci, V., Beckner, V., Lijewski, M.: Turbulence effects on cellular burning structures in lean premixed hydrogen flames. Combust. Flame **156**, 1035–1045 (2009)
10. de la Llave Plata, M., Cant, R.S.: On the application of wavelets to LES subgrid modelling. Quality and Reliability of Large Eddy Simulations (QLES07) (2007)

11. Deslauriers, G., Dubuc, S.: Symmetric iterative interpolation processes. Constr. Approx. **5**, 49–68 (1989)
12. Donoho, D.L.: Interpolating wavelet transforms. Presented at The NATO Advanced Study Institute conference on Wavelets and Applications, Il Ciocco, Italy (1992)
13. Echekki, T., Chen, J.H.: Unsteady strain rate and curvature effects in turbulent premixed methane-air flames. Combust. Flame **106**, 184–202 (1996)
14. Fernández, G., Periaswamy, S., Sweldens, W.: LIFTPACK: A software package for wavelet transforms using lifting. Wavelet Applications in Signal and Image Processing IV, Proc. SPIE, 396–408 (1996)
15. Fröhlich, J., Schneider, K.: An adaptive wavelet-vaguelette algorithm for the solution of PDEs. J. Comput. Phys. **130**, 174–190 (1997)
16. Fröhlich, J., Schneider, K.: Numerical simulation of decaying turbulence in an adapted wavelet basis. Appl. Comput. Harm. Anal. **2**, 393–397 (1995)
17. Fröhlich, J., Schneider, K.: A fast algorithm for lacunary wavelet bases related to the solution of PDEs. C. R. Math. Rep. Acad. Sci. Canada **16**, 83–86 (1996)
18. Klein, R.: Semi-implicit extension of a Godunov-type scheme based on low Mach number asymptotics I: One dimensional flow. J. Comput. Phys. **121**, 213–237 (1995)
19. Jameson, L.: On the Daubechies based wavelet differentiation matrix. NASA ICASE Report 93-95 (1993)
20. Jameson, L.: On the differentiation matrix for Daubechies-based wavelets on an interval. SIAM J. Sci. Comput. **17**, 498–516 (1996)
21. Mallat, S.G.: A theory for multiresolution signal decomposition: The Wavelet representation. IEEE Trans. Pattern Anal. Machine Intell. **11**, 674–692 (1989)
22. Meyer, Y.: Wavelets and Operators, Cambridge University Press (1992)
23. Newland, D.E.: Harmonic wavelet analysis. Proc. Roy. Soc. Lond. A **443**, 203–25 (1993)
24. Press, W.H., Teukolsky, S.A., Vetterling, W.T., Flannery, B.P.: Numerical Recipes in FORTRAN, Cambridge University Press (1992)
25. Prosser, R.: Numerical methods for the computation of cmbustion. Ph.D. Thesis, Cambridge University (1997)
26. Prosser, R.: An adaptive algorithm for compressible reacting flows using interpolating wavelets. Proc. Instn. Mech. Engrs. Part C **221**, 1397–1410 (2007)
27. Schröder, P., Sweldens, W.: Building your own wavelets at home. Lecture Notes in Earth Sciences Vol. 90, Springer (2000)
28. Sharpe, G.J., Falle, S.A.E.G.: Nonlinear cellular instabilities of planar premixed flames: numerical simulations of the reactive Navier-Stokes equations. Combust. Theor. Model. **10**, 483–514 (2006)
29. Sweldens, W.: The Lifting scheme: A custom design construction of biorthogonal wavelets. Appl. Comput. Harm. Anal. **3**, 186–200 (1996)
30. Sweldens, W.: The Lifting scheme: A construction of second generation wavelets. SIAM J. Math. Anal. **29**, 511–546 (1997)
31. Vasilyev, O.V., Kevlahan, N.K.R.: An adaptive multilevel wavelet collocation method for elliptic problems. J. Comput. Phys. **206**, 412–431 (2005)
32. Williams, F.A.: Combustion Theory - Second Edition, Addison Wesley (1985)

Part IV
Cross-Cutting Science

Chapter 15
Design of Experiments for Gaining Insights and Validating Modeling of Turbulent Combustion

A.R. Masri

Abstract This chapter addresses some key issues for consideration in the design and development of experiments that provide a deeper understanding of combustion physics and become benchmarks for the validation of calculations. Close interaction between numerical and experimental approaches has proven to be a key ingredient in advancing the predictive capabilities. A good experiment (i) addresses one or more specific scientific issues, (ii) has well-defined boundary conditions, (iii) is amenable to advanced diagnostics (iv) provides a range of conditions to test trends as well as absolute quantities, (v) makes the detailed data available including information about the accuracy of the measurements, and (vi) responds to the changing needs of modelers as computational approaches change and evolve. A regime diagram for turbulent combustion is first introduced followed by a section that details a series of important considerations in the design and conduct of combustion experiments. The last section provides details of three key burners that stabilize flames spanning most of the turbulent combustion regime from premixed to non- premixed. Highlights and pitfalls in the design of these burners are addressed in some detail.

15.1 Introduction

Progress in combustion science and engineering has been and continues to be evolutionary rather than revolutionary with incremental and somewhat slow advancement considering the long history of fires and flames. Developments in combustion theory, experiments and more recently numerical calculations have hardly been synchronized and do not necessarily occur in an ordered sequence. It is well known however, that combustion science has historically lagged behind engineering advancements that have largely relied on experience and expensive cut-and-try meth-

A.R. Masri
School of Aerospace, Mechanical and Mechatronic Engineering, University of Sydney, NSW 2006 Australia, e-mail: a.masri@usyd.edu.au

T. Echekki, E. Mastorakos (eds.), *Turbulent Combustion Modeling*,
Fluid Mechanics and Its Applications 95, DOI 10.1007/978-94-007-0412-1_15,

ods. It is now evident that combustion science is needed to guide further developments in engineering to overcome long-standing problems of high pollution, lower than optimal efficiency and diminishing reserves in fossil fuels. In the last century there have been significant advances in the understanding of, and the ability to compute thermo-chemistry, chemical kinetics and laminar flames; all of which are necessary platforms for furthering research in turbulent combustion which is the subject of this chapter.

The advent of computational methods [28] and advanced experimental diagnostics techniques [27, 47] has initiated a healthy competition between experimental and numerical scientists in combustion. In the second half of the twentieth century, computational fluid dynamics (CFD) has evolved developing numerical solutions for the Reynolds averaged Navier-Stokes (RANS) equations and hence reproducing the basic statistical features of turbulent flames such as mean and rms fluctuations of velocity, density, temperature and even composition fields [9]. It was not until years later that experimental methods were able to generate the necessary data for validating such calculations. For example, Raman scattering techniques capable of measuring the concentration of some stable species became available to turbulent combustion applications in the eighties [15, 16, 51]. The application of Raman scattering to hydrocarbon flames [20] has reversed the leadership roles and led to the discovery of local extinction and re-ignition events in non-premixed flames approaching global flame blow-off [57, 60, 61]. This feature of turbulence-chemistry interaction was not computed until a decade later [49, 50, 79].

The series of International Workshops on Measurements and Calculations of Turbulent Non- premixed Flames (TNF) which was initiated in the nineties [37] became a very successful venue for bringing experimental and numerical researchers together to advance the science of turbulent combustion. One of the basic ingredients for this success is the collective focus on generic test cases or model problems that embody one or more specific issues being researched. The attention of the first couple of workshops was directed to the study of turbulence-chemistry interactions in non-premixed flames with relatively simple flows. Extensive data for two burners with streaming flows, namely the simple jet flame and the piloted jet burner were made freely available including sufficient details on the boundary conditions [37, 55]. Calculations were then attempted by many groups using various approaches and the results were compared, analysed and discussed at subsequent workshops. This healthy exchange of information led to an enhancement in the capabilities of modelling approaches as well as to further sharpening in the design of burners and the presentation of experimental data. A clear outcome here is the breakthrough in the capabilities of the probability density function (PDF) methods to compute local extinction in turbulent non-premixed flames [49, 50, 79].

The lead of experiments in exploring new aspects of turbulence-chemistry interactions and hence further challenging modelers continued on many fronts. With non-premixed combustion, the next generation of burner designs, namely the bluff-body [18, 56, 59] and swirl [1–3, 41] burners were introduced bringing additional complexities to the flow. Auto-ignition in turbulent flows was investigated experimentally using two burners: one employs preheated air [54, 63] and the other uses hot combustion products in a vitiated co-flow [13]. A broadening in the scope of the TNF from non-premixed toward premixed flames has led to the introduction of the turbulent piloted premixed jet burner in vitiated co-flows [23]. Measurements in these highly sheared flames with distributed reaction zones revealed a gradual reduction in reactedness rather than sudden local extinction as was observed in non-premixed flames. This mode of turbulence-chemistry interaction needs to be accounted for in the modeling of premixed combustion. Partially premixed and stratified flames are now starting to receive attention and a few burners to study such flows are now being developed [4, 23, 72, 75]. It is only natural that the TNF approach of close interaction between modelers and experimentalists be extended to advancing knowledge in other fields of turbulent combustion. As a case-in-point, another workshop series, named the International Workshops on Turbulent Combustion of Sprays (TCS), has recently been initiated to replicate the successes of the TNF, initially in dilute, turbulent spray flames [38].

Good experiments should also benefit from the evolving requirements of advanced numerical approaches such as Large Eddy Simulations (LES) [7, 21, 44, 48, 71, 73] that is most likely to become a standard tool in the design of practical combustors. One such requirement is that means and rms fluctuations are no longer sufficient quantities to compare with LES which are producing time-sequences of the entire flow, temperature, and composition fields. In response to this, high-speed diagnostic methods have emerged to enable the imaging of velocity fields (using high-speed particle imaging velocimetry, HS-PIV) and selected reactive scalars (using high-speed laser induced fluorescence, HS-LIF) [8, 10, 78]. Admittedly, the imaging of true time-sequences are still limited by the repetition rates which are gradually improving with the evolution of higher speed lasers, intensifiers and cameras. The resulting time-dependent information, both numerical and experimental, pose new challenges not only in data storage and analysis but also in the development of smart methodologies for comparing measurements and calculations.

The improved capabilities of numerical approaches have gradually enabled the reliance on numerical experiments to complement the limited flexibility of laboratory experiments. On one extreme, Direct Numerical Simulation (DNS) offers the ultimate solution where no modeling is required but it is computationally prohibitive and is likely to remain limited to simple flows with modest Reynolds numbers. However, DNS is extremely useful in the development, testing and validation of specific physical processes or sub-models that are difficult to resolve with experiments. Two examples of such uses are the development of models for the scalar dissipation rate and its fluctuation and the DNS of auto-ignition to identify ignitable mixtures and resolve the composition of the radical pool prior to thermal run-away [26, 36, 64, 76]. DNS will continue to have an increasing and more effective role in the development and validation of turbulent combustion calculations.

The use of "numerical experiments" is also becoming accepted from closure-based approaches some of which are described in this book. However, care should be exercised here to limit these applications to areas where the level of confidence in the calculations is high. Such a healthy interaction between experiments and calculations can extend the applicability and capability to probing new areas of combustion science. As an example, the particle-based PDF approach has been used in conjunction with imaging experiments in auto-igniting flames of hydrocarbon fuels to identify transport processes as well as markers for auto-ignition [31, 32]. Similarly, LES methods are gradually advancing to enable the calculation and optimization of practical combustors [7, 21, 48, 73]. These enhanced linkages and the growing interdependence between experiments and calculations in turbulent combustion is a common theme in this book.

This chapter draws on years of experience in the design of burners that have advanced knowledge of turbulent combustion and provided deep insights into outstanding issues such as turbulence-chemistry interactions in flames. The data resulting from experiments on these burners formed an essential platform for the development and validation of calculations. Section 15.2 describes a regime diagram for turbulent combustion ranging from premixed to non-premixed. This diagram may also be applicable to combustion of sprays and forms the challenge space for modelers and experimentalists alike. Section 15.3 addresses key issues that need to be considered and addressed in the design of burners as well as related experiments. Section 15.4 gives an account of selected case studies highlighting major discoveries and advances as well as specific problems and issues that could be useful as potential guides for designers. The chapter closes with some concluding remarks.

15.2 The Turbulent Combustion Domain

The broad and historical classification of combustors into premixed and non-premixed is convenient but rather not representative of practical devices considering that in most modern engines flames crossover a multitude of regimes within the same combustion chamber. A typical example here is spray combustion where the broad classification is for non-premixed flames while in actual fact the droplets within the flow provide a source of fuel for partial premixing and stratification. Nonetheless, such broad classification is very useful as it enables separate physical sub-models to be developed independently for both premixed and non-premixed regimes before merging them together to account for the "mixed modes" of combustion where partial premixing dominates. A classical example of such merging is the development of the mixture fraction- reaction progress variable models which are now applied to partially premixed flows [39, 46, 70].

The use of mixture fraction and some measure of reaction progress to describe the regimes of turbulent combustion seems natural given the relevance of these concepts in the extremes of premixed and non-premixed flames. Figure 15.1 shows a realization of the regime diagram for turbulent combustion plotted in terms of the flow conditions at the inlet to the combustor in a two-stream configuration. One axis represents the inlet condition in stream 1 in terms of the overall fuel mass fraction (the balance being oxidant) while the second axis represents the mass fraction of oxidant in the second stream (the balance being fuel). Note here that the stoichiometric as well as lean, and rich flammability limits marked respectively by "St,

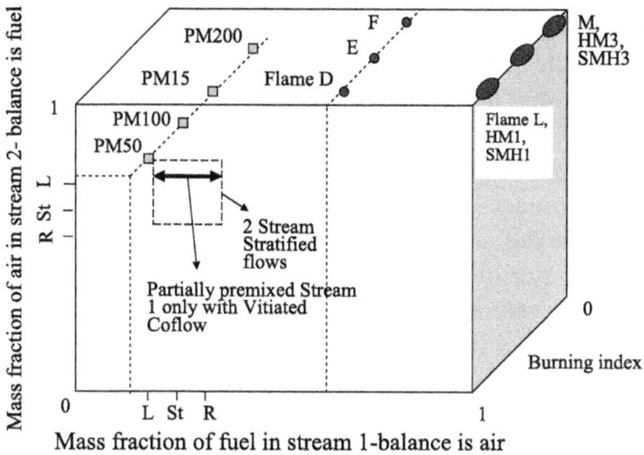

Fig. 15.1: Regime diagram for turbulent combustion in a two-stream flow. Piloted flames (L, B, M [57, 61] and D,E,F [5, 43]), bluff-body stabilised flames (HM1, HM2, HM3) [18, 59], swirl-stabilised flames (SMH1, SMH2 and SMH3) [1–3, 41] as well as piloted premixed vitiated coflow flames (PM50-PM200) [23, 25] are shown.

L and R" are only indicative and their actual location on the diagram varies with fuel mixture. The third axis represents overall departure from fast chemistry which may be indicated by an overall measure of burning index, B, reaction progress, C or a Damköhler number, Da. The overall burning index is used here as a parameter between zero and one where zero is for non-reacting and 1 is for fully reacting conditions at the fully burnt limits.

Most burners and flames that have been studied so far may be positioned on this regime diagram. Sydney's piloted flames L, B, and M have 100% fuel in stream 1, 100% air in stream 2 and range in burning index from high for flame L to almost zero in flame M [20, 57]. Piloted Sandia flames D, E and F are similar but partially premixed in the fuel stream [5, 43]. The complexity in the fluid mechanics is not reflected in this diagram so Sydney's bluff-body flames HM1, HM2 and HM3 [18, 59] and swirling flames with 100% fuel in the jet stream (flames SMH1, SMH2 and SMH3) [1–3, 41] would be on the same traverse as flames L, B, M. Flames positioned in a vitiated co-flow such as the Cabra and Dibble configuration [13] have, by definition a reduced amount of air in the co-flow as shown in Fig. 15.1 by the solid bold line with arrows that may extend further on the lean or rich sides. For example, the auto-igniting flames of Cabra et al. [13] and Gordon et al. [30] with partially premixed fuel would be positioned on the rich (right hand) side of the plot. The premixed flames of Dunn et al. [23, 25] are on the lean side and are indicated by PM-50 to PM-200 with a range of burning indices. The two-stream stratified flames introduced recently [4, 72, 75] would fit within the dashed box depending on the equivalence ratios used within each stream. Spray flames may also be represented on this diagram although the location may vary depending on the injection condition and the volatility of the parent liquid fuel. Direct injection of liquid into air yields diffusion-like flames initially while air-assisted injection may result in premixed or partially premixed depending on the volatility. Additionally, the evaporation of droplets within the flow may lead to some stratification.

Most modeling approaches are generally adaptable to one or more modes of combustion, such as premixed or non-premixed but not sufficiently general to apply to the entire spectrum described in this regime diagram. A "generalized combustion modeling approach" is one with a proven ability to compute the structure of turbulent gaseous flames over a broad range of conditions: premixed to non-premixed, low to high strain, thin to broad reaction zones, auto-igniting to flaming. This would be compounded by additional complexities associated with strong transients and instabilities such as those found in swirling flows and/or with the presence of droplets in sprays. These are significant challenges that may well be met by one or more of the approaches described in this book.

15.3 Basic Considerations

Good experiments are generally designed to explore specific scientific issues and advance knowledge to the point where the resulting findings can be used to challenge, develop and validate numerical approaches. Good interaction with modelers is essential here since an appreciation of the limitations and capabilities of numerical methods may result in improved burner designs that are more representative while still amenable to modeling. A few basic considerations are discussed in this section as aids rather than prescriptions for the design and conduct of future experiments.

15.3.1 Design Issues

The distinction is made at the outset that the concern here is with generic laboratory research burners rather than scaled-down versions of practical combustors. While scaling issues are not considered, attention should be given to the relevant dimensionless parameters to ensure that the relevant domain of the burners operation is satisfied. Here are the three main issues that govern the design of research burners:
Research focus: The principle here is well accepted and entails identifying and separating research issues that form the design focus of the burner. In the well-known piloted burners, the focus was on turbulence-chemistry and hence soot, liquid fuels and complex flows were initially eliminated. The latter condition was later relaxed in the bluff-body and swirl stabilized burners and, later, dilute sprays were introduced but in simple jet flows. The study of auto-ignition also uses rather simple gaseous flows in a hot vitiated co-flow as in the burner of Cabra et al. [13]. Turbulent stratified combustion requires a burner that allows flexibility in changing the mixture

fraction and mixture fraction gradients and preferably the attainment of high levels of u'/S_L. In premixed flames, most burner configurations, such as those stabilised by low swirl and bluff bodies [6, 14, 52, 53] are generally limited to values of u'/S_L of about 10. The provision of a hot vitiated co-flow enables much higher shear rates and hence a transition of the flames to the distributed reaction regime. Representative flames stabilized on these vitiated co-flow burners are located on the regime diagram of Fig. 15.1.

Boundary conditions: The issue of boundary conditions is of extreme importance and impacts directly on the final configuration of the burner. Modelers generally require as much information as possible but the level of detail and the meaning of boundary conditions depend on the numerical approach and location at which modelers start their calculations. For RANS type calculations, profiles of the mean and rms fluctuations would be sufficient, while time series may be required for LES. It is not unusual for modelers to commence their calculations well upstream of the exit plane of the burner. The case studies detailed later in this chapter will provide examples and highlight some of the associated issues. Figure 15.2 presents a sample of some basic details required for boundary conditions with profiles of means and rms fluctuations of velocity at the exit plane of three commonly used burners, namely the piloted, the bluff-body and the swirl stabilized burners.

Another issue of importance is the sensitivity to boundary conditions. This is harder to investigate experimentally and lends itself more easily to calculations once a certain level of confidence is reached. The boundary layer associated with jet flows may have impact on the flame. It has been shown for simple jet flames, that increasing the co-flowing velocity leads to a laminar to turbulent transition [58] in the external boundary layer hence affecting transport of the reactive scalars in the jet. Heat transfer to the burner may also be significant as is the case in bluff-body flows. Specifying the temperature of the bluff-body as well as the construction material will be important in accounting for such losses.

Stabilization with a pilot is simple and effective but adds a subtle level of complexity to the flow. The pilot is essentially another flow "stream" that needs to be accounted for. The level of difficulty associated with modeling the pilot stream is generally modest but depends on the approach. Accounting for the additional heat release from the pilot is generally easy. The composition of the pilot flame gases may be assumed to be at equilibrium or may be computed from laminar premixed flame codes or as a plug-flow reactor. When studying pilot-stabilized turbulent premixed flames, the stoichiometry of the pilot may be different from the rest of the flow and this tends to make the definition of the reaction progress variable no longer uniform as is the case with the piloted premixed burner in vitiated co-flows [25].

Optical access: This is a key requirement for experimental burners given that almost all advanced diagnostics methods will require some sort of optical access. Enclosed burners will use windows that are generally made of quartz if ultraviolet light transmission is needed. Flat windows are preferred to avoid reflection associated with curvature. Applications that involve high pressure are much harder to handle but these are outside the scope of this chapter.

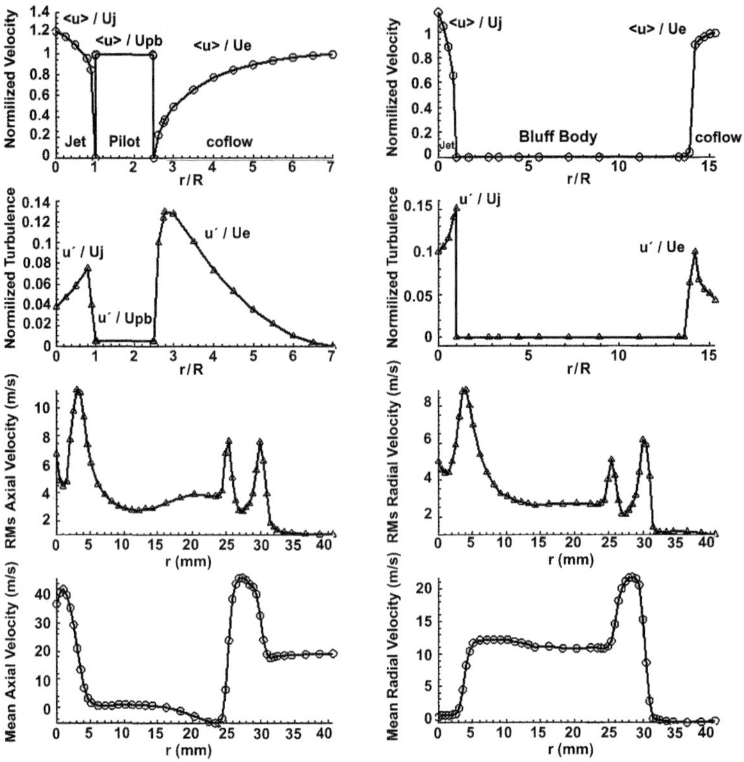

Fig. 15.2: Mean and rms fluctuations of the axial velocity measured at the jet exit plane of three burners: (Top LHS) piloted [61], (Top RHS) bluff-body [59], and (Bottom four plots) swirl [2] which also shows the tangential component of velocity.

15.3.2 Operational Envelopes

Understanding the burners operational envelope is very important since this dictates the limiting conditions and the combustion ranges that the burner can access. Three considerations are noted here:

Controlling parameters: Minor and major parameters that affect the flame and its stability need to be identified for fixed geometry parameters such as the size of the bluff-body or the jet diameter. Initial parametric studies may be needed to identify the relative importance of various parameters and this process is not always straight forward. In the swirl burner for example, the swirl number is essential but the ratio of the momentum of the co-flowing stream to the central stream is also very important. Note also that in bluff- body stabilised burners, the co-flow velocity is a major parameter but not so in the piloted flames.

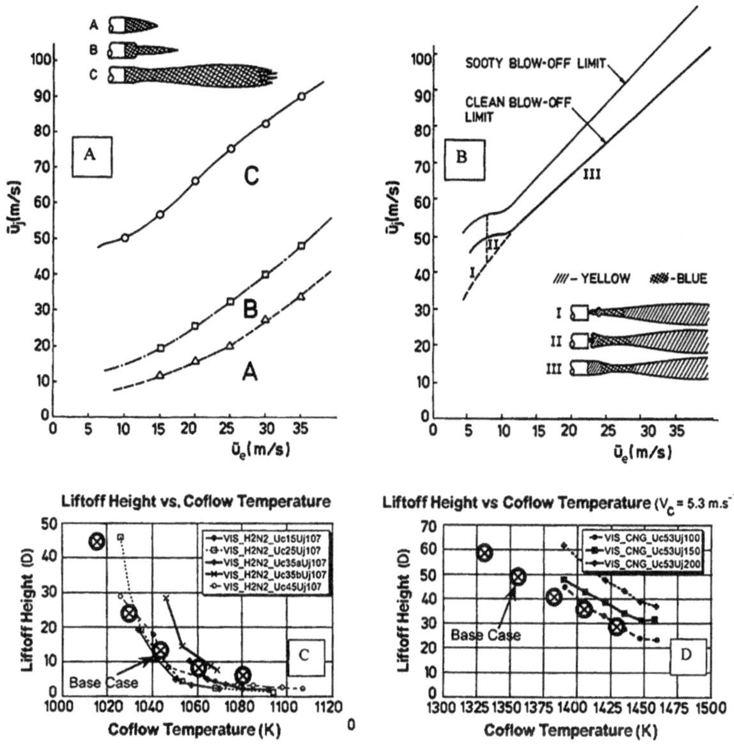

Fig. 15.3: Stability limits of the bluff-body burner [56] (top) and auto-ignition burner in vitiated coflows [31] (bottom).

Stability limits and test cases: While relatively easy to determine once the major controlling parameters are identified, these are important in setting the overall operational envelopes of the burner and hence selecting relevant flames for further investigation. Figure 15.3 shows typical stability limits for two burners: (i) the bluff-body burner [56] where the two controlling parameters are the velocities of the jet fuel, U_j and the co- flowing air, U_e, and, (ii) the auto-igniting flames in hot, vitiated coflow [31, 32] where the lift-off height is strongly affected by the temperature of the hot co-flowing stream. It is worth noting that in selecting the test cases, groups of flames are chosen in particular sequences to test the importance of each of the controlling parameters. For cxample, bluff-body flames HM1, HM2, and HM3 would approach blow-off by increasing the jet velocity at the same co-flow velocity (move vertically up in Fig.15.3). Similarly, piloted flame sequences L, B, M and D, E, F, as well as swirl flames SMH1, SMH2, SMH3 of methane/hydrogen fuel mixtures. Such sequences have proven to be extremely useful in showing relevant trends and hence in the validation of computational models.

Sensitivity to boundary conditions: In some flow configurations, the flames may be very sensitive to boundary conditions leading to broad variations when the same experiments are repeated at different times or different laboratories. A common example of this is the standard lifted flame where small variations in the surrounding air velocity or turbulence level causes significant changes in the lift-off height. Other illustrations of the sensitivity to boundary conditions are the auto-ignition burner which is strongly affected by the temperature of the co-flowing vitiated air [13], and the bluff-body burner which may be affected by the outer boundary layer on the air side of the bluff-body [18, 59]. Investigating such effects is possible experimentally but the difficulty depends on the parameters being studied. With sensitivity to the boundary layer, it may be easier to perform such studies numerically since changing the boundary layer experimentally is do-able but not trivial.

15.3.3 Experimental Considerations

Good planning of the experiments will improve the chances to strike a breakthrough in the understanding of the physics or at least to just provide a useful and reliable data base that will be used for model validation. Here is a brief description of the issues that are worth considering:

A non-reacting test case: Along with the sequence of test cases discussed earlier, the selection of a base case is extremely useful for model validation. This could be a simple flame or even a non-reacting case that possesses similar characteristics to the reacting cases. Modelers normally start with such cases to test the relatively simpler features of their calculations such as flow and mixing fields. In the investigation of sprays, it is extremely important to perform measurements in non-reacting case that will yield information about the dynamics and evaporation of droplets. Selecting test cases where the liquid fuels have different vapor pressures will also be extremely useful in optimizing the evaporation model.

Measurements:

- **Flow fields:** The most common techniques here are Laser Doppler Velocimetry (LDV) and Particle Imaging Velocimetry (PIV) both of which rely on seeding micron-sized of particles into the flow. LDV methods are well-developed but are inherently limited to single-point measurements. PIV methods are simple in principle and have also been well developed for low repetition rate. Recently, there have been two major developments with PIV: (i) High-speed PIV (HS-PIV) made possible by the development of high-speed lasers and cameras (around 20 kHz at the time of writing this chapter) and it is only a matter of time before the laser and camera technology develops even further to enable true time evolution measurements, (ii) Three dimensional PIV based on volume illumination and three-aperture-three-camera detection systems (V3V) [40]. This technology is relatively new and is currently limited to low repetition rates but is also expected to evolve to a stage where full 3D-measurements of the time evolution of the flow field becomes possible.

- **Mixing fields:** The use of combined Raman scattering and LIF to measure the concentration of a sufficient number of species to obtain a conserved scalar is limited by many factors including the complexity of the experiment and severe interferences on the Raman signals [61]. An alternative approach is to identify inert tracers that can fluoresce and withstand the flame temperatures without being affected. Despite testing a large number of tracers [74], this approach remains inconclusive for high temperature applications due to flame interferences resulting in the corruption of the original concentration of the tracer or due to significant quenching of fluorescence from the tracer. The use of noble gases as tracers is promising if the quenching issue is overcome since these species can easily survive the flame temperatures unscathed so they have a good chance of enabling a simple and reliable measure of mixture fraction.
- **Temperature and reactive species fields:** The use of Rayleigh scattering for temperature measurements is limited to non-sooting gaseous flames but still requires knowledge of the Rayleigh scattering cross-section, and hence a measure of species concentrations which is difficult due to the complexities associated with Raman methods. A relatively easy alternative here is to modify the fuel mixture such that the Rayleigh scattering cross section remains as uniform as possible across the entire mixture fraction space [19]. Combined imaging of Rayleigh-LIF is proving to be extremely useful in flames where Raman techniques are hard or inaccessible. Multi-species-LIF techniques are expensive since they require numerous laser sources depending on the species being detected but are extremely informative and do not suffer from the low signal levels associated with Raman measurements. High-speed LIF (HS-LIF) methods are now becoming more common albeit still limited to common species such as OH [8, 10, 78] and speeds of about 10 kHz. Such methods are extremely informative and are evolving quickly so it is only a matter of time before developments in the laser and camera technology enables higher speed and a broader selection of species that can be measured either solely or jointly.

Data presentation and availability: Two issues need urgent attention now that high-speed imaging methods are becoming commonplace along with the widespread use of LES for calculations: (i) presentation of huge quantities of space- and time-resolved data represented by time-sequenced images and (ii) relevant and tractable methods of processing and analyzing these images as well as providing meaningful comparisons with LES. Figure 15.4 shows sequences of high speed images (5 kHz, hence 200 μs separation) of LIF-OH collected at the base of a lifted flame of CNG with a jet diameter of 3.85mm and a Reynolds numbers of 4800. The top sequence illustrates the presence of what appears to be edge flames shaped in the form of hooks while the second shows blobs that grow at the same location indicating that these are coming into the page from another radial location at the base. Both modes appear to play a significant role in stabilising the flame. Thousands of these images are collected so that presenting the information in a sensible way is not straightforward. The other issue is the processing of the image sequences such that spatio-temporal information such as rate of change of spatial flow structures can be extracted in a sensible way. Proper orthogonal decomposition, (PoD) techniques

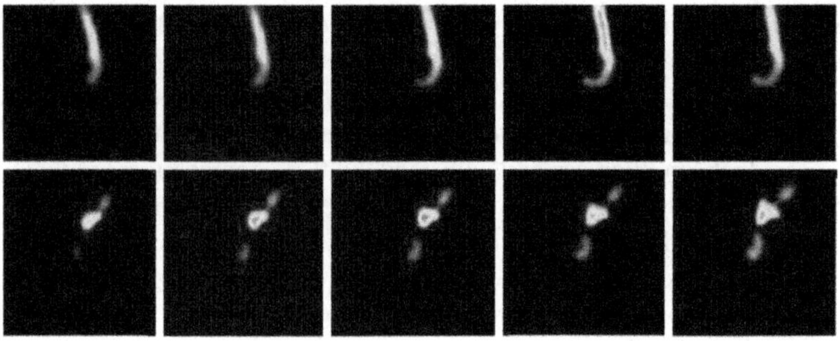

Fig. 15.4: Sequences of high-speed LIF-OH images collected at the base of a lifted methane flame. The fuel issues at 35 m/s into still air from a round nozzle with a diameter D=3.85 mm. The measurements are centred at x/D=30. The top sequence shows the edge flame "hooks" and the bottom sequence shows flame kernels likely to be flowing in or out of the page. From [67].

[17, 22] appear to be promising in this regard although significant development is needed before these are established as standard tools for data analysis.

15.3.4 Numerical Considerations

The focus on providing detailed and simple boundary conditions has already been identified as one of the benefits of the strong interactions between modelers and experimentalists. There are two others issues that are worth focusing in this section:
The starting point of the calculations: Commencing calculations at the exit plane of the jet requires details that enable the modelers to initiate the "same" condition for which data are available. Such information is generally in the form of means and rms profiles of velocity and of scalars such as, temperature and species mass fractions. Starting the calculation further upstream, requires details of the inner design of the burner but then only bulk flow conditions need be provided. Measurements provided at the exit plane can then be used to test and tune the calculations. Examples of such scenarios are given in the next section. Another consideration here is the need to compute the flow close to surfaces used to pipe the fluid. These are generally hard to reproduce particularly when spray flows are used. The interaction of droplets with solid surfaces is a difficult field that is not well modeled.
The requirements of LES: Large eddy simulations are providing an effective platform for modeling the entire spectrum of turbulent reacting flows. Relevant subgrid-scale models for combustion are being developed but a universally accepted model is not yet available. The use of LES imposes stringent requirement on experimental methods in two key areas: (i) more details are needed in the boundary conditions (such as time series of say velocity fields), and (ii) the provision of time-sequenced images of flame structure to compare with the calculations. This latter

requirement may now be satisfied thanks to the advent of high-speed imaging techniques [8, 10, 78] although methods to provide adequate and meaningful comparisons between the numerical and experimental data sets are yet to be devised.

15.4 Case Studies

This section presents three cases studies that span a broad region of the turbulent combustion diagram shown in Fig. 15.1. These are the swirl burner, the piloted premixed burner in vitiated co- flows and the piloted spray burner. While the first two are established burners in the TNF, the spray burner is now established in the new Turbulent Combustion in Sprays (TCS) [38] workshops that are attempting to duplicate the success of the TNF in sprays, starting with dilute sprays. While extensive details about these burners are well known from earlier publications, the focus of this section is to summarise the scientific advances and relay highlights and pitfalls in their design for future reference. A brief summary of the major insights and scientific advances resulting from each burner are mentioned at the beginning of each section.

15.4.1 The Swirl Stabilised Burner

This represents the third generation burner (after the piloted and bluff-body stabilised) in the TNF series of model problems aimed at studying turbulence-chemistry interaction in non- premixed flames. While the design maintains the features of a laboratory research configuration with simple, well-defined boundary conditions in an open flow, it is most representative of practical combustors with flow recirculation and instabilities. A comprehensive data bank is made available for a series of flames and three fuel mixtures as shown in Table 15.1. The results confirm earlier findings about the occurrence of local extinction in non-premixed flames even in such highly recirculating and unstable flows. There have been serious attempts to compute some of these flames [45, 62, 66] but with mixed success due to the complexity of the fluid dynamics. Figure 15.5 shows a detailed drawing of the full burner assembly. The following key features have not been addressed before in the literature and are worth discussing here for the record:

The start of the calculations at the exit plane of the burner requires details of the boundary conditions of the velocity time series in the central fuel jet and in the boundary layer. While the former may be assumed to conform with fully developed turbulent pipe flow, the co-flowing boundary layer is less certain. Using a quartz shroud for the swirl annulus as shown in Fig. 15.6, extensive measurements of the axial velocity were made just upstream of the exit plane. These have resulted in time series for the axial velocity a sample of which is also shown in Fig. 15.6. Also shown in Table 15.2 is a listing of the mean temperature of the ceramic on the

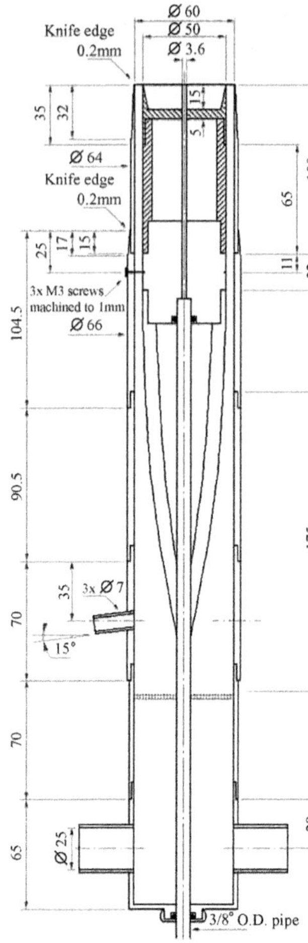

Fig. 15.5: Drawing of the swirl burner. Reprinted from [2] with permission from Taylor and Francis.

face of the bluff-body measured using a two-color pyrometer in the various swirling flames studied on this burner. To circumvent the problem of duplicating the measured conditions at the jet exit plane, some groups chose to start their calculations well upstream and hence the details shown in Fig. 15.5 become extremely useful for such calculations. The main difficulty with such approach for LES is the fact that fine resolution is required near the inner and outer surfaces to resolve the boundary layers. The benefit is that the calculations, when performed reliably, can yield extremely useful information about the flow sensitivity to the boundary layers.

Instabilities are inherent to swirling flows and at least two modes of instabilities have been identified here: (i) jet precession which is observed both in non-reacting and reacting flows, and (ii) puffing which is limited to flames. Figure 15.7 shows

Table 15.1: Flame conditions studied in the swirl burner.

Flame	Fuel Mixture (vol. ratio)	U_s (m/s)	W_s (m/s)	U_j (m/s)	Re_s (-)	Re_{jet} (-)	S_g (-)	U_{BO} (m/s)	L_f (m)	$\dot{W}$ (kW)
SM1	CNG	38.2	19.1	32.7	75900	7200	0.5	166	0.12	11.1
SM2		38.2	19.1	88.4	75900	19500	0.5	166	0.18	30
SMA1	CNG-air (1:2)	32.9	21.6	66.3	65400	15400	0.66	241	0.2	11.5
SMA2		16.3	25.9	66.3	32400	15400	1.59	216	0.23	11.5
SMA3		16.3	25.9	132.6	32400	30800	1.59	216	0.3	23
SMH1	CNG-H2 (1:1)	42.8	13.8	140.8	85000	19300	0.32	267	0.37	104.1
SMH2		29.7	16	140.8	59000	19300	0.54	281	0.4	104.1
SMH3		29.7	16	226	59000	31000	0.54	281	0.5	167.1

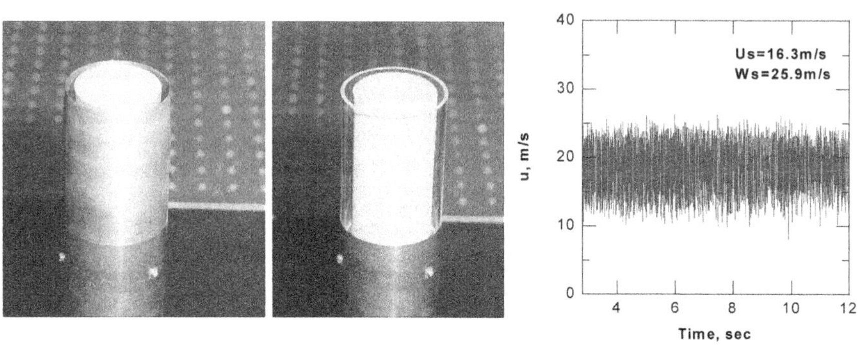

Fig. 15.6: Time series of axial velocity measured upstream of the exit plane of the swirl burner (using the quartz shroud shown). Reprinted from [3] with permission from the Combustion Institute.

the latter mode of puffing instability in flame SMH3. Such modes add a significant level of complexity to the task of computing these flows and impose an additional difficulty with measurements due to the need of phase locking. It should be noted that the data reported for these swirling flames were not phase-locked with any

Table 15.2: Measured mean temperature on the face of the ceramic for the various swirl flames investigated in Fig. 15.6 and Table 15.1.

Flame	SM1	SM2	SMA1	SMA2	SMA3	SMH1	SMH2	SMH3
Ave. temp, ◦C	**557**	**475**	**585**	**614**	**581**	**685**	**637**	**556**
Range, ±◦C	**10**	**19**	**35**	**35**	**33**	**24**	**11**	**28**

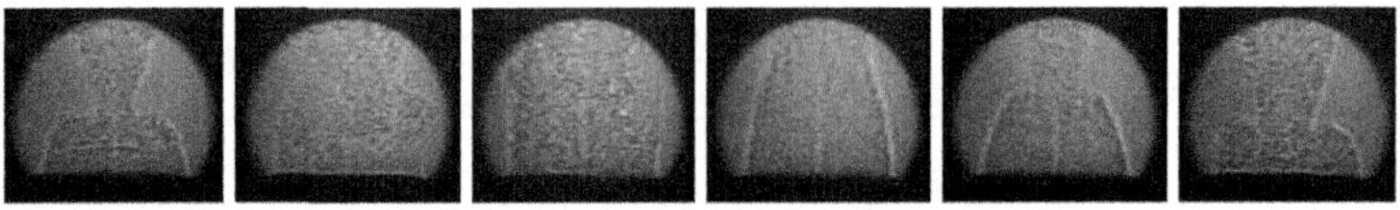

Fig. 15.7: Shadowgraph plots showing "puffing" instabilities in swirl stabilised flames of methane/hydrogen. This sequence is for flame SMH3 [3]. Reprinted from [3] with permission from the Combustion Institute.

frequency mode and were presented in the form of instantaneous scatter plots as well as means and rms fluctuations.

15.4.2 The Premixed Burner in Vitiated Coflows

There are two configurations for this burner (i) the unpiloted version which is essentially the basic layout as developed by Cabra and Dibble to study auto-ignition [13] and (ii) the piloted burner which involves an additional stoichiometric stream that maintains the flame connected to the exit plane. A layout of the piloted burner and its stability limits are shown in Fig. 15.8. Four piloted flames were selected for further studies and these are also shown in Fig. 15.8 along with further details in Table 15.3. A comprehensive data base exists for these flames which are now being computed [24].

Table 15.3: Details of premixed flames stabilized in hot coflowing pilot.

Flame	U_0 (m/s)	Re	q'/S_L	Re_t	Da
PM1-50	50	12500	40	720	0.0690
PM1-100	100	25000	230	3100	0.0089
PM1-150	150	37500	300	3700	0.0063
PM1-200	200	50000	390	5200	0.0053

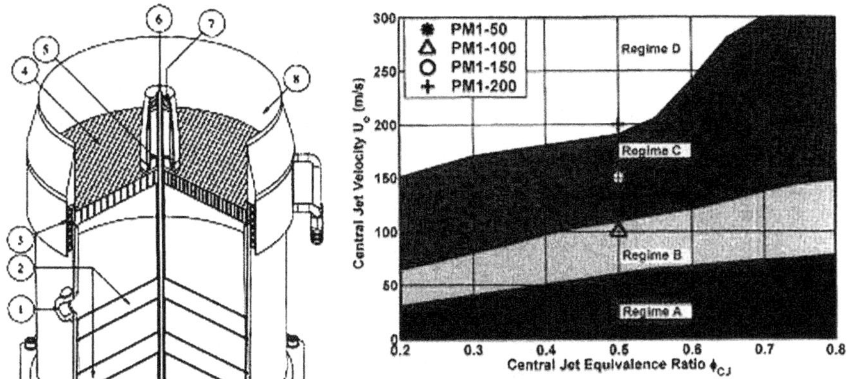

Balloon number	Description
1	Flashback over-pressure sensing port
2	Glass bead filled cavities
3	Cooling water coil
4	Coflow perforated baseplate
5	Pilot mixture feed exit
6	Central jet exit
7	Pilot perforated baseplate
8	Coflow collar

Fig. 15.8: The premixed piloted vitiated coflow burner and its stability limits. Also shown are the flames selected for further investigation (PM1-50, PM1-100, PM150, PM1200). Reprinted from [23, 25] with permission from the Combustion Institute.

The flames listed in Table 15.3 are gradually approaching blow-off and fit in the distributed reaction region of the premixed combustion regime. Note that q'/S_L increases from 40 to 390 while the Damköhler number, defined by the ratio of chemical and mixing time scales, decreases by more than an order of magnitude from flame PM1-50 to PM1-200 which is close to blow-off. Here q' is the fluctuation in the turbulent kinetic energy. A major finding revealed by the measurements is that the finite rate chemistry effects feature in a gradual reduction of reactedness rather than in localized extinction as was observed non-premixed flames [57, 60, 61]. Generalised models of turbulent combustion must account for this diversified behaviour. There are two special features in this burner design that deserve further elaboration:

- The extent of the hot coflow is a key consideration because the region of interest in the flames studied has to be fully within the hot co-flow and not affected by the outside air. Figure 15.9 shows mean temperature profiles measured at various axial locations in the co- flow which has a nominal temperature of 1500K. Flat-top profiles peaking at this temperature where measured down to x/D 50 (where D=4mm) and this corresponds to a "valid cone" that extends for about 200 mm downstream of the jet exit plane. Matching the co-flowing air velocity with the hot co-flow tends extends this region slightly.
- The additional pilot is essential for flame stabilization and for enabling sufficiently high rates of shear to be applied to the flame without inducing lift-off. For this to occur, a stoichiometric pilot composition was deemed necessary. The pilot has two main disadvantages: (i) its different stoichiometry and hence dif-

ferent temperature and composition fields in the pilot stream makes it difficult to define a universal reaction progress variable in terms of temperature or species mass fractions, (ii) the pilot adds another stream to the flow making it a three-stream problem. While this is not a problem for most modeling approaches, it may be so for others hence making this configuration less attractive. An alternative layout for this design is the use of bluff-body stabilisation which adds complexity to the flow but reduces the number of flow streams and eliminates the problems associated with the definition of the reaction progress variable.

15.4.3 The Piloted Spray Burner

Spray flows are now receiving significant attention and this is largely driven by the lack of progress in this area up to date, the continued interest in liquid fuels and the recent advances in computational and experimental capabilities. The piloted spray burner shown in Fig. 15.10 is designed to enhance current understanding of reacting and non-reacting dilute sprays and to provide a platform for the development and validation of codes. In this burner, the spray is prepared upstream of the jet exit plane and carried in a co-flowing stream of air or nitrogen to the jet exit plane. Studies reported to date [34, 77] revealed the following interesting but challenging features: (i) Jet spray flames using of acetone or ethanol fuel reveal different modes of combustion largely due to the different volatility of the fuel. Acetone flames are non-premixed near the jet exit plane due to the high vaporization rates while ethanol flames may have dual reaction zones with a premixed flame core and an outer diffusive layer. (ii) While single droplet combustion is occasionally observed in these flames, the general tendency is for combustion to occur around droplet clusters. Additional research is required to further explore these phenomena.

There are three underlying design principles in this burner: (i) simple streaming flows, (ii) well known boundary conditions and hence the preparation of the spray upstream of the exit plane (iii) ability to stabilize flames over a range of Damköhler numbers. While most of these objectives have been achieved in this burner design, there are a few issues that are worth addressing here:

Measurement of velocity and droplet fields: Phase Doppler Anemometry, PDA coupled with LDV methods are extremely useful here but are limited to single-point measurements and tend to fail when the concentration of droplets in the measurement probe is high. This tends to limit the LDV/PDA application to cases with relatively low droplet loadings and/or dilute regions of the flow. An alternative to LDV/PDA is high- speed PIV-based methods which have the potential of yielding true time evolution of the flow fields with the droplets being the scattering agent (instead of seeded particles). The limitation of the approach lies in the sparsity of droplets, particularly in dilute sprays. Classical PIV-processing tools expect large number of seed particles to produce valid cross-correlations. Expanding the window of interrogation is not really a useful option and hence the PIV processing tools

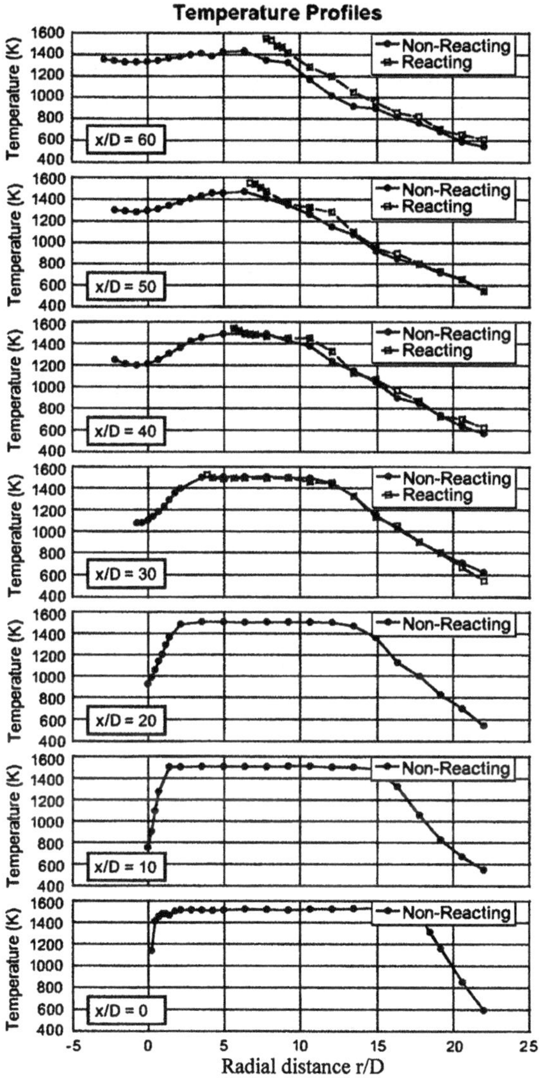

Fig. 15.9: Measured mean temperature profiles in the vitiated co-flow burner showing the extent of the "valid cone" marked by a region of uniform peak temperature. From [29].

need to be modified to handle small numbers of scattering elements in a reasonably small interrogation window.

Measurement of mixing, temperature, and reactive scalar fields: The existence of droplets limits the application of laser diagnostics methods such as Raman and Rayleigh scattering due to the inability to distinguish between gaseous and liquid

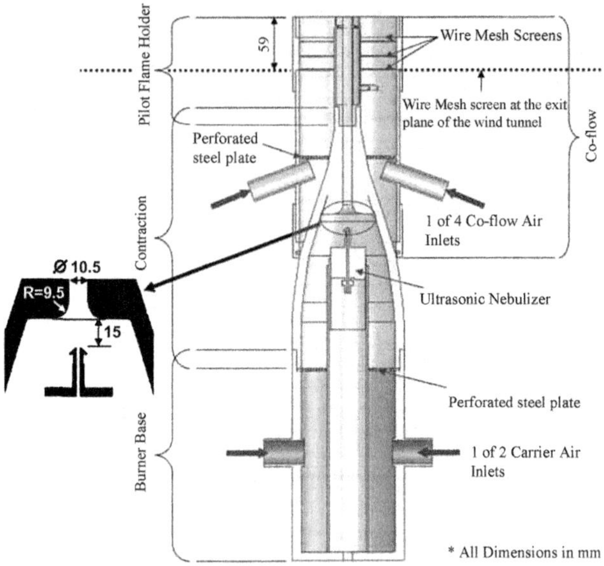

Fig. 15.10: Piloted spray burner showing details of spray inlet condition. Reprinted from [34] with permission from Taylor and Francis.

phases and the interferences of Mie scattering from particles. Thankfully, laser induced fluorescence works well here and these methods have been employed in spray flames to yield some very useful results for the measurements of reactive scalars such as OH, CH_2O and others. The potential of LIF-based methods to yield a measure of mixture fraction and temperature is a challenge that is not yet fully realised. This is either due to the inability of the tracer to survive higher temperatures without flame interference or due to significant quenching on the LIF signal. The use of NO as a tracer even in non-reacting sprays has proven to be problematic due to LIF-quenching or to Raman interference from droplets [35]. Figure 15.11 shows the fluorescence quenching of acetone and ethanol on NO. While n-hexane was identified as a non-quencher for NO, it had another problem in that the Raman interference from droplets on LIF-NO was significant, rendering it unviable. Other tracers such as dyes or even acetone may turn out to be more useful than NO. The measurement of gaseous temperature in reacting sprays is another serious challenge that is currently receiving attention. The use of elemental tracers such as indium [42, 65] or phosphorous [11, 12, 68, 69] appears to be promising although a lot more development is needed before such techniques can yield reliable quantitative information.

Tube wall effects: Measurements of velocity and turbulence levels at the jet exit plane have shown that when conditioned with the droplet size, the rms fluctuations of large droplets are higher that those of small droplets near the inner edge of the pipe. This is counter-intuitive and was referred to in earlier reports as the "pipe-flow anomaly". Scatter plots of velocity versus droplet size collected near the inner pipe

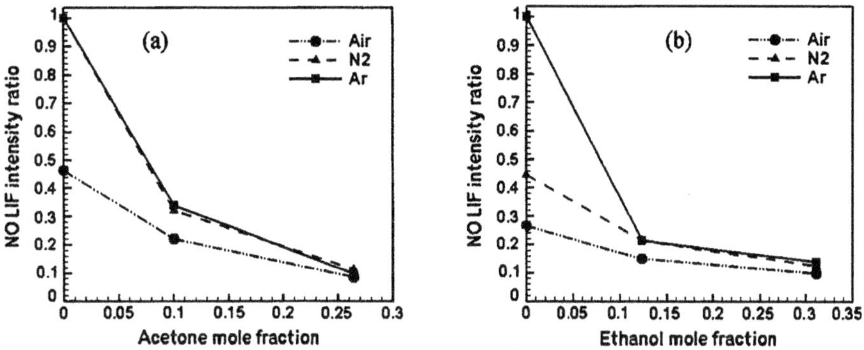

Fig. 15.11: Reduction in the intensity of LIF-NO due to quenching by (a) acetone and (b) ethanol vapor. Three profiles refer to different spray carries (air, nitrogen and agron). From [35].

wall have shown a bimodal distribution as seen in Fig. 15.12. The reason for such behaviour is now understood and explained as follows: droplets interact with the wall of the pipe and form a boundary layer of liquid. At the exit plane, such boundary layer sheds droplets in the form of ligaments which will quickly disintegrate into droplets as confimred by recent high- speed imaging of Mie scattering from droplets. Such droplets are relatively slow compared to other droplets of similar size that are not affected by this boundary layer and hence the bimodal distribution. The impact of such bimodality on the structure of the spray jet is not really known. Removing such bimodality requires a reduction in the length of pipe to about 30mm which is region where the droplets just start to impact on the wall and build the liquid boundary layer.

Non-reacting sprays and a variety of liquid fuels: Selecting a series of non-reacting cases is particularly important in this study given the importance of evaporation and the need to resolve uncertainties surrounding current models of evaporation. A reference case that represents a fuel with low evaporation, namely mineral turpentine which has a boiling point in the range of 140-190°C has proven to be useful. Measuring the liquid flux at the jet exit plane of a non-reacting turpentine spray ascertains whether the mass flux is measured correctly. Moreover, performing a series of measurements in non- reacting jets of sprays of various volatilities gives the necessary platform for validating the flow, droplet fields as well as the evaporation models.

15.5 Concluding Remarks

This chapter gives an overview of the principles associated with the design of generic laboratory burners that form a platform for exploring aspects of combustion science and gaining insights into outstanding research issues. The data resulting

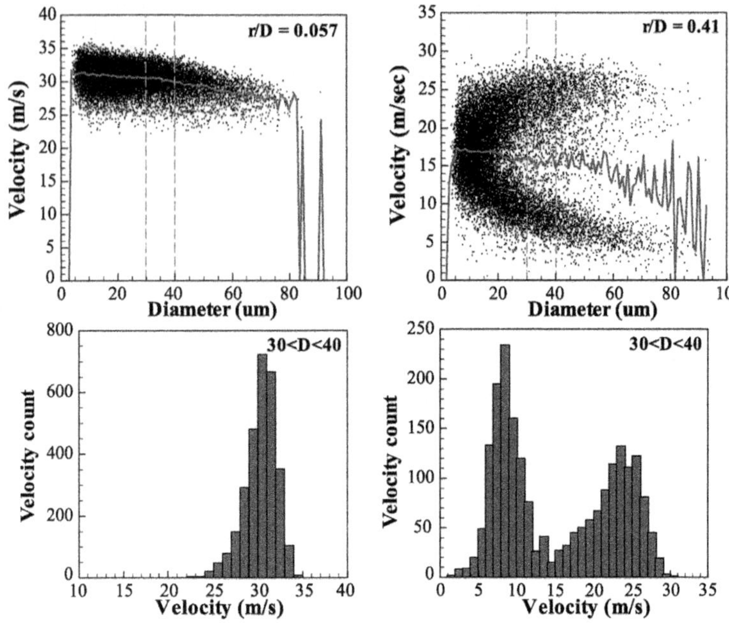

Fig. 15.12: Upper: Scatter plot of instantaneous velocity *vs.* diameter at $r/D = 0.057$ and 0.41. Lower: Velocity pdf for droplets in the diameter range $30 < d < 40$ μm. From [33].

from such experiments also form an international benchmark and validating and advancing modeling capabilities. A regime diagram that highlights a broad workspace in turbulent combustion is presented. Such a parameter space is huge and only representative sections of it can be studied experimentally. The challenge to the combustion community however, is to develop generalized modeling capabilities within this entire workspace. Examples discussed in this chapter cover scattered regions of this diagram and address a range of considerations and issues that would hopefully be of benefit in the design and conduct and future experiments in combustion. Good experiments may lead to major breakthroughs as was demonstrated in some of the case studies presented here. It is evident that continued interaction between modelers and experimentalists is extremely beneficial leading to a mutual understanding of needs and limitations, a better design of burners and experiments as well as enhanced computational methods. Other fields of combustion science are also benefiting from this approach.

Acknowledgements

The author is indebted to Professors R.W. Bilger and E. Mastorakos for their enlightening feedback on earlier drafts of this manuscript. The author acknowledges the support of Professor R.W. Dibble and Dr R.S. Barlow for enabling a long standing collaboration with Sandia's Combustion Research Facility at Livermore, CA. Thanks to Drs Sten Starner, Robert Gordon, Matthew Dunn and James Gounder as well as to Mr Mrinal Juddoo and William O'Loughlin for their help with the various experiments. Thanks also to the Australian Research Council for the continued support of his research.

References

1. Al-Abdeli, Y.M., Masri, A.R.: Recirculation and flow field regimes of unconfined swirling flows. Exp. Therm. Fluid Sci. **27**, 655–665 (2003)
2. Al-Abdeli, Y.M., Masri, A.R.: Stability characteristics and flow fields of turbulent swirling jet flows. Combust. Theor. Model. **7**, 731–766 (2003).
3. Al-Abdeli, Y.M., Masri, A.R., Marquez, G.R., Starner, S.H.: Time varying behaviour of turbulent swirling nonpremixed flames. Combust. Flame **126**, 200–214 (2006)
4. Anselmo-Filho, P., Hochgreb, S., Barlow, R.S., Cant, R.S.: Experimental measurements of geometric properties of turbulent stratified flames. Proc. Combust. Inst. **32**, 1763–1770 (2009)
5. Barlow, R.S., Karpetis, A.N.: Measurements of flame orientation and scalar dissipation in turbulent partially premixed methane flames. Proc. Combust. Inst. **30**, 665–672 (2005)
6. Bedat, B., Cheng, R.K.: Experimental study of premixed flames in intense isotropic turbulence, Combust. Flame **100**, 485–494 (1995)
7. Boileau, M., Staffelbach, G., Cuenot, B., Poinsot, T., Berat, C.: LES of an ignition sequence in a gas turbine engine. Combust. Flame **154**, 2–22 (2008)
8. Bohm, B., Heeger, C., Boxx, I., Meier, W., and Dreizler, A.: Time-resolved conditional flow field statistics in extinguishing turbulent opposed jet flames using simultaneous high-speed PIV/OH-PLIF. Proc. Combust. Inst. **32**, 1647–1654 (2009)
9. Borghi, R.: Turbulent combustion modelling. Prog. Energy Combust. Sci. **14**, 245–292 (1988)
10. Boxx, I., Heeger, C., Gordon, R., Bohm, B., Aigner, M., Dreizler, A., Meier, W.: Simultaneous three-component PIV/OH-PLIF measurements of a turbulent lifted, C3H8-Argon jet diffusion flame at 1.5kHz repetition rate, Proc. Combust. Inst. **32**, 905–912 (2009)
11. Burbach, J., Hage, M., Janicka, J., Dreizler, A.: Simultaneous phosphor and CARS thermometry at the wall-gas interface within a combustor. Proc. Combust. Inst. **32**, 855-8-61 (2009)
12. Brubach J., Patt A., Dreizler A.: Spray thermometry using thermographic phosphors. Appl. Phys. B – Las. Opt. **83**, 499–502 (2006)
13. Cabra, R., Myrvold, T., Chen, J.Y., Dibble, R.W., Karpetis, A.N., Barlow, R.S.: Simultaneous laser Raman-Rayleigh-LIF measurements and numerical modeling results of a lifted turbulent H_2/N_2 jet flame in a vitiated coflow. Proc. Combust. Inst. **29**, 1881–1888 (2002)
14. Chen, Y.-C., Bilger, R.W.: Experimental investigation of three-dimensional flame-front structure in premixed turbulent combustion II. Lean hydrogen/air Bunsen flames. Combust. Flame **138**, 155–174 (2004)
15. Cheng, T.S., Wehrmeyer, J.A., Pitz, R.W.: Simultaneous temperature and multispecies measurement in a lifted hydrogen diffusion flame. Combust. Flame **91**, 323–343 (1992)
16. Correa, S.M., Gulati, A.: Measurements and modeling of a bluff-body stabilized flame. Combust. Flame **89**, 195–213 (1992)

17. Cosadia, I., Boree, J., Dumont, P.: Coupling time-resolved PIV flow fields and phase invariant proper orthogonal decomposition for the description of the parameter space in a transparent diesel engine. Exp. Fluids **43**, 357–370 (2007)
18. Dally, B.B., Masri, A.R., Barlow, R.S., Fiechtner, G.J., Fletcher, D.F.: Instantaneous and mean compositional structure of bluff-body stabilised nonpremixed flames. Combust. Flame **114**, 119–148 (1998)
19. Dibble, R.W., Hollenbach, R.E.: Laser Rayleigh thermometry in turbulent flames. Proc. Combust. Inst. **18**, 1489–1499 (1981)
20. Dibble, R.W., Masri, A.R., Bilger, R.W.: The spontaneous Raman scattering technique applied to nonpremixed flames of methane. Combust. Flame **67**, 189–206 (1987)
21. Di Mare, F., Jones, W.P., Menzies, K.: Large eddy simulation of a model gas turbine combustor. Combust. Flame **137**, 278–294 (2004)
22. Druault, P., Guibert, P., Alizon, F.: Use of proper orthogonal decomposition for time interpolation from PIV data. Application to the cycle-to-cycle variation analysis of in-cylinder engine flows. Exp. Fluids **39**, 1009–1023 (2005)
23. Dunn, M.J., Masri, A.R., Bilger, R.W.: A new piloted premixed jet burner to study strong finite-rate chemistry effects. Combust. Flame **151**, 46–60 (2007)
24. Dunn, M.J., Masri, A.R., Bilger, R.W., Barlow, R.S.: Finite-rate chemistry effects in highly sheared turbulent premixed flames. Flow Turbul. Combust. In press (2010)
25. Dunn, M.J., Masri, A.R., Bilger, R.W., Barlow, R.S., Wang, G-H.: The compositional structure of highly turbulent piloted premixed flames issuing into a hot coflow. Proc. Combust. Inst. **32**, 1779–1786 (2009)
26. Echekki, T., Chen, J.H.: Direct numerical simulation of autoignition in non-homogeneous hydrogen-air mixtures. Combust. Flame **134**, 169–191 (2003)
27. Eckbreth, A.C.: Laser Diagnostics for Combustion Temperature and Species, Gupta, A.K. and Lilley, D.G. Eds, Abacus Press (1987)
28. Ferziger, J.H., and Peric, M.: Computational Methods for Fluid Dynamics, 2nd Ed. Springer (1999)
29. Gordon, R.L.: A numerical and experimental investigation of auto-ignition. PhD Dissertation, The University of Sydney (2007)
30. Gordon, R.L., Masri, A.R., Mastorakos, E.: Simultaneous Rayleigh temperature, OH- and CH2O-LIF imaging of methane jets in a vitiated coflow. Combust. Flame **155**, 181–195 (2008)
31. Gordon, R.L., Masri, A.R., Pope, S.B., Goldin, G.M.: A numerical study of auto-ignition in turbulent lifted flames issuing into a vitiated co-flow. Combust. Theor Model. **11**, 351–376 (2007)
32. Gordon, R.L., Masri, A.R., Pope, S.B., and Goldin, G.M.: Transport budgets in turbulent lifted flames of methane auto-igniting in a vitiated co-flow. Combust. Flame **151**, 495–511 (2007)
33. Gounder, J.D., Masri, A.R.: Flow field and mass flux measurements near the exit plane of spray jets. Presented at the ICLASS 2009, 11th Triennial International Annual Conference on Liquid Atomization and Spray Systems, Vail, Colorado, USA, July 2009
34. Gounder, J.D., Masri, A.R.: Turbulent spray flames of acetone and ethanol approaching extinction. Combust. Sci. Technol. **182**, 702–715 (2010)
35. Gounder, J.D., Juddoo, M., Masri, A.R., and Starner, S.H.: Difficulties associated with using laser induced fluorescence from NO as a conserved scalar in spray jets and flames, Proceedings of the Fifth Australian Conference on Laser Diagnostics in Fluid Mechanics and Combustion, O'Neill, P., and Thiagarajan, K. (Eds.), (ISBN 978-1-74052-177-2), The University of Western Australia, Perth, Australia, pp. 11–14 (2008)
36. Hilbert, R., Thevenin, D.: Autoignition of turbulent non-premixed flames investigated using direct numerical simulations. Combust. Flame **128**, 22–37 (2002)
37. International Workshop on Measurement and Computation of Turbulent Nonpremixed Flames, `http://www.ca.sandia.gov/tdf/Workshop.html` (2009)
38. International Workshop on Turbulent Combustion in Sprays `http://www.FloHeaCom.UGent.be/`

39. Ihme, M., Pitsch, H.: Prediction of extinction and reignition in nonpremixed turbulent flames using a flamelet/progress variable model 2. Application in LES of Sandia flames D and E. Combust. Flame **155**, 90–107 (2008)
40. Kajitani, L., Dabiri, D.: A full three-dimensional characterization of defocusing digital particle image velocimetry. Meas. Sci. Technol. **16**, 790–804 (2005)
41. Kalt, P.A.M., Al-Abdeli, Y.M., Masri, A.R., Barlow, R.S.: Swirling turbulent non-premixed flames of methane: flowfield and compositional structure. Proc. Combust. Inst. **29**, 1913–1919 (2002)
42. Kaminski, C.F., Engstrom, J., Alden, M.: Quasi-instantaneous two-dimenstional temperature measurements in a spark ignition engine using 2-line atomic fluorescence. Proc. Combust. Inst. **27**, 85–93 (1998)
43. Karpetis, A.N., Barlow, R.S.: Measurements of scalar dissipation in a methane/air jet flame. Proc. Combust. Inst. **29**, 1929–1936 (2002)
44. Kempf, A.M., Forkel, H., Sadiki, A., Chen, J.-Y., Janicka, J.: Large-eddy simulation of a counterflow configuration with and without combustion. Proc. Combust. Inst. **28**, 35–40 (2000)
45. Kempf, A., Malalsekera, W., Ranga-Dinesh, K.K.J., Stein, O.: Large eddy Simulations of swirling non-premixed flames with flamelet models: a comparison of numerical methods. Flow Turbul. Combust. **81**, 523–561 (2008)
46. Knudsen, E., Pitsch, H.: A general flamelet transformation useful for distinguishing between premixed and non-premixed modes of combustion. Combust. Flame **156**, 678–696 (2009)
47. Kohse-Hoinghaus, K., Jeffries, J.B., Eds: Applied Combustion Diagnostics, Hemisphere (2002)
48. Lacaze, J., Richardson, E., Poinsot, T.: Large eddy simulation of spark ignition in a turbulent methane jet. Combust. Flame **156**, 1993–2009 (2009)
49. Lindstedt, R.P., Louloudi, S.A., Vaos, E.M.: Joint probability density function modeling of pollutant formation in piloted turbulent jet diffusion flames with comprehensive chemistry. Proc. Combust. Inst. **28**, 149–156 (2000)
50. Lindstedt, R.P., Louloudi, S.A.: Joint scalar transport probability density function modeling of turbulent methanol jet diffusion flames. Proc. Combust. Inst. **29**, 2147–2154 (2002)
51. Magre, P., Dibble, R.W.: Finite chemical kinetic effects in a subsonic turbulent hydrogen flame. Combust. Flame **73**, 195–206 (1988)
52. Magre, P., Moreau, P., Collin, G., Borghi, R., Pealat, M.: Further studies by CARS of premixed turbulent combustion in a high velocity flow. Combust. Flame **71**, 147–168 (1988)
53. Mansour, M.S., Chen, Y.-C., Peters, N.: The reaction zone structure of turbulent premixed methane-helium-air flames near extinction. Proc Combust. Inst. **24**, 461–468 (1992)
54. Markides, C.N., Mastorakos, E.: An experimental study of hydrogen autoignition in a turbulent co-flow of heated air. Proc. Combust. Inst. **30**, 881–888 (2005)
55. Masri, A.R.: `http://www.aeromech.eng.usyd.edu.au/research/thermofluids/resources`
56. Masri, A.R., Bilger, R.W.: Turbulent diffusion flames of hydrocarbon fuels stabilised on a bluff body. Proc. Combust. Inst. **20**, 319–326 (1985)
57. Masri, A.R., Bilger, R.W., Dibble, R.W.: Turbulent Nonpremixed Flames of Methane Near Extinction: Probability Density Functions, Combust. Flame **73**, 261–285 (1988)
58. Masri, A.R., Bilger, R.W., Starner, S.H.: Transition and Transport in the Initial Region of a Turbulent Diffusion Flame. Dynamics of Flames and Reactive Systems, AIAA Progress in Astronautics and Aeronautics, (J.R. Bowen, N. Manson, A.K. Oppenheim and R.I. Soloukhin, Eds.) 1984, Vol. 95, pp. 293–304 (1984)
59. Masri, A.R., Dally, B.B., Barlow, R.S., Carter, C.D.: The structure of the recirculation zone of a bluff-body combustor. Proc. Combust. Inst. **25**, 1301–1308 (1994)
60. Masri, A.R., Dibble, R.W., Barlow, R.S., The Structure of Turbulent Nonpremixed Flames of Methanol over a Range of Mixing Rates, Combust. Flame **89**, 167–185 (1992)
61. Masri, A.R., Dibble, R.W., Barlow, R.S.: The structure of turbulent nonpremixed flames revealed by Raman-Rayleigh-LIF measurements. Prog. Energy Combust. Sci. **22**, 307–362 (1996)

62. Masri, A.R., Pope, S.B., Dally, B.B.: PDF computations of a strongly swirling nonpremixed flame stabilised on a new burner. Proc. Combust. Inst. **28**, 123–132 (2000).
63. Mastorakos, E.: Ignition of turbulent non-premixed flames. Prog. Energy Combust. Sci. **35**, 57–97 (2009)
64. Mastorakos, E., Baritaud, T.B., Poinsot, T.J.: Numerical simulation of autoignition in turbulent mixing flows. Combust. Flame **109**, 198–223 (1997)
65. Medwell, P.R., Chan, Q.N., Kalt, P.A.M., Alwahabi, Z.T., Dally, B.B., Nathan, G.J.: Development of temperature imaging using two-line atomic fluorescence. Appl. Opt. **48**, 1237–1248 (2009)
66. Oefelein, J.C., Sankaran, V., Drozda, T.G.: Large eddy simulation of swirling particle-laden flow in a model axisymmetric combustor. Proc. Combust. Inst. **31**, 2291–2299 (2007)
67. O'Loughlin, W., Juddoo, M., Masri, A.R.: High-speed LIF-OH imaging in the stabilization region of lifted flames. Proc. Australian Combust. Symp. 2009 (Klimenko, A.Y., Cleary, M.J., Feng, B., Rudolph, V., Boyce, R.R., Wandel, A.P., Clements, R., Eds.) (ISBN 978-1-864999802), The University of Queensland, Brisbane, Australia, pp. 91–94 (2009)
68. Omrane, A., Sarner, G., Alden, M.: 2D-temperature imaging of single droplets and sprays using thermographic phosphors. Appl. Phys. B – Las. Opt. **79**, 431–434 (2004)
69. Omrane, A., Juhlin, G., Ossler, F., Alden, M.: Temperature measurements of single droplets by use of laser-induced phosphorescence. Appl. Opt. **43**, 3523–3529 (2004)
70. Pierce, C.D., Moin, P.: Progress-variable approach for large-eddy simulation of non-premixed turbulent combustion. J. Fluid Mech. **504**, 73–97 (2004)
71. Pitsch, H., Lageneste, D.D.: Large–eddy simulation of premixed turbulent combustion using a level-set approach. Proc. Combust. Inst. **29**, 2001–2008 (2002).
72. Robin, V., Mura, A., Champion, M., Degardin, O., Renou, B., Boukhalfa, M.: Experimental and numerical analysis of stratified turbulent V-shaped flames. Combust. Flame **153** 288–315 (2008)
73. Sankaran, V., Menon, S.: LES of spray combustion in swirling flows. J. Turbulence **3**, 11–23 (2002)
74. Schultz, C., Sick, V.: Tracer-LIF diagnostics: quantitative measurement of fuel concentration, temperature and fuel/air ratio in practical combustion systems. Prog. Energy Combust. Sci. **31**, 75–121 (2005)
75. Seffrin, F., Fuest, F., Geyer, D., Dreizler, A.: Flow field studies of a new series of turbulent premixed stratified flames. Combust. Flame **157**, 384–396 (2010)
76. Sreedhara, H., Lakshmisha, K.N.: Assessment of conditional moment closure models of turbulent autoignition using DNS data. Proc. Combust. Inst. **29**, 2069–2077 (2002)
77. Starner, S.H., Gounder, J., Masri, A.R.: Effects of turbulence and carrier fluid on simple, turbulent spray jet flames. Combust. Flame **143**, 420–432 (2005)
78. Steinberg, A.M., Driscoll, J.F., Ceccio, S.L.: Temporal evolution of flame stretch due to turbulence and the hydrodynamic instability. Proc. Combust. Inst. **29**, 1713–1721 (2009)
79. Tang, Q., Xu, J., Pope, S.B.: Probability density function calculations of local extinction and NO production in piloted-jet turbulent methane/air flames. Proc. Combust. Inst. **28**, 133–140 (2000)

Chapter 16
Uncertainty Quantification in Fluid Flow

Habib N. Najm

Abstract This chapter addresses the topic of uncertainty quantification in fluid flow computations. The relevance and utility of this pursuit are discussed, outlining highlights of available methodologies. Particular attention is focused on spectral polynomial chaos methods for uncertainty quantification that have seen significant development over the past two decades. The fundamental structure of these methods is presented, along with associated challenges. We also discuss demonstrations of their use in a number of fluid flow applications covering a range of complexity that is inherent in turbulent combustion.

16.1 Introduction

Uncertainty Quantification (UQ) broadly refers to the quantitative estimation of uncertainty in computational modeling of physical processes. Recent years have seen increasing interest in this topic, particularly in the context of computational modeling of fluid systems [29, 52, 80, 125, 145]. This is due in part to the increased complexity of computational models and our increased reliance on computational predictions for engineering design, scientific discovery, and decision support. It is also due to the development, over the past couple of decades, of efficient and effective UQ methods based on probability theory. In this chapter, we outline the basics of these methods, as well as their key challenges, and discuss their utility and application in computations of a range of fluid systems involving various facets of the complexity inherent in turbulent reacting flow.

In general, uncertainty quantification is useful from both the engineering and scientific points of view. In the engineering context, confidence intervals on predicted system behavior are necessary for design optimization, for reliability assessment, for the determination of safety factors in engineered systems, and for decision

Habib N. Najm
Sandia National Laboratories, Livermore, CA 94550, USA, e-mail: hnnajm@sandia.gov

T. Echekki, E. Mastorakos (eds.), *Turbulent Combustion Modeling*,
Fluid Mechanics and Its Applications 95, DOI 10.1007/978-94-007-0412-1_16,

support purposes. In the scientific context, model validation with respect to experimental measurements requires quantification of uncertainty in both experimental measurements and computational predictions. In the absence of such uncertainty "error bars", it is not possible to state whether any disagreements between model predictions and experimental observations indicate lack of validity of a model, or whether they are within the inherent range of uncertainty in the predictions and the observations.

Further, from a broad perspective, UQ is an important element of the overall Verification and Validation (V&V) challenge in computational modeling [90, 92, 94]. V&V is of key importance for computational code certification [91, 121]. Starting from a given mathematical model of a physical system, any subsequent discretization of the governing equations for computational purposes necessarily introduces discretization errors. It is imperative that these errors be quantified and minimized such that their impact on computational predictions is "small". Once computational predictions are shown to be consistent with the underlying mathematical model, then the associated code is "verified" [14, 29, 91, 110]. This verification challenge is a significant undertaking, and there are many examples of substantial code-to-code scatter in computational predictions of fluid flow problems [18, 42]. However, verification is outside the scope of this UQ discussion. Rather, the UQ context of interest here is of relevance in "validation", where the focus is on the degree to which computational predictions agree with empirical observations [90, 94].

Potential sources of uncertainty in computational modeling include any inputs, initial/boundary conditions, and parameters that enter the governing equations, as well as those relevant in the specification of the geometry of the computational domain. In the present discussion, we refer to these broadly as model parameters. Other, non-parametric sources of uncertainty, such as modeling assumptions, some aspects of model structure, or choices of constitutive laws, can be dealt with using information theoretic methods, e.g. Bayesian model averaging [43], but are outside the present scope. In the following, we focus primarily on parametric uncertainty.

Moreover, we adopt the Bayesian view of probability [47], where probability is inherently the degree of belief in a proposition, and does not necessarily derive from sampling or observation. This viewpoint provides the means of handling both "epistemic" and "alleatoric" uncertainty [92] in the context of probability theory. Moreover, it provides the means of seamlessly merging subjective prior information and empirical data in probabilistic inference, and is a sound foundation for sequential learning. Particularly, in the absence of data, prior knowledge, *e.g.* based on expert elicitation, serves to provide the requisite probabilistic representation. The interested reader may consult numerous sources on alternative, non-probabilistic, viewpoints on this topic [41]. These include Evidence theory [8, 40, 93, 118], Possibility theory [22], Fuzzy set theory [23, 124], and Imprecise Probability theory [54]. Note that a key advantage of probabilistic UQ methods is the utility of the concept of *measure* in probability theory. In the following, we presume the existence of sufficient information/data about model parameters to allow assigning them PDFs and/or evaluating their statistics, enabling the use of probabilistic UQ methods.

Further, the present discussion is focused on the *forward* propagation of uncertainty in computational models, *i.e.* the propagation of uncertainty from parameters to outputs/observables of interest. On the other hand, the quantification of the parametric uncertainties themselves is an equally important task. When experimental data is used to estimate uncertain parameters, one is left with parametric uncertainty that reflects a number of sources of uncertainty in experimental measurements, *e.g.* limited data and instrument noise. Further, recall that experimental measurements rarely provide direct measurement of a given parameter. Rather, an experiment measures a given observable that is the result of some instrument/experiment forward model to which the parameter of interest is an input. Consequently, parameter estimation from empirical observations involves inference and the solution of an inverse problem. Given that inverse problems are frequently ill-conditioned and *a-causal* [48], this leads to significant challenges in parameter estimation and ensuing amplification of any noise effect, leading to potentially large parametric uncertainty. There are numerous strategies for parameter inference, and estimation of parametric uncertainties, including e.g. least-squares methods and Bayesian methods. If the chosen (forward) UQ strategy is a probabilistic one, requiring full specification of the uncertain parameters as stochastic quantities, then an inference method that provides full probabilistic characterization of the uncertain parameters, such as Bayesian inference, is necessary. Whether the resulting specification of uncertain parameters is in terms of confidence intervals or probability distributions, however, it is important to note that the solution of the inverse problem generally also identifies correlations between multiple model parameters. Such correlations, if they exist, need to be included in the forward UQ problem, for accurate evaluation of uncertainties in model predictions [81]. In the present context, we will presume that model parameters are specified probabilistically *a priori*, whether correlated or not, and will not discuss the inference problem. Bayesian methodologies have been used in general UQ studies [39, 49, 50, 72, 73, 88, 89], for model validation under uncertainty [109], and, more specifically, in computations of fluid systems [28].

There are in general a number of UQ strategies that have been employed in the literature. Before outlining available techniques however, it is useful to point out that the curse of dimensionality, and the complexity of the forward model, are significant challenges to all such strategies. Thus, a likely good first strategy before proceeding to quantification of uncertainty is a screening strategy [16] for identifying those parameters with negligible influence on the model predictions. In this approach, the forward model is run repeatedly with significant individual changes in each of the input parameters values, allowing a rough ranking of the parameters in terms of relative importance. This information can provide guidance as to what parameters may be safely dropped from any more detailed UQ studies of this model, given their negligible influence on the model outputs. With the likely set of "important" parameters identified, more precise UQ approaches can be subsequently pursued on a relatively small set of parameters. Of course, care must be taken using these methods in the context of models with significant non-linearity, and potential bifurcation behavior.

Historically, the most commonly used UQ approach is based on local sensitivity analysis and variance propagation [13, 16]. In this context, sensitivity analysis is done at the nominal value of model parameters, arriving at sensitivity coefficients quantifying the partial derivative of each model output with respect to each parameter. Given this information, and presuming sufficiently "small" degrees of uncertainty, the variances of the model output due to each parameter variation over its respected range are added to arrive at the total variance in model output [129]. This approach provides a reasonable representation of output uncertainties as long as the uncertainties in the input parmeters and the resulting model outputs are sufficiently small. Moreover, it does not allow for examination of bifurcation of the solution over the parameter uncertainty ranges. Further, it clearly does not make use of specific known probability distributions of model parameters, except in using the first and second moments; and does not provide probabilistic characterization of model outputs, again except in terms of the first and second moments. Walters and Huyse [125] discuss various UQ methods, including the propagation of error using sensitivity derivatives, and higher fidelity approaches such as moment methods, similarly based on Taylor series expansions about the mean input values, as used in CFD (see *e.g.* [105]).

Global sensitivity analysis techniques have also been extensively used in the UQ literature. These methods employ sampling of the model parameters, given their specified probability density functions (PDFs), and evaluate the output model prediction for each sample. The statistics of the model outputs provide the probabilistic characterization of their resulting uncertainty. These sampling-based methods fall squarely within the probabilistic context. We will discuss sampling methods below in relation to Polynomial Chaos (PC) UQ methods, and will address alternative sampling strategies. Beside this, however, we will not address other statistical sampling methods for global sensitivity analysis, and we refer the reader to the extensive literature in this area adequately reviewed by Saltelli [16]. Walters and Huyse [125] present an exposition of this approach, using random Monte Carlo (MC) sampling, with demonstrations in various fluid flow problems.

In the following, we begin with a discussion of the basics of the PC construction, its use in UQ, and associated challenges due to high dimensionality and nonlinearity. We then discuss the application of PC UQ methods in a range of fluid flow applications, starting with laminar incompressible flow, and proceeding through chemically reacting and compressible flow applications. These discussions lead up to the issue of UQ in turbulence, thereby rounding up the full range of complexity of the UQ problem in computations of turbulent combustion.

16.1.1 Polynomial Chaos

The key challenge with probabilistic UQ based on random sampling is that typical random sampling methods, which are largely variants of MC, have very slow convergence with respect to the number of samples. In other words, large numbers of samples are required to achieve adequate accuracy in computed moments, *e.g.* means and standard deviations, of uncertain quantities of interest. This is an insurmountable hurdle when the model is computationally expensive, requiring massively parallel large-scale computational effort for each sample. Significant progress has been made in this regard over the past couple of decades through the development of spectral methods for representation of random quantities, specifically PC methods. The underlying formulation for the PC representation of random variables is presented below.

Let $(\Omega, \mathfrak{G}, P)$ be a probability space, where Ω is an event space, $\mathfrak{G}$ is a σ-algebra on Ω, and P is a probability measure on $(\Omega, \mathfrak{G})$. Any RV $X : \Omega \rightarrow \mathbb{R}$, that has finite variance, *i.e.* is in $L^2(\Omega)$, can be represented as an expansion in terms of any complete set of functions that are orthogonal with respect to the probability measure on $L^2(\Omega)$. Given this, one introduces a set of n coordinates in $\mathbb{R}^n$, being the images of n standard RVs $\boldsymbol{\zeta} : \Omega \rightarrow \mathbb{R}^n$ with known distributions, to provide a convenient expansion in $\mathbb{R}^n$. Accordingly, any L^2 RV can be represented as an expansion in terms of a complete set of functions of $\boldsymbol{\zeta}$ that are orthogonal with respect to the density of $\boldsymbol{\zeta}$.

Wiener [130] presented the first use of this construction, employing Hermite polynomial functionals of standard normal RVs, for representation of Gaussian random processes, namely the Wiener process model for Brownian motion, terming the resulting expansion Polynomial Chaos (PC)*. Ghanem and Spanos [38] employed this Wiener-Hermite (WH) PC for representation of any L^2 random process $X(\omega, \boldsymbol{x}, t)$:

$$X(\omega, \boldsymbol{x}, t) = \sum_{k=0}^{\infty} X_k(\boldsymbol{x}, t) \Psi_k(\boldsymbol{\xi}) \tag{16.1}$$

where the Ψ_k are multivariate Hermite polynomials, and $\boldsymbol{\xi} = \{\xi_1, \ldots, \xi_n\} = \boldsymbol{\xi}(\omega)$ is a vector of n independent standard normal RVs. The PC expansion (PCE) is equally applicable to random processes/variables. For use in a computational setting, we truncate the PCE to finite order, giving

$$X \simeq \sum_{k=0}^{P} X_k \Psi_k(\boldsymbol{\xi}). \tag{16.2}$$

By construction, orthogonality of the Hermite polynomials with respect to the density of the $\boldsymbol{\xi}$-basis leads to

$$X_k = \frac{\langle X \Psi_k \rangle}{\langle \Psi_k^2 \rangle} = \frac{1}{\langle \Psi_k^2 \rangle} \int_{\Omega} X(\omega) \Psi_k(\boldsymbol{\xi}(\omega) \mathrm{d}P(\omega) = \frac{1}{\langle \Psi_k^2 \rangle} \int X(\boldsymbol{\xi}) \Psi_k(\boldsymbol{\xi}) p_{\boldsymbol{\xi}}(\boldsymbol{\xi}) \mathrm{d}\boldsymbol{\xi} \tag{16.3}$$

for $k = 0, \ldots, P$, where

* This PC representation of nonlinear functionals of the Wiener process was shown to be convergent in the limit of $n \rightarrow \infty$ [15]. Ogura [95] extended this proof for representation of nonlinear functionals of a Poisson process, employing expansions in terms of Charlier polynomials and Poisson-distributed RVs. These are the only two sets of orthogonal functions for which the convergence of PC in the infinite dimensional limit has been proven. In the present context we will handle finite n cases exclusively.

$$p_{\boldsymbol{\xi}}(\boldsymbol{\xi}) = \prod_{i=1}^{n} p_{\xi_i}(\xi_i) \tag{16.4}$$

and $\xi_i \sim N(0,1)$, *i.e.* $p_{\xi_i}(x) = \exp(-x^2/2)/\sqrt{2\pi}$. With this expansion, one can easily construct the PDF of X by generating random samples of $\boldsymbol{\xi}$, and binning the resulting X samples. Further, for any $f(X)$, we have

$$\langle f(X)\rangle = \int_{\Omega} f(X(\omega))\mathrm{d}P(\omega) = \int f(X(\boldsymbol{\xi}))p_{\boldsymbol{\xi}}(\boldsymbol{\xi})\mathrm{d}\boldsymbol{\xi}. \tag{16.5}$$

Thus, any $\langle f(X)\rangle$, including moments of interest, can be evaluated as long as $X(\boldsymbol{\xi})$ is known. Given the PCE for X, the integral can be evaluated analytically for simple moments. For example, $\mu_X = X_0$, and $\sigma_X^2 = \sum_{k=1}^{P} X_k^2 \langle \Psi_k^2 \rangle$.

To illustrate the use of PC for UQ, let us consider a model

$$F(u,x,t;\lambda) = 0 \tag{16.6}$$

where $x \in \mathbb{R}^3$ is the spatial coordinate, t is time, $\lambda \in \mathbb{R}$ is a parameter, and $u \in \mathbb{R}$ is the model output. The use of a single parameter and one model output is for convenience. The analysis applies equally to models with multiple outputs, and many uncertain parameters. Consider further that λ is uncertain, and is modeled as a random variable. Therefore the model output, u, is also a random variable. We are interested in evaluation of the distribution of u, its moments, or any $\langle f(u)\rangle$ of interest.

We presume that either the PDF p_λ of λ is known, or it can be estimated from available empirical observations of λ. In the MC context, we rely on some random sampling strategy to generate samples λ^i from p_λ. For each such sample, we solve the forward model to evaluate the corresponding u^i, from which we have

$$\langle f(u)\rangle = \frac{1}{N}\sum_{i=1}^{N} f(u^i). \tag{16.7}$$

Note that MC convergence is insensitive to dimensionality and does not rely on any smoothness of the integrand. However, this robustness is achieved at the cost of significant computational expense.

Employing PC, on the other hand, we begin by constructing the PCE for λ,

$$\lambda = \sum_{k=0}^{P} \lambda_k \Psi_k(\boldsymbol{\xi}) \tag{16.8}$$

where $\boldsymbol{\xi}$ is n-dimensional. This PCE can be constructed using the PDF of λ. Note that, in general, the PCE constructed from the PDF is not unique, since *e.g.* both expansions $\lambda = \lambda_0 \pm \lambda_1\xi$ have the same Gaussian PDF $N(\lambda_0, \lambda_1^2)$. For $n = 1$, the PCE coefficients can be simply found using

$$\lambda_k = \frac{1}{\langle \Psi_k^2 \rangle}\int_0^1 \Phi_\lambda^{-1}(y)\,\Psi_k(\Phi_\xi^{-1}(y))\,\mathrm{d}y \tag{16.9}$$

where $\Phi_\lambda()$ and $\Phi_\xi()$ are the cumulative distribution functions (CDFs) of λ, and ξ, respectively [141]. Doing this for $n > 1$ is typically achieved employing a fitting procedure, inferring the PCE coefficients that result in a PDF for λ that best approximates the target PDF. Finally, in the case where a joint PDF is known for a parameter vector $\boldsymbol{\lambda}$, the Rosenblatt transform [111] can be used to map $\boldsymbol{\lambda}$ to a vector of independent uniformly distributed RVs each on $[0,1]$, whose PCEs can be estimated per the above procedure.

Having determined the PCE for λ, the key objective is then to construct the PCE for u,

$$u = \sum_{k=0}^{P} u_k \Psi_k(\boldsymbol{\xi}) \tag{16.10}$$

by evaluating the mode strengths given by the Galerkin projection

$$u_k = \frac{\langle u\Psi_k \rangle}{\langle \Psi_k^2 \rangle} = \frac{1}{\langle \Psi_k^2 \rangle} \int u(\boldsymbol{\xi})\Psi_k(\boldsymbol{\xi}) p_{\boldsymbol{\xi}}(\boldsymbol{\xi}) \mathrm{d}\boldsymbol{\xi}. \tag{16.11}$$

This can be done generally in two ways, as follows.

16.1.1.1 Intrusive PC UQ

The first approach employs Galerkin projection of the governing equations,

$$\left\langle F\left[\left(\sum_{i=0}^{P} u_i\Psi_i\right), x, t; \left(\sum_{j=0}^{P} \lambda_j \Psi_j\right)\right] \Psi_k \right\rangle = 0, \quad k = 0,\ldots,P \tag{16.12}$$

thereby arriving at a reformulated set of governing equations for the u_k. This approach discards the original deterministic model code, requiring discretization, algorithms, and associated code development for the new set of equations, and has been consequently termed “intrusive”.

As a simple example, consider the ODE problem

$$\frac{\mathrm{d}u}{\mathrm{d}t} = \lambda u, \quad u(t=0) = u_0 \tag{16.13}$$

where $\lambda = \sum_{k=0}^{P} \lambda_k \Psi_k(\boldsymbol{\xi})$ is an uncertain parameter, with a chosen PCE dimensionality and order. Applying the above Galerkin projection for any mode k, leads to

$$\left\langle \Psi_k \frac{\mathrm{d}}{\mathrm{d}t} \sum_{l=0}^{P} u_l \Psi_l(\boldsymbol{\xi}) \right\rangle = \left\langle \Psi_k \left[\sum_{i=0}^{P} \lambda_i \Psi_i(\boldsymbol{\xi})\right] \left[\sum_{j=0}^{P} u_j \Psi_j(\boldsymbol{\xi})\right] \right\rangle \tag{16.14}$$

and since, by construction, $\langle \Psi_i \Psi_j \rangle = \delta_{ij} \langle \Psi_i^2 \rangle$, we have

$$\frac{\mathrm{d}u_k}{\mathrm{d}t} = \sum_{i=0}^{P}\sum_{j=0}^{P} \lambda_i u_j \frac{\langle \Psi_i \Psi_j \Psi_k \rangle}{\langle \Psi_k^2 \rangle}, \quad k = 0,\ldots,P \tag{16.15}$$

where $C_{ijk} = \langle \Psi_i \Psi_j \Psi_k \rangle / \langle \Psi_k^2 \rangle$ is a tensor that can be evaluated once and stored. Thus, we have a new, larger, ODE system to solve for PCE mode strengths $\{u_k\}_{k=0}^P$, which determine the PCE $u = \sum_{k=0}^P u_k \Psi_k(\boldsymbol{\xi})$. With this in hand, one can generate samples of u by sampling $\boldsymbol{\xi}$, to evaluate moments or distributions, or any other purpose, such as evaluating any $\langle f(u) \rangle$.

16.1.1.2 Non-intrusive PC UQ

Alternatively, one can evaluate the Galerkin projection integrals (Eq. 16.11) numerically. This can be done in a number of ways. To begin with, of course, one can do this using any variant of MC. There is no inherent computational advantage here, except that the projection of the statistics of u onto its PC modes immediately provides, for no additional cost, sensitivity information regarding the dependence of the uncertain u on any one of multiple model parameters, linked via derivatives on $\boldsymbol{\xi}$. Of course, relying on the above smooth *global* WH PC constructions, where $\boldsymbol{\xi}$ has infinite support, this approach does presume a smooth dependence $u(\boldsymbol{\xi})$, although the use of *local* PC methods, outlined further below, addresses this issue. Using this approach, with $\boldsymbol{\xi}$ sampled from its PDF, employing N samples of $\boldsymbol{\xi}$ and corresponding computed realizations of u, the integrals in Eq. 16.11 become

$$u_k = \frac{1}{\langle \Psi_k^2 \rangle} \frac{1}{N} \sum_{i=1}^{N} u^i \Psi_k(\boldsymbol{\xi}^i), \quad k = 0, \ldots, P \tag{16.16}$$

On the other hand, the key advantage of non-intrusive PC methods is to enable the use of *deterministic* sampling of u, based on quadrature formulae, to provide estimates of the projection integrals. The central problem is the evaluation of

$$\langle u \Psi_k \rangle = \int_{-\infty}^{\infty} \cdots \int_{-\infty}^{\infty} u(\boldsymbol{\xi}) \Psi_k(\boldsymbol{\xi}) \left[\prod_{i=1}^{n} p_{\xi_i}(\xi_i) \right] \mathrm{d}\xi_1 \cdots \mathrm{d}\xi_n \tag{16.17}$$

which can be done using n-dimensional Gauss-Hermite quadrature [1, 27, 64]. With $\tilde{m}$ quadrature points in each dimension, this integration is exact if $f(x)$ has a polynomial degree smaller than $2\tilde{m}$. For a PCE order p, the maximum degree of $(u\Psi_k)$ in each ξ_i is $2p$, such that a quadrature discretization in each dimension with $\tilde{m} = p+1$ is sufficient. Thus, the full tensor product Gauss-Hermite quadrature rule for the n-dimensional integral requires $m = (p+1)^n$ points. This exponential rise in the number of requisite evaluations of the forward model is quite detrimental in high dimensional problems, as is further discussed below.

16.1.1.3 Arbitrary basis

As already indicated, the above PC construction, with finite n, can be implemented using any set of orthogonal functions and their corresponding measure on $L^2(\Omega)$. In general, given a chosen random basis $\boldsymbol{\zeta}$ of arbitrary distribution, an associated set of orthogonal functions can be generated, which is then readily usable to construct an expansion to represent any RV in terms of $\boldsymbol{\zeta}$ [33, 120]. This has been demonstrated, for example, using a range of orthogonal polynomials, members of the Askey scheme of hypergeometric orthogonal polynomials [3, 115], and their associated random bases, and termed a generalized PC (gPC) [141, 142]. This set includes, besides the Hermite-Gaussian pair, Legendre-Uniform (LU), Laguerre-Gamma, Jacobi-Beta constructions for continuous RVs, and another set of polynomial-basis pairs for discrete valued RVs. The generation of tailored orthogonal functions has also been demonstrated in the construction of multiwavelet chaos expansions [61].

One advantage of this flexibility is the ability to choose an optimal basis under some conditions. In particular, given a RV Z whose distribution $p_Z()$ is *known*, then an optimal choice of the specific chaos expansion $X = \sum X_k \Phi_k(\boldsymbol{\zeta})$ is that whose basis $\boldsymbol{\zeta}$ has a PDF that is nearest to $p_Z()$ [12, 141, 142]. For example, the optimal representation of a Gaussian RV is in terms of WH PC, being exact employing a 1D first-order expansion. Similarly, a Uniform RV is best represented using LU PC. Of course, while the parameter PDFs may be known, allowing this selection, in general the distribution of an RV that is an output of a general model is not known *a priori*.

16.1.2 Challenges in PC UQ Methods

There are key challenges for the PC UQ approach associated with requirements for (a) high dimensionality n, and (b) high order p. These arise from a number of practical requirements in physical systems, and have motivated significant research in recent years. We outline these briefly in the following.

16.1.2.1 High dimensionality

The *necessary* dimensionality of the PC system in the UQ context represents the number of degrees of freedom that are required to represent the input space, *i.e.* the uncertain parameters, initial/boundary conditions, etc. Each independent uncertain parameter introduces at least one associated degree of freedom, while known dependences among parameters can constrain this number. Input parameters that are random processes, as opposed to random variables, can require many degrees of freedom to represent, according on their known correlation structure [38]. At the same time, the dimensionality is generally a modeling question. The *efficient* (*i.e.* low order for given accuracy) PC representation of the random variables/fields that are either inputs or components of the system state may well require introduction of a higher number of independent degrees of freedom. For example, by definition, the RV $Z = \sum_{k=1}^{n} \xi_k^2$, where the ξ_k's are independent standard normal RVs, has a chi-square distribution with n degrees of freedom, *i.e.* $Z \sim \chi_k^2$. Clearly, for any given $Y \sim \chi_k^2$, a second order WH PCE $Y = \sum_{k=0}^{2} Z_k \Psi_k(\xi_1, \ldots, \xi_n)$ is exact with the proper choice of coefficients, and is a more efficient model for Y than a PCE with $n < k$, for which higher order terms (> 2) would be required to approximate the exact model with high accuracy.

In general, the resulting dimensionality n of the PC system can be quite large. The limit of large n can lead to significant computational expense. To begin with, P, the overall size of the PCE increases with n, following

$$P+1 = \frac{(n+p)!}{n!p!} \tag{16.18}$$

where p is the order. Thus, a 4th order PCE with $n=4$ requires $P=69$, while $n=8$, requires $P=494$. In the intrusive context, the size of the Galerkin-projected equation system increases quickly with n. At the same time, in the non-intrusive context, dense full-tensor product quadrature formulae, *e.g.* employing Gauss-Hermite quadrature as outlined above, require a number of evaluations of the forward model that goes up exponentially with n, specifially $m=(p+1)^n$. With $p=4$, this requires $m=625$ for $n=4$, and $m=390,625$ for $n=8$. This can quickly become prohibitively expensive, more so than MC.

Recent developments have targeted these challenges. Reducing the number of terms in the expansion has been addressed using PCE constructions based on sparse tensor products [11, 30, 65, 116, 117, 123], and employing arbitrary adapted probability measures [120]. At the same time, reducing the number of forward model evaluations, in the non-intrusive/collocation context, has been addressed employing Smolyak [98–100, 119] or other sparse-quadrature/cubature integration formulae in general [9, 32, 34, 35, 77, 85–87, 140]. There has been significant progress in the application of these methods in UQ, such that integration of systems with substantial dimensionality is feasible, and more efficient than MC sampling [83, 84].

16.1.2.2 High order

The requisite PC order for representing a given RV corresponds to the highest necessary degree in the $\Psi_k(\boldsymbol{\xi})$ polynomials required to capture the inherent $u(\boldsymbol{\xi})$ dependence, with sufficiently small truncation error in high-order terms. This requisite order can be high, even with mild uncertainty in the input space, depending on the non-linearity of the forward model. Two important classes of problems exhibit this challenge, with particular relevance in fluid systems. These are systems exhibiting (1) long time horizon dynamics with large growth rate of phase-errors, *i.e.* strong rate of loss of correlation, or (2) sharp/discontinuous dependence on uncertain parameter values. These are considered in some detail below.

Long Time Horizon

A key characteristic of non-linear dynamics is the large rate of growth of small differences in initial conditions. This has direct consequences to the UQ problem. Small parametric/input uncertainties can translate to large and fast-growing phase-variances in the solution. In the PC context, representing the resulting uncertain system state as a function of the basis $\boldsymbol{\xi}$ used to characterize the input space, and maintaining a specific accuracy in the representation, requires progressively higher PC order as the time horizon of interest is increased. Pettit and Beran [101] provided a concise illustration of this challenge employing a sinusoidal function with an uncertain frequency, where the resulting oscillatory frequency of the system state as a function of $\boldsymbol{\xi}$ is observed to increase with the observation horizon.

This is a serious challenge for the application of PC methods in fluid systems exhibiting non-linear oscillatory dynamics or convective dynamics of vortex structures. This challenge applies to both intrusive and non-intrusive implementations as long as the system state observable of interest has a high degree of dependence on initial conditions, therefore exhibiting a large rate of loss of correlation, and growth of phase variances, in time. On the other hand, one easy resolution of this challenge is available conveniently in the non-intrusive context, by focusing on observables that are *well-behaved* with respect to initial conditions, even though the detailed system state is not. For example, in the context of DNS of turbulent flow, the instantaneous detailed state of the system, comprised of the velocity vector as a function of space and time, is highly sensitive to small changes in initial conditions. Representing this uncertain field over long time horizons, accounting for parametric/input uncertainties using PC, is a losing proposition, irrespective of the means employed for evaluating the associated PCE. On the other hand, the dependence of averaged quantities, such as mean/RMS profiles or energy spectra, is much smoother with respect to initial conditions or model parameters. Of course, without employing some means of turbulence modeling, there are no *a priori* equations for these smooth observables, so their use in the intrusive DNS context is not possible. However, they are usable in the non-intrusive context, where they, as opposed to the detailed system state, are expressed as PCEs. The associated dependence on $\boldsymbol{\xi}$ is smooth, and the requisite order is small. This is a clear argument for non-intrusive methods in this context. This approach has been employed for development of effective collocation-based long time-horizon PC and local Lagrange-interpolant UQ strategies for a range of transient oscillatory systems, including structural and/or flow-structure dynamics, employing smooth observables that parametrize the transient oscillatory system response [133–139].

Discontinuities

Another common feature of non-linear systems, frequently observed in fluid systems, is the existence of high-rates of change of the solution in space/time, discontinuities, and bifurcations over parametric space. Examples include shocks in compressible flow, ignition/flame-fronts in reacting flow, as well as flow instability and transitions between laminar/turbulent flow regimes. Discontinuity of the system state as a function of the input/parametric space, and therefore as a function of $\boldsymbol{\xi}$, is a serious challenge for the above introduced PC construction. Sharper dependence on $\boldsymbol{\xi}$, in principle requires higher PC order to maintain a given accuracy in the representation. However, as a discontinuous structure is approached, the above *global* spectral PC constructions, where the $\boldsymbol{\xi}$-basis has infinite support, will fail to maintain accuracy for any finite order. Obviously, a polynomial representation in terms of a smooth basis is not useful for representing a step-function, leading to the familiar Gibbs phenomenon. This challenge led to the development of *local* multiresolution analysis (MRA) PC methods employing multiwavelet bases with compact support [58, 61]. In this context, the input space is subdivided into a number of blocks/elements. Distinct PC representations are constructed over each element using local bases with compact support, and propagated through the forward model following conventional PC UQ procedures. Statistics and moments of the system state can be computed by combining associated elemental statistics/moments. In the simplest version of MRA, a PC construction comprised of Legendre polynomials along with first-level 1D wavelet details, both functionals of a local uniformly-distributed RV basis, are used on each element. The Legendre subset of the construction is the Legendre-Uniform gPC with a local basis, while the directional wavelet details provide useful information on the need for refinement of the element mesh in each stochastic dimension. The Multi-Element gPC (ME-gPC) construction employs the same above local LU PC representation, without the wavelet details [128]. Moreover, a range of stochastic finite element constructions have been developed employing local interpolants, as opposed to local PCEs, in numerous applications [2, 5–7, 25, 32, 71, 82, 133, 137]. In general, local constructions are useful and necessary in both the intrusive and non-intrusive contexts, where discontinuous/bifurcation behavior is present. Note, in particular, that even smooth observables can exhibit bifurcative behavior as a function of model parameters, *e.g.* due to flow instability and change of overall flow character at a critical parameter value.

16.2 Polynomial Chaos UQ in Fluid Flow Applications

There have been many demonstrations of the application of PC UQ methods in a range of fluid flow applications, the earliest involving flow in porous media [36, 37]. The discussion below addresses demonstrations in general incompressible, compressible, and reacting flow; with a last section devoted to turbulent flow.

16.2.1 Incompressible Flow

The incompressible Navier-Stokes equations have been solved in a number of flow contexts, using a range of numerical methods, employing PC UQ. We discuss these in the following, considering the general classes of Eulerian and Lagrangian schemes.

16.2.1.1 Eulerian methods

The intrusive PC reformulation of the incompressible Navier-Stokes equations in an Eulerian framework, employing Galerkin projection, leads to a set of coupled partial differential equations for the mode-strengths of the velocity field PCE, along with associated continuity constraints [60]. These are, for $k = 0,\ldots,P$,

$$\frac{\partial \mathrm{u}_k}{\partial t} + \sum_{i=0}^{P}\sum_{j=0}^{P} (\mathrm{u}_i \cdot \nabla)\mathrm{u}_j C_{ijk} = -\nabla \mathrm{p}_k + \sum_{i=0}^{P}\sum_{j=0}^{P} \nu_i \nabla^2 \mathrm{u}_j C_{ijk} \tag{16.19}$$

$$\nabla \cdot \mathrm{u}_k = 0, \tag{16.20}$$

where $\mathrm{u}_k, \mathrm{p}_k$, and ν_i are the PC modes for the velocity, pressure, and viscosity, respectively; and C_{ijk} has been already introduced. Using a projection scheme [20, 51] to solve these equations leads to a decoupled set of elliptic solves for the pressure field. Specifically, rewriting Eq. 16.19 as

$$\frac{\partial \mathrm{u}_k}{\partial t} = -\nabla \mathrm{p}_k + \mathscr{H}_k, \tag{16.21}$$

a projection scheme employing, *e.g.* a multi-step time integration strategy proceeds with a first fractional step at time t^n, giving

$$\frac{\mathrm{u}_k^* - \mathrm{u}_k^n}{\Delta t} = F(\mathscr{H}_k^n, \mathscr{H}_k^{n-1}, \ldots), \qquad k = 0,\ldots,P \tag{16.22}$$

where u_k^* are the provisional velocity modes. In the second fractional step, a pressure correction is applied to the provisional velocity in order to satisfy the divergence constraints, arriving at the velocity at time t^{n+1},

$$\frac{\mathrm{u}_k^{n+1} - \mathrm{u}_k^*}{\Delta t} = -\nabla \mathrm{p}_k, \qquad k = 0,\ldots,P \tag{16.23}$$

where the pressure fields p_k are determined so that the fields u_k^{n+1} satisfy the divergence constraints in (16.20), i.e.

$$\nabla \cdot \mathrm{u}_k^{n+1} = 0, \qquad k = 0,\ldots,P \tag{16.24}$$

Combining equations (16.23) and (16.24) results in the following system of decoupled Poisson equations:

$$\nabla^2 \mathfrak{p}_k = -\frac{1}{\Delta t} \nabla \cdot \mathfrak{u}_k^* \qquad k = 0, \ldots, P. \tag{16.25}$$

Each of the above independent Poisson equations is solved subject to Neumann boundary conditions, as in the conventional projection scheme. Presuming N mesh points over the physical domain, the solution of the decoupled pressure equations, requiring $(P+1)$ Poisson equation solutions, each involving an $N \times N$ matrix, is computationally efficient compared with the alternative, namely the solution of a fully coupled $(P+1)N \times (P+1)N$ system.

This construction, based on a global WH PC implementation and a finite difference discretization has been used to solve several variants of the laminar channel flow, including flows with uniform and temperature-dependent viscosity, and uncertain temperature boundary conditions [60]. It has also been used for computations of stable natural convective flow in a differentially heated cavity with specified uncertain wall temperature [63, 64]. When handling large temperature differences, in the non-Boussinesq limit, particular attention is necessary regarding mass conservation for ensuring numerical stability, requiring the use of a highly accurate stochastic inversion algorithm.

Application of these techniques in Rayleigh-Bénard flow, accounting for natural convection instabilities, requires adequate handling of bifurcations [58]. The stability of this flow is governed by the value of the Rayleigh number *Ra*, with a critical value Ra^* delineating the boundary between a stable regime with conductive heat transfer and an unstable regime with predominantly convective heat transfer. Thus, considering natural convection in a 2D cavity, when the specified uncertain bottom wall temperature results in an uncertain *Ra* spanning a range that straddles Ra^*, the spectral expansion has to represent a step-function dependence on the stochastic dimesion ξ, clearly infeasible for global polynomial expansions in terms of smooth functions. Studies of Rayleigh-Bénard flow have shown that the bifurcation at Ra^* challenges the utility of global LU PC, necessitating the use of of a local Wavelet-Uniform PC construction instead [58]. The local construction deals with this bifurcation effectively, exhibiting robust physical behavior.

The PC representation fits quite naturally in the context of finite or spectral element flow solution methods, and has been applied in this context for the study of a range of uncertain incompressible flows. Spectral element implementations have been applied, along with a high-order projection scheme, using global Wiener-Hermite PC, for studying flow-structure interactions in flow around a circular cylinder at low Reynolds number [143]. Global gPC implementations have also been used in this context, examining incompressible 2D channel and cylinder flows [142], and for studies of spatially developing shear layers [53]. Further, this construction has been used for computations of flow around fixed cylinders with multiple vibrational degrees of freedom, as well as oscillating cylinders where shedding mode switching resulting from noisy inflow was evident [67, 69, 70].

In the finite element context, global gPC UQ has been implemented with a variational multiscale stabilized finite element method for solving the incompressible Navier-Stokes equations in a number of flow configurations including channel, driven cavity, and cylinder flows [82]. The convergence rate of the GMRES solver used in this construction is reduced significantly in going from the deterministic problem to the stochastic one, suggesting the need for improved preconditioning strategies for the Galerkin-projected system of equations. This methodology has also been applied in studies of natural convection in a differentially heated cavity, where bifurcative behavior also necessitates use of local interpolant-polynomial constructions [4]. Non-intrusive adaptive sparse-grid collocation UQ approaches, employing local interpolants over stochastic space, have also been used in this natural convection context in the Boussinesq limit, considering differentially heated cavities with uncertain rough wall topology [32]. Surface roughness leads to significant enhancement of the flow and heat transfer in this system, and alteration of Rayleigh-Bénard convective dynamics. Different realizations of the random configuration of the rough-wall topology result in switching between two stable flow-patterns, and associated bimodal PDFs for the uncertain flow variables. Wall roughness and geometric uncertainty can also be handled using domain transformations, allowing flow solutions on a regular mesh [122, 144].

As indicated above, particular attention to accuracy needs to be considered in the context of UQ computations of oscillatory flow dynamics and unsteady vortical motions over long time horizons, due to the growth of phase errors in time. These issues are relevant in shear layer and bluff-body/recirculation wake flows, the latter including *e.g.* both cylinder flow and flow over an open cavity. Considerable amplification of the inflow uncertainty occurs in cylinder flows, highlighting the need for ensuring good accuracy in the high-order PC modes [82]. Analysis of both global and local gPC methods in spectral element solutions of incompressible flow over an open cavity at high Reynolds number indicate better efficiency with the local ME-gPC methods [127]. Studies of gPC in the context of a 1D oscillatory advection model illustrate that, in order to maintain a fixed accuracy, the requisite PC order of the global method must be increased linearly in time. On the other hand, the time horizon corresponding to a given error threshold and PC order is extended by a factor of N when a uniform mesh of N stochastic elements is used in the ME-gPC context [126]. Studies of vortex shedding around a circular cylinder indicate that, while gPC fails to converge at early integration times, ME-gPC captures the flow accurately over a significantly longer time horizon. The use of local methods, however, does not fully resolve the challenges with representing vortex dynamics over long time horizons, particularly for high-dimensional random inputs, as the increase in the number of elements can be prohibitive.

16.2.1.2 Lagrangian vortex methods

Lagrangian formulations of the incompressible Navier-Stokes equations have also been extended to account for uncertainty employing PC UQ methods. From an algorithmic point of view, the simplest construction in this regard models both particle locations and strengths as uncertain. Implementations along these lines have been demonstrated in flow around airfoils with uncertainty in the inflow velocity, where Lagrangian particles are used to represent the vorticity field. Both intrusive and non-intrusive PC constructions have been used in this context [57, 102]. However, the approach of representing particle locations as random variables is fraught with difficulty in general, due to the growth of phase errors and loss of correlation for the non-linear N-body problem. Alternate developments to deal with this issue have led to constructions employing deterministic particle locations with uncertainty lumped in particle strengths. This construction has been developed employing PC UQ with a stochastic hybrid particle-mesh vortex method for incompressible flow natural convection computations in the Boussinesq limit [59]. Vorticity diffusion is implemented using a particle strength exchange scheme with remeshing; an optimal choice given the uncertainty formulation. A mesh is introduced for the evaluation of the velocity field for a given vorticity distribution, as a means of accelerating the velocity computations, and as an alternative to Biot-Savart N-body computations. Particles are displaced with the mean velocity field, while their uncertain strengths are computed employing PC algebra due to both diffusion and convection terms.

16.2.2 Reacting Flow

Computations of reacting flow provide a strong case for the need for uncertainty quantification, primarily because of the proliferation of empirically-determined reaction rate and thermodynamic parameters in chemical kinetic models. A detailed chemical model for the oxidation of simple hydrocarbons, *e.g.* CH_4, involves literally hundreds of empirical parameters with significant associated uncertainties. Models of complex fuels involve $\mathcal{O}(10^4)$ uncertain parameters. Moreover, in general, there is the equally important uncertainty in the structure of the chemical reaction network itself.

One significant issue that has to be stressed in this context is the need for proper characterization of the input space. This is important in general of course, but it gains even added significance here because of the large number of uncertain inputs and their internal correlations. The large impact of the dependence relationship between the parameters on the resulting uncertainty in chemical model predictions has been illustrated in a simple ignition system with two uncertain parameters [80]. The fact that such parametric correlations, and in general the joint probabilistic distribution of uncertain chemical model parameters, is largely unavailable in the literature is a significant challenge for non-perturbative stochastic UQ studies in reacting flow. These distributions can be established based on experimental data, however raw data is largely unavailable. Recent efforts are making headway in terms of warehousing and accessibility of data on uncertainties and correlations in parameters of thermochemical models [31, 112].

Laying aside the unavailability of proper probabilistic characterization of the input space of reacting flow models, two other aspects of these systems present difficulties associated with the above-discussed challenges to PC UQ. Chemical models exhibit (1) high-dimensionality associated with many uncertain parameters; and (2) strong nonlinearity associated with the rates of chemical reactions, particularly those with high activation energy. The dimensionality challenge highlights the need for adaptive sparse tensor product constructions for representation of random variables, as discussed above, while also recalling the necessity of identifying the set of *important* uncertain parameters by sensitivity analysis or other means, before addressing the propagation of uncertainty. This should be done with care however, given the second challenge of strong nonlinearity and the associated potential of non-gaussian and multimodal distributions of uncertain solutions.

Strong nonlinearity is a hallmark of exothermic reaction systems and of paramount importance in models for oxidation of hydrocarbon fuels. Nonlinearity leads to a strong dependence on, and strong sensitivity to, small changes in initial conditions. It also leads to large rates of change of the solution over space/time, and similarly to bifurcations over parametric space. Arrhenius reaction rate expressions, and associated uncertain parameters, present a clear example of the inherent UQ challenge in these systems. Ignition computations with a single-step irreversible global reaction model for methane-air combustion, with small uncertainties in the Arrhenius rate expression pre-exponential constant and activation energy, highlight the strong dependence of ignition time on these parameters, the high amplification of input uncertainties, and the observed bimodal structure of the probability distribution of the temperature or other state variables [80]. These complexities clearly indicate the inadequacy of perturbative linear methods for UQ in these systems, and the need for full probabilistic accounting for uncertainty.

Probabilistic PC UQ was first implemented in chemical systems, using a collocation approach, in a zero-dimensional isothermal context, focusing on supercritical water oxidation (SCWO) chemistry [104]. This particular collocation approach, where $(P+1)$ linear algebraic equations are solved for the $(P+1)$ PCE mode strengths, while ensuring accuracy of the PC representation at the collocation points, says nothing about its accuracy elsewhere in the parametric space. This same chemical model has also been used in PC UQ studies of ignition and 1D premixed flames, employing non-intrusive spectral projection where MC Latin Hypercube Sampling (LHS) was used for evaluation of the projection integrals for the PC mode coefficients, allowing for 13 uncertain parameters [107]. The 1D flame structure exhibits high amplification of uncertainty due to the strong non-linearity of the model. As a result, particular attention has to be paid to convergence with PC order, and the number of LHS samples, especially when considering higher-order statistics [108]. By the same token, global intrusive PC UQ methods are severely challenged with time-stability concerns in these systems, even with implicit stiff ODE integrators [106] suggesting the development of positive eigenvalues due to poor spectral resolution.

These challenges motivated the development of MRA constructions [58, 61]. Intrusive MRA PC UQ with adaptive multi-block decomposition of parametric space has been demonstrated in isothermal SCWO chemical ignition [62]. More recently exothermal ignition employing this intrusive multiwavelet construction has also been demonstrated [80].

Isothermal reacting flow in microchannels, with electroosmotic pumping, has also been studied using intrusive WH PC [26]. This construction couples the Navier-Stokes equations with slip velocity due to the wall double-layer zeta potential, species conservation equations including the electrokinetic body force, and electrostatics. Both fast equilibrated electrolytic reactions and slow finite-rate reactions are used in the chemical model.

16.2.3 Compressible Flow

Compressible flow equations, being hyperbolic, present unique challenges to uncertainty quantification efforts. These include wave propagation phenomena, and associated characterstics, which dictate specific numerics, both in the bulk and at domain boundaries. They also include shocks, which, being discontinuities, also demand specialized numerics.

Shock discontinuities lead to significant difficulties in the application of intrusive global PC UQ methods. Studies in model 1D nozzle flow have been successful in the absence of strong discontinuities [76, 125]. The consequences of shocks are significant, however, to the utility of the global PC construction. The discontinuity in the primitive variables over physical space translates to a discontinuity of the PC modes dependence over stochastic space [19]. While numerical stability of the global construction is possible, employing suitable filters and collocation schemes, convergence can be slow, as would be expected for the representation of discontinuities using global polynomial expansions. Alternatively, local PC constructions employing non-intrusive stochastic collocation have been used for shocks in quasi-1D nozzle flow [74, 75]. Shocks in a 1D piston problem have also been studied employing a collocation UQ approach relying on local polynomial interpolants, adaptive decomposition of stochastic space into simplex elements, and Newton-Cotes quadrature [137].

In a 2D flow context, supersonic flow over a wedge, exhibiting an oblique shock, and that over an expansion corner, exhibiting a Prandtl-Meyer expansion wave, are canonical compressible flow problems. These problems have been studied using intrusive PC UQ, with uncertainty in the wedge/expansion angle, employing an implicit 2D finite volume solver for the Euler equations [97]. The supersonic wedge problem has also been studied using intrusive gPC/ME-gPC constructions employing a WENO scheme in physical space [66]. In this construction, the signs of the (uncertain) eigenvalues associated with characteristics are required to choose the proper spatial discretization stencils. The increased size of the hyperbolic governing equation system leads to an increased computational burden associated with the eigenproblem solution, which necessitates attention to optimal numerical strategies. Improved performance with the local ME-gPC construction has been reported in this context, which is consistent with better resolution of the discontinuity in stochastic space, and the improved performance of local PC methods in bifurcative flows in general. Both the supersonic wedge and expansion corner problems have also been studied using non-intrusive collocation PC methods [45]. Deterministic solutions were computed using NASA's CFL3D Euler code [56], presuming uncertainty in the wedge/corner angle. As might be expected, the wedge flow was found to be more challenging than the expansion corner problem, in terms of both PDF skewness and collocation error behavior.

Three-dimensional compressible flow around a transonic wing has been studied employing collocation PC UQ methods [44]. The compressible Euler equations were used, coupled with aeroelastic analysis of the wing structure, allowing for uncertainty in the free-stream Mach number and the angle of attack. Further, 2D and 3D transonic wing computations have been demonstrated, with uncertain free stream velocity, using adaptive collocation employing local polynomial interpolants on simplex elements [135, 137]. Coupled fluid-structure oscillations were studied in this context, using an Arbitrary Lagrangian-Eulerian formulation.

16.2.4 Turbulence

The application of polynomial chaos expansions in turbulent flow has a long history, although not in the UQ context. Wiener-Hermite PC expansions have been considered as potential means for representing turbulent motions. Initially, Wiener [131] had proposed the use of his chaos expansions for analyzing turbulence. This proposal did not get much further attention however till after it was further elaborated in [132]. The following decade saw significant work in this area. However, despite some encouraging signs, the idea eventually proved ineffectual. The central problem is that, while the infinite expansions can indeed be used to represent turbulent motions starting from some initial state, any truncated/finite expansion fails, except in a limited time window. Orszag and Bissonnette [96] pointed out early-on deficiencies of this approach in the context of the turbulent Burgers equation, and using Kraichnan's simple three-body problem [55]. In the latter case, it was found that non-constant time-correlation was the Achilles heel of this approach, even for stationary Gaussian processes, as non-linearities propagate energy into higher-order terms. Crow and Canavan [24] further highlighted the failure of the Wiener-Hermite expansions in representing the energy cascade of trubulent motions. Moreover, Canavan [17] showed that using the convected Gaussian basis, an idea originally proposed by Wiener, did not resolve the problem. Chorin [21] further analyzed the failure of these expansions, highlighting the significance of the high-order terms given the turbulent energy cascade, and suggesting that the construction may find some use only in the description of large-scale motions.

With the recent utilization of PC methods for UQ, this topic has regained further relevance prompting further inquiry. These issues have been examined via analysis of the 1D Burgers and 2D Navier-Stokes equations with a Brownian motion forcing term [46], employing Karhunen-Loève and PC representations with adaptive sparse tensor-product constructions [30, 65]. Despite significant demonstrations of interesting short-term dynamics, however, long-term dynamics remain a challenge. The necessary growth in the number of terms in the expansion, required to maintain a given accuracy threshold, frustrates the utility of the construction for long time horizons.

Limiting the scope to large-scale oscillatory dynamics, significant progress has been made, even though the time horizon challenge generally requires special treatment in the presence of non-linearities. Computations of vortical shedding behind circular cylinders accounting for parametric uncertainty have met with success over short time horizons [69, 82, 126, 143]. Further, vortex dynamics have been studied, using non-intrusive gPC UQ employing Gauss-Legendre quadrature, in a laminar 2D shear layer with uncertain inflow forcing modes [53]. The shear layer flow structure reveals distinct regions where the influence of the fundamental or subharmonic uncertain forcing modes dominates, largely coinciding with zones with high probability of vortex interactions. Further, localized regions of high uncertainty are evident, highlighting local flow features that are most affected by the uncertain inputs. As is typical of shear/recriculating flows, the most probable solution is more meaningful than the mean flow in this system, given that multiple dominant solutions have significant probability. Large-scale dynamics have also been studied using PC UQ methods in the context of airfoil limit-cycle oscillations [10, 78, 79, 101, 103] where local PC methods are found to adequately handle the Hopf bifurcation associated with the onset of flutter. In general, however, the accurate computation of the uncertain time evolution of oscillatory flow over arbitrarily long time horizons remains a challenge.

One means by which this challenge has been addressed, has been to use: (1) a non-intrusive stochastic collocation approach, and (2) a parameterization of the oscillatory response in terms of well-behaved observables, having smooth dependence on the uncertain inputs [133–139]. This approach has been developed and demonstrated on systems of increasing complexity, with recent demonstrations involving transient multifrequency aeroelastic response of 2D/3D structures in fluid flow [135]. It employs simplex elements with adaptive refinement in stochastic space and multi-frequency parameterization of oscillatory response functions. The resulting observables, being amplitudes, phases, and frequencies of sinusoidal functions capturing the transient oscillatory response, are themselves quite smooth and easily represented in terms of low order polynomials.

In the same spirit, focusing on smooth observables, non-intrusive gPC has been used effectively for investigation of the uncertainty in LES computations of statistical moments of decaying homogeneous isotropic turbulence, resulting from parametric uncertainty in the underlying Smagorinsky subgrid model constant C [68]. This uncertainty is found to have the largest impact on the smallest resolved scales in the flow. Of course the determination of the "correct" range and PDF of C is a key issue. This PDF can be inferred employing Bayesian methods based on experimental measurements, and/or DNS computations, of target smooth observables. Ultimately the utilization of robust models, guaranteeing a consistent level of accuracy over a range of conditions is of interest [114]. The validation of LES or RANS-LES computations, employing a range of useful observables, is increasingly feasible given computational resources. However, there has been little such validation taking into account uncertainties resulting from incomplete information in complex flow systems, the typically different time spans of experimental and computational results, and experimental measurement noise [113].

16.3 Closure

This chapter has addressed the topic of uncertainty quantification in fluid flow computations, focusing on the fundamentals and application of spectral Polynomial Chaos UQ methods. The availability of these methods, along with the continuing increase in computational capabilities, has led to many successful demonstrations of uncertainty quantification in a range of fluid flow computations. Ongoing algorithmic developments provide hope in the continued expansion of the range of difficulty of uncertain fluid problems that can be efficiently handled. Above all, challenges with dimensionality and nonlinearity demand ongoing attention, as do challenges with the accurate and self-consistent description of the uncertain input space.

Acknowledgement

This work was supported by the US Department of Energy (DOE), Office of Basic Energy Sciences (BES) Division of Chemical Sciences, Geosciences, and Biosciences. Sandia National Laboratories is a multiprogram laboratory operated by Sandia Corporation, a Lockheed Martin Company, for the United States Department of Energy under contract DE-AC04-94-AL85000.

References

1. Abramowitz, M., Stegun, I.: Handbook of Mathematical Functions. Dover, New York (1970)
2. Agarwal, N., Aluru, N.: A domain adaptive stochastic collocation approach for analysis of MEMS under uncertainties. J. Comput. Phys. **228**, 7662–7688 (2009)
3. Askey, R., Wilson, J.: Some basic hypergeometric polynomials that generalize jacobi polynomials. Memoirs Amer. Math. Soc. **319**, 1–55 (1985)
4. Asokan, B., Zabaras, N.: Using stochastic analysis to capture unstable equilibrium in natural convection. J. Comput. Phys. **208**, 134–153 (2005)
5. Babuška, I., Tempone, R., Zouraris, G.: Galerkin finite element approximations of stochastic elliptic partial differential equations. SIAM J. Numer. Anal. **42**, 800–825 (2004)
6. Babuška, I., Tempone, R., Zouraris, G.: Solving elliptic boundary value problems with uncertain coefficients by the finite element method: The stochastic formulation. Comput. Methods Appl. Mech. Engrg. **194**, 1251–1294 (2005)

7. Babuška, I., Nobile, F., Tempone, R.: A stochastic collocation method for elliptic partial differential equations with random input data. SIAM J. Num. Anal. **45**, 1005–1034 (2007)
8. Bae, H.R., Grandhi, R., Canfield, R.: Uncertainty quantification of structural response using evidence theory. AIAA J. **41**, 2062–2068 (2003)
9. Barthelmann, V., Novak, E., Ritter, K.: High-dimensional polynomial interpolation on sparse grids. Adv. Compu. Math. **12**, 273–288 (2000)
10. Beran, P.S., Pettit, C.L., Millman, D.R.: Uncertainty quantification of limit-cycle oscillations. J. Comput. Phys. **217**, 217–247 (2006).
11. Bieri, M., Schwab, C.: Sparse high order FEM for elliptic sPDEs. CMAME **198**, 1149–1170 (2009)
12. Boyd, J.: The rate of convergence of Hermite function series. Math. Comput. **35**, 1309–1316 (1980)
13. Cacuci, D. (ed.): Sensitivity and Uncertainty Analysis Theory, vol. 1. Chapman & Hall/CRC, Boca Raton, FL (2003)
14. Cadafalch, J., Pérez-Segarra, C., Cónsul, R., Oliva, A.: Verification of finite volume computations on steady-state fluid flow and heat transfer. ASME J. Fluids Eng. **124**, 11–21 (2002)
15. Cameron, R., Martin, W.: The orthogonal development of nonlinear functionals in series of Fourier-Hermite functionals. Annals Math. **48**, 385–392 (1947)
16. Campolongo, F., Saltelli, A., Sørensen, T., Tarantola, S.: Hitchhiker's guide to sensitivity analysis. In: A. Saltelli, K. Chan, E. Scott (eds.) Sensitivity Analysis. Wiley, Chicester (2000)
17. Canavan, G.: Some properties of a Lagrangian Wiener-Hermite expansion. J. Fluid Mech. **41**, 405–412 (1970)
18. Celik, I., Li, J.: Assessment of numerical uncertainty for the calculations of turbulent flow over a backward-facing step. Int. J. Numer. Meth. Fluids **49**, 1015–1031 (2005)
19. Chen, Q.Y., Gottlieb, D., Hesthaven, J.: Uncertainty analysis for the steady-state flows in a dual throat nozzle. J. Comput. Phys. **204**, 378–398 (2005)
20. Chorin, A.: A numerical method for solving incompressible viscous flow problems. J. Comput. Phys. **2**, 12–26 (1967)
21. Chorin, A.: Gaussian fields and random flow. J. Fluid Mech. **63**, 21–32 (1974)
22. de Cooman, G., Ruan, D., Kerre, E. (eds.): Foundations and Applications of Possibility Theory. World Scientific Publishing, Singapore (1995)
23. Cox, E. (ed.): The Fuzzy Systems Handbook: A Practitioner's Guide to Building, Using, and Maintaining Fuzzy Systems, 2 edn. AP Professional, Div. of Academic Press, San Diego, CA, USA (1999)
24. Crow, S., Canavan, G.: Relationship between a Wiener-Hermite expansion and an energy cascade. J. Fluid Mech. **41**, 387–403 (1970)
25. Deb, M.K., Babuška, I., Oden, J.: Solution of stochastic partial differential equations using Galerkin finite element techniques. Comput. Methods Appl. Mech. Engrg. **190**, 6359–6372 (2001)
26. Debusschere, B., Najm, H., Matta, A., Knio, O., Ghanem, R., Le Maître, O.: Protein labeling reactions in electrochemical microchannel flow: Numerical simulation and uncertainty propagation. Phys. Fluids **15**, 2238–2250 (2003)
27. Debusschere, B., Najm, H., Pébay, P., Knio, O., Ghanem, R., Le Maître, O.: Numerical challenges in the use of polynomial chaos representations for stochastic processes. SIAM J. Sci. Comput. **26**, 698–719 (2004)
28. DeVolder, B., Glimm, J., Grove, J., Kang, Y., Lee, Y., Pao, K., Sharp, D., Ye, K.: Uncertainty quantification for multiscale simulations. ASME J. Fluids Eng. **124**, 29–41 (2002)
29. Faragher, J.: Probabilistic methods for the quantification of uncertainty and error in computational fluid dynamics simulations. Tech. rep., Australian Gov., Dept. of Defense, Defense Sci. and Tech. Org., DSTO-TR-1633 (2004)
30. Frauenfelder, P., Schwab, C., Todor, R.: Finite elements for elliptic problems with stochastic coefficients. Comput. Methods Appl. Mech. Engrg. **194**, 205–228 (2005)
31. Frenklach, M.: Transforming data into knowledge–Process Informatics for combustion chemistry. Proc. Comb. Inst. **31**, 125–140 (2007). http://primekinetics.org

32. Ganapathysubramanian, B., Zabaras, N.: Sparse grid collocation schemes for stochastic natural convection problems. J. Comput. Phys. **225**, 652–685 (2007)
33. Gautschi, W.: On generating orthogonal polynomials. SIAM J. Sci. Stat. Comput. **3**, 289–317 (1982)
34. Gerstner, T., Griebel, M.: Numerical integration using sparse grids. Numerical Algorithms **18**, 209–232 (1998). (also as SFB 256 preprint 553, Univ. Bonn, 1998)
35. Gerstner, T., Griebel, M.: Dimension adaptive tensor product quadrature. Computing **71**, 2003 (2003)
36. Ghanem, R.: Probabilistic characterization of transport in heterogeneous media. Comput. Methods Appl. Mech. Engrg. **158**, 199–220 (1998)
37. Ghanem, R., Dham, S.: Stochastic finite element analysis for multiphase flow in heterogeneous porous media. Trans. Porous Media **32**, 239–262 (1998)
38. Ghanem, R., Spanos, P.: Stochastic Finite Elements: A Spectral Approach. Springer Verlag, New York (1991)
39. Ghanem, R.G., Doostan, A.: On the construction and analysis of stochastic models: Characterization and propagation of the errors associated with limited data. J. Comput. Phys. **217**, 63–81 (2006)
40. Guan, J., Bell, D.: Evidence Theory and its Applications, vol. I. Elsevier Science Publishers, Amsterdam, The Netherlands (1991)
41. Helton, J., Johnson, J., Oberkampf, W.: An exploration of alternative approaches to the representation of uncertainty in model predictions. Reliab. Eng. System Safety **85**, 39–71 (2004)
42. Hemsch, M.: Statistical Analysis of Computational Fluid Dynamics Solutions from the Drag Prediction Workshop. J. Aircraft **41**, 95–103 (2004)
43. Hoeting, J., Madigan, D., Raftery, A., Volinsky, C.: Bayesian Model Averaging: A Tutorial. Stat. Sci. **14**, 382–417 (1999)
44. Hosder, S., Walters, R., Balch, M.: Efficient uncertainty quantification applied to the aeroelastic analysis of a transonic wing. In: AIAA-2008-729, 46th AIAA Aerospace Sciences Meeting and Exhibit. Reno, NV (2008)
45. Hosder, S., Walters, R., Perez, R.: A non-intrusive polynomial chaos method for uncertainty propagation in cfd simulations. In: Paper AIAA 2006-0891, 44th AIAA Aerospace Sciences Meeting and Exhibit. Reno, NV (2006)
46. Hou, T.Y., Luo, W., Rozovskii, B., Zhou, H.M.: Wiener Chaos expansions and numerical solutions of randomly-forced equations of fluid mechanics. J. Comput. Phys. **216**, 687–706 (2006)
47. Jaynes, E.: Probability Theory: The Logic of Science, G.L. Bretthorst, Ed. Cambridge University Press, Cambridge, UK (2003)
48. Kaipio, J., Somersalo, E.: Statistical and Computational Inverse Problems. Springer (2005)
49. Kennedy, M., O'Hagan, A.: Predicting the output from a complex computer code when fast approximations are available. Biometrika **87**, 1–13 (2000)
50. Kennedy, M.C., O'Hagan, A.: Bayesian calibration of computer models. J. Royal Stat. Soc.: Series B **63**, 425–464 (2001)
51. Kim, J., Moin, P.: Application of a fractional-step method to incompressible Navier-Stokes equations. J. Comput. Phys. **59**, 308–323 (1985)
52. Knio, O., Le Maître, O.: Uncertainty propagation in cfd using polynomial chaos decomposition. Fluid Dyn. Res. **38**, 616–640 (2006)
53. Ko, J., Lucor, D., Sagaut, P.: Sensitivity of two-dimensional spatially developing mixing layers with respect to uncertain inflow conditions. Phys. Fluids **20**, 1–20 (2008)
54. Kozine, I.: Imprecise Probabilities Relating to Prior Reliability Assessments. In: 1st Int. Symp. on Imprecise Probabilities and their Applications. Belgium (1999)
55. Kraichnan, R.: Direct-interaction approximation for a system of several interacting simple shear waves. Phys. Fluids **6**, 1603 (1963)
56. Krist, S., Biedron, R., Rumsey, C.: CFL3D User's Manual (Version 5.0). Tech. rep., NASA TM-1998-208444, NASA Langley Res. Ctr., Hampton, VA (1998)

57. Le Maître, O.: Polynomial chaos expansion of a Lagrangian model for the flow around an airfoil. Comptes Rendus Mec. **334**, 693–699 (2006)
58. Le Maître, O., Ghanem, R., Knio, O., Najm, H.: Uncertainty propagation using Wiener-Haar expansions. J. Comput. Phys. **197**, 28–57 (2004)
59. Le Maître, O., Knio, O.: A stochastic particle-mesh scheme for uncertainty propagation in vortical flows. J. Comput. Phys. **226**, 645–671 (2007)
60. Le Maître, O., Knio, O., Najm, H., Ghanem, R.: A stochastic projection method for fluid flow I. Basic formulation. J. Comput. Phys. **173**, 481–511 (2001)
61. Le Maître, O., Najm, H., Ghanem, R., Knio, O.: Multi-resolution analysis of Wiener-type uncertainty propagation schemes. J. Comput. Phys. **197**, 502–531 (2004)
62. Le Maître, O., Najm, H., Pébay, P., Ghanem, R., Knio, O.: Multi-resolution-analysis scheme for uncertainty quantification in chemical systems. SIAM J. Sci. Comput. **29**, 864–889 (2007)
63. Le Maître, O., Reagan, M., Debusschere, B., Najm, H., Ghanem, R., Knio, O.: Natural convection in a closed cavity under stochastic, non-Boussinesq conditions. SIAM J. Sci. Comput. **26**, 375–394 (2004)
64. Le Maître, O., Reagan, M., Najm, H., Ghanem, R., Knio, O.: A stochastic projection method for fluid flow II. Random process. J. Comput. Phys. **181**, 9–44 (2002)
65. Li, R., Ghanem, R.: Adaptive polynomial chaos expansions applied to statistics of extremes in nonlinear random vibration. Prob. Engrg. Mech. **13**, 125–136 (1998)
66. Lin, G., Su, C.H., Karniadakis, G.: Predicting shock dynamics in the presence of uncertainties. J. Comput. Phys. **217**, 260–276 (2006)
67. Lucor, D., Karniadakis, G.: Noisy inflows cause a shedding-mode switching in flow past an oscillating cylinder. Phys. Rev. Lett. **92**, 154,501.1–154,501.4 (2004)
68. Lucor, D., Meyers, J., Sagaut, P.: Sensitivity analysis of large-eddy simulations to subgrid-scale-model parametric uncertainty using polynomial chaos. J. Fluid Mech. **585**, 255–279 (2007)
69. Lucor, D., Triantafyllou, M.: Parametric study of a two degree-of-freedom cylinder subject to vortex-induced vibrations. J. Fluids Struct. **24**, 1284–1293 (2008)
70. Lucor, D., Xiu, D., Su, C., Karniadakis, G.: Predictability and uncertainty in cfd. Int. J. Num. Meth. Fluids **43**, 483–505 (2003)
71. Ma, X., Zabaras, N.: An adaptive hierarchical sparse grid collocation algorithm for the solution of stochastic differential equations. J. Comput. Phys. **228**, 3084–3113 (2009)
72. Marzouk, Y.M., Najm, H.N.: Dimensionality reduction and polynomial chaos acceleration of Bayesian inference in inverse problems. J. Comput. Phys. **228**, 1862–1902 (2009)
73. Marzouk, Y.M., Najm, H.N., Rahn, L.A.: Stochastic spectral methods for efficient Bayesian solution of inverse probelms. J. Comput. Phys. **224**, 560–586 (2007)
74. Mathelin, L., Hussaini, M., Zang, T.: Stochastic Approaches to Uncertainty Quantification in CFD Simulations. Num. Algorithms **38**, 209–236 (2005)
75. Mathelin, L., Hussaini, M., Zang, T., Bataille, F.: Uncertainty propagation for turbulent, compressible flow in a quasi-1D nozzle using stochastic methods. In: AIAA-2003-4240, 16th AIAA Comput. Fluid Dyn. Conf. Orlando, FL (2003)
76. Mathelin, L., Hussaini, M., Zang, T., Bataille, F.: Uncertainty propagation for a turbulent, compressible nozzle flow using stochastic methods. AIAA J. **42**, 1669–1676 (2004)
77. Matthies, H.G., Keese, A.: Galerkin methods for linear and nonlinear elliptic stochastic partial differential equations. Comput. Meth. Appl. Mech. Eng. **194**, 1295–1331 (2005)
78. Millman, D., King, P., Beran, P.: Airfoil pitch and plunge bifurcation behavior with Fourier chaos expansions. J. Aircraft **42**, 376–384 (2005)
79. Millman, D., King, P., Maple, R., Beran, P., Chilton, L.: Uncertainty quantification with a B-spline stochastic projection. AIAA J. **44**, 1845–1853 (2006)
80. Najm, H.: Uncertainty quantification and plynomial chaos techniques in computational fluid dynamics. Ann. Rev. Fluid Mech. **41**, 35–52 (2009)
81. Najm, H., Debusschere, B., Marzouk, Y., Widmer, S., Le Maître, O.: Uncertainty quantification in chemical systems. Int. J. Num. Meth. Eng. **80**, 789–814 (2009)

82. Narayanan, V., Zabaras, N.: Variational multiscale stabilized FEM formulations for transport equations: stochastic advection-diffusion and incompressible stochastic Navier-Stokes equations. J. Comput. Phys. **202**, 94–133 (2005)
83. Nobile, F., Tempone, R., Webster, C.: A sparse grid stochastic collocation method for partial differential equations with random input data. SIAM J. Num. Anal. **46**, 2309–2345 (2008)
84. Nobile, F., Tempone, R., Webster, C.: An anisotropic sparse grid stochastic collocation method for partial differential equations with random input data. SIAM J. Num. Anal. **46**, 2411–2442 (2008)
85. Novak, E., Ritter, K.: High dimensional integration of smooth functions over cubes. Numerische Mathematik **75**, 79–97 (1996)
86. Novak, E., Ritter, K.: Simple cubature formulas with high polynomial exactness. Constructive Approximation **15**, 499–522 (1999)
87. Novak, E., Ritter, K., Schmitt, R., Steinbauer, A.: On an interpolatory method for high-dimensional integration. J. Comput. Appl. Math. **112**, 215–228 (1999)
88. Oakley, J., O'Hagan, A.: Bayesian inference for the uncertainty distribution of computer model outputs. Biometrika **89**, 769–784 (2002)
89. Oakley, J., O'Hagan, A.: Probabilistic sensitivity analysis of complex models: a Bayesian approach. J.R. Statist. Soc. B **66**, 751–769 (2004)
90. Oberkampf, W., Barone, M.: Measures of agreement between computation and experiment: Validation metrics. J. Comput. Phys. **217**, 5–36 (2006)
91. Oberkampf, W., Blottner, F.: Issues in computational fluid dynamics code verification and validation. AIAA J. **36**, 687–695 (1998)
92. Oberkampf, W., DeLand, S., Rutherford, B., Diegert, K., Alvin, K.: Error and uncertainty in modeling and simulation. Reliab. Eng. Sys. Safety **75**, 333–357 (2002)
93. Oberkampf, W., Helton, J.: Evidence Theory for Engineering Applications. In: E. Nikolaidis, D. Ghiocel, S. Singhal (eds.) Engineering Design Reliability Handbook, pp. 10.1–10.30. CRC Press (2005)
94. Oberkampf, W., Trucano, T.: Validation methodology in computational fluid dynamics. AIAA-2000-2549 (2000)
95. Ogura, H.: Orthogonal Functionals of the Poisson Process. IEEE Trans. Info. Theory **18**, 473–481 (1972)
96. Orszag, S., Bissonnette, L.: Dynamical properties of truncated Wiener-Hermite expansions. Phys. Fluids **10**, 2603–2613 (1967)
97. Perez, R., Walters, R.: An implicit polynomial chaos formulation for the euler equations. In: Paper AIAA 2005-1406, 43rd AIAA Aerospace Sciences Meeting and Exhibit. Reno, NV (2005)
98. Petras, K.: On the smolyak cubature error for analytic functions. Advances in Computational Mathematics **12**, 71–93 (2000)
99. Petras, K.: Fast calculation of coefficients in the smolyak algorithm. Numerical Algorithms **26**, 93–109 (2001)
100. Petras, K.: Smolyak cubature of given polynomial degree with few nodes for increasing dimension. Numerische Mathematik **93**, 729–753 (2003)
101. Pettit, C., Beran, P.: Polynomial chaos expansions applied to airfoil limit cycle oscillations. 45th AIAA/ASME/ASCE/AHS/ASC Structures, Struct. Dyn. and Mat. Conf., Palm Springs, CA, 2004 **AIAA-2004-1691** (2004)
102. Pettit, C., Hajj, M., Beran, P.: Gust loads with uncertainty due to imprecise gust velocity spectra. In: AIAA-2007-1965, 48th AIAA/ASME/ASCE/AHS/ASC Structures, Struct. Dyn., and Mat. Conf. Honolulu, Hawaii (2007)
103. Pettit, C.L., Beran, P.S.: Spectral and multiresolution wiener expansions of oscillatory stochastic processes. J. Sound Vibr. **294**, 752–779 (2006).
104. Phenix, B., Dinaro, J., Tatang, M., Tester, J., Howard, J., McRae, G.: Incorporation of parametric uncertainty into complex kinetic mechanisms: Application to hydrogen oxidation in supercritical water. Combust. Flame **112**, 132–146 (1998)

105. Putko, M., Taylor III, A., Newman, P., Green, L.: Approach for uncertainty propagation and robust design in CFD using sensitivity derivatives. ASME J. Fluids Eng. **124**, 60–69 (2002)
106. Reagan, M., Najm, H., Debusschere, B., Le Maître, O., Knio, O., Ghanem, R.: Spectral stochastic uncertainty quantification in chemical systems. Combust. Theory Model. **8**, 607–632 (2004)
107. Reagan, M., Najm, H., Ghanem, R., Knio, O.: Uncertainty quantification in reacting flow simulations through non-intrusive spectral projection. Combust. Flame **132**, 545–555 (2003)
108. Reagan, M., Najm, H., Pébay, P., Knio, O., Ghanem, R.: Quantifying uncertainty in chemical systems modeling. Int. J. Chem. Kin. **37**, 368–382 (2005)
109. Rebba, R., Mahadevan, S.: Model Predictive Capability Assessment Under Uncertainty. AIAA J. **44**, 2376–2384 (2006)
110. Roache, P.: Code verification by the method of manufactured solutions. ASME J. Fluids Eng. **124**, 4–10 (2002)
111. Rosenblatt, M.: Remarks on a multivariate transformation. Annals Math. Stat. **23**, 470–472 (1952)
112. Ruscic, B., Pinzon, R., Morton, M., von Laszewski, G., Bittner, S., Nijsure, S., Amin, K., Minkoff, M., Wagner, A.: Introduction to active thermochemical tables: Several "key" enthalpies of formation revisited. J. Phys. Chem. A **108**, 9979–9997 (2004)
113. Sagaut, P., Deck, S.: Large eddy simulation for aerodynamics: status and perspectives. Phil. Trans. Royal Soc. A **367**, 2849–2860 (2009)
114. Sagaut, P., Meyers, J., Lucor, D.: Uncertainty modeling, error charts and improvement of subgrid models. In: S.-H. Peng and W. Haase (Eds.): Adv. in Hybrid RANS-LES Modelling, NNFM 97. Springer-Verlag, Berlin (2008)
115. Schoutens, W.: Stochastic Processes and Orthogonal Polynomials. Springer (2000)
116. Schwab, C., Todor, R.: Sparse finite elements for stochastic elliptic problems. Numer. Math. **95**, 707–734 (2003)
117. Schwab, C., Todor, R.: Sparse finite elements for stochastic elliptic problems – higher order moments. Computing **71**, 43–63 (2003)
118. Shafer, G.: A Mathematical Theory of Evidence. Princeton University Press, Princeton, NJ (1976)
119. Smolyak, S.A.: Quadrature and interpolation formulas for tensor products of certain classes of functions. Soviet Math. Dokl. **4**, 240–243 (1963)
120. Soize, C., Ghanem, R.: Physical systems with random uncertainties: Chaos representations with arbitrary probability measure. SIAM J. Sci. Comput. **26**, 395–410 (2004)
121. Stern, F., Wilson, R., Shao, J.: Quantitative V&V of CFD simulations and certification of CFD codes. Int. J. Numer. Meth. Fluids **50**, 1335–1355 (2006)
122. Tartakovsky, D., Xiu, D.: Stochastic analysis of transport in tubes with rough walls. J. Comput. Phys. **217**, 248–259 (2006)
123. Todor, R., Schwab, C.: Convergence rates for sparse chaos approximations of elliptic problems with stochastic coefficients. IMA Journal of Numerical Analysis **27**, 232–261 (2007)
124. Unwin, S.: A fuzzy set theoretic foundation for vagueness in uncertainty analysis. Risk Analysis **6**, 27–34 (1986)
125. Walters, R., Huyse, L.: Uncertainty analysis for fluid mechanics with applications. Tech. rep., ICASE Report No. 2002-1; NASA/CR-2002-211449 (2002)
126. Wan, X., Karniadakis, G.: Long-term behavior of polynomial chaos in stochastic flow simulations. Comput. Meth. Appl. Mech. Eng. **195**, 5582–5596 (2006)
127. Wan, X., Karniadakis, G.: Multi-Element Generalized Polynomial Chaos for Arbitrary Probability Measures. SIAM J. Sci. Comput. **28**, 901–928 (2006)
128. Wan, X., Karniadakis, G.E.: An adaptive multi-element generalized polynomial chaos method for stochastic differential equations. J. Comput. Phys. **209**, 617–642 (2005)
129. Warnatz, J.: Resolution of gas phase and surface combustion chemistry into elementary reactions. Proc. Combust. Inst. **24**, 553–579 (1992)
130. Wiener, N.: The homogeneous chaos. Am. J. Math. **60**, 897–936 (1938)

131. Wiener, N.: The use of statistical theory in the study of turbulence. In: Fifth International Congress for Applied Mechanics. Wiley, New York, NY (1939)
132. Wiener, N.: Nonlinear Problems in Random Theory. MIT Press, Cambridge, MA (1958)
133. Witteveen, J., Bijl, H.: An alternative unsteady adaptive stochastic finite elements formulation based on interpolation at constant phase. Comput. Meth. Appl. Mech. Eng. **198**, 578–591 (2008)
134. Witteveen, J., Bijl, H.: An unsteady adaptive stochastic finite elements formulation for rigid-body fluid-structure interaction. Comp. Struct. **86**, 2123–2140 (2008)
135. Witteveen, J., Bijl, H.: Effect of randomness on multi-frequency aeroelastic responses resolved by Unsteady Adaptive Stochastic Finite Elements. J. Comput. Phys. **228**, 7025–7045 (2009)
136. Witteveen, J., Bijl, H.: Higher period stochastic bifurcation of nonlinear airfoil fluid-structure interaction. Math. Prob. Eng. 394387, 1–26 (2009)
137. Witteveen, J., Loeven, A., Bijl, H.: An adaptive stochastic finite elements approach based on Newton-Cotes quadrature in simplex elements. Comput. Fluids **38**, 1270–1288 (2009)
138. Witteveen, J., Loeven, A., Sarkar, S., Bijl, H.: Probabilistic collocation for period-1 limit cycle oscillations. J. of Sound Vib. **311**, 421–439 (2008)
139. Witteveen, J., Sarkar, S., Bijl, H.: Modeling physical uncertainties in dynamic stall induced fluid-structure interaction of turbine blades using arbitrary polynomial chaos. Comput. Struct. **85**, 866–878 (2007)
140. Xiu, D., Hesthaven, J.S.: High-order collocation methods for differential equations with random inputs. SIAM J. Sci. Comput. **27**, 1118–1139 (2005)
141. Xiu, D., Karniadakis, G.: The Wiener-Askey polynomial chaos for stochastic differential equations. SIAM J. Sci. Comput. **24**, 619–644 (2002)
142. Xiu, D., Karniadakis, G.: Modeling uncertainty in flow simulations via generalized polynomial chaos. J. Comput. Phys. **187**, 137–167 (2003)
143. Xiu, D., Lucor, D., Su, C.H., Karniadakis, G.: Stochastic modeling of flow-structure interactions using generalized polynomial chaos. ASME J. Fluids Eng. **124**, 51–59 (2002)
144. Xiu, D., Tartakovsky, D.: Numerical methods for differential equations in random domains. SIAM J. Sci. Comput. **28**, 1167–1185 (2006)
145. Zang, T., Hemsch, M., Hilburger, M., Kenny, S., Luckring, J., Maghani, P., Padula, S., Stroud, W.: Needs and opportunities for uncertainty-based multidisciplinary design methods for aerospace vehicles. Tech. rep., NASA/TM-2002-211462 (2002)

Chapter 17
Computational Frameworks for Advanced Combustion Simulations

J. Ray, R. Armstrong, C. Safta, B. J. Debusschere, B. A. Allan and H. N. Najm

Abstract Computational frameworks can significantly assist in the construction, extension and maintenance of simulation codes. As the nature of problems addressed by computational means has grown in complexity, such frameworks have evolved to incorporate a commensurate degree of sophistication, both in terms of the numerical algorithms that they accommodate as well as the software architectural discipline they impose on their users. In this chapter, we discuss a *component framework*, the Common Component Architecture (CCA), for developing scientific software, and describe how it has been used to develop a toolkit for simulating reacting flows. In particular, we will discuss why a component architecture was chosen and the philosophy behind the particular software design. Using statistics drawn from the toolkit, we will analyze the code structure and investigate to what degree the aims of the software design were actually realized. We will explore how CCA was employed to design a high-order simulation code on block-structured adaptive meshes, as well as a simulation capacity for adaptive stiffness reduction in detailed chemical models. We conclude the chapter with two reacting flow studies performed using the above-mentioned computational capabilities.

17.1 Introduction

Computational science has come to be regarded as the third leg of science, after theory and experimentation. With the advent of massively parallel computers, simulations have been used to investigate extremely challenging problems. However, as the problems have increased in complexity, the tools used to investigate them computationally – numerical algorithms and their software implementations – have developed a commensurate sophistication and intricacy. Software complexity, with

---

J. Ray, R. Armstrong, C. Safta, B. J. Debusschere, B. A. Allan and H. N. Najm
Sandia National Laboratories, Livermore, CA, e-mail: `[jairay,rob,csafta,bjdebus,baallan,hnnajm]@sandia.gov`

T. Echekki, E. Mastorakos (eds.), *Turbulent Combustion Modeling*,
Fluid Mechanics and Its Applications 95, DOI 10.1007/978-94-007-0412-1_17,

its detrimental impact on software maintainability and extensibility, is regarded as a large drain on time and effort and computational frameworks are intended to be a solution to this problem. One means of addressing complexity is to provide "shrink-wrapped" functionality, in essence transferring software and algorithmic complexity to the implementer of the framework from its user. A second way of addressing complexity is through some form of modularization. *Component frameworks* fall in the second category.

The word "framework" is an extremely overused and consequently, confusing term. It can refer to an *architecture*, a set of specifications that, when followed, imposes some standard on the software that adhere to it. By this token, a component framework is a set of specifications that permit software to be constructed by modular composition. The modules are referred to as *components*. Confusingly, "framework" can also refer to a software framework that is written to such a specification and is meant to locate, instantiate and compose components. The aim of component frameworks is to use *modularity* to divide and conquer complexity, by composing programs out of software building blocks. Studies have shown that the capability and complexity of a program is proportional to its size, which determines the number of software developers required to maintain the software [36, 63]. Component frameworks can be used to design an organizing principle for the efficient deployment of programming resources. Individual programmers have cognizance over one or more components of the overall program, which provides a natural division of responsibilities. It makes obvious that the interfaces between components are where programmers must negotiate. Interfaces exported to, or imported from components is functionality that is needed by, or provided to a programmer as part of his/her domain of responsibility. The structure provided by component-based software engineering is one of the main reasons for using it and defines the workflow in a large software project.

In this chapter, we will limit the discussion to the Common Component Architecture (CCA), a set of specifications for developing and composing component-based scientific simulation software. We will devote some time to the CCA frameworks used in scientific computings and provide examples of how CCA has been used for software for the simulation of flames.

17.2 Literature Review of Computational Frameworks

Scientific computational frameworks proliferate. Some were initially designed to assist in the creation of simulation codes in a certain scientific field and consequently have been adopted/adapted rather sporadically elsewhere. Others were intended to assist in nothing more than the implementation of a particular software design, irrespective of the nature/aim of the software being developed. It is impossible to review the myriads of scientific simulation frameworks that lie between these two extremes and this will not be attempted here. Since this chapter describes the design of a component-based simulation facility for reacting flows on adaptively refined meshes, we will devote the majority of our review to frameworks that assist in computations on block-structured adaptively refined meshes, and implementations of CCA that are relevant to scientific computing. We will, however, describe a basic categorization before proceeding with the details.

Scientific computing frameworks are application development frameworks which, at the very least, support parallel computing and pay particular attention to high performance in their design. This manifests itself in their ability to address operations on large arrays and the elimination/reduction of overheads that scale with the size of the data or with the number of processors on which the framework is expected to run. Further, they rarely contain any support for languages outside FORTRAN and C/C++, and make sparing use of remote method invocation. Within these confines, scientific simulation frameworks can be divided into three categories.

In the first category are frameworks that were originally designed to support a particular field of science. The Earth System Modeling Framework [30, 32] is one such example, developed for constructing climate simulation codes; Cactus [15, 37] is another, developed for numerical relativity studies while FLASH [31, 35], developed for astrophysical simulation, forms a third example. OpenFoam, developed initially for finite-volume simulation of fluid flow (but now containing extensions for solid mechanics and Direct Simulation Monte Carlo calculations as well) [80] is in widespread use. Such application-specific frameworks often contain numerical schemes and "utility" scientific models that find routine use. They are modular in design and implement many of the ideas (e.g., involving a separation of implementation from the interface of a module) that are formalized in CCA. Note that these frameworks do not intend to impose any design patterns on the resulting simulation code; rather one adopts the data structures and design patterns of the framework itself and reaps the benefits of using validated solvers and models in one's simulations. These frameworks allow one to rapidly develop simulation codes as long as the facilities provided by the framework (e.g. solvers and models) are profitably leveraged by the simulation.

The second category of frameworks consists of those that enable the use of a particular solution methodology e.g.. block-structured adaptive mesh refinement (AMR) in simulation codes. These frameworks are more general than those described in the previous paragraph – they provide data-structures and solvers (or interfaces to those implemented by external libraries) that are useful while developing simulation software. Again, they make no attempt to promote any particular design pattern. However, given that one makes heavy use of the framework-provided data structures (which are usually implemented in the form of objects), the design patterns employed for simulation codes bear a strong resemblance to those employed for the framework itself. We will review a few such frameworks/packages that enable the use of AMR in simulation codes.

Simulations using block-structured adaptively refined meshes are generally conducted where the spatial domain is rectangular/cuboid or can be logically described as such. One begins with a Cartesian mesh with a resolution that is insufficient to capture many of the features of the solution. Regions in the domain that require refinement are identified and collated into rectangles/cuboids and a finer mesh (usually, finer by a constant factor) is overlaid on the refined part of the initial mesh. This is performed recursively leading to a *mesh hierarchy* of a few (usually less than 10) levels. GrACE [5, 17, 25, 38, 67, 69, 70] and AmrLib [4] are a few frameworks that implement such an adaptive meshing strategy. These frameworks provide the infrastructure required to manipulate such meshes as well as data structures that allow fields to be described on them (on distributed memory computers, this involves very intricate book-keeping). The data structures also allow the use of *time-refinement* [13], a time-integration technique that allows one to mitigate the effect of having time-steps be CFL-constrained by the finest mesh in a mesh-hierarchy. While CHOMBO stores the data on a given level of the mesh hierarchy in a separate data-structure, both AMROC and GrACE present a data object that is described on the entire hierarchy. The three frameworks also largely automate the work of refining/coarsening the mesh hierarchy periodically (called regridding), based on a user-calculated "error metric" and perform load-balancing of the mesh and data after each regridding operation. While GrACE only provides a parallel block-structured adaptive mesh and the associated data object, CHOMBO and AMROC also provide a set of commonly used solvers.

All the three frameworks have been used for simulating reacting flows. AMROC has its origins in the simulation of shock-laden flows [26, 66], but has been extended to shock-fluid interactions [18, 28, 29] and shock-induced combustion [27]. CHOMBO (and its predecessors) have been used for an immense variety of simulations [69], including solutions of variable density formulation of the Navier-Stokes equations [3], embedded boundary methods [21], and fourth-order-in-space discretizations [9, 20] for AMR simulations. AmrLib, which has similar foundations as CHOMBO, has been used to develop algorithms for low Mach laminar flames simulations [24] as well as for turbulent flames [10, 11]. GrACE was initially developed to investigate load-balancers for block-structured adaptive meshes [54, 55], but has been used to develop efficient numerical schemes (e.g., extended stability time integrators [52] and fourth-order spatial discretizations [72]) for the simulation of flames [75]. These frameworks have been investigated for their scalability on parallel machines; see [61] for a discussion of scalability issues in GrACE and [78] for CHOMBO.

Overture [65] had its beginnings in simulations using overset meshes i.e. a gridding scheme, generally applied to complex geometries, where the domain is discretized using "patches" which could employ meshing schemes that were best suited to the geometry at hand. Thus it was possible, for example, to embed a small circular patch, employing a cylindrical mesh, within a larger mesh discretized in a Cartesian manner. It was used to develop fourth-order-in-space discretizations [39], as well as in simulations of high-speed reactive flows [40]. It has been extended to adaptive mesh refinement [41], multi-material flows [8] and has recently been parallelized and used for 3D calculations [42]. A full listing of Overture-related publications, including investigation of detonations etc. can be found on the Overture homepage [65].

In the third category are frameworks that primarily seek to assist software developers implement a certain design pattern. CCA defines one such design pattern; UINTAH [77] and CCAFFEINE [1, 14] are frameworks that implement it. Thus components designed to operate in the UINTAH and CCAFFEINE frameworks comply with the CCA architecture and bestow its advantages — modular design, reduction of software complexity, plug-and-play experimentation and multiple language interoperability — on simulation codes that follow the architecture. The framework is not required to provide any numerical or simulation capability (which renders the framework usable in diverse scientific applications) and indeed CCAFFEINE does not; however, UINTAH provides its users with a mesh and the associated data structures for describing fields on a discretized domain. A detailed discussion of CCA follows in Sect. 17.3.

17.3 The Common Component Architecture

The Common Component Architecture (CCA) specification defines a software standard allowing plug-and-play composition of scientific applications. Being component-based, it is necessarily object-oriented. Its development was driven by the benefits of modularization. The competitive advantages of modularization were recognized by the commercial establishment in the mid-1990s, which fashioned a solution in the form of component architecture; CORBA [22], Visual Basic [85] and Java Beans [33] are some industry-standard *component architectures*. These architectures employ certain subsets of object-oriented software design principles to realize modularity and interoperability of modules under a large variety of conditions. However, commercial component standards are ill-suited for scientific computing [1], the primary drawbacks being the lack of support for parallel computing. Starting from the concepts common to most component standards, the CCA retains most of the relevant characteristics while simultaneously allowing (but not stipulating) parallel computing in an SPMD fashion. CCAFFEINE [1] and UINTAH [77] are CCA-compliant frameworks that support parallel computing, while XCAT [87] does not.

Within CCA one considers *components*, modules (or objects) that implement a particular scientific or algorithmic functionality, and a *framework* e.g. CCAFFEINE or UINTAH, that "wires" components together into a functioning simulation code. Components implement interfaces (*Ports* in CCA parlance) through which the components provide their functionality; these ports are designed by the programmer implementing the component. Components are peers, i.e., they do not inherit implementation from other components, and are capable of being used independently. Since components implement interfaces, they can be developed without a tight coupling to a software development team. CCA stipulates that all components implement a particular interface (called the *Component* interface) through which individual components can interact with the framework (specifically, using the *Services* interface). Typically, this involves registering the Ports (interfaces/functionalities) that they provide (these interfaces are called the *ProvidesPorts* of the component) as well as the Ports that they use (i.e., the *UsesPorts* of the component). Driven by a user script, the framework matches the *ProvidesPort* and *UsesPort* registrations.

The individual components then fetch the matched-up Ports and use them by making calls on the methods on those interfaces The Port interface is a caller-callee boundary, not a data-flow abstraction.

17.3.1 Features of the Common Component Architecture

The initial development of the CCA specification took place using a compiler-independent, simple subset of C++ syntax (pure virtual classes) as the interface definition language. Subsequent CCA development (version 0.6 and beyond) use the Scientific Interface Definition Language (SIDL) and its compiler Babel [23] to enable language interoperability i.e. components written in the Babel-supported languages (C, FORTRAN, Python, or Java) can call each other transparently, without manually performing the data translations and language-bindings that such a mixture of programming languages would entail. When using Babel, the interface construct of the SIDL language replaces the pure virtual abstract class of C++ (of the initial design). In addition to language translation, SIDL/Babel provides modern software engineering functionality in each of the languages it supports e.g., passing and accessing multidimensional array data and C structs as references to preserve performance, support for exceptions, object-oriented design, tunably enforced programming-by-contract conditions when entering and leaving method calls etc. The C++ specification continues to exist and provide interfaces equivalent to those of the SIDL definition.

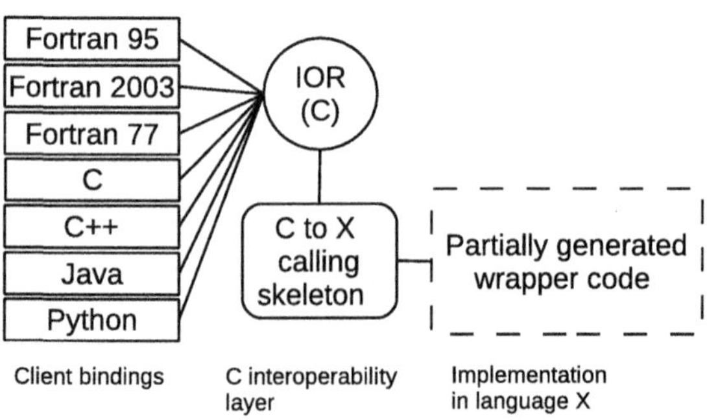

Fig. 17.1: Mechanics of any-to-any language calling with Babel. Client binding and interoperability layers (solid shapes) are entirely generated code. The implementation is partially generated in the implementor's choice of programming language, X, chosen from among the supported languages. The implementor fills in method bodies and private data structure definitions that are inaccessible from the clients. Lines indicate function calls.

Language interoperability: SIDL/Babel enables language-interoperability by interposing an Intermediate Object Representation (IOR) between components, with an independent client-side binding to each language. This is illustrated in Fig. 17.1.

In order that a component, written in language X, may be called from a host of other languages (in Fig. 17.1, these are various dialects of FORTRAN, C/C++, Java and Python), the programmer first expresses/writes the interface i.e., function calls, and their arguments (which may use the Babel-supported data-types like multi-dimensional arrays, exceptions etc. listed above) in a language-neutral manner using SIDL. The SIDL interface is then "compiled" using the Babel compiler to generate equivalent interfaces in C/C++, FORTRAN etc, collectively referred to as "client bindings" in Fig. 17.1 (left). It also generates a wrapper, in language X, (shown in dashed lines on the right side of Fig. 17.1) in which one implements one's components. The wrapper can also include calls to an external library, and in fact, provides a very simple route to making a library directly callable from a multitude of programming languages. The most significant, and intricate, piece in the cross-language orchestration is the IOR, shown in the middle of Fig. 17.1 ("C interoperability layer"). This IOR, essentially a C struct containing a function table and data pointers, maps function calls in the various "client" languages (left side of Fig. 17.1) into the equivalent functions/methods on the "server" side (the dashed structure on the right side of Fig. 17.1). The need for translation of data (e.g from FORTRAN multi-dimensional arrays on the "client" side to perhaps C on the "server"/implementation side), the throwing of exceptions and other functionalities that might be needed to enable cross-language operation are detected by Babel when "compiling" the SIDL interface, and is encoded into the IOR. There are a substantial number (as seen in Fig. 17.1) of code objects which are automatically derived from SIDL by Babel and then compiled. Most of the rote work needed to build and deploy SIDL-enabled applications has been automated with the Bocca tool [2], enabling entire component application structures to be prototyped within minutes.

Parallel computing: Since the use of parallel computing is critical to performance, the CCA standard is carefully crafted to avoid forcing the choice of any particular parallel programming model on the user. Component developers may use any communication model. A survey of the impact of using CCA on application performance is provided in Section 6 of [14]. The typical scientific application has a single-program-multiple-data structure (SPMD). This programming model is well supported by the CCAFFEINE framework [1] and default driver, which supports the component-based programming model analog, the single-component-multiple-data (SCMD) model. In the SCMD model, all components have a representative instance on all processes. For a single component, we call the group of parallel instances a cohort. Message passing, by whatever means, is restricted to exchanges within a cohort, as one component implementation cannot make any assumptions about the communication internals of a different component implementation. CCAFFEINE has also been used as a library to compose MCMD (multi-component-multiple-data) applications [46].

17.4 Computational Facility for Reacting Flow Science

The Computational Facility for Reacting Flow Science (CFRFS) is a toolkit for conducting high-fidelity simulations of laboratory-scaled flames. The toolkit adheres to the CCA specification (using C++, not SIDL/Babel as its interface language of choice) and is typically run using the CCAFFEINE framework. The CFRFS toolkit implements a set of novel numerical techniques, most of which were developed as the toolkit was being constructed. In this section we describe the toolkit and its numerical structure and discuss why a component architecture was necessary for its implementation. We conclude with an analysis, using data drawn from the toolkit, to determine to what degree the original aims of the software design have been realized.

17.4.1 Numerical Methods and Capabilities

The CFRFS toolkit is used to perform computations of flames using detailed chemical mechanisms. It solves the low Mach number approximation of the Navier-Stokes (NS) equations [86], augmented with evolution equations (convective-diffusive-reactive systems) for each of the chemical species in the chemical mechanism. It employs fourth-order finite-difference schemes for spatial discretizations within the context of AMR meshes [13]. It adopts an operator-split construction to enable the use of efficient integrators when time-integrating different physical processes in the system being solved. It uses extended-stability Runge-Kutta-Chebyshev (RKC) time-integration [76] for time-advancing the diffusive transport terms and an adaptive, backward difference stiff integrator for the chemical source terms. The numerical details of the high-order (spatial) methods can be found in [72] and those of RKC on block-structured adaptive meshes in [52]. The divergence constraint in the low Mach NS equations necessitates a projection scheme, which gives rise to a non-constant coefficient Poisson equation; this is solved using the conjugate-gradient method with a multigrid solver (employing high-order spatial stencils) as a preconditioner.

The CFRFS Toolkit, then, is an integration of many advanced numerical techniques. Many of them, e.g., RKC, had been demonstrated on uniform meshes [64] but had to be augmented to enable the solution of partial differential equations (PDEs) on AMR meshes. For example the RKC schemes had to be modified to preserve its order of accuracy when used with *time refinement* [52] on AMR meshes; further tests were required to determine various "free" parameters (in the RKC scheme) when time-advancing a convective-diffusive system [71]. The high-order spatial discretizations required appropriate interpolation schemes, and in certain cases, needed the solution to be filtered, to remove high-wavenumber content and prevent Gibbs phenomenon [72]; the correct pairings of discretization, interpolation and filter order were determined as a part of the implementation of the toolkit. The projection scheme adapts a fourth-order finite-volume formulation [44] for use in the context of a finite-difference approach [74]. Thus the construction of the high-order AMR simulation capability, as implemented in the CFRFS Toolkit entailed a significant amount of development of advanced numerical methods.

The CFRFS Toolkit makes copious use of external software. The adaptive meshing and load-balancing is currently provided by the GrACE package [38]; coupling to CHOMBO [17] is in progress. The stiff integration capability is provided by CVODE [19], while the elliptic solvers in Hypre [34] are used for the pressure solve. Legacy codes are used to provide implementations of various constitutive models (transport coefficients, gas-phase reaction and thermodynamics models etc). Being able to leverage existing, validated software (e.g., legacy codes) has saved much implementation effort, while numerical libraries (e.g. Hypre, CHOMBO) allow the CFRFS toolkit to take advantage of optimized and specialized capabilities in a facile manner.

17.4.2 The Need for Componentization

The goal of the CRFRS development effort was to develop a flexible, reusable toolkit. At the very outset it was expected that many of its advanced features would be contributed piecemeal by experts or incorporated using legacy software and the necessity of an extremely modular design was recognized very early.

The componentization in the CFRFS toolkit follows strictly along functional lines. Each component implements a physical model, a numerical scheme, or a computational capability like a "data object" that stores and manages domain-decomposed fields (e.g., a temperature field) on multiple levels on an AMR mesh. The functionality is expressed in the Port/interface design; components implement the functionality. As there are may different ways to provide a functionality e.g., one may time-integrate using many different algorithms or calculate transport properties using diverse models, a single Port may find disparate implementations. Each component is compiled into a shared library (also known as a dynamically loadable library); a simulation "code" is composed by loading a number of them into the CCA framework and "wiring them together". Figure 17.2 shows a wiring diagram, assembling approximately 40 components into a low-Mach number flame simulation, whose results are discussed in Section 17.5.1. The components in the wiring diagram implement flow models (fourth-order discretizations for convective and diffusive fluxes, detailed chemical models etc), numerical schemes (the pressure solution, sixth-order interpolation schemes), the AMR mesh and the associated data object and miscellaneous components for I/O etc. The components can be approximately collated into 3 sub-assemblies, responsible for scalar transport, momentum transport (including the projection required for solving the low-Mach number approximation of the Navier-Stokes equation) and for advancing reactive terms. The components are dynamically loadable, and so, the "code" is composed at runtime. The components and wiring connectivity are specified in an input file to the framework; components can be exchanged simply by changing a single line in this input file.

The aim of componentization was the taming of complexity. One measure of complexity is the pattern in which different components might use each other. A good design would exhibit modularity, where connectivity between components is sparse and connections are arranged in some regular manner e.g., if components are collected/connected into sub-assemblies, which are hierarchically composed into the simulation code. Figure 17.2 shows 3 separate sub-assemblies consisting of components that address the transport of species, chemical reactions and the momentum solve (including the pressure solution). The size of each component is a second measure of complexity; smaller components are easier to understand and maintain. In Fig. 17.3 (left), we plot a histogram of the size (lines of code) in a component; it is clear most components are small, less than 1000 lines of code. Since components implement Ports, this strongly suggests that individual Ports do not embody much complexity either i.e., they have few methods that need to be implemented. This is shown in Fig. 17.3 (right) where we plot a histogram of the number of methods in each port. Most of the ports have 10 or fewer functions/methods. Figures 17.4 (left) plots the histogram of the number of ProvidesPorts i.e, the number of Ports a given component implements. It is clear that the bulk of the components implement less than five ports each, which explains their small size; recall that most ports have fewer than 10 methods. Another measure of a component's complexity or importance is the number of UsesPorts it has. Components that link disparate sub-assemblies together tend to have many UsesPorts distributed among the sub-assemblies. In Fig. 17.4 (right) we plot the number of UsesPorts per component, which is proportional to the number of other components a given component requires to perform its functions. We see that almost all components have less than 10 UsesPorts. Since components provide, on an average, less than 5 ports each (see Fig. 17.4, left), a component is connected to approximately 2-3 other components, leading to sparse connectivity between components. These statistics show that a relatively sparse specification of functionalities and interconnections may suffice for the construction of quite complex scientific software.

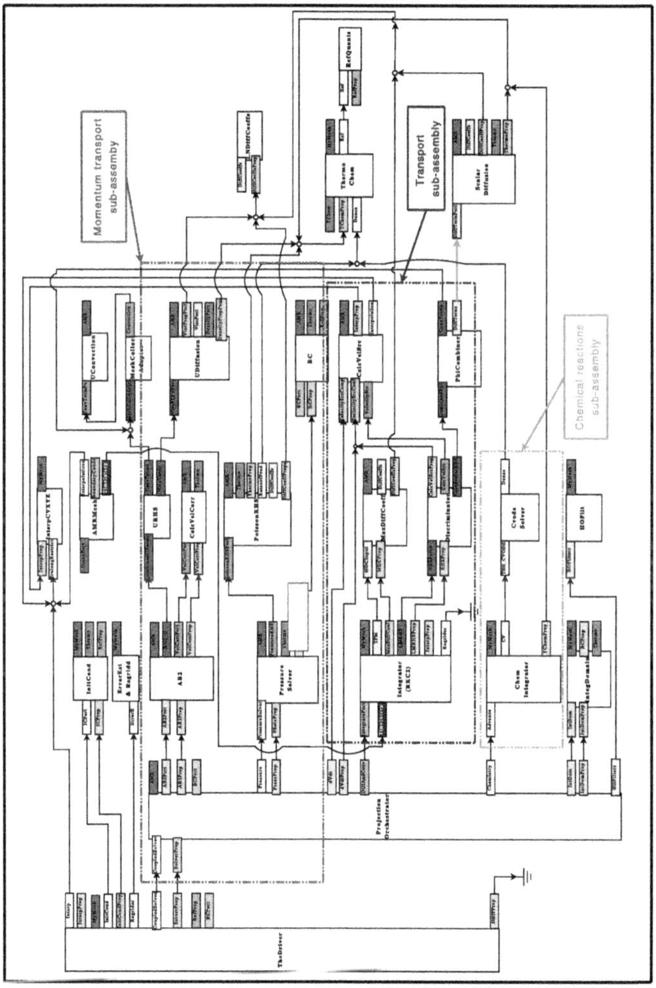

Fig. 17.2: Wiring diagram of the approximately 40 components that are used for simulating reactive flows in the CFRFS toolkit. The components are shown in white; different ports are colored. The ProvidesPorts appear on the left of a component; the UsesPorts on the right. The components are assembled, roughly, into 3 sub-assemblies, responsible for scalar transport (blue dash-dotted box), momentum transport (red dash-dotted box) and for modeling reactive processes (green dash-dotted box). The components can be strictly numerical (e.g., interpolation, pressure solver), models (detailed chemistry, diffusion coefficients) or "computer science" e.g., the AMR Mesh and the data object.

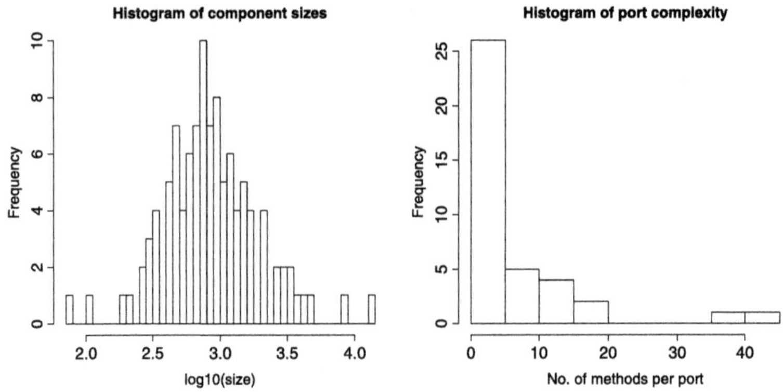

Fig. 17.3: Left: Histograms of the logarithm (to base 10) of component sizes, measured as the total number of lines, including comments and blank lines. The histogram clearly shows that half the components are less than 1000 lines long, and almost all are less than about 3000 lines. Components, therefore, are generally quite small. Right: Histogram of the number of functions/methods per port. We see that most ports have less than 10 methods each. There are a couple of ports, related to the mesh and the data object that have approximately 40 methods each. These statistics were extracted from a population of 100 components.

17.5 Computational Investigations Using CCA

In the previous section we described how CCA was used to architect and implement the CFRFS toolkit whose design philosophy stressed small, simple components, sparse connectivity between components and their hierarchical composition, via sub-assemblies, into functioning simulation code. In this section, we demonstrate two different ways in which the components of the the CFRFS toolkit are used.

The CFRFS toolkit consists of two sets of components. The first set, by far the bigger one, consists of components that address the numerical issues surrounding the use of fourth- (and higher) order spatially accurate methods on block-structured AMR meshes. These allow efficient resolution of fine flame structures without the necessity of overwhelming computational resources. The second set of components addresses the identification of low-dimensional manifolds in the chemical dynamics, so that chemical source terms may be tabulated and thus evaluated inexpensively within the context of spatially resolved flame simulations. This is done using Computational Singular Perturbation [51]. Many components, for example, those modeling chemical reactions, thermodynamics and constitutive models find use in both

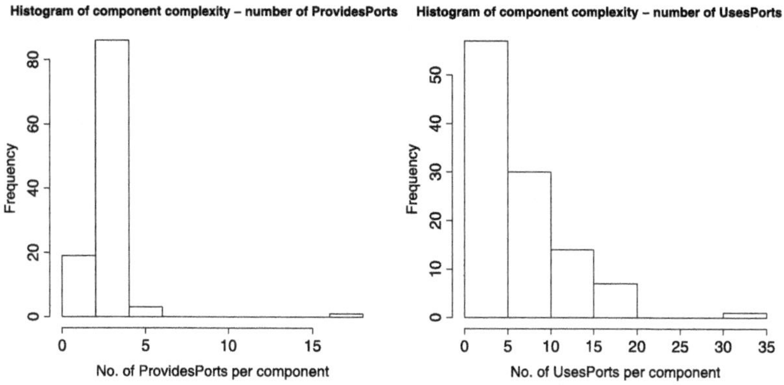

Fig. 17.4: Left: Histogram of the number of ProvidesPorts implements by the components in the CFRFS toolkit. We see that most of the components have 4 or fewer ProvidesPorts i.e., they implement very few ports. The distribution of the number of methods in each port is plotted in Fig. 17.3 (right). Right: Histogram of the number of UsesPorts a component uses. This is *one* measure of the complexity of the algorithm/functionality a component implements. Components that link sub-assemblies of components also tend to have many UsesPorts. We see that most components have less than 10 UsesPorts.

the efforts. The final goal is to replace/augment the reactive subsystem, consisting of a stiff-integrator and the chemical source terms (as described in Sec. 17.4), with an inexpensive tabulation scheme that would allow the toolkit to be used with large (and stiff) chemical mechanisms typically associated with higher hydrocarbons.

17.5.1 Fourth-order Combustion Simulations with Adaptive Mesh Refinement

Chemically reacting flow systems based on hydrocarbon fuels typically exhibit a large range of characteristic spatial and temporal scales. The complexity of kinetic models, even for simple hydrocarbon fuels, compounds this problem, making multidimensional numerical simulations difficult. This is true even for laboratory scale configurations.

These difficulties are commonly addressed in a variety of ways. For low speed flows, one may adopt a low Mach number approximation [58] for the momentum transport. This approximation assumes that acoustic waves travel at infinite speed, a justifiable assumption in many low-speed flows. One can also exploit the structure of the governing equations and adopt an operator-split mechanism, performing the transport and reactive time-advancement via specialized integrators [64]. In problems where fine structures exist only in a small fraction of the domain e.g., in laminar jet flames, one may employ AMR [13] to concentrate resolution only where needed [6, 12, 24, 73], while maintaining a coarse mesh resolution elsewhere.

The CFRFS toolkit implements a numerical model that can efficiently simulate flames with detailed chemical mechanisms. The use of AMR is not without its challenges, beyond just programming complexity. In order to reduce the number of grid points and the number of refinement levels in the computational mesh hierarchy we employ high-order stencils to discretize the governing equations and to interpolate between the computational blocks on adjacent mesh levels. A projection scheme is employed for the momentum transport. Since mesh adaptivity is driven by the narrow flame structure rather than the velocity field, we solve the momentum transport on the lowest level mesh in the AMR mesh hierarchy i.e., on a uniform mesh. This further enhances the efficiency of the model since the elliptic solver required by the pressure equation is more efficient on a uniform mesh, compared to a multilevel one [59]. The numerical approach and results obtained for canonical configurations are presented below.

17.5.1.1 Formulation

In the low-Mach number limit, the continuity, momentum and scalar transport equations for a chemically reacting flow system are written in compact form as

$$\nabla \cdot \mathbf{v} = -\frac{1}{\rho}\frac{D\rho}{Dt} \tag{17.1a}$$

$$\frac{\partial \mathbf{v}}{\partial t} = -\frac{1}{\rho}\nabla p + C_U + D_U \tag{17.1b}$$

$$\frac{\partial T}{\partial t} = C_T + D_T + S_T \tag{17.1c}$$

$$\frac{\partial Y_k}{\partial t} = C_{Y_k} + D_{Y_k} + S_{Y_k} \qquad k = 1, 2, \ldots, N_s. \tag{17.1d}$$

Here $\mathbf{v}$ is the velocity vector, ρ the density, T the temperature, Y_k the mass fraction of species k, p is the hydrodynamic pressure, and N_s is the number of chemical species. The $\frac{D}{Dt}$ operator in the continuity equation represents the material derivative, $\frac{D}{Dt} = \frac{\partial}{\partial t} + \mathbf{v} \cdot \nabla$. The system of equations is closed with the equation of state for an ideal gas. The thermodynamic pressure spatially uniform and is constant in time for an open domain in the low-Mach number limit. NASA polynomials are used to compute thermodynamic properties [60]. The transport properties are based on a mixture-averaged formulation and are evaluated using the DRFM package [68].

The equation of state is used to derive an expression for the right hand side of the continuity equation (Eq. 17.1a)

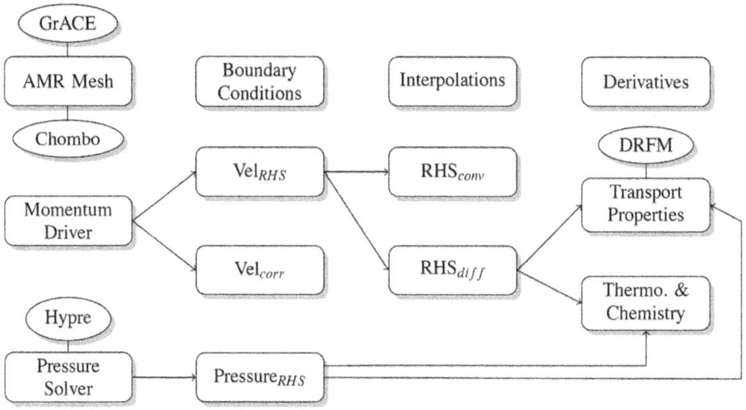

Fig. 17.5: Schematic of the momentum solver components in CFRFS.

$$\frac{DP_0}{Dt} = 0 \rightarrow \frac{1}{\rho}\frac{D\rho}{Dt} = -\frac{1}{T}(D_T + S_T) - \sum_{k=1}^{N_s} \frac{\bar{W}}{W_k}\left(D_{Y_k} + S_{Y_k}\right) \tag{17.2}$$

17.5.1.2 Implementation in the CCA Framework

The numerical integration of the system of equations (17.1a-17.1d) is performed in three stages. In the first stage, a projection scheme is used to advance the velocity field based on Eqs. 17.1a, 17.1b. Figure 17.5 shows the main CCA components involved in the momentum solver. The *Momentum Driver* component advances the velocity field to an intermediate value based on convection RHS_{conv} and diffusion RHS_{diff} contributions to the right-hand-side (RHS) term Vel_{rhs} of the momentum equation (17.1b). This is followed by an elliptic solve in the *Pressure Solver* component for the dynamic pressure p. The RHS values for the elliptic pressure equation are computed in $Pressure_{RHS}$. Transport and thermodynamic properties are provided by *Transport Properties* and *Thermo & Chemistry* components, respectively. The gradient of the pressure field is used to correct (Vel_{corr}) the intermediate velocities above to obtain a field that satisfies both the continuity and momentum equations (17.1a, 17.1b). The components shown at the top of Fig. 17.5 (AMR Mesh, Boundary Conditions, Interpolations and Derivatives) are generic components that handle the adaptive mesh refinement library, boundary conditions, interpolations, and derivatives. External libraries are shown in ellipses.

In the second stage, sketched in Fig. 17.6, temperature and species mass fractions are advanced using an operator split approach that separates the convection, C_T, C_{Y_k}, and diffusion, D_T, D_{Y_k}, contributions from the ones due to the chemical source terms, R_T, R_{Y_k}, in Eq. (17.1c, 17.1d). Symmetric Strang splitting is employed, beginning with the chemical source term contribution for half the time step, fol-

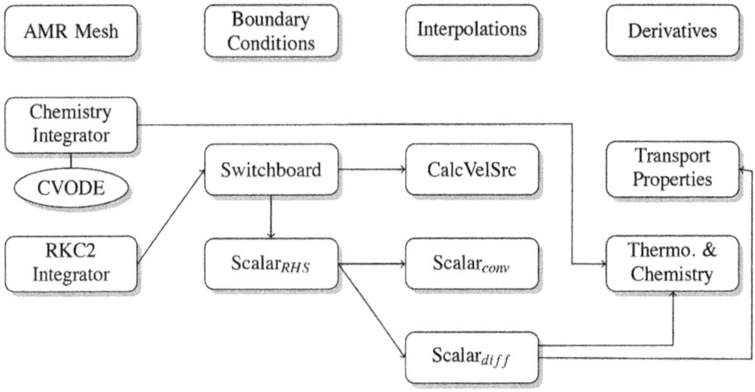

Fig. 17.6: Schematic of the scalar solver components in CFRFS.

lowed by the contributions from convection and diffusion terms for a full time step, and concluded by the remaining contribution from the reaction term for half the time step. The scalar advance due to the chemical source term is handled by *Chemistry Integrator*. The convection ($Scalar_{conv}$) and diffusion ($Scalar_{conv}$) contributions are combined by $Scalar_{RHS}$ component and provided to *RKC2 Integrator* which uses a Runge-Kutta-Chebyshev (RKC) algorithm [84] for time advancement. A *Switchboard* component is used to ensure that velocities are available at intermediate times during the multi-stage RKC integration.

The third stage repeats the projection algorithm from the first stage using the updated scalar fields from second stage. The overall algorithm is 4th-order accurate in space and 2nd-order in time.

Adaptive mesh refinement We employ an AMR approach where the computational domain is split into rectangular blocks. The advancement in time of the AMR solution is based on Berger-Colella time refinement [13, 52]. Figure 17.7 shows a schematic of this recursive time integration algorithm. Consider the solutions on levels L and $L+1$ at time t_n. Level L is first advanced to $t_n + \Delta t$, then the solution on $L+1$ is advanced in two half steps, $\Delta t/2$ to ensure numerical stability on the finer grid. During time advancement on $L+1$, boundary conditions are computed by interpolation using the solution on L. At $t_n + \Delta t$ the solution on $L+1$ is interpolated down to the corresponding regions on level L. In order to preserve the 4th-order spatial convergence of the numerical scheme, the interpolations between adjacent grid levels use 6th-order spatial stencils [72].

17.5.1.3 Application to Flame-Vortex Interaction

A canonical vortex-flame configuration [64] was chosen to explore the performance of the numerical construction. The computational domain is 1.5cm × 0.75cm. The

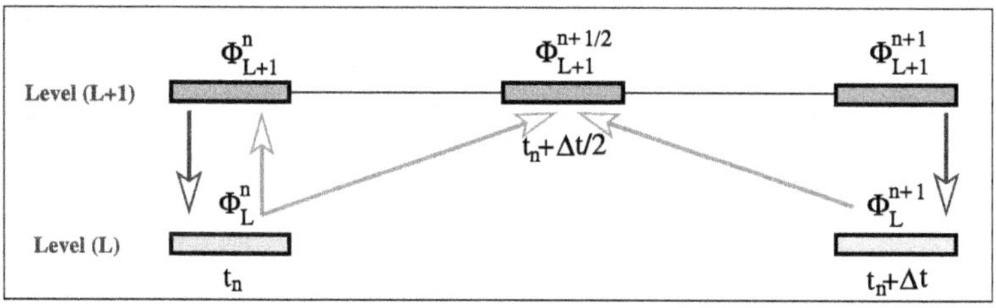

Fig. 17.7: Schematic of the time refinement in the context of AMR.

velocity field corresponding to a periodic row of counter-rotating Lamb-Oseen vortices is superimposed over the premixed 1D flame solution discussed above. A relatively coarse mesh was used for the base mesh, with a cell size of 50μm in each direction. Additional, finer, mesh levels were added in the flame region during the simulation.

A one-step, irreversible Arrhenius global reaction model is used in addition to a C1 kinetic model to study the vortex-flame interaction. Figure 17.8 shows freeze frames of the vorticity and heat release rate fields. The vortex pair is initially located 2mm upstream of the flame and propagates with approximately 10m/s towards it. As the vortex pair impinges into the flame, the flame intensity decreases on the centerline for the C1 model while the one-step solution shows little change in the interaction region. Similarly, at locations off-centerline the flame intensity for the C1 model decreases significantly as it stretched and rolled around the vortex pair. The last frames show a significantly contorted flame, and the relative increase in the overall burning rate is about about 50% more for the one-step reaction simulation compared to the simulation using the C1 model.

17.5.2 Computational Singular Perturbation and Tabulation

In the previous section, we described our experience with simulating flames using one-step and C1 chemical mechanisms. The primary challenge in going from simple one-step chemistry to a C1 mechanism was the steep increase in the stiffness of the dynamical system composed of the reactive processes; its main effect was to reduce the size of the time-step one could take without unacceptable splitting errors. Matters are further compounded when one considers C2 (or even more detailed) chemical mechanisms. This stiffness of detailed chemical mechanisms is due to the wide range of time scales that they model. It leads to considerable difficulties when time-advancing them in an efficient manner. Chemical model simplification and reduction strategies typically target these challenges by reducing the number of reactions and/or species in the model, with associated reduction in model complexity. When done properly, this strategy also reduces the system stiffness. Alternatively,

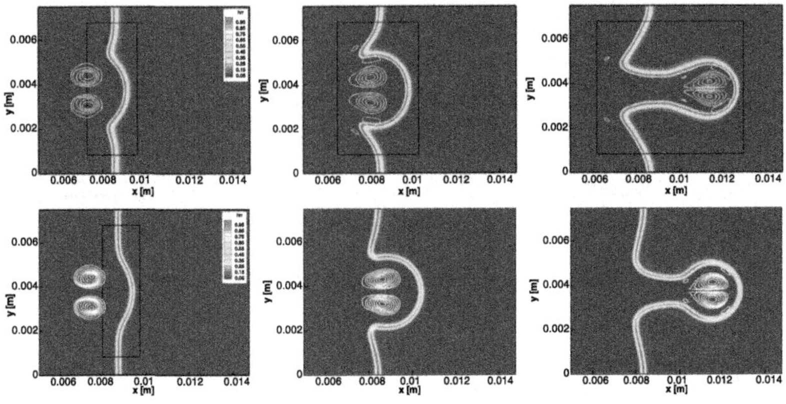

Fig. 17.8: Vorticity (white solid contours) and normalized heat release rate (hrr, shaded contours) for simulations using the one-step reaction model (upper row) and a C1 kinetic model consisting of 16 species and 46 reactions (lower row).

the Computational Singular Perturbation (CSP)-based time integration construction of [83] uses CSP analysis to project out the fast time scales from the detailed chemical source term, thereby rendering the equations non-stiff. The promise of this approach is that explicit time integrators can be used for large-time step integration of the resulting non-stiff source terms, and could potentially eliminate the need for operator-split time integration of reaction-diffusion source terms. More details are given in Chapter 9.

The key challenge with this time integration approach, however, is the large computational cost of solving for the requisite CSP information and the resulting projection matrices. An approach to mitigate this computational cost is tabulation. By adaptively storing and reusing the CSP information, the significant CSP overhead can be drastically reduced (by amortization), leading to an efficient overall implementation. Such a tabulation strategy has been explored for elementary model problems [50, 51]. This section describes the implementation of a CSP tabulation approach, relying on *k*d-trees [7] to efficiently store and retrieve CSP information along manifolds in the chemical configuration space.

In the following, the CSP time integration and tabulation approach is formulated, followed by a discussion of its implementation in the CCA framework for reacting flow simulation. Next, the approach is illustrated on the simulation of H_2 – air ignition.

17.5.2.1 Formulation and Implementation

Consider the chemical system described by $\mathrm{d}\boldsymbol{y}/\mathrm{d}t = \boldsymbol{g}(\boldsymbol{y})$, where $\boldsymbol{y} \in \mathbb{R}^N$, and $\boldsymbol{g}(\boldsymbol{y})$ is the chemical source term. The CSP basis vectors $\{\boldsymbol{a}_k\}_{k=1}^N$ and covectors $\{\boldsymbol{b}^k\}_{k=1}^N$, all in $\mathbb{R}^N$, enable the decoupling of the fast and slow processes, and the identification of low dimensional slow invariant manifolds (SIMs) [48]. Thus, we have

$$\frac{d\boldsymbol{y}}{dt} = \boldsymbol{g} = \boldsymbol{g}_{\text{fast}} + \boldsymbol{g}_{\text{slow}} = \boldsymbol{a}_1 f^1 + \boldsymbol{a}_2 f^2 + \cdots + \boldsymbol{a}_N f^N \tag{17.3}$$

where $f^i = \boldsymbol{b}^i \cdot \boldsymbol{g}$, for $i = 1, 2, \ldots, N$. In this equation, $\boldsymbol{g}_{\text{fast}}$ corresponds to the modes with fast transients, which are rapidly exhausted. After relaxation of fast transients, with M modes exhausted, $\boldsymbol{g}_{\text{fast}} = \sum_{r=1}^{M} \boldsymbol{a}_r f^r \approx 0$ and $\boldsymbol{g}_{\text{slow}} = \sum_{s=M+1}^{N} \boldsymbol{a}_s f^s = (I - \sum_{r=1}^{M} \boldsymbol{a}_r \boldsymbol{b}^r)\boldsymbol{g} = \boldsymbol{P}\boldsymbol{g}$. In practice, the number of exhausted modes is determined as the maximum M for which $\tau_{M+1} \sum_{r=1}^{M} \boldsymbol{a}_r f^r$ is less than a user-specified threshold, where τ_{M+1} is the time scale corresponding to the $(M+1)^{st}$ mode.

The CSP integrator [83] proceeds in each time step by first integrating the slow dynamics of the system, followed by a homogeneous correction (HC) to correct for the fast time scales:

$$\tilde{\boldsymbol{y}}(t + \Delta t) = \boldsymbol{y}(t) + \int_t^{t+\Delta t} \boldsymbol{P}\boldsymbol{g}\, dt' \tag{17.4}$$

$$\boldsymbol{y}(t + \Delta t) = \tilde{\boldsymbol{y}}(t + \Delta t) - \sum_{m,n=1}^{M} \boldsymbol{a}_m \tau_n^m |_t \hat{f}^n \tag{17.5}$$

$$\hat{f}^n = \boldsymbol{b}^n \cdot \boldsymbol{g}[\tilde{\boldsymbol{y}}(t + \Delta t)] \tag{17.6}$$

where τ_n^m is the inverse of λ_n^m, given by

$$\lambda_n^m = \left(\frac{d\boldsymbol{b}^m}{dt} + \boldsymbol{b}^m \boldsymbol{J} \right) \boldsymbol{a}_n \tag{17.7}$$

and $\boldsymbol{J}$ is the Jacobian of $\boldsymbol{g}$. The matrix τ_n^m is diagonal with entries the time scales $\{\tau_k\}_{k=1}^N$ when the CSP basis vectors are chosen to be the eigenvectors of $\boldsymbol{J}$ and the curvature of the SIM is neglected, *i.e.* $d\boldsymbol{b}^m/dt = 0$.

The procedure outlined above separates the fast, exhausted modes from the slow modes that drive the evolution of the system along the SIMs. As discussed in [47], CSP also identifies the species that are associated with these fast modes as *CSP radicals*. (These are the species whose concentration can be determined from the algebraic equations resulting from setting $f^i = 0,\ i = 1, \ldots, M$.) Accordingly, the species space can be separated into the CSP radicals and non-CSP radicals.

To improve the efficiency of the CSP integrator, a tabulation approach has been developed to enable reuse of the essential CSP quantities: the M fast CSP vectors and covectors, as well as the $M+1$ fastest time scales, which are sufficient to assemble the slow-manifold projector $\boldsymbol{P}$ needed for the HC and CSP integration, and to select the time step along the slow manifold. As the CSP vectors, covectors and time scales can be modeled as functions of the non-CSP radical species only, it is sufficient to tabulate these quantities in an $N - M$ dimensional table, rather than having to cover the full N-dimensional state space.

In the work presented here, a table with manifold conditions is constructed off-line, by performing full CSP analysis on a number of *design points* in state space. We first randomly sample a set of initial conditions over a range of initial temperatures, equivalence ratios and N_2 dilution factors (extra mole of N_2 per mole of air) and integrate them forward, with CVODE [19], using detailed reaction kinetics. A set of design points is constructed from the system states encountered during those simulations. For each of these design points, a CSP analysis is performed to identify associated SIMs. If a design point has exhausted modes, then successive HCs are applied to project that design point onto the corresponding SIM. Each SIM is characterized by a unique value of M and the associated CSP radicals.

For each identified SIM, the tabulation of the associated CSP information relies on a nonparametric regression approach. For this purpose, the CSP information is stored in kd-tree data structures over the the $N-M$ dimensional space of the non-CSP radical species. Note that, in order to give equal weight to all dimensions of the state vector in the computation of distance measures, all coordinates of the manifold points are first rescaled and shifted to range between 0 and 1 before being stored in the kd-trees. During time integration, the manifold that best corresponds to the current condition in the chemical configuration space is determined by finding the nearest neighbor, as measured by the Euclidean distance measure, in all of the manifolds in the table. If the manifold point that is closest to the current condition over all manifolds is within a maximum allowable distance d, then the associated manifold is assumed to be the one that is currently being followed by the system. The CSP information at the current condition is then approximated with the corresponding values at the nearest neighbor point in the table, which amounts to a 0th-order interpolation. Higher order interpolations, relying on interpolation between nearest neighbors or on polynomial response surfaces [49, 81], are the subject of ongoing work. In case none of the nearest neighbors in the tabulated manifolds are within the maximum allowable distance, then a full CSP analysis is performed on the current condition instead.

To implement the CSP integration approach, extensive use was made of existing components in the CCA framework. For example, the evaluation of the chemical kinetics source term and its Jacobian rely on the "AMR" set of components from CFRFS toolkit discussed in Sect. 17.4. Time integration relies on a CVODE component, part of the CFRFS toolkit. New components were developed to perform the CSP analysis as well as the table construction and interpolation for the tabulation approach. These components were joined together through the use of driver components that organize the overall algorithmic sequence of operations.

17.5.2.2 Application to H_2–air ignition system

The CSP integration method outlined in the previous section was applied to the simulation of ignition of a stoichiometric homogeneous H_2–air mixture at a temperature of $T = 1000$ K. The system is modeled using a 9 species reaction mechanism, resulting in a total state space dimension of $N = 10$ (9 species + temperature) [88].

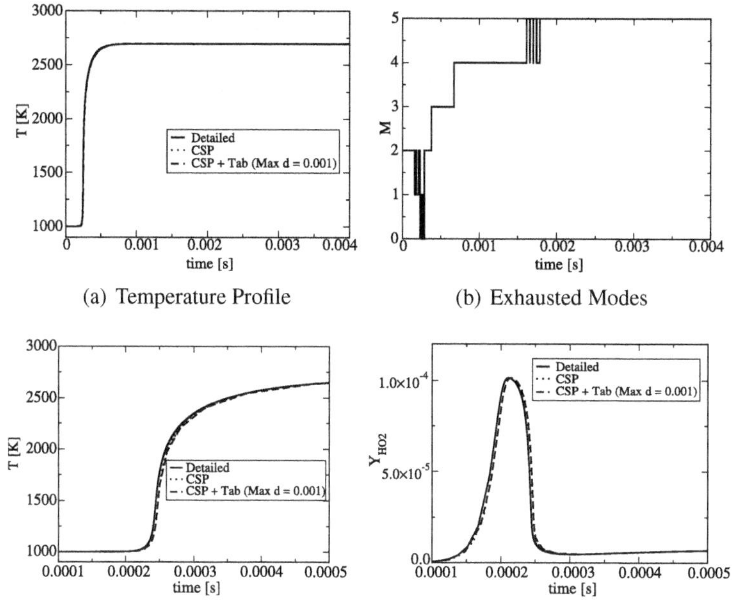

(a) Temperature Profile

(b) Exhausted Modes

(c) Temperature Detail

(d) HO_2 Detail

Fig. 17.9: a) Evolution of temperature in an igniting stoichiometric H_2–air system, simulated using the detailed reaction mechanism, the CSP solver, and the CSP solver with tabulation. b) Evolution of the number of exhausted modes, M, as obtained by CSP analysis. All approaches are in good agreement, except for minor differences in the ignition time delay, as shown in the close-up of the ignition zone for c) temperature and d) one of the trace species, HO_2. The initial conditions were: $T = 1000$ K, $Y_{H_2} = 0.0285$, $Y_{O_2} = 0.2264$ and $Y_{N_2} = 0.7451$.

Figure 17.9 compares the predicted temperature evolution obtained by integrating the detailed reaction kinetics (with the implicit solver CVODE), to the solution obtained with the CSP integrator (using the explicit fourth order Runge-Kutta (RK4) integration scheme), and with the CSP + tabulation approach.

For the tabulation approach a CSP table was constructed by sampling 100 initial conditions with Latin Hypercube Sampling over a range of equivalence ratios between 0.9 and 1, initial temperatures between 980 and 1020 K, and dilution factors between -0.005 and 0.005. From the design points extracted from these runs, close to 1 million states were identified on 9 different manifolds, with a number of exhausted modes ranging from 1 to 5.

The CSP integrated solution, both with and without tabulation is in good agreement with the full solution, except for a small difference in the ignition time delay, as is shown in detail in Fig. 17.9(c) and 17.9(d).

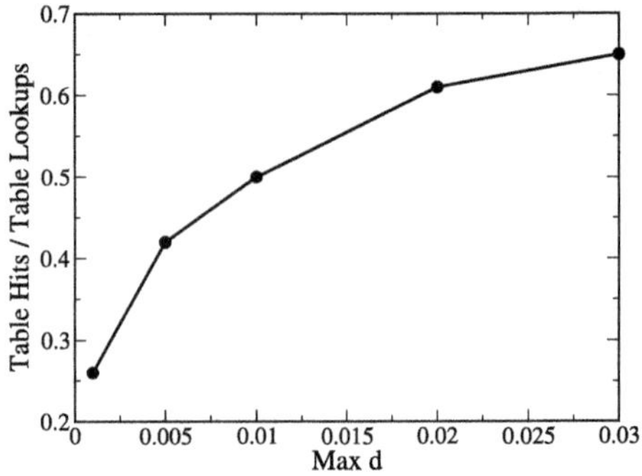

Fig. 17.10: The number of table hits, as a fraction of the total number of table lookups, increases as the maximum allowed distance to the nearest neighbor on a manifold is increased, resulting in more efficient usage of the tabulated data.

Note that, as the reaction progresses, the number of exhausted modes M, and the associated CSP radicals change according to the reaction dynamics. Figure 17.9(b) indicates that the system initially has two exhausted modes, followed by a time window during ignition where all modes are active, after which M gradually increases up to five at late time, as more and more modes become inactive. Accordingly, as the number of exhausted modes increases, the tabulation approach becomes more efficient in terms of storage and lookup times as the CSP information for each manifold is tabulated in $N-M$ dimensional kd-trees. For example, for the H_2–air system studied here, tabulation in a 5-dimensional table is sufficient for the section(s) of the 10-dimensional state space where 5 modes are exhausted (see Fig. 17.9(b)).

In terms of efficiency of the table usage, Fig. 17.10 shows the number of successful table hits as a fraction of the total number of table lookups. As a table lookup is performed in every time step, this number indicates how efficient the tabulation approach is at avoiding full CSP analyzes by providing tabulated CSP information instead. For the current table and initial condition, the table lookup success rate increases from 25 % to about 65 % as the maximum allowed nearest-neighbor distance is increased from 0.001 to 0.03, while the accuracy of the integration does not noticeably change (not shown here). Other numerical experiments indicate that this table hit success rate and the accuracy of the tabulation assisted simulations also depends on the density of the table in state space. A quantitative relationship between the table density, the maximum allowed distance in the nearest neighbor table lookup, and the accuracy of the CSP tabulation approach is the subject of ongoing work.

A comprehensive evaluation of the overall numerical performance of the CSP integration scheme with tabulation, as a function of table size, system and manifold dimensionality, degree of stiffness, and desired accuracy, is currently in progress. However, preliminary performance measurements show the CSP tabulation scheme to be competitive with direct CVODE integration for the cases studied in this paper.

To summarize, CCA provides a flexible framework for the implementation of several components for the integration of stiff chemical kinetics. The modularity of the framework allows easy reuse of components that were developed elsewhere in the project for operations such as time integration, or the evaluation of source terms or Jacobian. This flexibility allows the rapid development of codes to test various integration approaches and easy switching between them for method comparison and validation.

17.6 Research Topics in Computational Frameworks

The previous sections have described and demonstrated how modularity, obtained as a result of adopting a component-based design, may be used to mitigate the effects of software complexity. They have also shown the sophistication of the scientific software that can be designed using a component-based approach. However, the experience with CFRFS, as well as other componentization efforts [45, 46], have revealed a number of difficulties, Some solutions have recently been crafted, which we discuss below.

The Learning Curve: A significant challenge in adopting CCA has been the learning curve associated with using SIDL/Babel. While this was not encountered when developing the CFRFS toolkit (which uses the original C++-interface approach, not SIDL/Babel), it was observed in other componentization efforts [45, 46]. The process of generating client- and server-side code, as described in Sec. 17.3.1, is prone to error if performed manually, but can be automated. An integrated development environment, called Bocca [2], has been developed (see [16] for Bocca tutorials) for this purpose. Given the interfaces that a component uses and provides, Bocca automatically invokes Babel, creates the client- and server-side auto-generated code and constructs a build system to compile the resulting component skeleton. It minimizes what a CCA developer has to learn, enabling him/her to focus on more productive tasks.

Reluctance to Abandoning Working Software: The process of componentization could be significantly simplified if one could automatically *derive* components out of a non-component codebase. A concept, called OnRamp [43], is being investigated by CCA researchers to enable such an automated derivation of components. OnRamp is driven by annotations which are inserted into the codebase (indicating interfaces, code-blocks that will reside in components etc.), from which is it possible, under certain restrictions, to automatically generate components and its associated build system. This preserves the original code and most of the software development practices that the programmer is familiar with, while bestowing the benefits of componentization on the software in question.

Components also confer benefits beyond the fundamental requirement of constraining software complexity. Since components are a black box regarding implementation but adhere to a specified convention for communicating with the outside world, they are ideal for automating computation at a high level; CCA has focused on performance improvements as the aim of automation. Below, we mention some of the recent advances in this arena.

Automatic Proxy Generation: The collection of performance characteristics on a method-call basis is a required, but laborious task in performance modeling. In a component environment, this can be achieved quite easily by exploiting the fact that components publish the interfaces that they use and provide. The collection of performance data can be done by interposing a proxy component between an interface provider component and an interface using component. The proxy component serves to trap calls between components and switch-on/terminate the collection of performance metrics (elapsed time, cache misses, page faults etc) by a performance measurement tool e.g. TAU [79]. Such "performance-measurement" proxies can be generated automatically [82]. They can be used to collect performance data and identify bottlenecks; they can also be used to *monitor* an executing simulation and *optimize* it during runtime. This is described below.

Computational Quality of Service: Since the framework has a holistic view of the entire application, and proxy components can monitor the performance of individual components, it becomes possible to manipulate their behavior, with a view to ensuring robustness, celerity of computation etc. This is commonly referred to as "Computational Quality of Service" [62]. The manipulation of components may be performed by changing parameters that a component may provide or by replacing entire components [56, 57]. See [62] for how this may be performed without modifying any components; a working example, using a shock-hydrodynamics simulations in CCAFFEINE, can be found in [56].

17.7 Conclusion

In this chapter, we have investigated how a component architecture may be used to design and implement scientific simulation software. The Common Component Architecture was chosen because of its ability to accommodate parallel computing. Unlike many computational frameworks, a component framework does not require one to "marry into" a prescribed set of data and code structures; in many cases, such "marriages" lead to one's dependence on the framework for the integration of external libraries and/or legacy software. The peer nature of components (whereby all components are independent) prevent such dependencies from arising. However, it is to be noted that a software architecture merely lays down a few software development principles; their judicious use is a matter of software design. The design, in turn, is dictated by where one starts from (i.e., whether one starts with a *tabula rasa*, which was our case, or whether one starts componentizing a legacy code) and what one wishes to achieve with the particular design. Any lack of clarity regarding the second aspect invariably leads to an unsatisfactory end.

In Sect. 17.4 we described the particular ends that wished to achieve with our component-based design, particularly maintainabiliy and reduction in software complexity; the statistics drawn from the 100 components in our toolkit provide some confidence that we have largely succeeded. In Sect. 17.5 we showed how the toolkit is used, including a few example of component reuse. Though not described here, some of the AMR components used in Sect. 17.5.1 have also been used to simulate problems in shock-hydrodynamics [53]. Thus there is some evidence to indicate that the plug-and-play promise of component software, widely realized in non-scientific software, may be replicated in our field too.

This chapter has drawn examples from the CFRFS effort, which emphasized small, manageable components designed without any constraints imposed legacy software. There are other efforts where legacy software has dictated both the aims and the course of componentization (see [45] and references within), and still others, usually involving the componentization of libraries, where users played a role (see Sect. 11 in [14]). Component-based design, and CCA in particular, is a versatile methodology for designing and developing maintainable software, and is most profitably used when one has a clear idea of *why* one wishes to use it. Unlike many frameworks developed to enable the *rapid prototyping* of codes, its attractiveness lies in the long term.

Acknowledgements: The work documented in this chapter was funded by the Department of Energy, under its Scientific Discovery through Advanced Computing (SciDAC) program. Some of the computations were performed at the National Energy Research Supercomputing Center (NERSC) in Oakland, CA. The work was performed in Sandia National Laboratories, CA. Sandia is a multiprogram laboratory operated by Sandia Corporation, a Lockheed Martin company, for the United States Department of Energy's National Nuclear Security Administration under Contract DE-AC04-94-AL85000.

References

1. Allan, B.A., Armstrong, R.C., Wolfe, A.P., Ray, J., Bernholdt, D.E., Kohl, J.A.: The CCA core specifications in a distributed memory SPMD framework. Concurrency-Pract. Ex. **14**, 323–345 (2002)
2. Allan, B.A., Norris, B., Elwasif, W.R., Armstrong, R.C.: Managing scientific software complexity with Bocca and CCA. Sci. Program. **16**, 315–327 (2008)
3. Almgren, A., Bell, J., Colella, P., Howell, L., Welcome, M.: A conservative adaptive projection method for the variable density incompressible Navier-Stokes equations. J. Comput. Phys. **142**, 1–46 (1998)
4. AmrLib Homepage. Available from (accessed October 2009) `https://ccse.lbl.gov/Software/index.html`
5. AMROC Homepage. Available from (accessed October 2009) `http://amroc.sourceforge.net/`
6. Anthonissen, M.J.H., Bennett, B.A.V., Smooke, M.D.: An adaptive multilevel local defect correction technique with application to combustion. Combust. Theory Model. **9**, 273–299 (2005)
7. Arya, S., Mount, D.M., Netanyahu, N.S., Silverman, R., Wu, A.Y.: An optimal algorithm for approximate nearest neighbor searching in fixed dimensions. J. Assoc. Comput. Mach. **45**, 891–923 (1998)

8. Banks, J.W., Schwendeman, D.W., Kapila, A.K., Henshaw, W.D.: A high-resolution Godunov method for compressible multi-material flow on overlapping grids. J. Comput. Phys. **223**, 262–297 (2007)
9. Barad, M., Colella, P.: A fourth-order accurate local refinement method for Poisson's equation. J. Comput. Phys. **209**, 1–18 (2005)
10. Bell, J.B., Day, M.S., Grcar, J.F., Lijewski, M.J., Driscoll, J.F., Filatyev, S.F.: Numerical simulation of a laboratory-scale turbulent slot flame. Proc. Combust. Inst. **31**, 1299–1307 (2007)
11. Bell, J.B., Day, M.S., Shepherd, I.G., Johnson, M., Cheng, R.K., Grcar, J.F., Beckner, V.E., Lijewski, M.J.: Numerical simulation of a laboratory-scale turbulent V-flame. Proc. Natl. Acad. Sci. USA **102**, 10,006–10,011 (2005)
12. Bennett, B.A.V., Smooke, M.D.: Local rectangular refinement with application to axisymmetric laminar flames. Combust. Theory Model. **2**, 221–258 (1998)
13. Berger, M., Colella, P.: Local adaptive mesh refinement for shock hydrodynamics. J. Comput. Phys. **82**, 64–84 (1989)
14. Bernholdt, D.E., Allan, B.A., Armstrong, R., Bertrand, F., Chiu, K., Dahlgren, T.L., Damevski, K., Elwasif, W.R., Epperly, T.G.W., Govindaraju, M., Katz, D.S., Kohl, J.A., Krishnan, M., Kumfert, G., Larson, J.W., Lefantzi, S., Lewis, M.J., Malony, A.D., McInnes, L.C., Nieplocha, J., Norris, B., Parker, S.G., Ray, J., Shende, S., Windus, T.L., Zhou, S.: A component architecture for high-performance scientific computing. Intl. J. High-Perf. Computing Appl. **20**, 162–202 (2006)
15. Cactus Homepage. Available from (accessed October 2009) `http://en.wikipedia.org/wiki/Cactus_Framework`
16. CCA Tutorials Hands-On Guide. Available from (accessed October 2009) `http://www.cca-forum.org/tutorials/`
17. CHOMBO. Available from (accessed October 2009) `http://seesar.lbl.gov/anag/chombo/`
18. Cirak, F., Deiterding, R., Mauch, S.P.: Large-scale fluid-structure interaction simulation of viscoplastic and fracturing thin shells subjected to shocks and detonations. Comput. Struct. **85**, 1049–1065 (2006)
19. Cohen, S.D., Hindmarsh, A.C.: CVODE, a stiff/nonstiff ODE solver in C. Comput. Phys. **10**, 138–143 (1996)
20. Colella, P., Dorr, M., Hittinger, J., Martin, D.F., McCorquodale, P.: High-order finite-volume adaptive methods on locally rectangular grids. J. Phy.: Conf. Ser. **180**, 012,010 (5pp) (2009)
21. Colella, P., Graves, D.T., Keen, B.J., Modiano, D.: A Cartesian grid embedded boundary method for hyperbolic conservation laws. J. Comput. Phys. **211**, 347–366 (2006)
22. CORBA Component Model Webpage. Available from (accessed October 2009) `http://www.omg.org`
23. Dahlgren, T., Epperly, T., Kumfert, G., Leek, J.: Babel User's Guide. CASC, Lawrence Livermore National Laboratory, Livermore, CA (2005)
24. Day, M.S., Bell, J.B.: Numerical simulation of laminar reacting flows with complex chemistry. Combust. Theory Model. **4**, 535–556 (2000)
25. Deiterding, R.: Object-oriented design of an AMR algorithm for distributed memory computers. In: 8th Int. Conf. on Hyperbolic Problems. Magdeburg (2000)
26. Deiterding, R.: Detonation structure simulation with AMROC. In: L.T. Yang (ed.) High Performance Computing and Communications, no. 3726 in Lecture Notes in Computer Science, pp. 916–927. Springer, Berlin Heidelberg (2005)
27. Deiterding, R.: A parallel adaptive method for simulating shock-induced combustion with detailed chemical kinetics in complex domains. Comput. Struct. **87**, 769–783 (2009)
28. Deiterding, R., Cirak, F., Mauch, S., Meiron, D.: A virtual test facility for simulating detonation- and shock-induced deformation and fracture of thin flexible shells. Int. J. Multiscale Computational Engineering **5**, 47–63 (2007)
29. Deiterding, R., Radovitzky, R., Mauch, S.P., Noels, L., Cummings, J.C., Meiron, D.I.: A virtual test facility for the efficient simulation of solid material response under strong shock and detonation wave loading. Eng. Comput. **22**, 325–347 (2006)

30. Drake, J.B., Jones, P.W., Carr Jr., George R., Overview of the software design of the community climate system model. Intl. J. High-Perf. Computing Appl. **19**, 177–186 (2005)
31. Dubey, A., Antypas, K., Ganapathy, M.K., Reid, L.B., Riley, K., Sheeler, D., Siegel, A., Weide, K.: Extensible component based architecture for FLASH, A massively parallel, multiphysics simulation code. Parallel Comput. (2009). Submitted, preprint at `http://arxiv.org/pdf/0903.4875`
32. Earth Systems Modeling Framework Homepage. Available from (accessed October 2009) `http://www.esmf.ucar.edu/`
33. Englander, R., Loukides, M.: Developing Java Beans (Java Series). O'Reilly and Associates (1997)
34. Falgout, R., Yang, U.: Hypre: a library of high performance preconditioners, in Computational Science. In: P.M.A. Sloot, C. Tan, J.J. Dongarra, A.G. Hoekstra (eds.) Lecture Notes in Computer Science, vol. 2331, pp. 632–641. Springer-Verlag (2002)
35. Fryxell, B., Olson, K., Ricker, P., Timmes, F.X., Zingale, M., Lamb, D.Q., MacNeice, P., Rosner, R., Truran, J.W., Tufo, H.: FLASH: an adaptive mesh hydrodynamics code for modeling astrophysical thermonuclear flashes. Ap. J. Supplement Series **131**, 273–334 (2000)
36. Godfrey, M.W., Tu, Q.: Evolution in open source software: A case study. In: Proceedings of the International Conference on Software Maintenance, pp. 131–142 (2000)
37. Goodale, T., Allen, G., Lanfermann, G., Mass, J., Radke, T., Seide, E., Shalf, J.: The Cactus framework and tolkit: Design and applications. In: Proceedings of Vector and Parallel Processing - VECPAR 2002 (2002)
38. GrACE Homepage. Available from (accessed October 2009) `http://www.caip.rutgers.edu/TASSL/`
39. Henshaw, W.D.: A fourth-order accurate methods for the incompressible Navier-Stokes equations on overlapping grids. J. Comput. Phys. **113**, 13–25 (1994)
40. Henshaw, W.D., Schwendeman, D.W.: An adaptive numerical scheme for high-speed reactive fow on overlapping grids. J. Comput. Phys. **191**, 420–447 (2003)
41. Henshaw, W.D., Schwendeman, D.W.: Moving overlapping grids with adaptive mesh refinement for high-speed reactive and non-reactive flow. J. Comput. Phys. **216**, 744–779 (2006)
42. Henshaw, W.D., Schwendeman, D.W.: Parallel computation of three-dimensional Flows using Overlapping Grids with Adaptive Mesh Refinement. J. Comput. Phys. **227**, 7469–7502 (2008)
43. Huelette, G.C., Sottile, M.J., Armstrong, R., Allan, B.: OnRamp: Enabling a New component-based development paradigm. In: Proceedings of Component-Based High Performance Computing (2009)
44. Kadioglu, S., Klein, R., Minion, M.: A fourth-order auxiliary variable projection method for zero-Mach number gas dynamics. J. Comput. Phys. **227**, 2012–2043 (2008)
45. Kenny, J.P., Janssen, C.L., Valeev, E.F., Windus, T.L.: Components for integral evaluation in quantum chemistry. J. Comput. Chem. **29**(4), 562–577 (2008)
46. Krishnan, M., Alexeev, Y., Windus, T.L., Nieplocha, J.: Multilevel parallelism in computational chemistry using common component architecture and global arrays. In: SC '05: Proceedings of the 2005 ACM/IEEE Conference on Supercomputing, p. 23. IEEE Computer Society, Washington, DC, USA (2005)
47. Lam, S.: Using CSP to understand complex chemical kinetics. Combust. Sci. Technol. **89**, 375–404 (1993)
48. Lam, S., Goussis, D.: Understanding complex chemical kinetics with computational singular perturbation. Proc. Combust. Inst. **22**, 931–941 (1988)
49. Lee, J., Najm, H., Lefantzi, S., Ray, J., Frenklach, M., Valorani, M., Goussis, D.: A CSP and tabulation based adaptive chemistry model. Combust. Theory Model. **11**(1), 73–102 (2007)
50. Lee, J., Najm, H., Lefantzi, S., Ray, J., Goussis, D.: On chain branching and its role in homogeneous ignition and premixed flame propagation. In: K. Bathe (ed.) Computational Fluid and Solid Mechanics 2005, pp. 717–720. Elsevier Science (2005)
51. Lee, J.C., Najm, H.N., Lefantzi, S., Ray, J., Frenklach, M., Valorani, M., Goussis, D.: A CSP and tabulation based adaptive chemistry model. Combust. Theory Model. **11**, 73–102 (2007)

52. Lefantzi, S., Ray, J., Kennedy, C.A., Najm, H.N.: A component-based toolkit for reacting flows with high order spatial discretizations on structured adaptively refined meshes. Prog. Comput. Fluid Dy. **5**, 298–315 (2005)
53. Lefantzi, S., Ray, J., Najm, H.N.: Using the common component architecture to design high performance scientific simulation codes. In: Proceedings of the International Parallel and Distributed Processing Symposium. Nice, France (2003)
54. Li, X., Parashar, M.: Hierarchical partitioning techniques for structured adaptive mesh refinement applications. J. Supercomput. **28**, 265–278 (2004)
55. Li, X., Parashar, M.: Hybrid runtime management of space-time heterogeneity for parallel structured adaptive applications. IEEE Transactions on Parallel and Distributed Systems **18**, 1202–1214 (2007)
56. Liu, H., Parashar, M.: Enabling self-management of component-based high-performance scientific applications. In: Proceedings of the 14th IEEE International Symposium on High Performance Distributed Computing (HPDC-14). Research Triangle Park, NC (2005)
57. Liu, H., Parashar, M.: Accord: A programming rramework for autonomic applications. IEEE Transaction on Systems, Man, and Cybernetics **36**, 341–352 (2006). Special issue on Engineering Autonomic Systems, Editors: R. Sterritt and T. Bapty
58. Majda, A., Sethian, J.: The derivation and numerical solution of the equations for zero Mach number combustion. Combust. Sci. Technol. **42**, 185–205 (1985)
59. Martin, D.F., Colella, P.: A cell-centered adaptive projection method for the incompressible euler equations. J. Comput. Phys. **163**, 271–312 (2000)
60. McBride, B.J., Gordon, S., Reno, M.: Coefficients for calculating thermodynamic and transport properties of individual species. Tech. Rep. TM-4513, NASA (1993)
61. McInnes, L.C., Allan, B.A., Armstrong, R., Benson, S.J., Bernholdt, D.E., Dahlgren, T.L., Diachin, L.F., Krishnan, M., Kohl, J.A., Larson, J.W., Lefantzi, S., Nieplocha, J., Norris, B., Parker, S.G., Ray, J., Zhou, S.: Parallel PDE-based simulations using the common component architecture. In: Numerical Solution of Partial Differential Equations on Parallel Computers, pp. 327–384. Springer (2006)
62. McInnes, L.C., Ray, J., Armstrong, R., Dahlgren, T.L., Malony, A., Norris, B., Shende, S., Kenny, J.P., Steensland, J.: Computational quality of service for scientific CCA applications: Composition, substitution, and reconfiguration. Tech. Rep. ANL/MCS-P1326-0206, Argonne National Laboratory (2006). `ftp://info.mcs.anl.gov/pub/tech_reports/reports/P1326.pdf`
63. Meir "Manny" Lehman's FEAST project. Available from (accessed October 2009) `http://www.doc.ic.ac.uk/~mml/feast`
64. Najm, H., Knio, O.: Modeling lw Mach number reacting flow with detailed chemistry and transport. J. Sci. Comp. **25**, 263–287 (2005)
65. Overture Homepage. Available from (accessed October 2009) `https://computation.llnl.gov/casc/Overture/`
66. Pantano, C., Deiterding, R., Hill, D.J., Pullin, D.I.: A low numerical dissipation patch-based adaptive mesh refinement method for large-eddy simulation of compressible flows. J. Comput. Phys. **221**, 63–87 (2007)
67. Parashar, M., Browne, J.C.: System engineering for high performance computing software: The HDDA/DAGH infrastructure for implementation of parallel structured adaptive mesh refinement. In: D.B.G.S.B. Baden, M.P. Chrisochoides, M.L. Norman (eds.) Structured Adaptive Mesh Refinement, *IMA*, Vol. 117. Springer-Verlag (2000)
68. Paul, P.H.: DRFM: A new package for the evaluation of gas-phase-transport properties. Sandia Report SAND98-8203, Sandia National Laboratories, Albuquerque, New Mexico (1997)
69. Publications from the Applied Numerical Algorithms Group. Available from (accessed October 2009) `http://seesar.lbl.gov/anag/publication.html`
70. Publications Using AMROC and Virtual Test Facility. Available from (accessed October 2009) `http://www.csm.ornl.gov/~r2v/html/pub.htm`
71. Ray, J., Kennedy, C., Steensland, J., Najm, H.N.: Advanced algorithms for computations on block-structured adaptively refined meshes. J. Phys.: Conf. Ser. **16**, 113–118 (2005)

72. Ray, J., Kennedy, C.A., Lefantzi, S., Najm, H.N.: Using high-order methods on adaptively refined block-structured meshes - derivatives, interpolations, and filters. SIAM J. Sci. Comp. **29**, 139–181 (2007)
73. Ray, J., Najm, H.N., Milne, R.B., Devine, K.D., Kempka, S.: Triple flame structure and dynamics at the stabilization point of an unsteady lifted jet diffusion flame. Proc. Combust. Inst. **28**, 219–226 (2000)
74. Safta, C.: (2009). Personal Communication
75. Safta, C., Ray, J., Najm, H.: A high-order projection scheme for AMR computations of chemically reacting flows. In: Proceedings of the 2009 Fall Meeting of the Western States Section of the Combustion Institute, Irvine, CA (2009)
76. Sommeijer, B.P., Shampine, L.F., Verwer, J.G.: RKC: An explicit solver for parabolic PDEs. J. Comp. Appl. Math. **88**, 315–326 (1998)
77. de St. Germain, J.D., McCorquodale, J., Parker, S.G., Johnson, C.R.: UINTAH: A massively parallel problem solving environment. In: HPDC '00: Ninth IEEE International Symposium on High Performance and Distributed Computing (2000)
78. van Straalen, B., Shalf, J., Ligocki, T., Keen, N., Yang, W.S.: Scalability challenges for massively parallel AMR applications. In: Proceedings of the 23rd IEEE International Symposium on Parallel and Distributed Processing (2009)
79. TAU: Tuning and Analysis Utilities. Available from (accessed November 2009) `http://www.cs.uoregon.edu/research/paracomp/tau/`
80. The OpenFOAM Homepage. Available from (accessed October 2009) `http://www.opencfd.co.uk/openfoam/`
81. Tonse, S., Moriarty, N., Brown, N., Frenklach, M.: PRISM: Piecewise reusable implementation of solution mapping. An economical strategy for chemical kinetics. Israel J. Chem. **39**, 97–106 (1999)
82. Trebon, N., Morris, A., Ray, J., Shende, S., Malony, A.D.: Performance modeling using component assemblies. Concurr. Comp.-Pract. E. **19**, 685–696 (2007)
83. Valorani, M., Goussis, D.: Explicit time-scale splitting algorithm for stiff problems: Auto-ignition of gaseous-mixtures behind a steady shock. J. Comput. Phys. **169**, 44–79 (2001)
84. Verwer, J.G., Sommeijer, B.P., Hundsdorfer, W.: RKC time-stepping for advection-diffusion-reaction problems. J. Comput. Phys. **201**, 61–79 (2004)
85. Visual Basic Webpage. Available from (accessed October 2009) `http://msdn.microsoft.com/en-us/vbasic/default.aspx`
86. Williams, F.: Combustion Theory, 2nd edn. Addison-Wesley, New York (1985)
87. XCAT Homepage. Available from (accessed October 2009) `http://www.extreme.indiana.edu/xcat/`
88. Yetter, R., Dryer, F., Rabitz, H.: A comprehensive reaction mechanism for carbon monoxide/hydrogen/oxygen kinetics. Combust. Sci. Technol. **79**, 97–128 (1991)

Chapter 18
The Heterogeneous Multiscale Methods with Application to Combustion

Weinan E, Björn Engquist and Yi Sun

Abstract The framework of the heterogeneous multiscale methods (HMM) is briefly reviewed. Both the original HMM and the seamless HMM are discussed. Applications to interface capturing and flame front tracking are presented.

18.1 The Heterogeneous Multiscale Method

The heterogeneous multiscale method (HMM) is a general framework for designing multiscale methods for a wide range of applications [8, 10]. HMM is a top-down strategy: it starts with an incomplete macroscale model, the missing data in the macroscale model is estimated on the fly using the microscale model. As such, it offers a platform for designing numerical algorithms that makes use of the knowledge we have at both the macro- and micro-scales. Here we will briefly review the main ideas, its advantages and limitations. We will also discuss possible applications to combustion.

Weinan E
Department of Mathematics and Program in Applied and Computational Mathematics, Princeton University, e-mail: weinan@math.princeton.edu

Björn Engquist
Department of Mathematics and Institute for Computational Engineering and Sciences, University of Texas, Austin, e-mail: engquist@ices.utexas.edu

Yi Sun
Statistical and Applied Mathematical Sciences Institute, e-mail: yisun@samsi.info

T. Echekki, E. Mastorakos (eds.), *Turbulent Combustion Modeling*,
Fluid Mechanics and Its Applications 95, DOI 10.1007/978-94-007-0412-1_18,

18.1.1 The Basic Framework

The general setting is as follows. At the macroscopic level, we make an assumption about the form of the macroscale model, which can be expressed abstractly as

$$F(U,D)=0, \tag{18.1}$$

where D denotes the data needed for the macroscale model to be complete. This data is estimated using a microscopic model, say in the form:

$$\partial_t u = \mathscr{L}(u;U), \quad Qu(0)=U_0. \tag{18.2}$$

or

$$f(u,d)=0, \quad d=d(U). \tag{18.3}$$

Here the macroscale variable U may enter the system as constraint, in order to make sure that the data does correspond to the macroscale state we are interested in, Q is the compression (or projection) operator that maps micro-state variables to macro-states variables [8]. Once the form of the macroscale model and the detailed microscale models are chosen, we select a macroscale solver. We also need a constrained microscale solver and a way of estimating the needed macroscale data from the results of the microscale simulation. Therefore the three main ingredients of HMM are (Figure 18.1):

1. macroscale solver
2. constrained microscale solver
3. data estimator

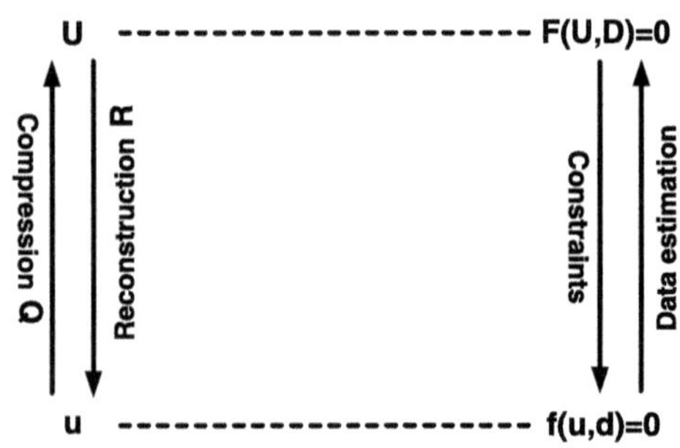

Fig. 18.1: Schematics of the HMM framework.

Take the example of incompressible fluids for which the macroscale model is a continuum model for the macroscale velocity field **U** in the form:

$$\partial_t \mathbf{U} + (\mathbf{U} \cdot \nabla)\mathbf{U} = \nabla \cdot \tau,$$
$$\nabla \cdot \mathbf{U} = 0.$$

These are simply statements of the conservation of momentum and mass. The unknown data is the stress τ: $D = \tau$.

Let us assume that the micro model is a molecular dynamics model for the particles that make up the fluid:

$$m_i \frac{d^2 \mathbf{y}_i}{dt^2} = \mathbf{f}_i, \quad i = 1, 2, \cdots, N. \tag{18.4}$$

Here $m_i, \mathbf{y}_i$ are respectively the mass and position of the i-th particle, $\mathbf{f}_i$ is the force acting on the i-th particle.

Given that the macroscale model is in the form of an incompressible flow equation, it is natural to select the projection method as the macro-solver [4]. In the implementation of the projection method, we will need the values of τ at the appropriate grid points. These are the data that need to be estimated. At this point, we have to make an assumption about what τ depends on, and this enters in the constraints that we put on the microscale model. Let us assume that

$$\tau = \tau(\nabla \mathbf{U}). \tag{18.5}$$

We will constrain the molecular dynamics in such a way that the average strain rate is given by the value of $\nabla \mathbf{U}$ at the relevant grid point. In general, implementing such constraints is the most difficult step in HMM. For the present example, one possible strategy is discussed in [27].

From the results of the microscale model, we need to extract the values of the needed components of the stress. For this purpose, we need a formula that expresses stress in terms of the output of the molecular dynamics. This can be obtained by modifying the Irving-Kirkwood formula [27]. These details can be found in [27].

In this example, the three ingredients are:

1. macro-solver, the projection method;
2. micro-solver, the constrained molecular dynamics;
3. data estimator, the modified Irving-Kirkwood formula.

We can write down the HMM procedure formally as follows. At each macro time step,

1. Given the current state of the macro variables U_n, reinitialize the micro-variables:

$$u_{n,0} = RU_n; \tag{18.6}$$

2. Evolve the micro variables for some micro time steps:

$$u_{n,m+1} = \mathscr{S}_{\delta t}(u_{n,m}; U_n), \quad m = 0, \cdots, M-1; \tag{18.7}$$

3. Estimate D:

$$D_n = \mathscr{D}_M(u_{n,0}, u_{n,1}, \cdots, u_{n,M}); \tag{18.8}$$

4. Evolve the macro-variables for one macro time step using the macro-solver:

$$U_{n+1} = S_{\Delta t}(U_n; D_n). \tag{18.9}$$

Here R is some reconstruction operator that reinitializes the micro model in a way that is consistent with the current state of the macro variables, $\mathscr{S}_{\delta t}$ is the micro solver, which also depends on U_n through the constraints, as indicated. $\mathscr{D}_M$ is some data processing operator which in general involves spatial/temporal/ensemble averaging. This is sometimes referred to as the data estimator. Finally $S_{\Delta t}$ is the macro solver.

There are two important time scales that we need to consider. The first, denoted by t_M, is the time scale for the dynamics of the macro-variables. The second, denoted by τ_ε, is the relaxation time for the microscopic model. We will need to distinguish two different cases. The first is when there is no time scale separation, i.e. $\tau_\varepsilon \sim t_M$. In this case, from the viewpoint of numerical efficiency, there is not much room to play with as far as time scales are concerned. We just have to evolve the microscale model along with the macroscale model. The second case is when $\tau_\varepsilon \ll t_M$. This is the case we will focus on. The general guideline in this case is to choose Δt to accurately resolve the t_M time scale, and to choose M such that $M\delta t$ covers sufficiently the τ_ε time scale for equilibration to take place in the micro model.

One main difference between HMM and other multiscale methods such as systematic upscaling and the "equation-free" approach [2, 19] is that HMM starts with a macroscale model. This does represent a compromise in the sense that we need to work with a preconceived form of the macroscale model. If the correct macroscale model is a stochastic differential equation, then clearly an HMM based on an assumption that the macroscale model is a deterministic equation leads to wrong results. On the other hand, there are two important reasons for starting from a macroscale model. The first is that even for problems for which adequate closed-form macroscale models are available, designing a stable and accurate numerical method for that macroscale model might be a significant task. The second is that for most problems of practical interest, we often have already accumulated some knowledge about the macroscale behavior of the problem. We should make use of such information in designing multiscale methods. In cases when one makes a wrong assumption, one can still argue that HMM produces an "optimal approximation" for the macroscale behavior of the solution in the class of models considered. In this sense, HMM is a way of addressing the following question: What is the best we can do given the knowledge we have about the problem at all scales? This was indeed the underlying designing principle for HMM.

18.1.2 The Seamless Algorithm

The version of HMM described above requires converting back and forth between the macro- and micro-states of the system. This can become rather difficult in actual implementations, particularly when constructing discrete micro-states (needed for example in molecular dynamics) from continuous macro variables. The seamless strategy proposed in [12] (see also [11, 16, 26]) is intended to bypass this difficult step.

To motivate the seamless algorithm, let us consider the trivial example of stiff ODEs:

$$\begin{aligned} \frac{dx}{dt} &= f(x,y), \\ \frac{dy}{dt} &= -\frac{1}{\varepsilon}(y-\varphi(x)). \end{aligned} \tag{18.10}$$

One simplest idea is to change the small parameter ε to a bigger value ε', the size of which is determined by the accuracy consideration:

$$\begin{aligned} \frac{dx}{dt} &= f(x,y), \\ \frac{dy}{dt} &= -\frac{1}{\varepsilon'}(y-\varphi(x)). \end{aligned} \tag{18.11}$$

This is then solved using standard methods.

We can look at this differently. Instead of changing the value of ε, we may change the clock for the microscale model, i.e. if we use $\tau = \varepsilon t/\varepsilon'$ in the second equation in (18.11), then (18.11) can be written as:

$$\begin{aligned} \frac{dx}{dt} &= f(x,y) \\ \frac{dy}{d\tau} &= -\frac{1}{\varepsilon}(y-\varphi(x)). \end{aligned} \tag{18.12}$$

If we discretize this equation using standard ODE solvers but with different time step sizes for the first and second equations in (18.12), we obtain the following algorithm:

$$y^{n+1} = y^n - \frac{\delta\tau}{\varepsilon}(y^n - \varphi(x^n)), \tag{18.13}$$

$$D^{n+1} = y^{n+1}, \tag{18.14}$$

$$x^{n+1} = x^n + \tilde{\Delta}t f(x^n, D^{n+1}). \tag{18.15}$$

Here $y^n \sim y(n\delta\tau)$ and $x^n \sim x(n\tilde{\Delta}t)$. The value of $\delta\tau$ is the time step size we would use if we attempt to solve (18.10) accurately. The choice of $\tilde{\Delta}t$ is more tricky. It should not only resolve the macro time scale, but also allow the micro state to relax, i.e. to adjust to the changing macroscale environment. For example, if Δt is the time step size required for accurately resolving the macroscale dynamics and if τ_ε is the relaxation time of the microscopic model, then we should choose $\tilde{\Delta}t = \Delta t/M$ where $M > \tau_\varepsilon/\delta\tau$.

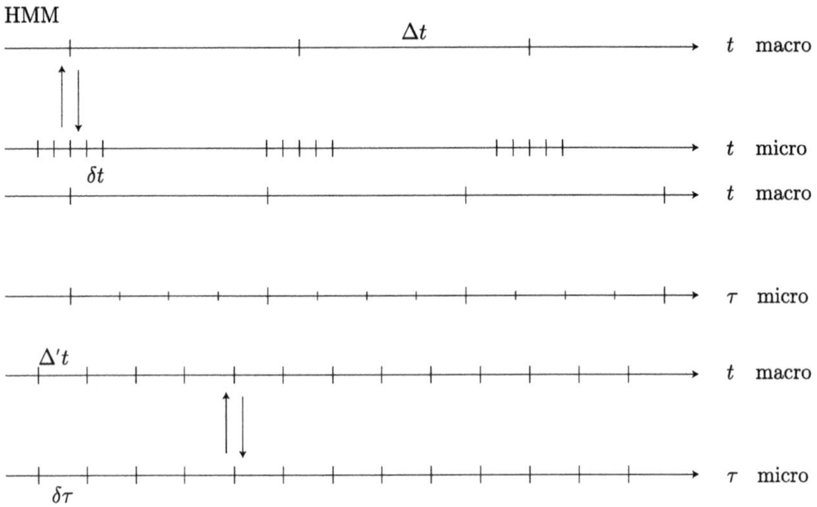

Fig. 18.2: Illustration of HMM (upper panel) and the seamless algorithm (lower panel). Middle panel: rescaling the micro time scale.

In a nutshell, we can formulate the strategy as follows:

1. Run the micro solver using its own time step $\delta\tau$.
2. Run the macro solver on a slower pace than a standard macro model: $\tilde{\Delta}t = \Delta t/M$.
3. Exchange data between the micro- and macro-solvers at every step.

The intuitive idea is illustrated in Figure 18.2. What one does is to force the microscale model to accommodate the changes (here the change in x) in the macroscale environment at a much faster pace. For example, assume that the characteristic macro time scale is 1 second and the micro time scale is 1 femtosecond $= 10^{-15}$ second. In a brute force calculation, the micro model will run 10^{15} steps before the macroscale environment changes. HMM makes use of the separation of the time scales by running the micro model only until it is sufficiently relaxed, which requires much fewer (say M) than 10^{15} steps, and then extracts the data in order to evolve the macro system over a macro time step of 1 second. That means that HMM skips $10^{15} - M$ micro steps of calculation, and it exchanges data between the macro and micro solvers after every 10^{15} micro time step interval. The price one has to pay is that one has to reinitialize the microscale solver at each macro time step, due to the gap created by skipping $10^{15} - M$ micro steps. One can take a different viewpoint. If we define

$$\tilde{y}(k\tilde{\Delta}t + t^n) = y^{n,k}, \quad k = 1, \cdots, M, \tag{18.16}$$

where $\tilde{\Delta}t = \Delta t/M$ and $y^{n,k}$ is the k-th step solution to the microscale model at the n-th macro step, the variable $\tilde{y}$ is defined everywhere in the new rescaled time axis. By doing so, we change the clock for the micro model, but we no longer need to reinitialize the micro model every macro time step, since the gap mentioned above no longer exists. Because the cost of the macro solver is typically very small compared with the cost of the micro solver, we may as well run the macro solver using a smaller time step (i.e. $1/M$ sec) and exchange data every time step. In this way the data exchanged tend to be more smooth.

Using the setup discussed earlier, we can write the seamless algorithm as follows:

1. Given the current state of the micro variables $u(\tau)$ and the macro variables $U(t)$, evolve the micro variables for one time step

$$u(\tau+\delta\tau) = \mathscr{S}_{\delta\tau}(u(\tau);U(t)); \tag{18.17}$$

2. Estimate D:

$$D = \mathscr{D}(u(\tau+\delta\tau)); \tag{18.18}$$

3. Evolve the macro variables

$$U(t+\tilde{\Delta}t) = S_{\tilde{\Delta}t}(U(t);D). \tag{18.19}$$

In this algorithm, we alternate between the macro- and the micro-solvers, each running with its own time step (therefore the micro- and macro-solvers use different clocks). At every step, the needed macroscale data is estimated from the results of the micro-model (at that step) and is supplied to the macro-solver. The new values of the macro-state variables are then used to constrain the micro-solver.

Example: SDEs with multiple time scales. Consider the stochastic ODE:

$$\begin{aligned} \frac{dx}{dt} &= f(x,y), \\ \frac{dy}{dt} &= -\frac{1}{\varepsilon}(y-\varphi(x)) + \sqrt{\frac{1}{\varepsilon}}\dot{w}. \end{aligned} \tag{18.20}$$

where $\dot{w}(t)$ is the standard white noise. The averaging theorems suggest that the effective macroscale equation should be of the form of an ODE:

$$\frac{dx}{dt} = F(x). \tag{18.21}$$

The original HMM with forward Euler as the macro-solver proceeds as follows:

1. Initialize the micro-solver, e.g. $y^{n,0} = y^{n-1,M}$;
2. Apply the micro-solver for M micro steps:

$$y^{n,m+1} = y^{n,m} - \frac{\delta t}{\varepsilon}(y^{n,m} - \varphi(x^n)) + \sqrt{\frac{\delta t}{\varepsilon}}\xi^{n,m}, \tag{18.22}$$

for $m = 0, 1, \cdots, M-1$. Here $\{\xi^{n,m}\}$ are independent normal random variables with mean 0 and variance 1;

3. Estimate $F(x)$:

$$F^n = \frac{1}{M} \sum_{m=1}^{M} f(x^n, y^{n,m}); \tag{18.23}$$

4. Apply the macro-solver:

$$x^{n+1} = x^n + \Delta t\, F^n. \tag{18.24}$$

In contrast, the seamless HMM with forward Euler scheme is simply:

$$y^{n+1} = y^n - \frac{\delta\tau}{\varepsilon}(y^n - \varphi(x^n)) + \sqrt{\frac{\delta\tau}{\varepsilon}}\xi^n, \tag{18.25}$$

$$x^{n+1} = x^n + \tilde{\Delta}t\, f(x^n, y^{n+1}), \tag{18.26}$$

where $\{\xi^n\}$ are independent normal random variables with mean 0 and variance 1. Note that for HMM, we have $x^n \sim x(n\Delta t)$, but for the seamless algorithm, we have $x^n \sim x(n\tilde{\Delta}t) = x(n\Delta t/M)$.

18.1.3 Stability and Accuracy

One of the main advantages of HMM is that it comes with a general framework for conducting error analysis. The basic idea, as was explained in [8], is to compare the HMM solution with the solution of the selected macroscale solver for the effective macroscale model. Their difference is caused by an additional error in the HMM solution due to the error in the data estimation process. This new error term is called the HMM error, denoted by $e(\text{HMM})$. Assume that both the HMM and the macro-solver for the effective macroscale model can be expressed in the form (note that this is not necessarily the forward Euler scheme):

$$U^{n+1}_{\text{HMM}} = U^n_{\text{HMM}} + \Delta t F^{\varepsilon}(U^n_{\text{HMM}}, U^{n-1}_{\text{HMM}}, \cdots), \tag{18.27}$$

$$\bar{U}^{n+1}_H = \bar{U}^n_H + \Delta t \bar{F}(\bar{U}^n_H, \bar{U}^{n-1}_H, \cdots). \tag{18.28}$$

Note that

$$||Qu^{\varepsilon} - U_{\text{HMM}}|| \le ||Qu^{\varepsilon} - \bar{U}|| + ||\bar{U}_H - \bar{U}|| + ||U_{\text{HMM}} - \bar{U}_H||, \tag{18.29}$$

where $\bar{U}$ is the solution of the macroscale model, and $\bar{U}_H$ is the numerical solution to the effective macroscale model computed using (18.28). The first term on the right hand side of (18.29) is due to the error of the effective model; the second term is due to the error in the macroscale solver; the third term is the HMM error, due to the error in the estimated data. Normally we expect that estimates of the following type hold:

$$||Qu^{\varepsilon} - \bar{U}|| \leq C\varepsilon^{\alpha}, \tag{18.30}$$

$$||\bar{U} - \bar{U}_H|| \leq C(\Delta t)^k, \tag{18.31}$$

where k is the order of accuracy of the macro-solver. In addition, define

$$e(\text{HMM}) = \max_U ||\bar{F}(U^n, U^{n_1}, \cdots) - F^{\varepsilon}(U^n, U^{n_1}, \cdots)||. \tag{18.32}$$

Then under suitable stability conditions, one can show that [8]:

$$||U_{\text{HMM}} - \bar{U}_H|| \leq Ce(\text{HMM}) \tag{18.33}$$

for some constant C. Therefore, we have

$$||Qu^{\varepsilon} - U_{\text{HMM}}|| \leq C(\varepsilon^{\alpha} + (\Delta t)^k + e(\text{HMM})). \tag{18.34}$$

The key in getting concrete error estimates and thereby giving guidelines to designing multiscale methods lies in the estimation of $e(\text{HMM})$. This is specific to each problem. Explicit examples can be found in [10].

18.2 Capturing Macroscale Interface Dynamics

In this section we discuss how the HMM philosophy can be applied to the study of interface motion in a multi-scale setting. This discussion follows that of [3]. Our main interest is to capture the macroscale dynamics of the interface in cases where the velocity is not explicitly specified. Instead, it has to be extracted from some underlying microscale model.

18.2.1 Macroscale Solver: The Interface Tracking Methods

We can select as the macroscale solver any conventional methods for interface dynamics such as the level set method [25], the front tracking method [17], or the segment projection method [30]. Here we will follow [3, 29] and focus on the level set method and the front tracking method.

In the framework of the level set method, an interface is described as the zero level set of a globally defined function, ϕ, called a level set function. All operations, in particular evolution, are then performed on this function in place of the interface of interest. The level set function ϕ satisfies the PDE:

$$\phi_t + V \cdot \nabla\phi = 0,$$

where V is a globally defined velocity field. On the interface, this is equivalent to

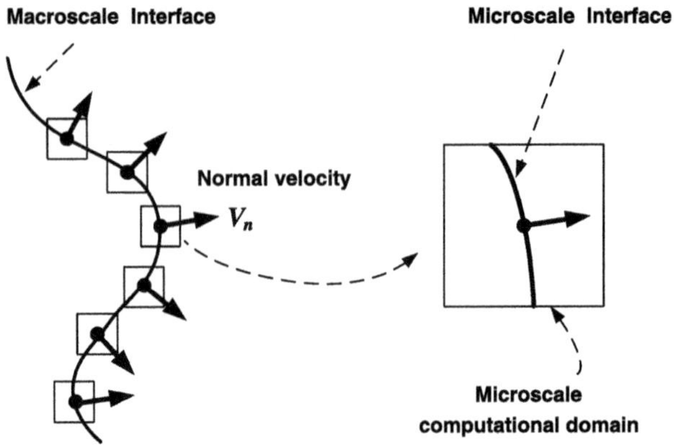

Fig. 18.3: A pictorial description of the HMM setup showing the link between the macroscopic and microscopic levels.

$$\phi_t \pm V_n |\nabla \phi| = 0, \tag{18.35}$$

which involves the normal velocity only (Figure 18.3). In many problems, though, the velocity is only naturally specified at the interface.

The front tracking method is based mainly on the point set definition, in which an interface is simply defined by a set of discrete points, together with an interpolation rule connecting these points (Figure 18.3). The motion of the interface is done by moving each point individually as a Lagrangian marker. Given a velocity field $\mathbf{v}(\mathbf{x},t)$ by which the interface should be moved, each discrete point $\mathbf{x}_i$ is advected by $\mathbf{v}(\mathbf{x},t)$,

$$\frac{d\mathbf{x}_i}{dt} = \mathbf{v}(\mathbf{x},t), \tag{18.36}$$

After the points have been moved, a new parametric connection need to be calculated. Moreover, any connectivity data structure must be updated continuously to avoid clustering or depletion of points. New points can be added where necessary by using the calculated parameterization, and crowded points on other parts of the interface can be removed.

18.2.2 Estimating The Macroscale Interface Velocity

The data that need to be estimated from the microscale models are the normal velocities of the macroscale interface at each macroscale grid point. The first step is to locally reconstruct the interface. If the normal velocity of the macroscale interface is known to only depend on the orientation of the local tangent plane of the interface then we may approximate the interface locally by a hyperplane. On the other hand, if the normal velocity is also known to depend on the local curvature, a quadratic approximation is needed.

As illustration, we first consider the homogenization problem. The level set representation of the microscale model in this case takes the form

$$\phi_t + c\left(x, \frac{x}{\varepsilon}\right) |\nabla \phi| = 0, \tag{18.37}$$

which describes motion in the normal direction at speed c. In this case, we can use the hyperplane reconstruction. Cheng and E [3] suggests working on a transformed coordinate through a change of variables so that this hyperplane coincides with the $\{x_n = 0\}$-plane. The original microscale model (18.37) is then solved in a domain Δ in the transformed space, rectangular with sides orthogonal to the coordinate axes, that should be larger than the size of the periodic cell or the correlation length of c. Periodic boundary conditions are imposed in the $x_1, x_2, \ldots, x_{n-1}$-directions and a periodic jump condition can be imposed in the x_n-direction.

To extract the quantity of interest, at each microscale time step, the microscale Hamiltonian is averaged in the central region $\tilde{\Delta}$ of the domain to reduce spurious effects that may arise from the boundary and the chosen boundary conditions. Denote this value by $h^{\varepsilon}(t)$

$$h^{\varepsilon}(t) = \frac{1}{|\tilde{\Delta}|} \int_{\tilde{\Delta}} c^{\varepsilon}(x) |\nabla \phi(x,t)| dx, \tag{18.38}$$

the velocity of the front at the particular location is obtained from

$$V_n = \frac{1}{|\nabla \phi|} \int_0^{t^*} h^{\varepsilon}(t) K\left(\frac{t}{t^*}\right) dt. \tag{18.39}$$

Here K is an averaging kernel, as discussed earlier. Having obtained V_n at all the macroscale grid points at the interface, the interface can be evolved using standard procedures in the level set method.

Cheng and E [3] also considered the case where the microscale model is described by a phase-field equation:

$$u_t = \nabla \cdot \left(b\left(x, \frac{x}{\varepsilon}\right) u\right) + \varepsilon \nabla \cdot \left(a\left(x, \frac{x}{\varepsilon}\right) \nabla u\right) - \frac{1}{\varepsilon} \frac{\partial V}{\partial u}(u), \tag{18.40}$$

where $a(x,y) > 0$ and $b(x,y)$ are smooth functions that are either periodic in y or stationary random in y with rapidly decaying correlation at large distances and V is a double well potential with minima at $u = \alpha, \beta$. For the details concerning this example, as well as the numerical results, we refer to [3].

In [29], the authors used the HMM framework to design a multiscale method for solving the numerical difficulty due to rapid microscale transition at the interface, which is modeled by a stiff reaction-diffusion-convection equation [15, 21]

$$\frac{\partial u}{\partial t} + \mathbf{f}(x,y) \cdot \nabla u = \psi(u) + \varepsilon \Delta u \tag{18.41}$$

with the source term

$$\psi(u) = -\frac{1}{\varepsilon}u(u-1)\left(u-\frac{1}{2}\right). \tag{18.42}$$

When the numerical resolution is not sufficiently high, the interface or discontinuity may propagate at a nonphysical velocity. This problem rises from the smearing of the discontinuity caused by the transport, which introduces a nonequilibrium state into the calculation. Then the stiff source terms turn on and immediately restore equilibrium, thus shifting the discontinuity to a grid boundary. To overcome this difficulty, we can evolve the model locally around the interface on a much finer grid δx with a time step δt that resolves the smallest scales enforced by ε. As shown in [29], since the velocity quickly relaxes to a quasi-stationary value after τ_ε, which is much shorter than the macro step t_M, we may estimate the velocity after the relaxation time τ_ε and use it to move the interface in the front tracking method on the macroscale. In this way we can save much computational cost. For the details and the numerical results, we refer to [29].

In summary, here are the steps of the HMM technique for interface tracking:

1. The initial macroscale interface location $\Gamma(t)$ represented by a level set function ϕ or a set of marker points $\{X_i(t)\}$ in the front tracking method is given.
2. For each of these macroscale points, choose the microscale domains $\mathscr{D}_i$ and reconstruct the initial states for the microscale model.
3. Evolve the microscale model in the domains $\mathscr{D}_i$ for some micro time steps. Time or ensemble averaging may be used to estimate the velocities v_i.
4. Move the interface for one macroscale time step by taking $\{v_i\}$ as the normal velocity of the interface.
5. Solve the macroscale system for one macroscale time step with the level set method or the front tracking method. Perform the reintialization of the level set function or the redistribution of the marker points in the front tracking method if needed.
6. Stop if the desired time is reached in the macroscale model. Otherwise, repeat from step 2 using the newly evolved interface location.

This approach is closely related to the general philosophy of the adaptive mesh refinement (AMR) methods [1]. As in AMR methods, the end result is to work with a mesh which is locally refined at the interface and coarsened away from the interface. However, there are some important differences on how the mesh refinement is carried out. The first is that AMR would refine everywhere near the interface, whereas HMM would refine only around the macroscale grid points on the interface. The second is that AMR methods aim at computing accurately the local solutions for all times, and for that purpose local time stepping is carried out for all times. In contrast, HMM aims only at computing a macroscale quantity, the velocity of the interface, and this does not require following the microscale solution for all times.

18.3 HMM Interface Tracking of Combustion Fronts

HMM has been applied to several combustion models in [29], such as 1-D Majda's model and the reactive Euler equations in one and two dimensions. In these applications, the goal is to capture the front dynamics on the macroscale with the velocity of the front computed in the microscale region.

18.3.1 Majda's Model

We consider a simplified combustion model derived by Majda in [23]. This model is a 2×2 system, which couples Burgers' equation to a chemical kinetics equation:

$$U_t + \left(\frac{1}{2}U^2 - q_0 Z\right)_x = 0, \tag{18.43}$$

$$Z_x = K(U)Z, \tag{18.44}$$

where U is a variable with some features of pressure or temperature. Z is the mass fraction of unburnt gas, where $Z = 1$ describes the unburnt gas and $Z = 0$ the completely burnt state. $q_0 > 0$ is the heat release and $K(U)$ is the reaction rate. If U represents the temperature T, the function $K(T)$ is typically given by the Arrhenius form

$$K(T) = K_0 e^{-A/T}, \tag{18.45}$$

where K_0 is the rate constant and A is the activation energy [24]. The reaction rate is negligible at low temperature and grows exponentially fast if the temperature is high enough. For computational purposes, the reaction rate (18.45) may be replaced by a discrete ignition temperature kinetics model

$$K(T) = \begin{cases} K_0 & \text{if} \quad T \geq T_{ign}, \\ 0 & \text{if} \quad T < T_{ign} \end{cases} \tag{18.46}$$

where T_{ign} is the ignition temperature. The numerical difficulty in solving this model problem comes from the stiffness of the chemical source term when $K_0 = \frac{1}{\varepsilon}$ is very large.

Colella, Majda, and Roytburd [5] applied the Godunov's method and its high resolution extension [6] to this problem, using a fractional step method. Given (U_i^n, Z_i^n), in the first step, we compute Z_i^{n+1} via numerical integration of the ODE (18.44) with the formula

$$Z_{i-1}^{n+1} = Z_i^{n+1} \exp\left(-\Delta x \frac{K(U_i^n) + K(U_{i-1}^n)}{2}\right) \tag{18.47}$$

and the initial condition $Z_i^{n+1} = 1$ for i large enough. In the second step, we compute U_i^{n+1} from $\{U_i^n\}, \{Z_i^{n+1}\}$ by applying the Engquist–Osher (EO) scheme [13] to (18.43). The formula for computing U_i^{n+1} is given by

$$U_i^{n+1} = U_i^n - \frac{\Delta t}{\Delta x}\left[(f_+(U_i^n) - f_+(U_{i-1}^n)) + (f_-(U_{i+1}^n) - f_-(U_i^n)) \right] + q_0 \frac{\Delta t}{\Delta x}(Z_i^{n+1} - Z_{i-1}^{n+1}), \tag{18.48}$$

with $f_+(U) = \begin{cases} 0 & \text{if } U \le 0, \\ U^2/2 & \text{if } U \ge 0, \end{cases}$ and $f_-(U) = \begin{cases} U^2/2 & \text{if } U \le 0, \\ 0 & \text{if } U \ge 0. \end{cases}$

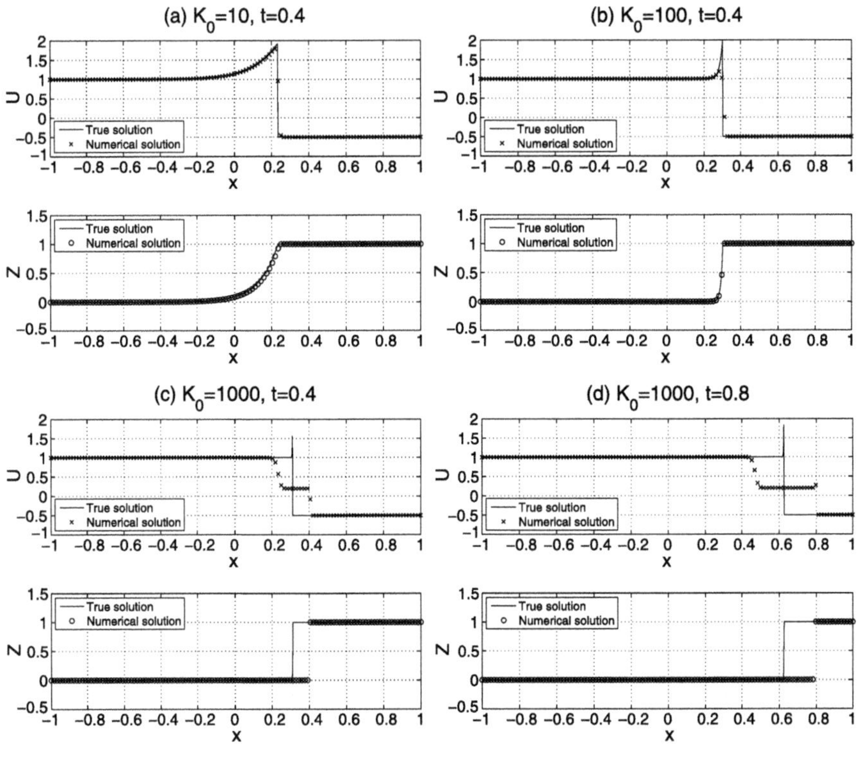

Fig. 18.4: Numerical results of Majda's model using the fractional step method (18.47) and (18.48) with the initial data in (18.49) for different values of the reaction rate constant K_0. We take $T_{ign} = 0$, $q_0 = 0.8$, $\Delta x = 0.01$, and CFL number $= 0.5$. Reprinted from [29] with permission. Copyright © 2006, Society for Industrial and Applied Mathematics.

Figure 18.4 shows numerical results of $q_0 = 0.8$ and different values of the reaction rate constant K_0 with the initial data

$$U(x,0) = \begin{cases} U_l = 1.0 & \text{if} \quad x \le 0, \\ U_r = -0.5 & \text{if} \quad x > 0. \end{cases} \tag{18.49}$$

As shown in Figure 18.4(a), for small value of $K_0 \Delta t$ $(= 0.05)$, the fractional step method gives the results of the detonation profiles of U and Z that match the true solutions well. For larger $K_0 \Delta t$ $(= 0.5)$, the spike of the detonation profile of U drops down due to the underresolution around it, but the location of the detonation is still correct, as shown in Figure 18.4(b). However, for the case $K_0 \Delta t = 5$, shown in Figures 18.4(c) $(t = 0.4)$ and 18.4(d) $(t = 0.8)$, the scheme produces a nonphysical solution, which is propagating with a velocity of one grid per time step as mentioned before.

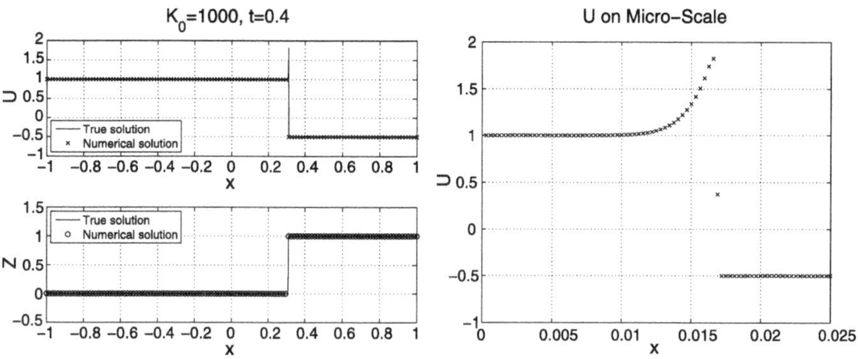

Fig. 18.5: Numerical results of Majda's model at time $t = 0.4$ using HMM with initial data (18.49) for the reaction rate constant $K_0 = 1000$. We take $T_{ign} = 0$, macroscale grid size $\Delta x = 0.01$, microscale grid size $\delta x = 0.01/50 = 2 \times 10^{-4}$, and a fixed CFL number 0.5. Reprinted from [29] with permission. Copyright © 2006, Society for Industrial and Applied Mathematics.

Figure 18.5 shows numerical results at time $t = 0.4$ using the HMM interface tracking technique introduced in previous section with the initial data in (18.49) for the reaction rate constant $K_0 = 1000$. The locations of detonation match the true solution quite well. We used 200 macroscale grids on the interval $[-1, 1]$. The refinement ratio between the macroscale and microscale grid sizes is 50. When $K_0 = 10^4$, a higher refinement ratio of 500 can achieve the same expected result. As shown in the right panel in Figure 18.5, the spiked strong detonation profile is resolved on the microscale grids. The interface condition on the macroscale enters into the flux evaluation in the EO scheme (18.48) only by inserting the new extrapolated values.

18.3.2 Reactive Euler Equations

In this section, we extend our HMM technique to 2-D reactive Euler equations. The results for 1-D multispecies reactive Euler equations can be found in [29]. The system of equations in 2-D is given by

$$\frac{\partial U}{\partial t} + \frac{\partial F(U)}{\partial x} + \frac{\partial G(U)}{\partial y} = S(U) \tag{18.50}$$

with

$$U = \begin{pmatrix} \rho \\ \rho u \\ \rho v \\ E \\ \rho Z \end{pmatrix}, F(U) = \begin{pmatrix} \rho u \\ \rho u^2 + p \\ \rho uv \\ u(E+p) \\ \rho u Z \end{pmatrix}, G(U) = \begin{pmatrix} \rho v \\ \rho uv \\ \rho v^2 + p \\ v(E+p) \\ \rho v Z \end{pmatrix}, S(U) = \begin{pmatrix} 0 \\ 0 \\ 0 \\ 0 \\ -K(T)\rho Z \end{pmatrix}$$

where u is the velocity of the gas in the x-direction and v is the y-component of the velocity. The total energy is given by the equation of state

$$E = \frac{p}{\gamma - 1} + \frac{1}{2}\rho(u^2 + v^2) + q_0 \rho Z.$$

where q_0 is the heat release and γ is the ratio of specific heat for ideal gas. The reaction rate $K(T)$ is described by the discrete ignition temperature kinetics model (18.46). The temperature is given by $T = p/\rho\mathscr{R}$, where $\mathscr{R}$ is the specific gas constant.

The fractional step method for approximating the solutions of (18.50) is given by

$$U^{n+1} = L_{\mathrm{ODE}}^{\Delta t} L_{y\mathrm{Roe}}^{\Delta t} L_{x\mathrm{Roe}}^{\Delta t} U^n. \tag{18.51}$$

where $L_{x\mathrm{Roe}}^{\Delta t}$ and $L_{y\mathrm{Roe}}^{\Delta t}$ denote Roe's scheme [28] for solving the nonreactive system in the x- and y-directions, respectively. $L_{\mathrm{ODE}}^{\Delta t}$ denotes the ODE-solver for the source term.

Here we consider a radially symmetric detonation wave in two dimensions. The initial values consist of totally burnt gas inside of a circle with radius 0.3 and totally unburnt gas everywhere outside of this circle. The burnt and unburnt states are chosen as follows (same as Example 3 in [18]):

$$\begin{cases} \rho_b = 1.4, \quad u_b = v_b = 0, \quad p_b = 1, \quad Z_b = 0 \quad \text{if } x^2 + y^2 \le 0.3^2, \\ \rho_u = 0.8876, \quad \begin{cases} u_u = -0.5774\cos\theta, \\ v_u = -0.5774\sin\theta, \end{cases} \quad p_u = 0.1917, \quad Z_u = 1 \quad \text{elsewhere.} \end{cases} \tag{18.52}$$

where θ is the angle in polar coordinates. The ignition temperature is set to $T_{ign} = 0.26$. The other parameters are set to $\gamma = 1.4$, $q_0 = 1$, and $\mathscr{R} = 1$. The computational domain is $[0,1] \times [0,1]$.

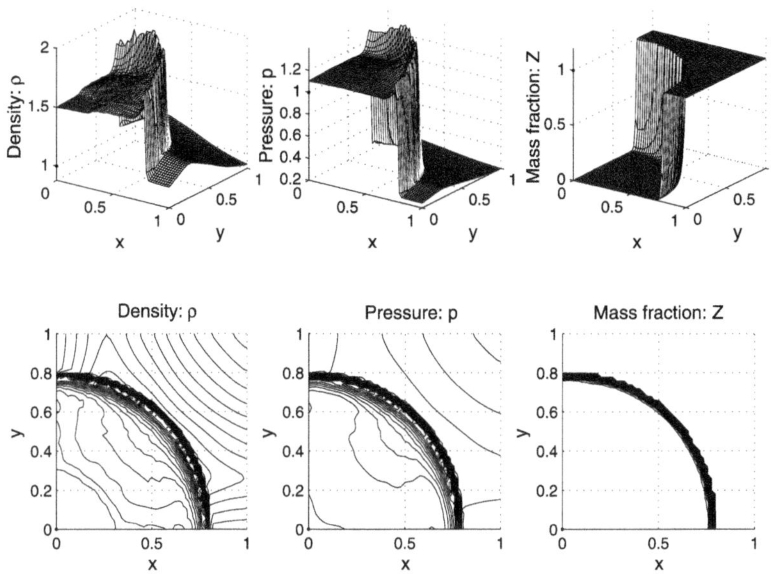

Fig. 18.6: Numerical solutions of 2-D reactive Euler equations at time $t = 0.5$ using fractional step method (18.51) with Roe's scheme and the initial data in (18.52) for $K_0 = 100$ and $T_{ign} = 0.26$. We take $\Delta x = \Delta y = 0.01$ and CFL number 0.5. Reprinted from [29] with permission. Copyright © 2006, Society for Industrial and Applied Mathematics.

Figure 18.6 shows numerical results at time $t = 0.5$ for the reaction rate constant $K_0 = 100$. The combustion front is a quarter of a circle that should expand as the system evolves. The distance from the center to the front should be 0.8 at this time. However, for a larger $K_0 = 1000$, the combustion front moves too fast, and there are nonphysical intermediate states for the density and the pressure, as evidenced by the top panel of Figure 18.7. Moreover, the circular geometry cannot be resolved, as shown in the bottom panel.

By using the HMM interface tracking technique introduced in previous section, we resolve the system (18.50) with the initial data in (18.52) at time $t = 0.5$ for the reaction rate constant $K_0 = 1000$. We used 100×100 macroscale grids on the domain $[0,1] \times [0,1]$. The refinement ratio between the macroscale and microscale grid sizes is 50. We note that we evolve the 1-D system along the normal line instead of 2-D system (18.50) as our microscopic solver. This idea can achieve both accuracy and efficiency. The spiked strong detonation profiles of all components are

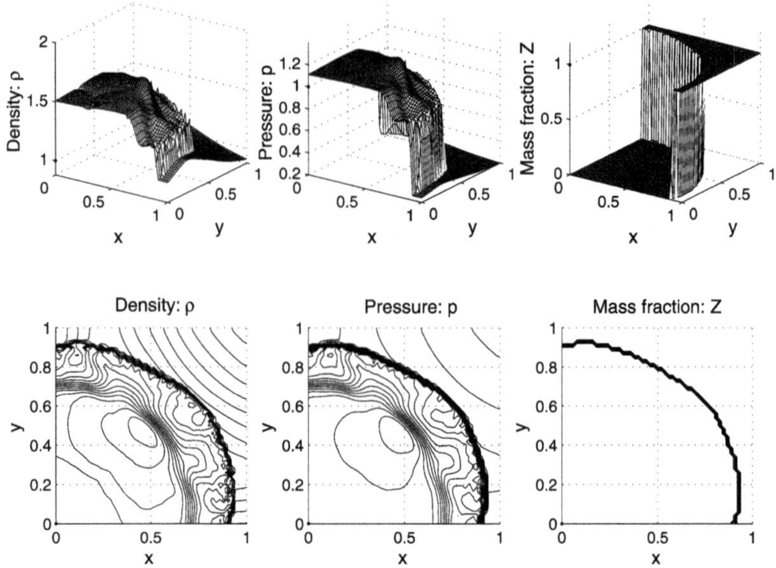

Fig. 18.7: Numerical solutions of 2-D reactive Euler equations at time $t = 0.5$ using fractional step method (18.51) with Roe's scheme and the initial data in (18.52) for $K_0 = 1000$ and $T_{ign} = 0.26$. We take $\Delta x = \Delta y = 0.01$ and CFL number 0.5. Reprinted from [29] with permission. Copyright © 2006, Society for Industrial and Applied Mathematics.

resolved on the microscale grids that are similar to the profile shown in the right panel of Figure 18.5. Figure 18.8 shows that the combustion front moves with the correct velocity and the circular geometry is resolved quite well on the grid. The small oscillation close to the front is typical of high resolution shock capturing with steep front. The accuracy of the whole method is first order since it is based on the time-splitting method and the upwind scheme.

18.4 Conclusions

The examples presented here are admittedly quite simple. However, they do show the promise of HMM. In principle, the philosophy also applies to more complex situations. In practice, many other issues have to be dealt with, such as the handling of intermediate scales. It is likely that HMM should be combined with more traditional techniques such as AMR in order to produce numerical algorithms that are truly effective for problems of practical interest.

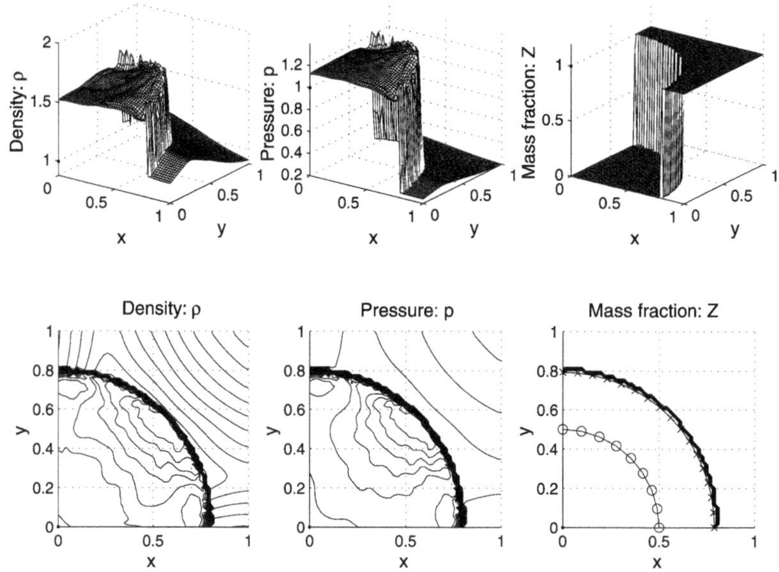

Fig. 18.8: Numerical solutions of 2-D reactive Euler equations at time $t = 0.5$ using HMM with Roe's scheme and the initial data in (18.52) for $K_0 = 1000$ and $T_{ign} = 0.26$. We take macroscale grid size $\Delta x = \Delta y = 0.01$, microscale grid size $\delta x = \delta y = 0.01/50 = 2 \times 10^{-4}$, and a fixed CFL number 0.5. In the contour plot of mass fraction Z, we also show two sets of marker points at different times. There are 8 segments between the points represented by "$\circ$" at time $t = 0.2$ and 16 segments between the points represented by "$\times$" at time $t = 0.5$ since we perform a redistribution of the marker points after time $t = 0.2$ due to the increasing distance between the neighboring points. Reprinted from [29] with permission. Copyright © 2006, Society for Industrial and Applied Mathematics.

Acknowledgements W. E is supported in part by ONR Grant N00014-01-0674 and DOE Grant DE-FG02-03ER25587. B. Engquist is supported in part by NSF Grant DMS-0714612. Y. Sun is supported by NSF Joint Institutes' Postdoctoral Fellowship through SAMSI.

References

1. Berger, M.J., Oliger, J.: Adaptive mesh refinement for hyperbolic partial differential equations. J. Comput. Phys. **53**, 484–512 (1984)
2. Brandt, A.: Multiscale scientific computation: Review 2001. In: Barth, T.J., et al. (eds.) Multiscale and Multiresolution Methods: Theory and Applications, Yosemite Educational Symposium Conf. Proc., 2000, Lecture Notes in Comp. Sci. and Engrg. **20**, pp. 3–96. Springer-Verlag, New York (2002)

3. Cheng, L.-T., E, W.: The heterogeneous multiscale method for interface dynamics. In Cheng, S.Y., et al. (eds.) Recent Advances in Scientific Computing and Partial Differential Equations. Contemp. Math. **330**, pp. 43–53. Amer. Math. Soc., Providence (2003)
4. Chorin, A.J.: A numerical method for solving incompressible viscous flow problems. J. Comput. Phys. **2**, 12–26 (1967)
5. Colella, P., Majda, A., Roytburd, V.: Theoretical and numerical structure for reacting shock waves. SIAM J. Sci. Statist. Comput. **7**, 1059–1080 (1986)
6. Colella, P., Woodward, P.: The piecewise-parabolic method (PPM) for gas-dynamical simulations. J. Comput. Phys. **54**, 174–201 (1984)
7. E, W.: Analysis of the heterogeneous multiscale method for ordinary differential equations. Comm. Math. Sci. **1**, 423–436 (2003)
8. E, W., Engquist, B.: The heterogeneous multi-scale methods. Comm. Math. Sci. **1**, 87–133 (2003)
9. E, W., Engquist, B.: Multiscale modeling and computation. Notices of the AMS, **50**, 1062–1070 (2003)
10. E, W., Engquist, B., Li, X., Ren, W., Vanden-Eijnden, E.: Heterogeneous multiscale methods: A review. Comm. Comput. Phys. **3**, 367–450 (2007)
11. E, W., Lu, J.: Seamless multiscale modeling via dynamics on fiber bundles. Comm. Math. Sci. **5**, 649–663 (2007)
12. E, W., Ren, W., Vanden-Eijnden, E.: A general strategy for designing seamless multiscale methods. J. Comput. Phys. **228**, 5437–5453 (2009)
13. Engquist, B., Osher, S.: Stable and entropy satisfying approximations for transonic flow calculations. Math. Comp. **34**, 45–75 (1980)
14. Engquist, B., Tsai, R.: Heterogeneous multiscale method for a class of stiff ODEs. Math. Comp. **74**, 1707–1742 (2005)
15. Fan, H., Jin, S.: Front motion in muti-dimensional viscous conservation laws with stiff source terms driven by mean curvature and front thickness. Quart. Appl. Math. **61**, 701–721 (2003)
16. Fatkullin, I., Vanden-Eijnden, E.: A computational strategy for multiscale systems with applications to Lorenz 96 model. J. Comput. Phys. **200**, 605–638 (2004)
17. Glimm, J., McBryan, O., Menikoff, R., Sharp, D.H.: Front tracking applied to Rayleigh-Taylor instability. SIAM J. Sci. Statist. Comput. **7**, 230–251 (1986)
18. Helzel, C., LeVeque, R.J., Warnecke, G.: A modified fractional step method for the accurate approximation of detonation waves. SIAM J. Sci. Comput. **22**, 1489–1510 (2000)
19. Kevrekidis, I.G., Gear, C.W., Hyman, J.M., Kevrekidis, P.G., Runborg, O., Theodoropoulos, C.: Equation-free, coarse-grained multiscale computation: enabling microscopic simulators to perform system-level analysis. Comm. Math. Sci. **1**, 715–762 (2003)
20. LeVeque, R.J.: Numerical Methods for Conservation Laws. Birkhäuser-Verlag, Boston (1990)
21. LeVeque, R.J., Yee, H.C.: A study of numerical methods for hyperbolic conservation laws with stiff source terms. J. Comput. Phys. **86**, 187–210 (1990)
22. Lions, P.L., Papanicolaou, G., Varadhan, S.R.S.: Homogenization of Hamilton-Jacobi equations. unpublished.
23. Majda, A.: A qualitative model for dynamic combustion. SIAM J. Appl. Math. **41**, 70–93 (1981)
24. Oran, E.S., Boris, J.P.: Numerical Simulation of Reactive Flow. Elsevier, New York (1987)
25. Osher, S., Sethian, J.A.: Fronts propagating with curvature-dependent speed: algorithms based on Hamilton-Jacobi formulations. J. Comput. Phys. **79**, 12–49 (1988)
26. Ren, W.: Seamless multiscale modeling of complex fluids using fiber bundle dynamics. Comm. Math. Sci. **5**, 1027–1037 (2007)
27. Ren, W., E, W.: Heterogeneous multiscale method for the modeling of complex fluids and microfluidics. J. Comput. Phys. **204**, 1–26 (2005)
28. Roe, P.L.: Approximate Riemann solvers, parameter vectors and difference schemes. J. Comput. Phys. **43**, 357–372 (1981)

29. Sun, Y., Engquist, B.: Heterogeneous multiscale methods for interface tracking of combustion fronts. SIAM Multiscale Model. Simul. **5**, 532–563 (2006)
30. Tornberg, A.-K., Engquist, B.: The segment projection method for interface tracking. Comm. Pure Appl. Math. **56**, 47–79 (2003)
31. Vanden-Eijnden, E.: Numerical techniques for multiscale dynamical systems with stochastic effects. Comm. Math. Sci. **1**, 385–391 (2003)
32. Vanden-Eijnden, E.: On HMM-like integrators and projective integration methods for systems with multiple time scales. Comm. Math. Sci. **5**, 495–505 (2007)

Chapter 19
Lattice Boltzmann Methods for Reactive and Other Flows

Christos E. Frouzakis

Abstract The lattice Boltzmann method (LBM) is receiving increasing attention in recent years as an alternative approach for computational fluid dynamics. Through its kinetic theory origin, the method inherits the physically appealing particle picture that can be adapted to simulate multiscale and multiphysics systems with sizes ranging from the microscale (where the continuum hypothesis may break down) to macroscale applications. The method is characterized by its straightforward implementation in complex geometries and the fact that it involves only nearest neighbor interactions without global operations, making LBM algorithms ideally suited for parallelization. However, the method in general employs a larger number of degrees of freedom per grid point than classical CFD approaches, and parallel implementation may be essential in order to meet the higher memory requirements. In this chapter, an overview of the method and its applications is presented focusing on recent model developments for the description of the averaged macroscopic behavior of isothermal and non-isothermal, single- and multi-component and reactive flows.

19.1 Introduction

The conventional approach for modeling most scientific and engineering flows is based on the continuum description of macroscopic behavior formulated as partial differential equations for a few fields (e.g. continuity and Navier-Stokes equations for the density and momentum, respectively, in the case of isothermal flows). A numerical technique (finite difference/volume/element, spectral or spectral element method) is then employed to obtain the discretized set of equations on a topologically connected mesh that can be solved numerically on the computer.

Christos E. Frouzakis
Aerothermochemistry and Combustion Systems Laboratory, Swiss Federal Institute of Technology, Zurich, Switzerland, e-mail: frouzakis@lav.mavt.ethz.ch

T. Echekki, E. Mastorakos (eds.), *Turbulent Combustion Modeling*,
Fluid Mechanics and Its Applications 95, DOI 10.1007/978-94-007-0412-1_19,

Alternatively, since on the microscopic level fluids consist of discrete particles (atoms or molecules interacting via classical or quantum mechanics potentials and following the Newton equations of motion), one can opt for a description based on physical particles and obtain the evolution of macroscopic variables as collective averages over an enormous number of individual trajectories. Introduced about fifty years ago, molecular dynamics (MD) is probably the most widely used family of particle methods [2]. It provides the most detailed description by solving Newton's equations of motion to track the position and velocity of each atom or molecule in the system. While MD yields the correct description of fluids on the microscopic as well as the hydrodynamic scales, the typical length and time scales that can be simulated in practice are of the order of a few tens of nanometers and a few hundred nanoseconds, respectively.

A large variety of (physical as in MD or notional) particle methods have been developed for the simulation of fluid flows at different scales. They include Monte Carlo and direct simulation Monte Carlo methods (DSMC) [17] at the microscopic level, lattice gas models [41], dissipative particle dynamics [36, 53] at the mesoscopic, and smoothed particle hydrodynamics [67], vortex particle methods [34], and the fluid particle model [37] at the macroscopic level.

This chapter focuses on the lattice Boltzmann method (LBM), a mesoscopic approach where notional particles exist on a set of discrete points spaced at regular intervals to form a lattice. Time is also discrete and during a time step the particles move between neighboring (conventionally, nearest neighbor) lattice sites along predefined directions, and then scatter according to carefully chosen kinetic rules that satisfy local conservation laws.

The lattice Boltzmann methods evolved from the FHP model [41], the first accurate lattice gas cellular automaton (LGCA) to simulate fluid flow proposed in 1986 by Frisch, Hasslacher and Pomeau, who realized the importance of the lattice symmetry for the recovery of the Navier-Stokes equation. Unfortunately, the advantage of the exact representation of the system state through Boolean variables marking the presence or absence of a particle at a certain lattice location with a certain velocity were overshadowed by major drawbacks: a relatively high fixed viscosity, lack of Galilean invariance, and the intrinsic stochastic noise which necessitates averaging over long time and large areas. The suggestion of McNamara and Zanetti [63], and Higuera and Jimenez [52] to replace the particles (represented by a Boolean number) by the single-particle probability distribution function led to the lattice Boltzmann model (LBM) which turned out to be suitable for the efficient simulation of a broad class of flows. For a detailed description of LGCA and for the development of the LBM from LGCA the reader is referred to [82, 90].

Due to its capability to describe different physics in an algorithmically simple and elegant way, the lattice Boltzmann method is attracting increasing interest from the application, the implementation and the theoretical sides alike. Additional advantages include the linear and exact description of the advection term, eliminating some of the difficulties plaguing conventional CFD solvers, the local character of the update rules, resulting in nearly ideal amenability to parallel implementation, and the ease with which complex geometries can be set up.

This chapter reviews some recent developments of stable and efficient models addressing the physico-chemical processes relevant to turbulent combustion (turbulent flow, mixing of multicomponent systems, strongly non-isothermal effects and reaction). It is practically impossible to present fully the method, its foundation and review the rapidly expanding literature on lattice Boltzmann methods within the limits of this chapter. Additional information on the method and its applications can be found in the review of Chen and Doolen [22], in the books of Wolf-Gladrow [90], Succi [82], Sukop and Thorne [87], and the references therein.

19.2 The Boltzmann Equation

19.2.1 Basic Considerations

The microscopic description of flows is based on moving and colliding particles. Instead of following the enormous number of individual particles, Boltzmann proposed to track the temporal evolution of the one-particle distribution function $f(\mathbf{x},\mathbf{c},t)$ prescribing the probability of finding the center of a particle with velocity $\mathbf{c}$ at location $\mathbf{x}$ at time t. For a sufficiently dilute gas in the presence of an external force $\mathbf{F}$ acting on the particles, the temporal evolution of $f(\mathbf{x},\mathbf{c},t)$ is described by taking into account only two processes: the free flight of the particles (second term in the left hand side) and their collisions (right hand side)

$$\frac{\partial f}{\partial t} + \mathbf{c}\cdot\nabla_{\mathbf{x}} f + \mathbf{F}\cdot\nabla_{\mathbf{c}} f = Q(f). \tag{19.1}$$

Assuming only binary collisions and that the velocity of a particle is uncorrelated with its position (the molecular chaos assumption), Boltzmann expressed the collision integral $Q(f)$ as a nonlinear integral.

The resulting nonlinear integro-differential Boltzmann equation is difficult to solve and various simplifications have been proposed, resulting in *kinetic models* that preserve certain features of the Boltzmann equation. The most commonly used one is the single-relaxation time model BGK model proposed by Bhatnagar, Gross and Krook

$$Q_{BGK} = -\frac{1}{\tau}(f - f^{eq}), \tag{19.2}$$

describing the relaxation of f to the *local Maxwellian distribution function*

$$f^{eq} = \frac{\rho}{(2\pi RT)^{D/2}} \exp\left[-\frac{(\mathbf{c}-\mathbf{u})^2}{2RT}\right] \tag{19.3}$$

with a characteristic time scale τ. Local Maxwellians are parametrized by the values of the fluid density ρ, mean velocity $\mathbf{u}$, and temperature T (D is the spatial dimension, and R is the gas constant.)

H theorem. One of the most important contributions of Boltzmann is the celebrated *H theorem*, which roughly states that entropy does not decrease in time. Related to entropy by $S = -H$, the H-function

$$H(t) = \int f \ln f \, d\mathbf{x} d\mathbf{v}$$

therefore satisfies

$$\frac{dH}{dt} \leq 0,$$

with the equality holding at equilibrium. The H-theorem then dictates that the relaxation to the unique global Maxwellian distribution must be accompanied by a monotonic increase of the entropy. The local Maxwellian can also be specified as the maximizer of the Boltzmann entropy function, subject to the constraints of local conservation laws, and is also called the local equilibrium distribution function.

Link to hydrodynamics. From the point of view of the Boltzmann equation, the macroscopic hydrodynamic fields are the first few moments of the distribution function

$$\rho = \int f d\mathbf{c}, \qquad \rho\mathbf{u} = \int \mathbf{c} f d\mathbf{c}, \qquad \frac{\rho DRT}{2} = \int \frac{(\mathbf{c}-\mathbf{u})^2}{2} f d\mathbf{c}. \tag{19.4}$$

Methods of reduced description. The closed set of equations for the macroscopic hydrodynamics fields can be obtained by reducing the Boltzmann equation using one of the classical reduction methods: the Hilbert method, the Chapman-Enskog method, and the Grad moment method, as described briefly in e.g. [42].

The Chapman-Enskog (CE) method approximates the solution of the Boltzmann equation as a formal series expansion in powers of the Knudsen number Kn, the ratio of the mean free path to the scale of variations of the macroscopic hydrodynamic fields. It starts from the known local equilibrium, and proceeds to derive macroscopic equations at different orders of truncation: the Euler hydrodynamics at Kn^0, the Navier-Stokes hydrodynamics at Kn^1, the Burnett hydrodynamics at Kn^2, etc. The equations past the Navier-Stokes description become ill-conditioned making it difficult to extend the hydrodynamic description into the highly non-equilibrium regime. In addition to enabling the derivation of macroscopic equations without *a priori* guessing, the method allowed to express transport coefficients in terms of particle interactions (see, for example, [19, 20]).

Grad's method is based on the assumption of the decomposition of the moments into a set of slowly evolving moments M' that does not change significantly during a time of order τ (slow dynamics) in comparison to the rest of the moments M'' (fast dynamics); M' includes density, momentum, energy and a subset of higher-order moments. Towards the end of the fast evolution, the values of M'' become functions of M', and, for $t \gg \tau$, the dynamics are fully determined by the evolution of the slow moments. Grad [44] approximated the solution to the Boltzmann equation as a series expansion in terms of Hermite orthogonal polynomials in velocity space and kept thirteen of the leading terms corresponding to the five macroscopic variables (density, momentum and energy) and their fluxes.

The Chapman-Enskog and Grad's methods are also employed in the derivation of lattice Boltzmann schemes.

19.2.2 Lattice Boltzmann Model

19.2.2.1 Original Lattice Boltzmann

The lattice Boltzmann model is based on the realization that the continuum of microscopic velocities can be reduced to a finite set without sacrificing the accurate description of the macroscopic dynamics. Lattice Gas and Lattice Boltzmann approaches construct computationally efficient discrete kinetic models obtained by discretizing the velocity space on a set of q discrete velocities,

$$\mathbf{C} = \{\mathbf{c_1}, \mathbf{c_2}, \ldots, \mathbf{c_q}\},$$

where q is as small as possible. The distinguishing characteristic of these methods is that the discrete velocities define the links of a lattice on which notional particles are allowed to propagate. The task is to find a set that recovers the target macroscopic hydrodynamic behavior.

Denoting by $f_i(\mathbf{x},t)$ the populations corresponding to the discrete velocities $\mathbf{c_i}$, $i = 1, \ldots, q$, the continuous in space and time 'standard' lattice Boltzmann model adopts the BGK model and assumes that during collision the populations relax to their equilibrium value with a single relaxation time τ

$$\frac{\partial f_i}{\partial t} + c_{i\alpha}\frac{\partial f_i}{\partial x_\alpha} = Q_i(f_i) = -\frac{1}{\tau}(f_i - f_i^{eq}) + F_i, \qquad i = 1, \ldots, q, \tag{19.5}$$

where $\alpha = 1, \ldots, D$ is the index for the spatial dimensionality of the system (D=1,2, or 3 for one-, two-, or three-dimensional geometries, respectively) over which summation is assumed. The source term F_i represents the force term of the Boltzmann equation and provides an elegant way to construct models for complex problems (fast flows, heat transfer, chemical reactions, phase transitions etc. [82]).

The easiest way to discretize the configuration (i.e. physical) space is to use a regular lattice so that for every grid node $\mathbf{x}$, $\mathbf{x} + \mathbf{c}_\alpha \delta t$ is also a grid node. This way, the information at all grid nodes is automatically known at the next time step. The lattices used in LBM are typically characterized as $DdQq$, where d denotes the spatial dimension and q the number of discrete velocities. Commonly used lattices for one- ($D1Q3$), two- ($D2Q9$) and three-dimensional simulations ($D3Q19$) are shown in Fig. 19.1.

In the original 'heuristic' LB formulations, the equilibrium function was chosen as low-order (typically second) polynomials of the hydrodynamic fields (e.g. fluid density and velocity in the isothermal case), and the polynomial coefficients

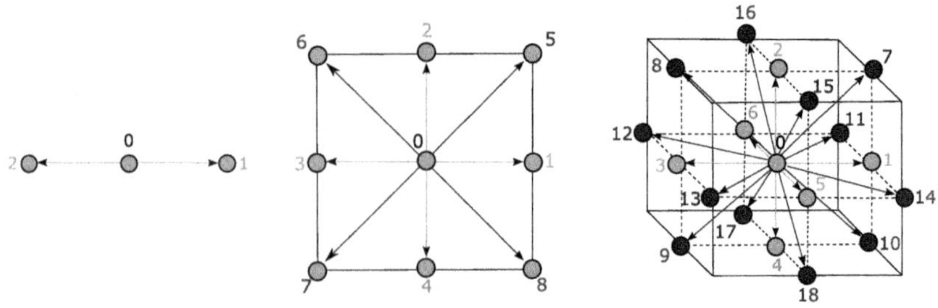

(a) $D1Q3$ (b) $D2Q9$ (c) $D3Q19$

Fig. 19.1: Commonly used lattices for (a) 1-D, (b) 2-D, and (c) 3-D simulations.

	$\mathbf{c}_i$	w_i
$D1Q3$	$\{0,1,-1\}$	$\{\frac{4}{6},\frac{1}{6},\frac{1}{6}\}$
$D2Q9$	$\{(0,0),$ $(1,0),(0,1),(-1,0),(0,-1),$ $(1,1),(-1,1),(-1,-1),(1,-1)\}$	$\{\frac{16}{36},$ $\frac{4}{36},$ $\frac{1}{36}\}$
$D3Q19$	$\{\quad (0,0,0),$ $(1,0,0),(0,1,0),(-1,0,0),(0,-1,0),(0,0,1),(0,0,-1),$ $(1,1,0),(-1,1,0),(-1,-1,0),(1,-1,0),$ $(1,0,1),(-1,0,1),(-1,0,-1),\{1,0,-1\},$ $(0,1,1),(0,-1,1),(0,-1,-1),(0,1,-1)\quad\}$	$\{\frac{12}{36},$ $\frac{2}{36},$ $\frac{1}{36},$ $\frac{1}{36}$ $\frac{1}{36}\}$

Table 19.1: Discrete velocities grouped in 'energy levels' that carry the same weights for the $D1Q3, D2Q9$ and $D3Q19$ LB models.

were determined *a posteriori* using the Chapman-Enskog procedure so that the correct target hydrodynamics (e.g. the Navier Stokes description) could be recovered in the large-scale long-time limit [82]. The way of constructing kinetic models is not unique, and different equilibrium functions and/or discrete velocity sets can be used. It was found, however, that not all models constructed this way result in stable numerical schemes. While for isothermal hydrodynamics relatively stable schemes emerged essentially by trial and error, the stability problem was particularly severe when temperature was allowed to vary.

LBM was 'freed' from its LGA origin when it was shown that it can be obtained directly from a special finite-difference discretization of the continuous BGK form of the Boltzmann equation [1, 48, 49]. The discrete velocities were related to the roots of Gauss-Hermite quadratures, which ensure accurate evaluation of the low-order moments of the distribution function defining the desired macroscopic hydrodynamics, while the equilibrium function was taken to be a second-order expansion of the Maxwellian in terms of low fluid velocity,

$$f_i^{eq} = \rho w_i \left[1 + \frac{\mathbf{c}_\alpha \cdot \mathbf{u}}{c_s^2} + \frac{(\mathbf{c}_\alpha \cdot \mathbf{u})^2}{2c_s^4} - \frac{\mathbf{u} \cdot \mathbf{u}}{2c_s^2} \right],$$

with $c_s = \sqrt{RT}$ being the 'speed of sound' of the model, and w_i related to the weights of the quadrature [48]. The discrete velocities and the corresponding weights for three commonly used LB lattices are given in Table 19.1. It was further shown that the resulting f_i^{eq} was the one obtained by the 'heuristic' approach.

The second-order expansion of the equilibrium function restricts application to hydrodynamics at the Navier-Stokes level. Shan and He [78] realized that if discrete velocities are constructed from the zeros of the Hermite polynomials, the method of discrete velocity is essentially equivalent to Grad's moment method [44]. They proposed expanding the equilibrium to higher orders and adopting higher degree Gauss-Hermite quadratures (i.e. lattices with larger number of discrete velocities), so that higher-order approximations to the Boltzmann equation important for non-isothermal and more complex flows can be constructed. A recent presentation of this systematic approach can be found in [79]. For orders higher than three, however, the ratios of the roots of the Hermite polynomials defining the discrete velocities are not in general integers and do not coincide with the lattice nodes such that the exact space discretization of the advection step offered by the lattice has to be abandoned.

As in the continuous Boltzmann equation case, the hydrodynamic fields are the first few moments of the populations:

$$\rho = \sum_{i=1}^{q} f_i, \qquad \rho u_a = \sum_{i=1}^{q} f_i c_{i\alpha}, \qquad \rho DT + \rho u^2 = \sum_{i=1}^{q} f_i c_i^2. \tag{19.6}$$

19.2.2.2 Entropic Lattice Boltzmann

In kinetic theory, the equilibrium distribution function is the maximum entropy state, and Boltzmann's H-theorem, which ensures the increase of entropy, also ensures stability. Entropic Lattice Boltzmann models are constructed with the target of restoring the H-theorem that is not satisfied by the standard LBM schemes discussed so far, thereby ensuring numerical stability of the scheme [18, 54].

From the perspective of kinetic theory, local equilibria minimize the convex H-function subject to the constraints of the locally conserved hydrodynamic fields. In the entropic construction of LB models, evaluation of Boltzmann's H-function by the Gauss-Hermite quadrature defines the discrete form of the H-function

$$H = \sum_{i=1}^{q} f_i \ln\left(\frac{f_i}{w_i}\right), \tag{19.7}$$

where the weights w_i associated with the discrete velocities $\mathbf{c}_i$ must be chosen appropriately. The equilibrium populations f_i^{eq} are then found as the minimizers of the H function under the constraints of the locally conserved quantities, e.g.

$$\sum_{i=1}^{q} f_i^{eq} = \rho, \quad \sum_{i=1}^{q} c_{i\alpha} f_i^{eq} = \rho u_\alpha, \quad \sum_{i=1}^{q} c_i^2 f_i^{eq} = \rho DT + \rho u^2$$

Analytical solutions for f^{eq} have been found in some cases [5]. For example, the isothermal 1-D Navier-Stokes description at some reference temperature T_0 was obtained for the $D1Q3$ model. In two and three dimensions, the discrete velocities and the weights of the $D2Q9$ and $D3Q27$ models were constructed by taking tensor products of the one-dimensional lattice, and the explicit solution of the constrained minimization problem in D dimensions is the product of D one-dimensional solutions [4, 5]

$$f_i^{eq} = \rho w_i \prod_{\alpha=1}^{D} \left(2 - \sqrt{1+u_\alpha^2}\right) \left(\frac{\frac{2}{\sqrt{3}} u_\alpha + \sqrt{1+u_\alpha^2}}{1 + u_\alpha \sqrt{3}}\right)^{c_{i\alpha}/\sqrt{3}c_s}, \tag{19.8}$$

where the speed of sound is $c_s = \sqrt{T_0}$ in lattice units and the fluid velocity is non-dimensionalized by c_s. It is worth noting that the second-order Taylor series expansion of the equilibria (19.8) coincides with the polynomial equilibria used in the standard $D2Q9$ LBM. This implies that this model possesses an H-function and explains its stability in comparison to other LB models. However, the equilibrium expressions (19.8) are preferable because they approximate more accurately higher-order moments needed to establish isothermal hydrodynamics. A tutorial presentation of the entropic lattice Boltzmann method for 1-D hydrodynamics can be found in [56].

More generally, the solution to the minimization problem can be expressed in product form as

$$f_i^{eq} = w_i \, \chi \prod_{\alpha=1}^{D} \zeta_\alpha^{c_{i\alpha}}, \tag{19.9}$$

where $\chi(\rho, \mathbf{u}), \zeta_\alpha(\rho, \mathbf{u})$ are Lagrange multipliers associated with the mass and momentum conservation constraints, respectively, and can be obtained analytically by perturbation around the zero-velocity solution $\chi(\rho, \mathbf{0}) = \rho, \zeta_\alpha(\rho, \mathbf{0}) = 1$ to any desired order in velocity [28]. The product form is particularly useful in reducing the computational complexity of LB models with large number of discrete velocities significantly.

In addition to the constraints imposed by local conservation, f^{eq} must also respect several other higher-order non-conserved moments. Failure to do so results in incomplete Galilean invariance, and the macroscopic hydrodynamic description cannot be completely recovered. In the non-isothermal case, the high-order moments of interest are the stress tensor, the energy flux tensor, and the rate of energy flux tensor

$$
\begin{aligned}
P^{eq}_{\alpha\beta} &= \sum_{i=1}^{q} f_i^{eq} c_{i\alpha} c_{i\beta} = \rho T_0 \delta_{\alpha\beta} + \rho u_\alpha u_\beta + P'_{\alpha\beta} \\
Q^{\rm eq}_{\alpha\beta\gamma} &= \sum_{i=1}^{q} f_i^{eq} c_{i\alpha} c_{i\beta} c_{i\gamma} = \rho T_0 (u_\alpha \delta_{\beta\gamma} + u_\beta \delta_{\alpha\gamma} + u_\gamma \delta_{\alpha\beta}) + \rho u_\alpha u_\beta u_\gamma + Q'_{\alpha\beta\gamma} \\
R^{\rm eq}_{\alpha\beta\gamma\delta} &= \sum_{i=1}^{q} f_i^{eq} c_{i\alpha} c_{i\beta} c_{i\gamma} c_{i\delta} = (D+2)\rho T_0^2 \delta_{\alpha\beta} + \rho T_0 u^2 \delta_{\alpha\beta} + (D+4)\rho T_0 u_\alpha u_\beta + \\
&\qquad \rho u^2 u_\alpha u_\beta + R'_{\alpha\beta\gamma\delta},
\end{aligned}
$$

respectively, where the right-most sides are the target moment expressions obtained from kinetic theory when the primed terms become equal to zero.

It should be stressed that it is impossible to exactly match the Maxwell-Boltzmann expressions and achieve full Galilean invariance with LBM. What can be achieved is to approximate these expressions to such high orders in velocity, that numerically the deviations become negligible. This can be achieved by using high-order LB models, i.e. models with a large number of discrete velocities, but the use of quadratures of higher order [79] faces the difficulties associated with off-lattice velocities and cannot, in general, guarantee stable numerical schemes. A systematic approach to construct stable approximate models that are on the lattice was proposed recently in the framework of the entropic construction [29, 30].

Stability can be further enhanced by employing either the entropic time stepping instead of the the time stepping of standard LB schemes [4], or the Multi Relaxation Time (MRT) formulation described below.

19.2.2.3 Lattice Boltzmann algorithm

The LBGK scheme is derived by integrating eqn. (19.5) using a second-order trapezoidal scheme to approximate the time integral of the collision term and a transformed set of populations, $g_i(\mathbf{x},t) = f_i(\mathbf{x},t) - \frac{\delta t}{2\tau} Q_i(f_j(\mathbf{x},t))$ satisfying $g_i^{eq} = f_i^{eq}$ to obtain an explicit second-order integration scheme [47, 56]

$$
\begin{aligned}
g_i(\mathbf{x}+\mathbf{c_i}\delta t, t) &= g_i(\mathbf{x},t) + (\frac{2\delta t}{2\tau+\delta t})\, Q_i(g_j(\mathbf{x},t)) \\
&= g_i(\mathbf{x},t) + \omega \left[g_i(\mathbf{x},t) - g_i^{eq}(g_i(\mathbf{x},t)) \right].
\end{aligned}
\tag{19.10}
$$

Here, $\omega(\tau, \delta t) = \frac{2\delta t}{2\tau+\delta t}$ is the discrete inverse relaxation time which, for fixed δt, is restricted in the linear stability interval $[0,2]$: $\omega \to 0$ when $\tau \to \infty$; $\omega \to 2$ when $\tau \to 0$. The relaxation time is related to the kinematic viscosity $\nu = \mu/\rho$ of the fluid via

$$
\tau = \frac{\nu}{c_s^2}.
$$

LBM operates on a highly efficient stream-along-links-and-collide-at-nodes schedule, making the method almost ideally suited for parallel implementations. The algorithm shown in Fig. 19.2 uses intermediate populations f_α^* to store the result of the collision step, doubling the memory requirements. By carefully combining the collision and the streaming steps, a single set of populations can (and should) be used.

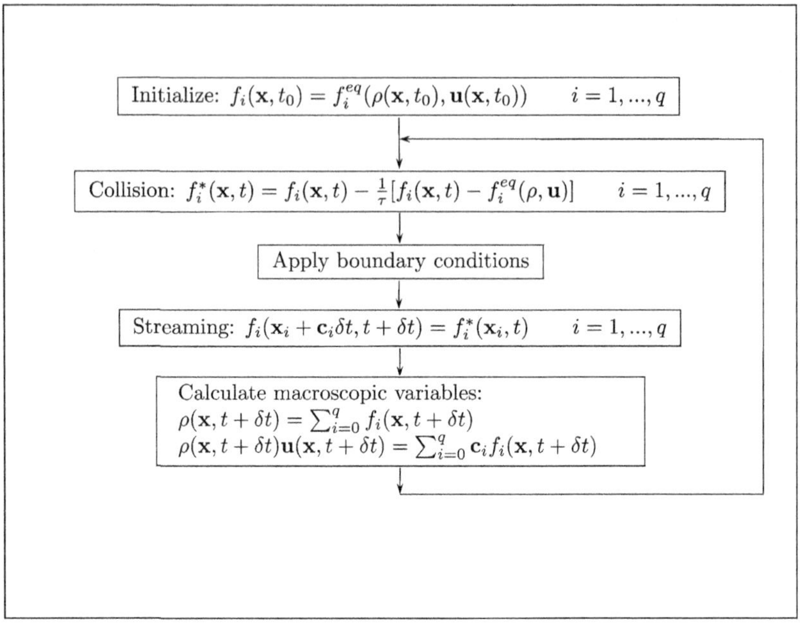

Fig. 19.2: Stream-along-links-and-collide-at-nodes LBGK algorithm.

19.2.3 Variations on the LBM Theme

LBM can be viewed as a reduced kinetic model that can reproduce targeted macroscopic behaviors. In this sense, many different approaches have been (and continue to be) developed that differ in the form of the collision term, the type of the lattice (choice of discrete velocities), the equilibrium function f_i^{eq}, or any combination of the above. The choice of the discrete velocities and of f_i^{eq} was discussed in the previous section. Here, we turn to some of the options available for the collision operator.

Matrix form of the collision operator. The most common form of the collision term is the single relaxation time BGK form, but other forms have also been proposed. The LB formulation was actually born in matrix, multi-time relaxation form [51, 52, 82],

$$\frac{\partial f_i}{\partial t} + c_{i\alpha}\frac{\partial f_i}{\partial x_\alpha} = \sum_j A_{ij}(f_j - f_j^{eq}), \qquad i = 1,\dots,q.$$

The components of the scattering matrix A_{ij} were considered as free parameters that could be tuned to match the macroscopic fluid viscosity [82].

Multi-relaxation time. The simplicity of the single relaxation time (SRT) employed in the popular BGK model comes at a price: bulk and shear viscosity are equal and the Prandtl number cannot be independently controlled. Driven from a systematic stability analysis, alternative formulations known as multi-relaxation-time (MRT) models map at each time step the distribution function f_i from the discrete velocity space to the moment space $\mathbf{m} = M \cdot \mathbf{f}$, where the velocity-moment transformation matrix M is a critical component of MRT. Relaxation is done on these moments using different relaxation rates

$$\mathbf{f}(\mathbf{x} + \mathbf{c}_i\delta t, t + \delta t) = \mathbf{f}(\mathbf{x}, t) - M^{-1} \cdot S \cdot (\mathbf{m} - \mathbf{m}^{eq}),$$

S being the $q \times q$ diagonal matrix of relaxation rates s_i which determine the transport coefficients of the system. Before performing the streaming step, the values are transformed back to discrete velocities and the propagation and boundary condition implementation is performed as in the SRT model. In addition to more flexibility in describing richer physics through individual control of all moments, MRT models can achieve better numerical stability properties in comparison to the original BGK model by appropriate tuning of the eigenvalues of the transformation matrix. Additional details can be found in e.g. [50].

Quasi-equilibrium LBM. A number of LB models for non-isothermal and multi-component flows have been derived based on a representation of the fast-slow decomposition of moment relaxations near a quasi-equilibrium (QE) state. Here, distribution functions approaches equilibrium in two steps: first from the initial state to the QE state, f^*, and then from the QE to the equilibrium, f^{eq} each with its own relaxation time [9]:

$$\frac{\partial f_i}{\partial t} + c_{i\alpha}\frac{\partial f_i}{\partial x_\alpha} = -\frac{1}{\tau_1}(f_i - f_i^*) - \frac{1}{\tau_2}(f_i^* - f_i^{eq}), \qquad i = 1,\dots,q,$$

Although a single distribution function is considered in the previous equation, additional such functions can be introduced to describe flows where different transport processes take place. In addition to the set of conserved fields, the QE LBM assumes that there is a set of quasi-conserved slow fields, and the QE state is defined as the minimizer of the H-function under the constraints of local conservation and of fixed values of the slow fields. The choice of the constraints depends on the particular problem [9, 12, 69, 73].

Force terms. Due to insufficient isotropy of the lattices, additional physics cannot be included in the equilibrium terms. By introducing force terms to the right hand side of the LB equation, a number of new fluid phenomena, including non-isothermal and multiphase flows, can be simulated [82].

Lattice Boltzmann methods without a lattice. Attempts were also made to generalize the lattice to unstructured grids and employ a finite difference of finite volume method for the spatial discretization. The experience with these approaches were "not so positive, since the methods lose their essential advantage of simple algorithm and high accuracy" [46]. In addition to the increased numerical dissipation introduced by the interpolation of the populations on the lattice, stability imposes a significantly shorter integration time step [16].

19.2.4 Initial and Boundary Conditions

In LBM, initial and boundary conditions must be specified for the whole set of distribution functions. What is available from the macroscopic description are the conserved low-order moments of the distribution functions (density, momentum etc.), while the non-conserved ones (stresses and other high-order fluxes) are unknown. In many cases, the higher-order moments will relax and become slaved to the low-order ones that determine the dynamics after a short transient. It then suffices to initialize populations at equilibrium, which is a function only of the macroscopic fields. For incompressible flows, Mei et al. [64] proposed an iterative procedure to generate consistent initial conditions that takes into account the equivalent set of moments that can be generated by the populations.

There is no generally accepted formulation for the boundary conditions and different approaches exist. At the inflow and outflow boundaries, the incoming and outgoing populations, respectively, are unknown and have to be approximated by other means (see, for example, Chikatamarla et al. [29] for a method based on Grad's moment closure). On flat walls, the simplest and most commonly used boundary condition is the bounce-back scheme that was taken from the LG method [22] and exists in different forms, which can influence the accuracy and stability of the computation.

LBM has received increasing attention recently for microfluidic simulations [3, 8, 75, 76, 81, 83]. The need for a boundary condition, that can describe properly the flow at non-vanishing Knudsen numbers, led to the introduction of the diffusive wall boundary condition [3]. Originating from the continuum kinetic theory, this boundary condition appears to be more appropriate for microfluidic simulations [8].

LB is particularly well-suited for complex geometries which can be easily described by marking the character of each lattice node as fluid or solid. For curved boundaries this results in jagged (staircase) boundaries, which can introduce additional errors. More accurate descriptions can be obtained by grid refinement [38], or by extrapolating the non-equilibrium part of the distribution function at the wall nodes [45].

19.2.5 Computational Cost

Most of the computational effort in LBGK simulations arises from the evaluation of the equilibrium expressions. Other computational costs include evaluation of the local conserved quantities, and the application of the BGK collision step. One of the biggest advantages of LBM is the locality of the collision step, making the computations local and the scheme ideally suited for parallelization. In addition, since the method is explicit, *a priori* estimation of the computational costs is possible, and domain decomposition for parallel computations can be easily and efficiently implemented. The effort of evaluating the equilibrium, however, grows non-linearly with the order of equilibrium (powers of velocity) and linearly with the number of discrete velocities involved. The non-linear effort involved in evaluating the equilibrium evaluation can be circumvented for low Mach numbers using the product form of the equilibrium (eqn. (19.9) [28].) In this case, the Lagrange multipliers are first evaluated to high powers of velocity and the equilibrium is then obtained as a product of these Lagrange multipliers. Since the Lagrange multipliers are few (equal to the number of local conservation laws) and independent of the lattice used, the computational costs of higher-order lattices with large number of discrete velocities can be reduced by more than an order of magnitude [28].

19.3 Applications

Lattice Boltzmann models continue to be developed not only to address new physics, but also to overcome shortcomings (mainly numerical instability and lack of Galilean invariance) of currently available models. At the same time, the rapidly accelerating list of applications cover ‘classical’ CFD areas (incompressible, non-isothermal, single- and multi-phase, single- and multi-component flows etc.), as well as nano- and micro-fluidics, biological flows, magnetohydrodynamics, in computer graphics and visualization, etc. Model development for other applications will continue to expand the range of applicability of the method.

19.3.1 Isothermal Flows

19.3.1.1 Turbulent flows

The LBM has been shown to be an efficient and accurate alternative CFD method for turbulent flows, and has been employed in direct and large eddy simulations as well as for the construction of turbulence models.

Direct numerical simulations using LBGK models and comparison with results obtained with a spectral solver were performed quite early after the introduction of the method (e.g. [21, 62]). In a careful study of a 2-D turbulent shear layer [62], for example, the time histories of global quantities and wavenumber spectra were found to be very close to those obtained with spectral methods, although discrepancies in the small scale features of the flow (spatial locations of vortical structures) were observed at late stages. Similar works up to the mid nineties are listed in the review paper of Chen and Doolen [22].

More recent applications include the DNS of grid-generated turbulence [35], fully-developed turbulence in incompressible plane channel flow [60], and the detailed comparison of the results obtained with a high-order entropic LB simulation and a spectral element solver of a 3-D Kida flow [30, 32]. Using a second-order $D3Q19$ LBGK model with second-order bounce back boundary conditions to simulate plane channel turbulence at $Re_\tau = 180$, Lammers et al. [60] concluded that their parallel LB code produces single-point statistics of the same quality as pseudo-spectral methods at comparable resolutions and with a competitive computational cost.

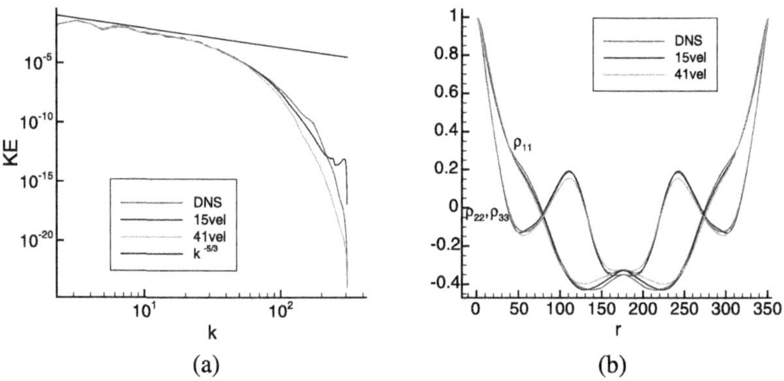

Fig. 19.3: Simulation of the Kida flow for $\mathrm{R}e = 4000$. Comparison of (a) the energy spectrum, (b) two-point longitudinal and transverse correlation functions (DNS: spectral element solution, 15v and 41v LBM solutions using the $D3Q15$ and $D3Q41$ models). From [30].

Chikatamarla et al. [32] used entropic $D3Q15$ and $D3Q41$ LB models to perform direct numerical simulation of the Kida flow in a triply periodic box. Both models were obtained by systematically reducing ("prunning") a $D3Q125$ model by removing discrete velocities contained in the same energy level [31]. The $D3Q41$ lattice (a high-order lattice with respect to the discretization of the velocity space) ensures Galilean invariance of the stress ($P^{eq}_{\alpha\beta}$) and energy flux ($Q^{eq}_{\alpha\beta\gamma}$) tensors, whereas the lower-order LB model recovers $Q^{eq}_{\alpha\beta\gamma}$ only up to $O(u)$. The remaining $O(u^3)$ deviation of the $D3Q15$ model results in a spurious dependence of the kinematic viscosity on the fluid velocity and limits its range of applicability. It is worth noting here that the efficient evaluation of the equilibrium populations using the product form becomes essential for efficiency as the order of the lattice is increased [28]. Comparison of enstrophy, kinetic energy, energy spectrum, skewness factor, second and higher-order structure functions and two-point correlation functions with those obtained with a spectral element solver, showed that DNS-quality results can be obtained by LB models (Fig. 19.3, [32]).

Large eddy simulations and turbulence models. Smagorinsky-type large eddy simulation LB models can be easily obtained by modifying the relaxation time using an effective viscosity obtained from some version of Smagorinsky's eddy viscosity model instead of the fluid viscosity. Model derivations and recent applications can be found in [6, 22, 58, 65, 74, 92] and the references therein. Similarly, standard turbulence models like the algebraic or two-equation models can be incorporated into LB models to compute the eddy viscosity and model high Reynolds number turbulent flows (see, e.g. [23, 82, 88].)

19.3.1.2 Microfluidics and porous media

Typical flows in microdevices are characterized by (a) Knudsen numbers, Kn, ranging from $Kn \ll 1$ (continuum flows) to $Kn \sim 1$ (weakly rarefied flows), and (b) highly subsonic velocities. As Kn is increased the continuum (Navier-Stokes) description breaks down, first in the slip-flow regime ($0.01 \lesssim Kn \lesssim 0.1$), and then in the transition regime ($0.1 \lesssim Kn \lesssim 10$). There is no well established and reliable macroscopic equation for the description of non-equilibrium physics beyond the Navier-Stokes regime, and LB with its kinetic origin is in principle better suited to model flows until the early transition regime [3, 8, 75, 76, 81, 83].

For a correct microscopic description, the diffusive wall boundary condition appears to be more appropriate. In contrast to the bounce-back boundary condition, in essence it allows the particles to 'forget' the incoming direction after reaching the wall by redistributing the populations in a way that is consistent with the mass-balance, the normal-flux conditions, and the Maxwellian distribution at the wall [3, 7]. A direct comparison with DSMC and several convergence analyses clearly indicate that the capability of the LB models to describe microflows is not an 'artifact of numerics' or a 'discretization error', as some authors have claimed. Analytic solutions obtained for flows, like the micro-Couette flow, suggest that the accuracy of the method in the transition regime is improved when higher-order lattices are used [8, 57].

Combining the microflow capability with the flexibility to handle very complex geometries, the method is suitable for the study of flows through porous media like the electrodes used in solid oxide fuel cells (SOFCs). Figure 19.4 shows the computed velocity contours of a methane/water mixture flowing from left to right through a 2-D slice extracted from the 3-D geometry of a real SOFC anode.

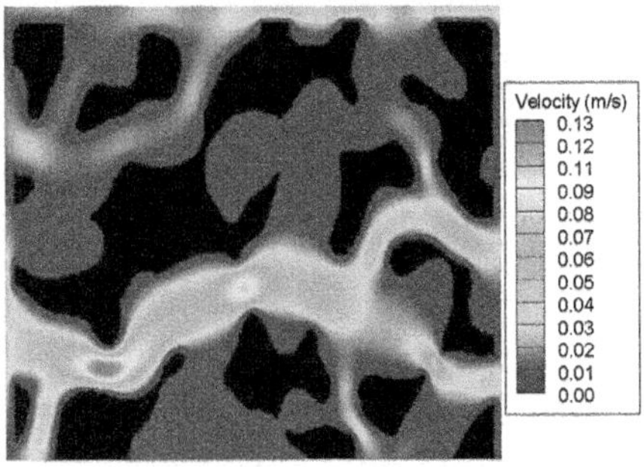

Fig. 19.4: Flow through a porous medium. Velocity contours in m/s are plotted for a square domain of 100×100 μ m (Courtesy of N. Prasianakis and J. Mantzaras).

19.3.2 Non-Isothermal Flows

Flows with variable temperature proved much more difficult to tame with LB models, exhibiting severe numerical instabilities. In order to account for variable temperature, the equilibrium populations must satisfy in addition to the $P^{eq}_{\alpha\beta}$ and $Q^{eq}_{\alpha\beta\gamma}$ moments for any temperature, the $R^{eq}_{\alpha\beta\gamma\delta}$ moments. The D2Q9 lattice, due to geometric constraints, cannot fully satisfy even the $Q^{eq}_{\alpha\beta\gamma}$ moments for a constant temperature. In addition, when the single-relaxation BGK collision model is used, all the higher-order moments relax towards equilibrium at the same rate. Since the relaxation rate of the higher-order moments controls the kinematic viscosity and the thermal conductivity, the Prandtl number has a constant value $Pr = 1$. For all these reasons, new degrees of freedom have been introduced leading to a multitude of approaches for heat transfer simulations. The most important ones are the following:

Extended lattices. Although it was shown theoretically that it is possible to simulate thermal flows using additional discrete velocities and including higher-order velocity terms in the equilibrium distribution function, early implementations suffered from severe numerical instabilities and the range of temperature variations that could be tolerated was narrow [66]. Recent results towards this direction are more encouraging (see, for example, [68, 80]). Even though the stability compared to their predecessors is improved, the range of allowable temperature variations remains quite limited. Also, the increased number of velocities and the complexity of the equilibrium expressions inevitably lead to less efficient schemes.

Hybrid models employ an LB model to compute the flow and couple it to a finite difference or finite element solver for the heat equation (e.g. [59]).

Passive-scalar approaches employ a separate distribution function, independent of the distribution f_i for the density and the momentum fields to simulate the temperature, which is treated as a passive scalar. The two distributions are coupled through a buoyancy forcing term added to the equation for f_i. Compressible flows where density, momentum and energy are tightly coupled cannot be simulated with this approach [77].

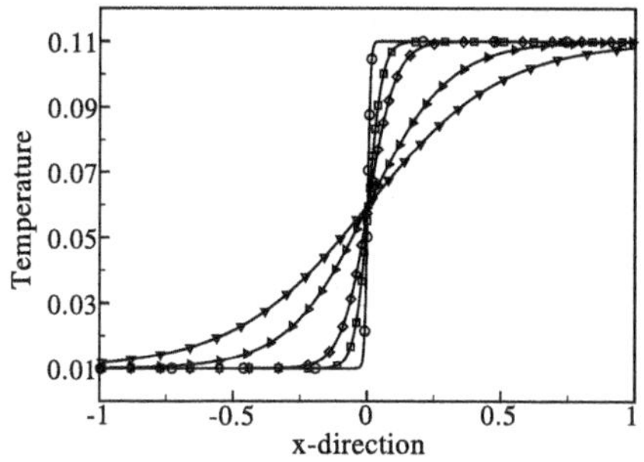

Fig. 19.5: Temperature profile of a model premixed flame using the enhanced entropic thermal LB model (lines: analytic solution for decreasing 'flame thicknesses'; symbols: LBM results). Temperature ratios of more than ten can be achieved. Reprinted from [72] with permission. Copyright © 2008, American Physical Society.

Multiple distribution functions also employ two distributions: one for the density and momentum and another for the internal energy [47]. The standard isothermal model is used on the first lattice and the coupling is through force terms designed to recover the correct viscous heat dissipation and the correct compression work done by the pressure.

Enhanced entropic LB. The entropic construction was also used to construct stable and thermodynamically consistent LB schemes which are off-lattice (e.g. the D1Q4 model in [5]), or on standard lattices (D2Q9 in [10]). The Prandlt number was allowed to vary by improving the collision process [9, 70]. In comparison to the standard lattices, the method has been enhanced in two ways. First, the accuracy of the equilibrium populations was improved by including a larger set of minimization constraints, leading to higher accuracy in higher-order moments. Second, forcing terms were introduced in the lattice BGK equation in order to cancel the effect of the remaining terms in the higher-order moments of the equilibrium populations that deviate from the corresponding Maxwell-Boltzmann expressions [69, 71]. The resulting model offers some highly desirable properties without increasing significantly the computational cost. The speed of sound is described correctly for a large range of temperatures, and shock tube simulations verified the stability of the algorithm. The model was used to solve a 1-D problem mimicking premixed flame propagation [89], showing that large temperature and density variations pertinent to combustion applications can be considered and that heat sources and sinks can easily be included (Fig. 19.5, [72]). Finally, the model is Galilean invariant thus relaxing the low Mach number restriction as long as the flow remains subsonic [73].

19.3.3 Multicomponent Mixtures

Combustion of gas mixtures typically involves a large number of chemical species, even for relatively simple fuels. Many attempts have been made to extend LB models to account for both the self collisions between particles of the same type and the cross collisions involving different particles in multicomponent mixtures (see, for example, the references in [12, 14]). Multiple fluid approaches using different BGK collision terms and allowing the species to relax to equilibrium with different relaxation times are more suitable than single-fluid based models [14]. However, most of the models fail to satisfy important physical properties such as the indifferentiability principle (i.e. they do not reduce to the single-component BGK fluid when the species become identical), or the H-theorem, facing numerical instabilities at large molecular mass ratios [12].

The entropy-based quasi-equilibrium LB approach was employed to construct isothermal models for binary [11] and multicomponent mixtures [12] with different Schmidt numbers and arbitrary molecular weights. The kinetic equations for species j in a mixture with M components take the form

$$\frac{\partial f_{ji}}{\partial t} + c_{ji\alpha}\frac{f_{ji}}{\partial x_\alpha} = -\frac{1}{\tau_{1j}}(f_{ji} - f_{ji}^*) - \frac{1}{\tau_{2j}}(f_{ji}^* - f_{ji}^{eq}) + F_{ji}, \quad j = 1,\ldots M, i = 0,\ldots q,$$

where the associated two relaxation times τ_{1j}, τ_{2j} are related to the viscosity of each component and to the mixture-averaged diffusion coefficient, respectively. F_{ji} is a forcing term acting on species j in order to maintain the momentum balance when the mixture-average approximation for the diffusion coefficients is used [12]. The different molecular weights of the species result in different lattice speeds, and populations corresponding to species with lowest molecular weight are streamed on the lattice, while heavier components with lower speeds stream off-lattice. The corresponding lattice values are calculated by interpolating the off-lattice neighbors, and the interpolation scheme becomes a critical component of the model in terms of accuracy as well as computational efficiency. The models recover the Navier-Stokes and the Stefan-Maxwell diffusion equations at the hydrodynamic limit assuming that the diffusion coefficients of the species are given by the mixture-averaged formulation. The results of the simulations of a mixing layer, an opposed-jet flow, and a micro-Couette flow were found to be in good agreement with continuum and direct simulation Monte Carlo [12]. In the opposed-jet mixing case, viscosities and species diffusivities computed with CHEMKIN were used to evaluate the local relaxation times τ_{1j} and τ_{2j}. Figure 19.6(a) shows the hydrogen mole fraction distribution between opposed jets ejecting different mixtures of H_2, O_2, N_2, and H_2O at a fixed temperature of 300 K; along the axis of symmetry the LB results agree very well with the 1-D solution obtained with the OPPDIF package of CHEMKIN [61].

An MRT lattice Boltzmann model for mixtures was also recently derived from the linearized Boltzmann equations and was used to simulate active and passive scalar mixing in 3-D homogeneous decaying isotropic turbulence [15].

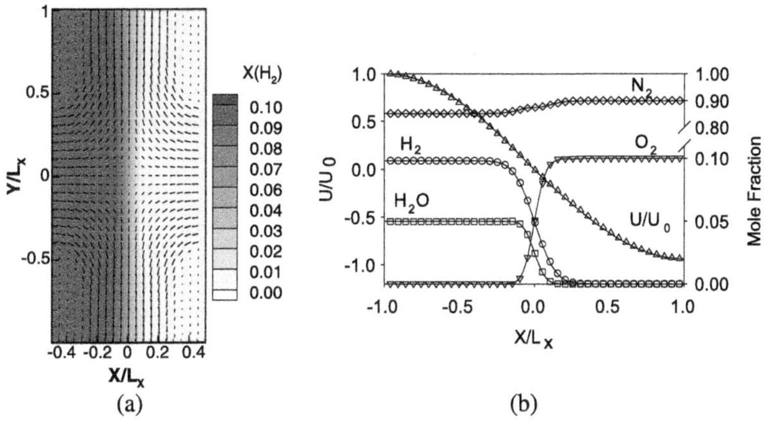

Fig. 19.6: (a) Hydrogen mole fraction distribution obtained by mixing a stream with mole fractions $X_{H_2} = 0.1, X_{N_2} = 0.85, X_{H_2O} = 0.05$ (left boundary) and $X_{N_2} = 0.90, X_{O_2} = 0.1$ (right boundary) in an opposed-jet configuration. (b) Comparison of the LB (symbols) with the OPPDIF solution (lines). Reprinted from [12] with permission. Copyright © 2007, American Physical Society.

19.3.4 Reactive Flows

In principle, once lattice Boltzmann models can properly account for large temperature variation and mixing, extension to reactive flows essentially involves adding appropriate source terms in the species equations to account for the reaction rate. However, no reactive LB model has been proposed so far that capitalizes on the latest developments.

Existing models can be classified in two groups: (a) hybrid models where the flow is solved by an LB solver coupled to a classical CFD solver for the macroscopic energy and species equations [40, 46], and (b) reduced model descriptions either via a tabulated mixture fraction formulation [84, 91], or by using a reduced mechanism obtained by the method of invariant grids [27].

Succi et al. [84] proposed the first extension of LBM for combustion applications by considering mixture fraction and temperature as passive scalars to construct non-premixed combustion models for infinitely fast reaction and constant density. The constant density assumption was also employed by Yamamoto et al. [91] in $D2Q9$ LBGK models using separate populations for the incompressible flow field, the temperature, and each of the species concentrations fields to simulate premixed flames in an opposed-jet setup with constant transport properties and single-step kinetics. The results agreed well with the solution of an incompressible macroscopic model, but, as expected, significant deviations were found when compressibility was taken into account.

The presence of a large number of chemical species in detailed combustion mechanisms results in high computational requirements for any LB model when each species is taken into account by a different population. The requirements become more demanding in higher dimensions and for higher-order lattices (i.e. lattices with larger number of discrete velocities like the common $D3Q27$, or the more recently proposed $D3Q41$ models).

The hybrid models proposed by Filippova and Hänel [39, 40] and Hänel et al. [46] for the simulation of low Mach number reactive flows employ additional free parameters and correction terms to couple LBGK models used to solve for the flow field to a finite difference solver for the macroscopic energy and species conservation equations. These models are not only more memory-conscious in dealing with large number of species, but also remove the constant density limitations of other LB approaches. Recently, the low Mach number formulation was also used by Chen et al. [24] to construct purely LB models for low speed combustion.

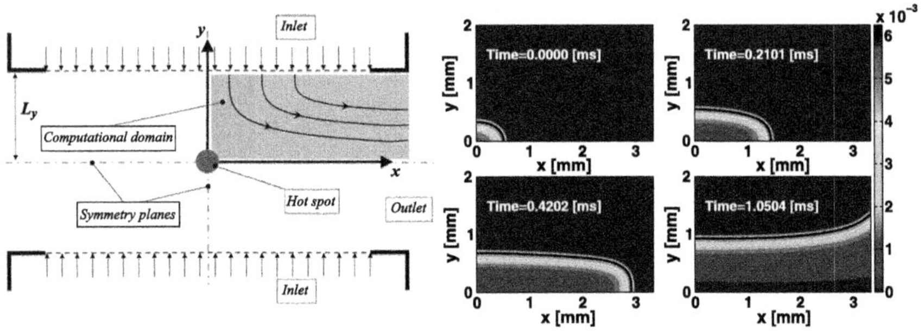

Fig. 19.7: Ignition and stabilization of a H_2/air premixed gas in an opposed-jet geometry: (a) Schematic of the computational setup, (b) temporal evolution of the distribution of the O mass fraction profile. Courtesy of E. Chiavazzo [25].

Reduced kinetic schemes offer another way to lower the computational requirements. The LB model of Yamamoto et al. [91] was employed to simulate the propagation of a 1-D H_2/air premixed flame propagation [27] and ignition/stabilization of a premixed flame in a 2-D opposed-jet geometry [25] (Fig. 19.7). Detailed as well as a reduced two-dimensional invariant manifold description [26] obtained with the method of invariant grids [42, 43], were used to describe the kinetics. The reduced description requires that only two additional equations for the variables parameterizing the manifold need to be solved by the LB model. After each LB step consisting of collision, streaming and reaction, the species populations are projected back to the invariant manifold along the local fast directions. In the premixed flame propagation case, the laminar flame speeds obtained with the full and the reduced scheme were found to agree well with the measured value. Incorporation of accurate reduced chemistry description into more elaborate multicomponent [12] and thermal

LB models [71] will pave the way to the simulation of combustion applications using the lattice Boltzmann method.

Since LB models can easily represent complex geometries, they are particularly well suited for the simulation of reactive flows in porous media and catalyst pores. An exploratory study of complex physics in catalytic devices was presented in [85]. More recently, the multicomponent model [12] was extended to simulate catalytic surface reactions described via appropriate diffusive-wall boundary conditions [13]. Coupling the code with surface CHEMKIN [33] allowed for the easy implementation of realistic surface kinetics. The results of the simulation of a single-step catalytic oxidation of methane in a planar channel at a constant temperature of 1200 K and atmospheric pressure were in very good agreement with a steady-state finite volume macroscopic equation solver (Fig. 19.8).

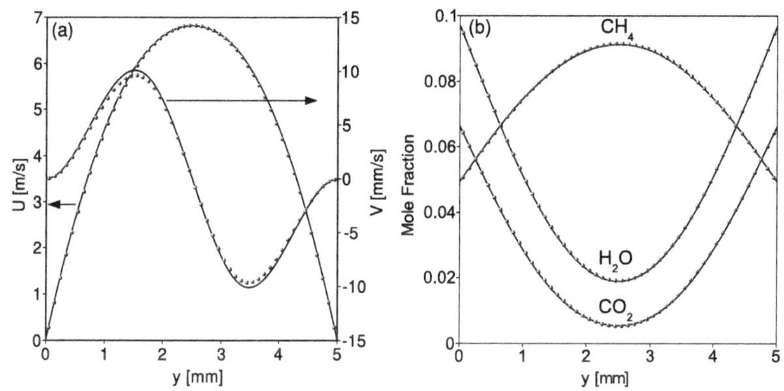

Fig. 19.8: Comparison of the velocity and species profiles obtained with the LB model (symbols) and a CFD solver (lines) in the wall-normal direction in an isothermal catalytic channel. Reprinted from [13] with permission. Copyright © 2008, American Physical Society.

19.4 Conclusions

The lattice Boltzmann method offers a theoretically sound and flexible basis for the construction of multiscale and multi-physics models that are simple to implement into numerical tools. Since the operations are local, LBM is ideally suited for parallel implementation and for the large-scale computation of complex flows at different scales.

Since its inception about 20 year ago, LBM has gone a long way to become an accurate and competitive fluid dynamics solver for complex flows within as well as beyond the regime of the Navier-Stokes hydrodynamics and has already proven its versatility in modeling a wide range of problems. Recently, stable and accurate lattice Boltzmann models have been developed to simulate the different processes relevant to turbulent combustion (turbulent flows, multi-component mixing, strongly non-isothermal and multiphase flows which were not covered here). Particularly interesting are models that can cope with large temperature variations pertinent to combustion without suffering from numerical instabilities, and accurate models for multi-component mixing.

For detailed chemistry reactive flow simulations, LBM is confronted with the high computational cost associated with the large number of chemical species present in combustion kinetics. Already a problem for 'conventional' CFD solvers, this is magnified in LBM since each additional species introduces a new set of populations. Particularly for the high-order lattices that are currently considered to address stability and lack of Galilean invariance issues, the computational toll can then become excessive. When passive scalar approaches are not an option, hybrid models coupling an LBM solver for the flow and a conventional solver for the energy and species lead to significant savings, but sacrifice the algorithmic simplicity of the method. Another option is offered by reduced kinetic descriptions which have already been shown to work in reactive constant density LB models.

The lattice Boltzmann method is still in rapid development. Additional systematic validation studies, similar to the ones that other CFD approaches have undergone, are required before recently proposed approaches can be combined into models capable of simulating reactive flows. The impressive progress to date supports that "reaching full maturity is just a matter of time and labour, no conceptual hurdles in sight" [82] holds not only for the method in general, but also for its extension for the simulation of turbulent reactive flows.

Acknowledgements

Many thanks to I.V. Karlin for enlightening discussions about the lattice Boltzmann method and the opportunity to follow developments in the field during the past six years. Special thanks to N. Prasianakis, S.S. Chikatamarla, S. Arcidiacono, and J. Mantzaras for the input they provided for this manuscript.

References

1. Abe, T.: Derivation of the Lattice Boltzmann Method by Means of the Discrete Ordinate Method for the Boltzmann Equation. J. Comp. Phys. **131**, 241–246 (1997)
2. Alder, B.J., Wainright, T.E.: Studies in molecular dynamics: I. General methods. J. Chem. Phys. **32**, 459–466 (1959)
3. Ansumali, S., Karlin, I.V.: Kinetic boundary conditions in the lattice Boltzmann method. Phys. Rev. E **66**, 026311 (2002)
4. Ansumali, S.: Minimal kinetic modeling of hydrodynamics. PhD thesis, ETH Dissertation No 15534, Swiss Federal Institute of Technology, Zurich, Switzerland (2004)

5. Ansumali, S., Karlin, I.V., Öttinger, H.C.: Minimal entropic kinetic models for simulating hydrodynamics. Europhys. Lett. **63**, 798–804 (2003)
6. Ansumali, S., Karlin, I.V., Succi, S.: Kinetic theory of turbulence modeling: smallness parameter, scaling and microscopic derivation of Smagorinsky model. Physica A **338**, 379–394 (2004)
7. Ansumali, S., Karlin, I.V., Frouzakis, C.E., Boulouchos, K.: Entropic lattice Boltzmann method for microflows. Physica A **359**, 289–305 (2005)
8. Ansumali, S., Karlin, I.V., Arcidiacono, S., Abbas, A., Prasianakis, N.: Hydrodynamics beyond Navier-Stokes: Exact solution to the lattice Boltzmann hierarchy. Phys. Rev. Lett. **98**, 124502 (2007)
9. Ansumali, S., Arcidiacono, S., Chikatamarla, S.S., Prasianakis, N.I., Gorban, A.N., Karlin, I.V.: Quasi-equilibrium lattice Boltzmann method. Eur. Phys. J. **56**, 135–139 (2007)
10. Ansumali, S., Karlin, I.V.: Consistent Lattice Boltzmann Method. Phys. Rev. Lett. **95**, 260605 (2005)
11. Arcidiacono, S., Mantzaras, J., Ansumali, S., Karlin, I.V., Frouzakis, C.E., Boulouchos, K.: Simulation of binary mixtures with the lattice Boltzman method. Phys. Rev. E **74**, 056707 (2007)
12. Arcidiacono, S., Karlin, I.V., Mantzaras, J., Frouzakis, C.E.: Lattice Boltzmann model for the simulation of multicomponent mixtures. Phys. Rev. E **76**, 046703 (2007)
13. Arcidiacono, S., Mantzaras, J., Karlin, I.V.: Lattice Boltzmann simulation of catalytic reactions. Phys. Rev. E **78**, 046711 (2008)
14. Asinari, P.: Viscous coupling based lattice Boltzmann model for binary mixtures. Phys Fluids **17**, 067102 (2005)
15. Asinari, P., Luo, L.-S.: A consistent lattice Boltzmann equation with baroclinic coupling for mixtures. J. Comp. Phys. **227**, 3878–3895 (2008)
16. Bardow, A., Karlin, I.V., Gusev, A.A.: General characteristic-based algorithm for off-lattice Boltzmann simulations. Europhys. Lett. **75**, 434–440 (2006)
17. Bird, G.A.: Molecular Gas Dynamics and the Direct Simulation of Gas Flows, Oxford Universiy Press (1994)
18. Boghosian, B.M., Yepez, J., Coveney, P.V., Wagner, A.: Entropic Lattice Boltzmann Methods. Proc. Roy. Soc. Lon. A **457**, 717–766 (2001)
19. Cercignani, C.: The Boltzmann equation and its applications. In Applied Mathematical Sciences, Vol. 61, Springer-Verlag, New York (1988)
20. Chapman, S., Cowling, T.G.: The Mathematical Theory of Non-Uniform Gases, 3rd edition, Cambridge University Press, London (1970)
21. Chen, S., Wang, Z., Shan, X., Doolen, G.D.: Lattice Boltzmann computational fluid dynamics in three dimensions. J. Stat. Phys. **68**, 379–400 (1992)
22. Chen, S., Doolen, G.D.: Lattice Boltzmann methods for fluid flows. Annu. Rev. Fluid Mech. **30**, 329–364 (1998)
23. Chen, S., Kandasamy, S., Orszag, S., Shock, R., Succi, S., Yakhot, V.: Extended Boltzmann Kinetic Equation for Turbulent Flows. Science **301**, 633–646 (2003)
24. Chen, S., Liu, Z., Zhang, C., He, Z., Tian, Z., Shi, B., Zheng, C.: A novel coupled lattice Boltzmann model for low Mach number combustion simulation. Appl. Math. Comp. **193**, 266–284 (2007)
25. Chiavazzo, E.: Invariant manifolds and lattice Boltzmann method for combustion, PhD Thesis, ETH Dissertation No. 18233, Swiss Federal Institute of Technology, Zurich, Switzerland (2009)
26. Chiavazzo, E., Karlin, I.V., Frouzakis, C.E., Boulouchos, K.: Method of invariant grid for model reduction of hydrogen combustion. Proc. Combust. Inst. **32**, 519–526 (2009)
27. Chiavazzo, E, Karlin, I.V., Gorban, A.N., Boulouchos, K.: Combustion simulation via lattice Boltzmann and reduced chemical kinetics. J. Stat. Mech. P06013 (2009)
28. Chikatamrla, S.S., Ansumali, S., Karlin, I.V.: Entropic lattice Boltzmann models for hydrodynamics in three dimensions. Phys. Rev. Lett. **97**, 010201 (2006)

29. Chikatamrla, S.S., Ansumali, S., Karlin, I.V.: Grad's approximation for missing data in lattice Boltzmann simulations. Europhys. Lett. **74**, 215–221 (2006)
30. Chikatamarla, S.S.: Hierarchy of Lattice Boltzmann Models for Fluid Mechanics, PhD Thesis, ETH Dissertation No. 17893, Swiss Federal Institute of Technology, Zurich, Switzerland
31. Chikatamarla, S.S., Karlin, I.V.: Lattices for the lattice Boltzmann method. Phys. Rev. E **79**, 046701 (2009)
32. Chikatamarla, S.S., Frouzakis, C.E., Karlin, I.V., Tomboulides, AG, Boulouchos, K., Lattice Boltzmann method for direct numerical simulation of turbulent flows. J. Fluid Mech. (submitted)
33. Coltrin, M.E., Kee, R.J., Rupley, F.M.: Sandia National Laboratories, Report No. SAND90-8003C (1996)
34. Cottet, G.-H., Koumoutsakos, P.: Vortex Methods, Theory and Practice, Cambridge University Press (2000)
35. Djenidi, L.: Lattice-Boltzmann simulation of grid-generated turbulence. J. Fluid Mech. **552**, 13–55 (2006)
36. Español, P., Warren, P.B.: Statistical-mechanics of dissipative particle dynamics. Europhysics Lett. **30**, 191–196 (1995)
37. Español, P.: Fluid particle model. Phys. Rev. E **57**, 2930–2948 (1998)
38. Filippova, O., Hänel, D.: Grid refinement for lattice-BGK models. J. Comput. Phys. **147**, 219–228 (1998)
39. Filippova, O., Hänel, D.: Lattice-BGK model for low Mach number combustion. Int. J. Mod. Phys. C **9**, 1439–1445 (1998)
40. Filippova, O., Hänel, D.: A novel Lattice BGK approach for low Mach number combustion. J. Comp. Phys. **158**, 139–160 (2000)
41. Frisch, U., Hasslacher, B., Pomeau, Y.: Lattice-gas automata for the Navier-Stokes equation, Phys. Rev. Lett. **56**, 1505–1508 (1986)
42. Gorban, A.N., Karlin, I.V.: Invariant Manifolds for Physical and Chemical Kinetics. Lect. Notes Phys. 660, Springer, Berlin, Heidelberg
43. Gorban, A.N., Karlin, I.V.: Method of invariant manifold for chemical kinetics. Chem. Eng. Sci. **58**, 4751 (2003)
44. Grad, H.: On the kinetic theory of rarefied gases. Commun. Pure Appl. Math. **2**, 331–407 (1949)
45. Guo, Z., Zheng, C., Shi, B.: An extrapolation method for boundary conditions in lattice Boltzmann method. Phys. Fluids **14**, 2007–2010 (2002)
46. Hänel, D., Lantermann, U., Kaiser, R., Wlokas, I.: Generalized lattice-BGK concept for thermal and chemically reacting flows at low Mach numbers. Int. J. Num. Meth. Fluids **51**, 351–369 (2006)
47. He, X., Chen, C., Doolen, G.D.: A novel thermal model for the Lattice Boltzmann method in incompressible limit. J. Comp. Phys. **146**, 282–300 (1998)
48. He, X., Luo, L.-S.: A priori derivation of the lattice Boltzmann equation. Phys. Rev. E **55**, R6333–R6336 (1997)
49. He, X., Luo, L.-S.: Theory of the lattice Boltzmann method: From the Boltzmann equation to the lattice Boltzmann equation. Phys. Rev. E **56**, 6811–6817 (1997)
50. d'Humieres, D., Ginzburg, I., Krafczyk, M., Lallemand, P., Luo, L.S.: Multiple relaxation-time lattice Boltzmann models in three-dimensions. Phil. Trans. R. Soc. **360**, 437 (2002)
51. Higuera, F.J., Succi, S., Benzi, R.: Lattice gas dynamics with enhanced collisions. Europhys. Lett. **9**, 345 (1989)
52. Higuera, F.J., Jimenez, J.: Boltzmann approach to lattice gas simulations. Europhys. Lett. **9**, 663–668 (1989)
53. Hoogerbrugge, P.J., Koelman, J.M.V.A.: Simulating microscopic hydrodynamic phenomena with dissipative particle dynamics. Europhysics Lett. **19**, 155–160 (1992)
54. Karlin, I.V., Gorban, A.N., Succi, S., Boffi, V.: Maximum entropy principle for lattice kinetic equations. Phys. Rev. Lett. **81**, 6–9 (1998)

55. Karlin, I.V., Ferrante, A, Öttinger, H.C.: Perfect entropy functions of the lattice Boltzmann method. Europhys. Lett. **47**, 182–188 (1999)
56. Karlin, I.V., Ansumali, S., Frouzakis, C.E., Chikatamarla, S.S.: Elements of the lattice Boltzmann method I: Linear advection equation. Comm. Comp. Phys. **1**, 616–655 (2006)
57. Kim, S.H., Pitsch, H., Boyd, I.D.: Accuracy of higher-order lattice Boltzmann methods for microscale flows with finite Knudsen numbers. J. Comp. Phys., 8655 (2008)
58. Krafczyk, M., Tölke, J., Luo, L.-S.: Large-eddy simulations with a multiple-relaxation-time LBE model. Int. J. Mod. Phys. B **17**, 33–39 (2003)
59. Lallemand, P., Luo, L.-S.: Theory of the lattice Boltzmann method: Acoustic and thermal properties. in two and three dimensions, Phys. Rev. E **68**, 036706 (2003)
60. Lammers, P., Beronov, K.N., Volkert, R., Brenner, G., Durst, F.: Lattice BGK direct numerical simulation of fully developed turbulence in incompressible plane channel flow. Comp. Fluids **35**, 1137–1153 (2006)
61. Lutz, A.E., Kee, R.J., Grcar, J.F., Rupley, F.M.: OPPDIF: A Fortran Program for Computing Opposed-Flow Diffusion Flames, Sandia National Laboratories, Sandia (1997)
62. Martinez, D.O., Matthaeus, W.H., Chen, S.: Comparison of spectral methods and lattice Boltzmann simulations of two-dimensional hydrodynamics. Phys. Fluids **6**, 1285–1298 (1994)
63. McNamara, GR, Zanetti, G.: Use of the Boltzmann Equation to Simulate Lattice-Gas Automata. Phys. Rev. Lett. **61**, 2332–2335 (1988)
64. Mei, R., Luo, L.-S., Lallemand, P., d'Humières, D.: Consistent initial conditions for lattice Boltzmann simulations. Comp. & Fluids **35**, 855–862 (2006)
65. Menon, S., Soo, J.-H.: Simulation of vortex dynamics in three-dimensional synthetic and free jets using the large-eddy lattice Boltzmann method. J. Turbul. **5**, 2–26 (2004)
66. McNamara, G., Alder, B.: Analysis of the lattice Boltzmann treatment of hydrodynamics. Physica A **194**, 218–228 (1993)
67. Monaghan, J.J.: Smoothed particle hydrodynamics. Rep. Prog. Phys. **68**, 1703–1759 (2005)
68. Philippi, P.C., Hegele, L.A., Dos Santos, L.O.E., Surmas, R.: From the continuous to the lattice Boltzmann Equation: The discretization problem and thermal models. Phys. Rev. E **73**, 056702 (2006)
69. Prasianakis, N.: Lattice Boltzmann method for thermal compressible flows, PhD Thesis, ETH Dissertation No. 17739, Swiss Federal Institute of Technology, Zurich, Switzerland
70. Prasianakis, N., Boulouchos, K.: Lattice Boltzmann method for simulation of weakly compressible flows at arbitrary Prandtl number. Int. J. Mod. Phys. C **18**, 602 (2007)
71. Prasianakis, N., Karlin, I.V.: Lattice Boltzmann method for thermal flow simulation on standard lattices. Phys. Rev. E **76**, 016702 (2007)
72. Prasianakis, N., Karlin, I.V.: Lattice Boltzmann method for simulation of compressible flows on standard lattices. Phys. Rev. E **78**, 016704 (2008)
73. Prasianakis, N., Karlin, I.V., Mantzaras, J., Boulouchos, K.: Lattice Boltzmann method with restored Galilean invariance. Phys. Rev. E **79**, 066702 (2009)
74. Premnatha, K.N., Pattison, M.J., Banerjee, S.: Dynamic subgrid scale modeling of turbulent flows using lattice-Boltzmann method. Physica A **388**, 2640–2658 (2009)
75. Sbragaglia, M., Succi, S.: Analytical calculation of slip flow in lattice Boltzmann models with kinetic boundary conditions. Phys. Fluids **17**, 093602 (2005)
76. Sbragaglia, M., Succi, S.: A note on the lattice Boltzmann method beyond the Chapman-Enskog limits. Europhys. Lett. **73**, 370 (2006)
77. Shan, X.: Simulation of Rayleigh-Bénard convection using a lattice Boltzmann method. Phys. Rev. E **55**, 2780–2788 (1997)
78. Shan, X.W., He, X.Y.: Discretization of the velocity space in the solution of the Boltzmann equation. Phys. Rev. Lett. **80**, 65–68 (1998)
79. Shan, X., Yuan, X.-F., Chen, H.: Kinetic theory representation of hydrodynamics: a way beyond the Navier–Stokes equation. J. Fluid Mech. **550**, 413–441 (2006)
80. Shan, X., Chen, H.: A General Multiple-Relaxation-Time Boltzmann Collision Model. Int. J. Mod. Phys. C **18**, 635 (2007)

81. Sofonea, V., Sekerka, R.F.: Boundary conditions for the upwind finite difference Lattice Boltzmann model: Evidence of slip velocity in micro-channel flow. J. Comp. Phys. **207**, 639 (2005)
82. Succi, S.: The Lattice Boltzmann Equation for Fluid Dynamics and Beyond, Oxford University Press, Oxford
83. Succi, S.: Mesoscopic modeling of slip motion at fluid-solid interfaces with heterogeneous catalysis. Phys. Rev. Lett. **89**, 064502 (2002)
84. Succi, S., Bella, G., Papetti, F.: Lattice kinetic theory for numerical combustion, J. Sci. Comp. **12**, 395–408 (1997)
85. Succi, S., Smith, G., Kaxiras, E.: Lattice Boltzmann simulation of reactive microflows over catalytic surfaces. J. Stat. Phys. **107**, 343–366 (2004)
86. Sterling, J.D., Chen, S.: Stability analysis of lattice Boltzmann methods. J. Comp. Phys. **123**, 196–206 (1996)
87. Sukop, M.C., Thorne, D.T.: Lattice Boltzmann Modeling: An Introduction for Geoscientists and Engineers, Springer (2007)
88. Teixeira, C.M.: Incorporating turbulence models into the lattice-Boltzmann method. Int. J. Mod. Phys. C **9**, 1159–1175 (1998)
89. Tomboulides, A.G., Orszag, S.A.: A quasi-two-dimensional benchmark problem for low Mach number compressible codes. J. Comput. Phys. **146**, 691–706 (1998)
90. Wolf-Gladrow, D.A.: Lattice-Gas Cellular Automata and Lattice Boltzmann Models, Springer (2000)
91. Yamamoto, K., He, X., Doolen, G.D.: Simulation of combustion field with lattice Boltzmann method. J. Stat. Phys. **107**, 367–383 (2002)
92. Yu, H., Luo, L.-S., Girimaji, S.S.: LES of turbulent square jet flow using an MRT lattice Boltzmann model. Comp. Fluids **35**, 957–965 (2006)

Index

T. Echekki, E. Mastorakos (eds.), *Turbulent Combustion Modeling*,
Fluid Mechanics and Its Applications 95, DOI 10.1007/978-94-007-0412-1,

Printed by Printforce, the Netherlands